ADVANCE 1

Education in the Creative Economy

"At a time when new media and a host of new technologies are reshaping every aspect of life and society, there has been a lack of analyses capable of integrating and critically engaging these new forces. We not only live in an age of constant information flows and technological breakthroughs, but we also inhabit a society in which individual and social agency are under attack. Innovation is stripped from matters of equity and social costs and the results look less like a step forward for humanity than a step backwards into what Bertolt Brecht once called 'dark times.' Technologies and their diverse modes of knowledge do not merely overwhelm us, they increasingly erase matters of equity, justice, and social responsibility. What *Education in the Creative Economy* does brilliantly is to remind us that learning is not merely the precondition for innovation, technological progress, and the over-hyped rise of an information economy, it is also the precondition for imagination, civic courage, and collective struggles that expand and deepen the process of democratization. *Education in the Creative Economy* is a breakthrough book that critically engages a range of complex issues regarding the brave new world in which new technologies and knowledge production redefine the nature of the present. But it does more. In breathtaking fashion it dresses cogently and with great insight what this new regime of production, consumption, and learning might mean for a future in which democratic values, social justice, and political agency once again matter."

—Henry Giroux, Global Television Network Chair in the Department of English and Cultural Studies, McMaster University

"A brilliant collection in substance, scale and variety. It is like a website in print. Michael A. Peters and Daniel Araya have done us a fine service in assembling all of this material in a single volume—one that is destined to be first portal on education and creativity for some time to come. The book is a lattice of diverse messages and interpretations, but all of the contributors share an engaged and enthusiastic realism, in which 'culture' and 'economy' are no longer separated into different boxes but are continually used to interrogate each other."

—Simon Marginson, Professor of Higher Education, Centre for the Study of Higher Education, University of Melbourne

"This book brings together multiple perspectives on and creates a compelling argument for the role of creativity in economic growth and development. It examines how an eclectic mix of technology, talent and educational policy are vital for innovation and economic growth, and argues the need for new learning paradigms and for better links across academic disciplines and between universities and business. Increased and accelerating globalization in all areas of the economy has never made creativity more important to innovation than now, so it is good to see a book that tackles the importance of creativity from a range of perspectives and from across the world. The book starts by defining creativity in the context of economic growth and globalization, and the chapters that follow under the headings of educational policy, technology and economy, and culture and curriculum indicate the challenges that need to be addressed in order for growth to occur and for developing countries to rise up. As curricula in schools continue to be locked down and a narrow range of skills are defined and measured with levels of accountability that drive out creativity, this book is a warning to policymakers to stop and review their directions, and to consider the implications of a testing regime that fails to encompass the need for demonstrations of creativity in its design."

—Michelle Selinger, Director, Education Practice, Internet Business Solutions Group, Cisco Systems

"At a time when educational improvement is increasingly seen as a prerequisite for economic prosperity, this volume centers on vital, underemphasized objectives for twenty-first century learning, teaching, and policy."

—Christopher Dede, Timothy E. Wirth Professor in Learning Technologies, Harvard University

"The edited collection *Education in the Creative Economy* brings together top educators to document various dimensions of the current crisis of education and the economy, and ways to jump-start a stalled economy and reinvigorate education. Spanning a broad range of issues from educational policy and the role of the university in today's economy and society to how new technologies are dramatically transforming our culture, economy, and schooling, the authors show the need to rethink and reconstruct education for the challenges of the twenty-first century."

—Douglas Kellner, George F. Kneller Chair in the Philosophy of Education at UCLA and author of Guys and Guns Amok: Domestic Terrorism and School Shootings from the Oklahoma City Bombings to the Virginia Tech Massacre

"The trick with creative education is to enable curious, playful and aspirational minds to make connections hitherto unimagined. That's what this book does. The impressive international array of authors reaches way beyond the traditional arts to encompass creative innovation and renewal in public policy, business strategy, creative practice and scholarly domains. Their work makes a compelling case for the widest possible application of creative education for an open future. For the curious reader, the book adds up to a creative education in itself."

—John Hartley, Research Director, ARC Centre of Excellence for Creative Industries & Innovation, Queensland University of Technology.

"Oh brave new world that has such essays in it! The words 'innovative' and 'comprehensive' don't often go together, but for a world where innovation and creativity are the key to economic success, here is a definitive look at what creativity means—and how people learn it—in the digital age."

—David Williamson Shaffer, Professor of Educational Psychology, University of Wisconsin-Madison

"This collection makes a seductive and intriguing read. Of course, learning in and around the creative economy has a fundamental role to play in the evolution of new economies in this third millennium, but one is left realising that the creative and ingenious minds evidenced by these papers share the more fundamental role of guiding education in re-examining the ways that we might learn together: if the creative industries cannot reinvent and invigorate learning, then who can?

This collection offers a small but unmissable insight into the future of learning. If we think we might be able to mend the world with learning, and I do think so, then this collection might be a very useful place to start the debate about how."

—Professor Stephen Heppell, Centre for Excellence in Media Practice, Bournemouth University

Education in the Creative Economy

This book is part of the Peter Lang Education list.
Every volume is peer reviewed and meets
the highest quality standards for content and production.

PETER LANG
New York • Washington, D.C./Baltimore • Bern
Frankfurt • Berlin • Brussels • Vienna • Oxford

Education in the Creative Economy

Knowledge and Learning in the Age of Innovation

Edited by Daniel Araya & Michael A. Peters

PETER LANG
New York • Washington, D.C./Baltimore • Bern
Frankfurt • Berlin • Brussels • Vienna • Oxford

Library of Congress Cataloging-in-Publication Data

Education in the creative economy: knowledge and learning
in the age of innovation / edited by Daniel Araya, Michael A. Peters.
p. cm.
Includes bibliographical references and index.
1. Educational technology. 2. Education—Effect of technological
innovations on. 3. International education. 4. Education and globalization.
I. Araya, Daniel. II. Peters, Michael A.
LB1028.3.E326 338.4'337—dc22 2010026814
ISBN 978-1-4331-0745-0 (hardcover)
ISBN 978-1-4331-0744-3 (paperback)

Bibliographic information published by **Die Deutsche Nationalbibliothek.**
Die Deutsche Nationalbibliothek lists this publication in the "Deutsche
Nationalbibliografie"; detailed bibliographic data is available
on the Internet at http://dnb.d-nb.de/.

The paper in this book meets the guidelines for permanence and durability
of the Committee on Production Guidelines for Book Longevity
of the Council of Library Resources.

29 Broadway, 18th floor, New York, NY 10006
www.peterlang.com

Printed in the United States of America

Contents

Foreword: Education in the Creative Economy

John Seely Brown

The last thing I wanted to do was to write another foreword, but once I saw the breadth and scope of the topics covered herein I sensed that this book would be both important—as a discourse setter—and timely. The one fact that everyone in our current economic crisis seems to agree upon is that innovation is going to be the ultimate means to rebuild and reinvigorate our economy. Fine, one might say, but look deeper; nearly all the discussion is about importance of science, technology, engineering, and mathematics (STEM) to this effort. As the former Chief Scientist of Xerox and director of its famous (or infamous) Palo Alto Research Center (PARC), I, of course, appreciate the importance of these fields, but the real key to innovation and the creative economy extends way beyond the hard sciences. It turns on imagination; it turns on understanding social systems, and it turns on agency. In addition, each of these concepts needs to be informed by the arts and humanities as much as by the sciences.

This fact was driven home to me by Rich Gold, the person at PARC responsible for our PARC Artists in Residence (PAIR) program, which sought to pair avant-garde artists, artists who were used to challenging the *status quo,* with scientists in a one-on-one engagement that could last up to a year or more. This program most definitely helped keep PARC at the forefront. Behind it was a conceptual framework that Rich created and that still continues to influence many of us well after Rich's early death. Let me portray this most succinctly by a simple diagram, one that I have always thought of as our "secret recipe," secret only in the sense that no matter how many times we talked about it in public, its full force seemed always elusive.

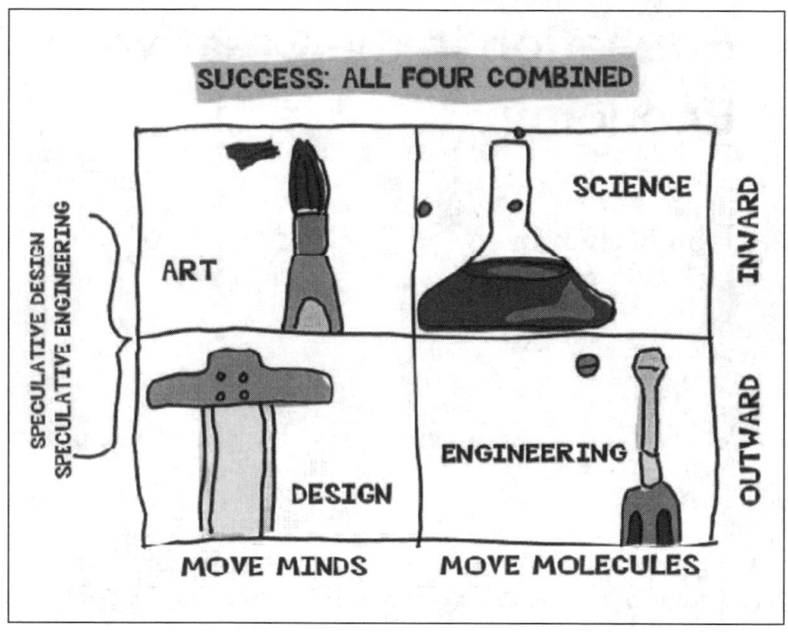

Creativity requires interdisciplinarity and synthesis

Key to understanding this paradigm is to first recognize that engineering can quickly become stale without the constant infusion of ideas flowing in from the sciences or at least from a rich participation with the sciences. Similarly, design grows stale without its interaction with the arts, humanities, and social sciences. At PARC we had all these disciplines jostling around, bumping up against one another, thus setting the stage for productive, and often provocative, interactions that challenged our approaches and techniques.

When Rich first created this framework, I thought that some of these interactions would flow naturally. The artists and designers would want to interact given that, at a high level of abstraction, both communities were involved in moving minds. Similarly, the scientists and engineers would want to interact since both were involved in moving molecules, metaphorically speaking. Alas, my belief was dead wrong! The most natural interactions developed among the artists and scientists, on the one hand, and between the designers and engineers, on the other hand. Why? Because the artists and scientists were driven by an imagination stimulated by looking inward, and the designers and engineers were driven by an imagination stimulated by looking outward. Show me a designer/engineer who does not pay attention to focus groups, ethnographic analyses, or user specifications, and most likely that designer will miss the mark. Said more dramatically, Picasso didn't traffic in focus groups, good designers do.

It would be tempting to assert that innovation emerges from waving a magic wand over this 2 by 2 and getting folks from these different quadrants to work together. However, such a neat confluence of the minds never directly happens. Each quadrant reflects different sensibilities, different ways of making meaning, and even making sense. Even so, an appreciation for the deep craft that lies behind high performance in each of these areas rises to the surface. Moreover, a near universal appreciation for deep craft helps to bridge and meld the different sensibilities of the different epistemic/artistic cultures. It is the seductive power of the problem and a desire to get to its root, unpacking the surface problem, and discovering the problem beneath the problem which, said simply, is a universal passion of each. In so doing, innovation ecologies are created that constantly and profoundly surprise. It is this ability to create meaningful surprises, continually and ubiquitously, that so much of this book is about.

As the many talented authors in this volume observe, we are living in an age of continuous innovation. Curiosity is the motivating force for innovation, but it is often blunted by the rigid educational systems we have today. As my colleague Douglas Thomas and I suggest, the educational needs of the twenty-first century pose a number of serious problems for current educational practices. First and foremost, the twenty-first century is characterized by constant change. Educational practices that focus on the transfer of static knowledge simply cannot keep up with the rapid rate of change. Educational practices that focus on adaptation or reaction to change fare better, but they are still finding themselves outpaced by an environment that requires content to be updated almost as fast as it can be taught. What is required in education today are learning systems that are responsive to constant flux.

For much of the twentieth century, learning focused on the acquisition of skills or transmission of information or what we define as "learning about." Near the end of the twentieth century, learning theorists started to recognize the value of "learning to be," of putting learning into a situated context that deals with systems and identity as well as the transmission of knowledge. In this volume the authors suggest that now even that is not enough. Although learning about and learning to be worked well in a relatively stable world, in a world of constant flux, we need to embrace a theory of *learning to become*. Where most theories of learning see becoming as a transitional state toward becoming something, the twenty-first century requires us to think of learning as a practice of becoming over and over again. In order to understand both what that means and how it might be achieved, we need to examine some of the new modes of learning that have emerged in the twenty-first century. In particular, we need to consider the social, distributed, and networked dimensions of learning. More than this, we need to consider the broader economic and technological landscape in which these new modes of learning are forming. That is precisely what this book sets out to accomplish.

John Seely Brown
Former Chief Scientist, Xerox Corporation
Director Emeritus, Xerox PARC

Introduction
The Creative Economy: Origins, Categories, and Concepts

Michael A. Peters & Daniel Araya

A creative economy is the fuel of magnificence.

—Ralph Waldo Emerson, *English Traits,* 1856.

The Economic Nature of Cultural Enterprise and Creativity as the Generation of Innovation

In *The Creative Economy: A New Definition,* a report prepared for the New England Foundation for the Arts (NEFA), Douglas DeNatale and Gregory H. Wassall (2007) note that creative economy research consists of two separate models: "one emphasizes the production of cultural goods and services—however defined—as a valuable contributor to society; the other emphasizes the role of intellectual innovation as an economic driver of particular value during periods of societal transition" (p. 5). They go on to write: "Definitions of the creative economy diverge at the point of whether "creative" should be interpreted as culturally based or ideational in nature, using "creative" as shorthand for cultural expression on the one hand, and intellectual invention on the other" (p. 5).

NEFA's model identifies three primary and interrelated components: the Creative Cluster (industry), the Creative Workforce (occupation) and Creative Communities (geography). NEFA's new definition comprises the "cultural core" conceived in concentric circles, surrounded by "cultural periphery" and "creative industries." The "cultural core" "includes occupations and industries that focus on the production and distribution of cultural goods, services and intellectual property." The beauty of this definition is that it constitutes an operational definition that allows researchers to use U.S. federal sources of data to list and count industries and occupations according to explicit and standard categories.[1]

As the nexus of innovation increasingly moves from labor-intensive "smokestack" industries to "mind work," creativity is becoming critical to policy discussions on economic growth. Creativity and knowledge-based innovation are now increasingly seen as powerful engines driving the global economy. In the white paper "Creative Economy Research in New England: A Reexamination," for example, DeNatale and Wassall (2006) trace New England's regional experience with the creative economy framework to the New England Creative Economy Initiative in 1998, noting that economists have devoted little time to "creative industries." New England's initiation of creative economy research is the oldest in the United States, and DeNatale and Wassall acknowledge that the first statewide survey of creative industries dates back to 1973, followed by an arts impact study in 1978 (*The Arts and the New England Economy*, 1980). These and subsequent studies formed the basis for the NEFA tripartite model. Most of this work was classificatory rather than conceptual or theoretical.

As DeNatale and Wassall (2006) go on to write: "Richard Florida's 2002 book, *The Rise of the Creative Class*, brought the concept of a creative economy to a national audience." Florida's use of the term "creative economy" is more expansive than many previous approaches, springing from a growing understanding of the way digital production has changed the dissemination and control of intellectual property, rather than simply the economics of the cultural and creative industries. The economics of the cultural enterprise is a strand of economics that began in the early 1970s around the *Journal of Cultural Economics*, established in 1973, and the Association of Cultural Economics International, informally organized in 1979.[2] This form of analysis tended to be concerned with the economic impact of non-profit cultural organizations such as galleries, theatres, television, and cultural tourism.[3] DeNatale and Wassall (2006) suggest that Richard Caves's 2000 book, *Creative Industries: Contracts Between Art and Commerce* is the standard work defining the field.[4] According to Caves (2000), creative industries are characterized by seven economic properties:

1. *Nobody knows principle:* Demand uncertainty exists because the consumers' reaction to a product are neither known beforehand nor easily understood afterward.

2. *Art for art's sake:* Workers care about originality, technical professional skill, harmony, etc. of creative goods and are willing to settle for lower wages than offered by "humdrum" jobs.

3. *Motley crew principle:* For relatively complex creative products (e.g., films), the production requires diversely skilled inputs. Each skilled input must be present and perform at some minimum level to produce a valuable outcome.

4. *Infinite variety:* Products are differentiated by quality and uniqueness; each product is a distinct combination of inputs leading to infinite variety options (e.g., works of creative writing, whether poetry, novel, screenplays, or otherwise).

5. *A list/B list:* Skills are vertically differentiated. Artists are ranked on their skills, originality, and proficiency in creative processes and/or products. Small differences in skills and talent may yield huge differences in (financial) success.

6. *Time flies:* When coordinating complex projects with diversely skilled inputs, time is of the essence.

7. *Ars longa:* Some creative products have durability aspects that invoke copyright protection, allowing a creator or performer to collect rents.

The Creative Industries and Economic Policy

At the national policy level, the most widely cited definition for the creative economy comes from the UK Government Department for Culture, Media and Sport (DCMS), referring to "those industries which have their origin in individual creativity, skill and talent . . . through the generation and exploitation of intellectual property" (DCMS, 2001, p. 4). These industries include advertising; architecture; art and antiques markets; computer and video games; crafts; design; designer fashion; film and video; music; performing arts; publishing; software; television and radio.[5] More recently, the UN *Creative Economy Report 2008* identifies creative industries with "the creation, production and distribution of goods and services that use creativity and intellectual capital as primary inputs [to] produce tangible goods and intangible intellectual or artistic services" (p. 4).

The Creative Industries Task Force (CITF) set up under Tony Blair's new Labor government identified six core issues underlying creative industries: export promotion; skills and education; access to finance; taxation and regulation; intellectual property rights; and regional issues and commissioned the Creative Industries Mapping Documents (2001) which identified the activities and economic performance of each creative industry; their potential for growth; and the key barrier to growth. The taskforce also initiated a series of mapping exercises.[6]

It is clear that the twin policy imperatives for the UK and New England developed first, and then came the more theoretical and general accounts even if it is the case that the emergence of these policy notions and measurements of the "creative economy" took place within a changing theoretical environment that, in particular, included new understandings such as the "knowledge economy" (OECD, 1996), endogenous growth theory (Romer, 1990), and the "learning economy" (Lundvall & Johnson, 1994). The "creative economy" is in fact a conceptual outgrowth of a changed understanding based on work completed on the "knowledge economy" during the 1990s and early 2000s (Carafa, 2008) as a development of earlier theories in sociology and management concerning the "information society" and "post-industrial society" (Peters et al., 2009; Peters, 2007; Peters & Besley, 2006). In the U.K.'s white paper developed by the Department for Trade and Industry (1998) *Our Competitive Future: Building the Knowledge-Driven Economy*, a knowledge-based economy is defined as: " . . . one in which the generation and the exploitation of knowledge has come to play the predominant part in the creation of wealth. It is not simply about pushing back the frontiers of knowledge; it is also about the more effective use and exploitation of all types of knowledge in all manner of activity."[7] It is suggested that knowledge is more than just information and cites a distinction between codified and tacit knowledge. Codifiable knowledge can be written down and easily transferred to others, whereas tacit knowledge is "often slow to acquire and much more difficult to transfer." Knowledge was included by the World Bank as a theme in its 1998 *World Development Report*: "For countries in the vanguard of the world economy, the balance between knowledge and resources has shifted so far towards the former that knowledge has become perhaps the most important factor determining the standard of living. . . . Today's most technologically advanced economies are truly knowledge-based."[8] In this white paper, the Department for Trade and Industry (1998) also notes that the OECD has drawn attention to the growing importance of knowledge, indicating that the emergence of knowledge-based economies has significant policy implications for the organization of production and its effect on employment and skill requirements. The report emphasizes "new growth theory," charting the ways in that education and technology are now viewed as central to economic growth. Neoclassical economics is limited in that it does not specify how knowledge accumulation occurs and thus cannot acknowledge externalities. It also fails to consider human capital, suggesting that education has no direct role. In contrast, new growth theory has highlighted the role of education in the creation of human capital and in the production of new knowledge (see Solow 1956, 1994). On this basis, it has explored the possibilities of education-related externalities. Not only do research and development expenditures contribute to productivity growth, but education is also important in explaining the growth of national income.

As Chapter 1 of the UK report *Creative Britain* (2008) makes clear, new modes of education are critical to this discussion:

> Ideas are the raw material of the creative industries. But unlike those for traditional products, we cannot dig them out of the ground or pick them off trees. Ideas are generated through individual and collective talent and innovation. That is why it is essential to prepare the ground thoroughly—to give every child and young person the opportunity to develop their creative talent to the full. So we will establish the "Find Your Talent" program for children and young people—to ignite a desire for creativity in the next generation, and provide additional means to discover talent where it might otherwise have lain hidden.

The report suggests that "there is a growing recognition of the need to find practical ways of nurturing creativity at every stage in the education system: from the nursery through to secondary school; whether in academic or vocational courses; on apprenticeships or at university."[9]

Aspects of the Creative Economy

The *Creative Economy Report* (2008), an initiative of the partnership between UNCTAD and the UNDP Special Unit for South-South Cooperation bringing together five United Nations Bodies (UNCTAD, UNDP, UNESCO, WIPO, and ITC), begins its overview with the following:

> In the contemporary world, a new development paradigm is emerging that links the economy and culture, embracing economic, cultural, technological and social aspects of development at both the macro and micro levels. Central to the new paradigm is the fact that creativity, knowledge and access to information are increasingly recognized as powerful engines driving economic growth and promoting development in a globalizing world. "Creativity" in this context refers to the formulation of new ideas and to the application of these ideas to produce original works of art and cultural products, functional creations, scientific inventions and technological innovations. There is thus an economic aspect to creativity, observable in the way it contributes to entrepreneurship, fosters innovation, enhances productivity and promotes economic growth. (p. 3)

The report notes that the creative economy is an evolving concept that can "foster income generation, job creation and export earnings while promoting social inclusion, cultural diversity and human development"; that "[i]t is a set of knowledge-based economic activities with a development dimension" with creative industries at the heart, embracing "economic, cultural and social aspects interacting with technology, intellectual property and tourism objectives" (p. 4). The report defines "creative industries" as "cycles of creation, production and distribution of goods and services that use creativity and intellectual capital as primary inputs. They comprise a set of knowledge-based activities that produce tangible goods and intangible intellectual or artistic services with creative content, economic value and market objectives" (p. 4).

Furthermore, it usefully defines creativity in relation to artistic, scientific, and economic creativity, all of which involve technological creativity:

- artistic creativity involves imagination and a capacity to generate original ideas and novel ways of interpreting the world, expressed in text, sound and image;

- scientific creativity involves curiosity and a willingness to experiment and make new connections in problem solving;

- economic creativity is a dynamic process leading towards innovation in technology, business practices, marketing, etc., and is closely linked to gaining competitive advantages in the economy (p. 9).

The report also makes a distinction between cultural goods (and services) and creative goods, suggesting that the former are a subset of the latter, indicating different models of "creative industries" depending on the model accepted. The UN report here, against DeNatale and Wassall (2006), gives priority to the "creativity as innovation" model over "cultural economics." This is a decisive shift that moves away from taxonomy and classification of creative industries and its measurement to entertaining new theoretical frameworks for the investigation of why creativity and innovation occupy the central position in economic, social, and cultural life.

Creativity and the Global Knowledge Economy

There is now widespread agreement among economists, sociologists, and policy analysts that creativity and innovation are at the heart of the global knowledge economy: together creativity and innovation define knowledge capitalism and its ability to continuously reinvent itself. Together and in conjunction with new communications technologies they give expression to the essence of digital capitalism—the "economy of ideas"—and to new architectures of mass collaboration that distinguish it as a new generic form of economy different in nature from industrial capitalism. The fact is that knowledge in its immaterial digitized informational form as sequences and value chains of 1s and 0s—ideas, concepts, functions, and abstractions—approaches the status of pure thought. Unlike other commodities, it operates expansively to defy the law of scarcity that is fundamental to classical and neoclassical economics and to the traditional understanding of markets. A generation of economists have expressed this truth by emphasizing that knowledge is (almost) a global public good: it is non-rivalrous and barely excludable. It is non-rivalrous in the sense that there is little or marginal cost to adding new users. In other words, knowledge and information, especially in digital form, cannot be consumed. One's use of knowledge or information as digital goods can be distributed and shared at no extra cost, and the distribution and sharing are likely to add to its value rather than to deplete it or use it up. This concept reflects the essence of the economics of file-sharing; it is also the essence of new forms of distributed creativity, intelligence, and innovation in an age of mass participation and collaboration. Only when knowledge is codified, when it is converted from its immateriality at the ideation stage to become embodied as a physical resource—a book, CD, tape, image, movie, or code—can it easily become a form of property and therefore be bought and owned. Taken together, the ubiquitous presence of digital technologies and the impact of networked collaboration constitute a new mode of information production that is reshaping societies around the world.

In *Creativity and the Global Knowledge Economy* (Peters, Marginson, & Murphy, 2009), we described this trend using the following vocabulary. Today there is a strong renewal of interest by politicians and policy-makers worldwide in the related notions of creativity and innovation, especially in relation to terms like *the creative economy, knowledge economy, enterprise society, entrepreneurship,* and *national systems of innovation.*[10] In its rawest form, the notion of the creative economy emerges from a set of claims that suggests that the Industrial Economy is giving way to the Creative Economy based on the growing power of ideas and virtual value—the turn from steel and hamburgers to software and intellectual property.[11] In this context, policy increasingly latches on to the issues of copyright as an aspect of IP, piracy, distribution systems, network literacy, public service content, the creative industries, new interoperability standards, the WIPO, and the development agenda, WTO and trade, and means to bring creativity and commerce together.[12] At the same time, this focus on creativity has exercised strong appeal to policy-makers who wish to link education more firmly to new forms of capitalism emphasizing how creativity must be taught, how educational theory and research can be used to improve student learning in mathematics, reading, and science, and how different models of intelligence and creativity can inform educational practice.[13] Under the spell of the creative economy discourse, there has been a flourishing of new accelerated learning methodologies together with a focus on giftedness and the design of learning programs for exceptional children.[14] One strand of the emerging literature highlights the role of the creative and expressive arts, of performance, of aesthetics in general, and the significant role of design as an underlying infrastructure for the creative economy.[15]

Rise of the Creative Economy

In the last twenty years, we have moved from the postindustrial economy to the information economy to the digital economy to the knowledge economy to the "creative economy." The notion of creative economy, pioneered in different ways by Charles Landry, John Howkins, Richard Florida, and Charles Leadbeater early in this decade, increasingly has become associated with postmarket notions of open source public space, democratized creativity, and intellectual property law that has been relativized to the cultural context emphasizing the socio-cultural conditions of creative work. Alongside this development, the notion of entrepreneurship, as interpreted originally by Schumpeter, breaks out of its business origins, becoming a rubric for larger transformation, and a set of infrastructural conditions enabling creative acts. Likewise the endogenous growth theory developed simultaneously by Paul Romer and others in economics has opened a space for the primacy of ideas and installed continuous innovation as mainstream OECD economic policy. These moves have brought to the forefront forms of knowledge production based on the commons and driven by ideas not profitability per se; and have posed the question of not just "knowledge management" but the design of "creative institutions" embodying new patterns of work.

Education in the Creative Economy

We seem to be moving into a different world now; a world in which the raw materials are no longer coal and steel produced by machines but creativity and meaning produced by the human imagination. Beyond conventional discussions on the knowledge economy, many scholars suggest that creative work and a rising "creative class" are fomenting shifts in advanced economies from mass production to creative innovation. Emerging along several paths, Charles Landry, John Howkins, and Richard Florida have been pioneers in understanding these dynamics. The publication of Landry's *The Creative City* (2000), Howkins's *The Creative Economy*

(2001) and Florida's *The Rise of the Creative Class* (2002) has catalyzed a rich discourse on the value and importance of creativity to the global economy. Laying the foundations, John Howkins offered the first account of this new economy, even as Charles Landry explored the possibility of developing benchmarks for stoking creative cities. Most recently, Richard Florida has offered an empirical account of the logic and dynamism of the creative economy.

We know that creativity and innovation have become critical to understanding the complex challenges facing us in the twenty-first century. In this volume we examine the contours of the creative economy discourse and consider its implications for education. Bringing together eminent scholars and practitioners from around the world, we consider the need for new modes of education that respond to the growing importance of creativity to a global economy and society.

John Dewey once said that education is the foundation for an ever-evolving economy and culture. This vision has clearly become reality today. Much as the assembly line shifted the key factor of production from labor to capital, computer networks are now shifting the key factor of production from capital to innovation. It seems increasingly clear that information and communications technologies (ICTs) are restructuring global production so that innovation is now anchored to social networks that criss-cross nations, cultures, and peoples.

In *Education in the Creative Economy*, we want to explore the need for new modes of education that can effectively tap the collective intelligence that powers these social networks. One of the major questions that we explore through this book is "What systems, policies and structures are most conducive to making it possible for the largest number of people in a society to participate in the creation and development of new cultural forms?" Creativity has become the economic engine of the twenty-first century and this volume explores the changing landscape in which creativity and innovation are now embedded.

The Organization of This Book

Part One: Educational Policy

In Part One, Educational Policy, we examine the contours of the creative economy in the context of educational policy, looking particularly at the rising importance of education for creativity and innovation. Tracing current socio-economic discourse on the creative economy, we examine the underlying logic of creativity and innovation and consider the factors of production that are linked to national systems of education.

Chapter 1 begins with Araya's examination of the creative economy discourse in the context of education and the changing dynamics of the global economy. In his view, digital networks serve as platforms for collective intelligence and should be seen as the key to renewing systems of education in advanced countries.

In Chapter 2, Cunningham and Jaaniste explore the policy milestones that mark the evolution of creativity in public policy. As they conclude, education policy today must support a rapprochement of the arts and sciences in order to better coordinate disciplines and engender the necessary human capital for an emergent creative economy.

In Chapter 3, Florida, Knudsen, and Stolarick offer an empirical study of the economic role of universities through the lens of Florida's 3Ts (technology, talent, and tolerance) of economic development. They suggest that the university's role in the first T, technology, while important, is overemphasized by most theories on innovation; they contrast the trend by examining the role of universities in attracting and mobilizing talent, and in establishing a diverse social climate.

In Chapter 4, Flew examines dynamic trends in economic geography and considers their implications for universities. As he concludes, universities that see their future development as linked to creative clusters will need to make serious commitments to the social environments in which they are embedded.

In Chapter 5, Hearn and Bridgstock pose the question: "to what extent do current education theory and practice prepare graduates for the creative economy?" They go on to explore innovation, transdisciplinarity, and networks as the core of the creative economy and examine the need for redesigning educational policy and practice for this changing milieu.

In Chapter 6, Brown and Lauder critique the technocratic account of the knowledge economy. Challenging the dominant theories on education in a global knowledge-based economy, they argue that supra-national forces, including transnational corporations, have exploited the digital revolution to organize and standardize global production under a kind of "digital Taylorism."

In Chapter 7, Pitroda explores the need for a paradigm shift in education. Outlining the goals behind India's National Knowledge Commission, he elaborates on a blueprint for reform of knowledge institutions and infrastructure to support India's knowledge economy. As he suggests, education is critical to India's future.

In Chapter 8, Lundvall, Rasmussen, and Lorenz consider the constant need for new competencies in an age in which innovation makes knowledge obsolete. Looking at learning and education from the context of Europe's learning economy, they argue that educational policy should focus on collaboration and interdisciplinarity in order to prepare people for participation in a learning economy and society.

Finally in Chapter 9, Rooney argues that a lack of adequate conceptual frameworks for knowledge production keeps policymakers unnecessarily anchored to an instrumentalist logic. Instead, he links knowledge and creativity to wisdom and values and explores the complex adaptive systems out of which creativity and wisdom emerge.

Part Two: Technology and Economy

In Part Two, Technology and Economy, we look closely at information and communications technologies and their relationship to collaboration in the creative economy. Beyond the command-and-control systems characteristic of industrial society, digital technologies have become fundamental to a network society. Technology is now so critical to such a wide range of overlapping industries and disciplines that conventional boundaries seem to be breaking down. Underlying this socioeconomic restructuring is the critical importance of digital networks as platforms for collaborative innovation.

Chapter 10 begins with Peters's exploration of openness and creativity from the perspective of decentralized networked communications and a global knowledge economy. As he points out, digitization transforms all aspects of cultural production and consumption. New digital logics alter the organization of knowledge, education, and culture and spawn new technologies as a condition of open innovation.

In Chapter 11, Aigrain, Chan, Guédon, Willinsky, and Benkler reflect on the notions of peer production introduced in Benkler's landmark book, *The Wealth of Networks* (2006). Aigrain begins by asking, "How does the growth of information commons and related non-market activities interact with the monetary economy?" Chan explores a parallel question in terms of human development and poverty alleviation. Guédon, in turn, considers the broader anthropological questions introduced by peer production. Lastly, Willinsky compares Benkler's work with Adam Smith's *Wealth of Nations* (1776/1910) and ponders the implications of peer production for transforming education. In response, Benkler attempts to answer these difficult questions in terms of the emergent logic of peer production itself.

In Chapter 12, Howkins considers the tension between intellectual property (as the lynch-pin of a creative ecology) and the growing importance of commons-based peer production. What, he asks, is the right way to regulate the ownership of ideas in the twenty-first century?

In Chapter 13, Fitzgerald and Shi examine copyright issues emerging with the evolutionary dynamics of networked innovation and ask the question "to what extent should copyright law allow copyright owners the right to control reproduction and communication to the public?" In their view, copyright law should not only facilitate the opportunity to create but also make possible the opportunity to distribute and communicate creative material to the broadest possible audience.

Following on Fitzgerald and Shi, Pasquinelli makes a dynamic argument for defending the commons against capitalist exploitation. In his view, the grammar of sabotage has become the *modus operandi* of the multitudes captive to the network society of cognitive capitalism. Simply put, sabotage has become the only possible gesture to defend the commons.

In Chapter 15, Bauwens argues that peer production represents a revolutionary mode of political economy that transcends capitalism. Beyond the recent economic crisis, he explores the possibilities of a phase transition into a postcapitalist era centered on peer production.

In Chapter 16, Murphy examines Schumpeter's notions of innovation in light of the recent global recession and the ongoing competition between various economic and social dogmas. As he suggests, creativity is born of paradox and contradiction; cultures that can internalize and integrate opposing views are the crucibles of peak creation.

In Chapter 17, Landry summarizes his notion of the Creative City and attempts to assess how creative thinking and new forms of learning might play a role in the creative ecology of cities. As he suggests, the Creative City is ultimately driven by learning because learning and education are central to the creative milieu.

Lastly, in Chapter 18, Nederveen Pieterse offers a critical review of the challenges facing the United States and other advanced capitalist countries under the spell of innovation rhetoric. The main problem for the United States, he concludes, is that American corporations have become complacent, dependent upon low-wage, low-tax, and low-regulation environments.

Part Three: Culture and Curriculum

In Part Three, Culture and Curriculum, we examine the growing importance of cultural production and explore the interface between innovation and design in the context of educational renewal. While a society based on industrial production could once effectively deliver a single, standardized curriculum in support of a Fordist economy, it is becoming obvious that the twenty-first century requires a different model of education.

Today, public education systems desperately need to be redesigned to embrace tools and practices that tap the indigenous talents of students. Part Three begins with Balsamo's examination of the rapid technological changes impacting knowledge and learning. In this context, she explores the notion of the Singularity and considers the key institutional elements necessary to cultivate the "technological imagination."

In Chapter 20, Whitney considers the fundamental economic and technological challenges facing schools today. As he suggests, schools need to become creative hubs at the center of networks of learning and innovation.

Following Whitney, McWilliam, Tan, and Dawson examine the challenges of embedding ICTs in contemporary public schools. As they suggest, the nexus between creativity and digitality has become critical to the educational sector, and yet schools appear to be unable to make the necessary cultural and pedagogical shift to meet this challenge.

In Chapter 22, Besley considers the growing questions surrounding youth identity in a digital age. Looking particularly at recent empirical research on youth identity, she examines the creativity of youth in the construction of emergent subjectivities while engaging and negotiating social media. In her view, creativity has become fundamental to a post-Fordist age, and yet schools and universities seem to actively discourage its development.

In Chapter 23, Cormier asks "What is the curriculum for creativity and innovation?" He suggests that the answer lies with the community as curriculum. That is, community as a distributed learning network in which learning is collaboratively generated and shared.

In Chapter 24, McCulloch-Lovell explores the "creative campus" movement and asks the question, "Are colleges and universities truly fostering the conditions in which innovation and creativity flourish?" In her view, valuing creativity means developing systems, measures, and even budgets that encourage creativity.

In Chapter 25, Parsons examines the importance of art education in educating for creativity. He suggests that art is the only subject where creativity is an inherent value of the subject as a conceptual structure. While the current discourse of social and educational policy stresses the importance of nurturing creative scientists, mathematicians, and technologists, he suggests that art and design have a special relation to creativity and should therefore be an explicit target of teaching.

In Chapter 26, de la Fuente observes that art is now so fully integrated with economics that the global economy increasingly functions as if art were the model for the whole of the market. In his view, one of the key institutions in this transformation has been the emergence of the art school, and its blending of the bohemian with the entrepreneur.

Following de la Fuente, Holden explores the need to democratize the arts in order to ensure that they are not the preserve of a cultural elite. In his view, a community of self-governing citizens, a *demos*, understands, creates, and reinvigorates itself through culture. He argues that the aim of a democratic society is to release the talents of all its citizens and not just an elite few.

In Chapter 28, Cope and Kalantzis consider the importance of design to the creative economy. They examine design as both a discipline and a process of meaning making, and explore a future-oriented agenda for design vocations.

Finally in Chapter 29, Strand asks the questions: "What are our images of creativity? And how do these images relate to our ways of seeing workplace learning within the new and globalized symbolic economy?" In this chapter, she addresses these questions through three philosophical discourses that metaphorize creativity as "expression," "production," and "reconstruction" in the context of workplace learning.

References

Baumol, W. J. (2002). The free-market innovation machine: Analyzing the growth miracle of capitalism. Princeton, NJ: Princeton University Press.

Baumol, W. J., & Bowen, W. G. (1966) *Performing Arts: The Economic Dilemma. The economics of the arts.* London, Hartin Robinson, 42–57.

Benkler, Y. (2006). *The wealth of networks: How social production transforms markets and freedom.* New Haven: Yale University Press.

Blaug, M. (2002). Where are we now on cultural economics? *Journal of Economic Surveys, 15*(2), 123–143.

Blythe, M. (2000). Creative learning futures: Lterature review of training & development needs in the creative industries. Retrieved from http://www.cadise.ac.uk/projects/creativelearning/New_Lit.doc.

Carafa, A. (2008). The creative sector and the knowledge economy in Europe: The Case of the United Kingdom's creative economy programme. Retrieved from http://papers.ssrn.com/sol3/papers.cfm?abstract_id=1403465

Caves, R. E. (2000). *Creative industries: Contracts between art and commerce.* Cambridge: Harvard University Press.

Cowen, T. (2002). *Creative destruction: How globalization is changing the world's cultures.* Princeton, NJ: Princeton University Press.

Davenport, T., & Beck, J. (2001). *The attention economy: Understanding the new economy of business.* Cambridge, MA: Harvard Business School Press.

DeNatale, D., & Wassall, G. H. (2006). *Creative economy research in New England: A reexamination.* Retrieved from http://www.bu.edu/artsadmin/news/symposium/denatale_paper.pdf.

DeNatale, D., Wassall, G. H. (2007). *The creative economy: A new definition.* The New England Foundation for the Arts (NEFA). Retrieved from http://www.nefa.org/sites/default/files/ResearchCreativeEconReport2007.pdf.

Department for Trade and Industry UK. (1998). *Our competitive future: Building the knowledge-driven economy.* London.Retrieved from http://www.stats.bis.gov.uk/competitiveness5/Past%20Indicators/UKPC1999.pdf.

Department for Culture Media and Sport (2008). Creative Britain: New talents for the new economy. Retrieved from: http://www.culture.gov.uk/reference_library/publications/3572.aspx/

Florida, R. (2002). *The rise of the creative class.* New York: Basic Books.

Frey, B. (2000). Arts and economics: Analysis and cultural policy. New York: Springer.

Frey, B. S., & Pommerehne, W. W. (1989). *Muses and markets: Explorations in the economics of the arts.* Cambridge, MA: Blackwell.

Ginsburgh, V. A., & Menger, P.-M. (Eds.). (1996). *Economics of the arts.* Amsterdam: North Holland.

Gordon, W. J. (1996). A property right in self-expression: Equality and individualism in the natural law of intellectual property. *Yale Law Journal, 102,* 1533, 1568–1572.

Heilbrun, J., & Gray, C. M. (2001). *The economics of art and culture* (2nd ed.). New York: Cambridge University Press.

Hesmondhalgh, D. (2002). *The cultural industries.* London: Sage.

Howkins, J. (2001). *The creative economy: How people make money from ideas.* London: Allen Lane.

Hughes, J. (1977) The philosophy of intellectual property. 77. *GEO. L.J.* 287, 337–344.

Landry, C. (2000). *The creative city: A toolkit for urban innovators.* London: Earthscan.

Lash, S., & Urry, J. (1994). *Economies of signs and space.* Thousand Oaks, CA: Sage.

Lemley, M. A. (2005). Property, intellectual property, and free riding. 83 *Texas Law Review,* 1031.

Lundvall, B., & Johnson, B. (1994). The learning economy. *Industry & Innovation, 1*(2), 23–42.

Mt. Auburn Associates. (2000). *The creative economy initiative: The role of the arts and culture in New England's economic competitiveness.* Prepared for the New England Council.

NEFA. (1980). The arts and the New England economy. Cambridge, Mass.: New England Foundation for the Arts.

OECD. (1996). *Employment and growth in the knowledge-based economy.* Paris: The Organisation.

Peters, M. A. (2007). *Knowledge economy, development and the future of higher education.* Rotterdam: Sense Publishers.

Peters, M., Marginson, S. & Murphy, P. (2009). *Creativity and the global knowledge economy.* New York: Peter Lang.

Peters, M. A., & Besley, T. (A. C.) (2006). *Building knowledge cultures: Education and Development in the Age of Knowledge Capitalism.* Lanham, MD: Rowman & Littlefield.

Romer, P. (1990). Endogenous technological change. *Journal of Political Economy,* 98(5), 71–102.

Shapiro, C., & Varian, H. (1998). *Information rules: A strategic guide to the network economy.* Cambridge, MA: Harvard Business School Press.

Solow, R. (1956). A contribution to the theory of economic growth. *Quarterly Journal of Economics,* 70, 65–94.

Solow, R. (1994). Perspectives on growth theory. *Journal of Economic Perspectives,* 8, 45–54.

United Kingdom Department for Culture, Media, and Sport. (1998). Creative industries mapping. London: Author.

United Kingdom Department for Culture, Media, and Sport. (2001). Creative industries mapping. London: Author.

United Nations (2008) *Creative economy report 2008. The challenge of assessing the creative economy: Towards informed policy making.* Retrieved from *http://www.unctad.org/en/docs/ditc20082cer_en.pdf.*

Wagner, R. P. (2003). Information wants to be free: Intellectual property and the mythologies of control. 103 *Columbia Law Review,* 995, 1001–1003.

Weinstock Netanel, N. (1996). Copyright and a democratic civil society. 106 *Yale Law Journal, 283,* 347–362.

Weinstock Netanel, N. (1998). Asserting copyright's democratic principles in the global arena. 51 *Vanderbilt Law Review,* 217, 272–276.

World Bank, the (1998) 1998 *World Development Report.* Retrieved from http://econ.worldbank.org/WBSITE/EXTERNAL/EXTDEC/EXTRESEARCH/EXTWDRS/EXTWDR19981999/0,,ImgPagePK:64202988~entityID:000178830_98111703550058~pagePK:64217930~piPK:64217936~theSitePK:766720,00.html.

Notes

1. See the Appendix for the NEFA classification system based on the definition for enterprises and occupations at http://www.nefa.org/sites/default/files/ResearchCreativeEconReport2007.pdf.
2. From the journal's Web site at http://www.culturaleconomics.org/history.html: "In 1973, Professor William Hendon of the University of Akron, Ohio, founded the *Journal of Cultural Economics,* and he organized the first international conference on cultural economics, at Edinburgh in 1979. He also started an Association for Cultural Economics (ACE) that held conferences in collaboration with host organizations in Maastricht, Netherlands (1982), Akron, Ohio, USA (1984), Avignon, France (1986), Ottawa, Canada (1988), Umea, Sweden (1990) and Fort Worth, Texas, USA (1992). In 1993, the ACE was transformed into the presently organized ACEI as a membership society with the election for officers and the adoption of the constitution."
3. DeNatale and Wassall (2006) mention the path-breaking work of William J. Baumol and William G. Bowen (1966) *Performing Arts: The Economic Dilemma* and Baumol's "cost-disease"—the problem of financing the performing arts in the face of ineluctably rising unit costs.

4. See Mark Blaug's (2002) "Where Are We Now on Cultural Economics?" where he addresses nine topics that cover the field (1) taste and taste formation, (2) demand and supply studies, (3) the media industries, (4) the art market, (5) the economic history of the arts, (6) the labor market for artists, (7) Baumol's cost disease, (8) non-profit arts organizations, and (9) public subsidies to the arts. See also Ruth Towse (2003) *A Handbook of Cultural Economics.*
5. See *The Creative Industries Fact File* at http://www.culture.gov.uk/PDF/ci_fact_file.pdf.
6. See http://www.culture.gov.uk/reference_library/publications/4740.aspx/(1998) and http://www.culture.gov.uk/reference_library/publications/4632.aspx/ (2001).
7. See http://www.dti.gov.uk/comp/competitive/main.htm.
8. See http://www.dti.gov.uk/comp/competitive/main.htm.
9. See http://www.culture.gov.uk/images/publications/CEPFeb2008.pdf.
10. For example, see: William J. Baumol, *The Free-Market Innovation Machine: Analyzing the Growth Miracle of Capitalism* (Princeton, NJ: Princeton University Press, 2002); Tyler Cowen. *Creative Destruction: How Globalization Is Changing the World's Cultures* (Princeton, NJ: Princeton University Press, 2002); Scott Lash and John Urry, *Economies of Signs and Space* (Thousand Oaks, CA: Sage, 1994).
11. See Richard Florida *The Rise of the Creative Class* (New York: Basic Books, 2002); John Howkins, *The Creative Economy: How People Make Money from Ideas* (London: Allen Lane, 2001); Charles, Landry, *The Creative City: A Toolkit for Urban Innovators* (London: Earthscan, 2000).
12. See Cowen, *Creative Destruction;* Carl Shapiro, and Hal Varian, *Information Rules: A Strategic Guide to the Network Economy* (Cambridge, MA: Harvard Business School Press, 1998); Thomas Davenport and John Beck, *The Attention Economy: Understanding the New Economy of Business* (Cambridge, MA: Harvard Business School Press, 2001); Justin Hughes, "The Philosophy of Intellectual Property," GEO. L.J. 287, 337–44 (1977); Neil Weinstock Netanel, "Asserting Copyright's Democratic Principles in the Global Arena," 51 *Vanderbilt Law Review*, 217 (1998): 272–76; Neil Weinstock Netanel, "Copyright and a Democratic Civil Society," 106 *Yale Law Journal* 283 (1996): 347–62; Wendy J. Gordon, "A Property Right in Self-Expression: Equality and Individualism in the Natural Law of Intellectual Property," 102 *Yale Law Journal,* 1533 (1996): 1568–72; Mark A. Lemley, "Property, Intellectual Property, and Free Riding," 83 *Texas Law Review,* 1031 (2005): 1031; R. Polk Wagner, "Information Wants to Be Free: Intellectual Property and the Mythologies of Control," 103 *Columbia Law Review,* 995 (2003): 1001–1003.
13. Mark Blythe. 2000. "Creative Learning Futures: Literature Review of Training & Development Needs in the Creative Industries." http://www.cadise.ac.uk/projects/creativelearning/New_Lit.doc.
14. See The Center for Accelerated learning at http://www.alcenter.com/; see, e.g., *The Framework for Gifted Education* at http://education.qld.gov.au/publication/production/reports/pdfs/giftedandtalfwrk.pdf.
15. Richard E. Caves, *Creative Industries: Contracts between Art and Commerce* (Cambridge, Mass: Harvard University Press, 2000); Bruno Frey, "Arts and Economics: Analysis and Cultural Policy (New York: Springer, 2000); Bruno S. Frey and Werner W. Pommerehne, *Muses and Markets: Explorations in the Economics of the Arts* (Cambridge, Mass: Blackwell, 1989); Victor A. Ginsburgh and Pierre-Michel Menger, eds.,

Economics of the Arts (Amsterdam: North Holland, 1996); James Heilbrun and Charles M. Gray, *The Economics of Art and Culture* (2nd ed.). (New York: Cambridge University Press, 2001); Desmond Hesmondhalgh, *The Cultural Industries* (Thousand Oaks, CA: Sage, 2002).

Part One: Educational Policy

1 Educational Policy in the Creative Economy

Daniel Araya

No social order ever perishes before all the productive forces for which there is room in it have developed; and new, higher relations of production never appear before the material conditions of their existence have matured in the womb of the old society itself.

—Karl Marx, *Contribution to the Critique of Political Economy*

We can no longer succeed—or even tread water—with an education system handed down to us from the industrial age, since what we no longer need is assembly-line workers. We need one that instead reflects and reinforces the values, priorities, and requirements of the creative age. Education reform must, at its core, make schools into places where human creativity is cultivated and can flourish.

—Richard Florida, *Flight of the Creative Class*

Structural changes in capitalist production linked to digital technologies have been widely apparent since at least the early 1970s. Institutional reconfigurations in the management of industrial production, coupled to technological shifts in industrial systems, have led to a wide-range of forecasts regarding a coming post-industrial economy. Much as skilled labor was the dominant social and political force in the twentieth century, knowledge workers are predicted to become the dominant social and political force in the twenty-first century. Beyond conventional discussions on the knowledge economy, however, many scholars today are now pointing to the growing importance of innovation and a *creative economy* (Rifkin, 2000; Leadbeater, 2000).

Richard Florida (2002), for example, argues that an emergent "creative class" is fomenting a shift in advanced economies from mass production to cultural innovation. Transcending and including knowledge workers (researchers, engineers, scientists), he suggests, is a growing segment of "cultural creatives" (writers, artists, producers, and designers) who form the vanguard of a coming creative age.

In this chapter, I examine the contours of the creative economy discourse and consider the implications of cultural production for institutions of learning and education. Beyond established arguments for a global knowledge economy, I argue that creativity is critical to the renewal of advanced capitalist countries. One major reason for the growing importance of creativity, I contend, is the convergence of creative industries and digital technologies in the context of a network-driven global economy.

Beyond the Knowledge Economy

Over the past half century, theorists like Peter Drucker (1966, 1985, 1993), Daniel Bell (1973), and Alvin Toffler (1970, 1980, 1990) have argued that the future of advanced capitalist countries is intimately connected to the exploitation of knowledge and information: Just as agricultural society was transformed by industrialization, so is industrial society being transformed by knowledge-based innovation. In his book, *A Whole New Mind* (2005), Dan Pink offers a cogent critique of this approach. He writes,

> For a nearly a century, Western society in general, and American society in particular, has been dominated by a form of thinking and an approach to life that is narrowly reductive and deeply analytical. Ours has been the age of the "knowledge worker," the well-educated manipulator of information and deployer of expertise. But that is changing. Thanks to an array of forces—material abundance that is deepening our nonmaterial yearnings, globalization that is shipping white collar work overseas, and powerful technologies that are eliminating certain kinds of work altogether—we are entering a new age. (p. 2)

In Pink's view, the knowledge-based economy has already peaked in advanced capitalist countries and is now migrating to Asia and elsewhere. Much as the routine mass production work that went before it, knowledge-based services in software, accounting, finance, telecommunications, and healthcare are increasingly shifting to newly industrializing countries (NICs). This is not to say that advanced economies no longer need knowledge workers but that knowledge-based labor is migrating downstream to developing countries or simply becoming embedded in information and communication technologies (ICTs).

As a rising tide of knowledge workers continues to grow outside of rich countries, predictions of a coming Western-biased knowledge economy look increasingly naive. What is clear is that globalization is reconfiguring *all* parts of the global economy including "left-brain" knowledge industries. Perhaps even more problematic is the growing importance of computer automation. While early industrial machines simply leveraged physical labor, ICTs have begun to displace human labor entirely.

Since the earliest days of mass manufacture, technology has been a necessary instrument in the industrialization of capitalist production. With the introduction of computers, however, this process has accelerated dramatically (Zuboff, 1988). The use of industrial robotics in construction and assembly has both advanced productivity and enhanced quality control while significantly reducing the need for human labor. At the same time, the evolution of computerized control systems is now increasingly enabling various industries to manage production with greater ease and precision. This technological shift is having a disconcerting impact on labor. Clear evidence of this is seen in falling rates of workers in the manufacturing and service sectors worldwide (Hilsenrath & Buckman, 2003).

Globalization 2.0

Advanced industrialized countries now appear bereft of a coherent post-industrial economic model. The increasing ascendancy of Korean technology, Indian software, and Chinese mass manufacture appears to be reordering the organization and distribution of global economic power. Consider, for example, China's impact on the Asian region:

> China's production chains are now the focal point around which the Asian regional economy spins. Replacing both Japan and the US, China has become the largest manufacturer and trading partner in an interregional market that hit US$722.2 billion in 2001 and had the fastest rate of growth in the world since 1985. Recently, intraregional trade accounted for the majority of Asia's export growth, with much of the increase flowing to China. China is now both Japan's and South Korea's largest trading partner. In fact, much of Japan's growing recov-

> ery depends on goods going to China. Previously idle capacity in construction machinery, steel and shipbuilding is now running at full stretch. Over the last year, Japan's exports to China have grown by 33.8 per cent while exports to the US have fallen by 5.4 per cent. (Harris, 2008, p. 174)

In 2009, China became the second-largest economy in the world after the US. Within three decades, China is predicted to become the world's largest economy and India, the third largest. The simultaneous economic ascendance of China and India (two nations that together account for one-third of the world's population) is nothing less than astounding. Given their current trajectory, most economists predict that India and China will likely account for half of global output by the middle of this century.

If the twentieth century was the American century, then the twenty-first century is likely the Chinese century. By every measure (consumer markets, investment, domestic savings, energy use, global exports, rate of growth) China is becoming a global super-power. Globalization itself seems to have entered a new phase in which NICs have considerable competitive advantage (Frank, 1998). The spread of global value chains, for example, has created a new level of complexity in international markets that is unprecedented. Steady deindustrialization of advanced capitalist countries alongside the rapid industrialization of countries in Asia has made the Asian region the center of industrial manufacturing. This trend will only grow deeper as education and skills development in NICs continues to improve.

The recent financial crisis in the U.S. and Europe has only served to exaggerate this global economic restructuring. The crash of 2008 inflicted profound damage to Western countries and their dominance over global trade and finance. The IMF (International Monetary Fund) estimates that loan losses for global financial institutions will eventually reach $1.5 trillion (Altman, 2009). In an article for *Foreign Affairs*, Altman outlines some of the geopolitical consequences of the recent crisis. As he concludes, the Western dominated international financial system has been devastated. While the U.S. share of world GDP has been declining for the past seven straight years, its geopolitical authority has now been significantly crippled. In response to the financial crisis, central banks in the United States and Europe have injected a total of 2.5 trillion dollars of liquidity into the credit markets (by far the biggest monetary intervention in history). What is obvious is that the credibility of Anglo-American laissez-faire capitalism has collapsed. As Altman concludes, Western governments simply "have neither the resources nor the economic credibility to play the role in global affairs that they otherwise would have played" (p. 1).

China, on the other hand, has been relatively insulated from the Western financial contagion. While experiencing its own economic slowdown, China's financial system was largely undamaged. Its foreign exchange reserves now approach $2 trillion, making it the world's strongest country in terms of liquidity. In financial terms China was little affected by the financial crisis:

> [China's] entire financial system plays a relatively small role in its economy, and it apparently has no exposure to the toxic assets that have brought the U.S. and European banking systems to their knees. China also runs a budget surplus and a very large current account surplus, and it carries little government debt. Chinese households save an astonishing 40 percent of their incomes. And China's $2 trillion portfolio of foreign exchange reserves grew by $700 billion last year, thanks to the country's current account surplus and foreign direct investment. (Altman, 2009, p. 5)

Largely driven by domestic demand, the IMF forecasts Chinese GDP to continue to grow at a rate of 8.5 percent. As China and the Association of Southeast Asian Nations (ASEAN) move closer to building the world's largest free-trade area, China has the opportunity to solidify its strategic advantage as a global power. With China's GDP projected to become the largest in the world, East Asia's geopolitical importance has become undeniable.

Economic Policy in the Creativity Economy

One increasingly important question that advanced economies must now seriously consider is "what next?" *What remains after we have mechanized agriculture, industry, and messaging technologies* (Lèvy, 1997)? Alongside discourse on a knowledge economy, many economists are now pointing to the increasing importance of *creativity* and a creative economy. Florida (2002), in particular, has argued that a new creative class made up of intellectuals, artists, and designers is an ascendant force today that is reshaping advanced capitalist countries. He elaborates,

> In 1900, creative workers made up only about 10% percent of the U.S. workforce. By 1980, that figure had risen to nearly percent. Today, almost 40 million workers—some 30 percent of the workforce—are employed in the creative sector. . . . When we divide the economy into three sectors—the creative, manufacturing and service sectors—and add up all the wages and salaries paid, the creative sector accounts for nearly half of all wage and salary income in the United States. That's nearly $2 trillion, almost as much as manufacturing and services combined. (Florida, 2007, pp. 29–30)

Commentators have debated changes to the industrial economy for several decades now. Beginning with Peter Drucker's (1959, 1966) predictions of a rising class of knowledge workers in the 1960s and continuing through Daniel Bell's (1973) explorations of a coming postindustrial society in the 1970s, contemporary discussions on the creative economy are only the most recent waypoint in this cultural migration (Healy, 2002). Linking discussions on the creative economy to broad structural mutations in the technologies underlying capitalist production, there are at least four common threads linking this discourse:

1. The diffusion of ICTs and consequent transformations in Fordist production.

2. The growing significance of a global market and globally fragmented production systems.

3. The increasing importance of highly educated workers or human capital within continuous cycles of creative innovation.

4. The rise of alternative centers of production outside advanced industrial countries.

Generally speaking there are two heavily overlapping modalities for understanding what is meant by the "creative economy." The first modality argues that *creative industries* and the cultural sector more broadly, represent a highly energized and growing portion of the broader economy. The second modality explores *creativity as an axial principle* underlying postindustrial shifts linked to globalization. Looking at both modalities in detail we see significant differences in their definitions of the "creative economy."

1) Creative Industries

The first modality for defining the creative economy is linked to discussions on "creative industries" as a growing sector. These industries include publishing, music, visual/performing arts, film, media, architecture, advertising and design. Since the 1990s, policymakers have developed fairly elaborate definitions of creative industries in the context of broader national innovation strategies. The idea of creative industries has existed for some time, however. Beginning with Adorno and Horkheimer's (1944/1977) early neo-Marxist critiques of mass media and the "culture industry" and evolving into a complex, though highly contested discourse on the nature and function of art and culture in the global market.

Creative industries today are estimated to be growing globally at an average rate of 8.7% per year (UNCTAD, 2008, p. 24). U.S. creative industries (defined in terms of arts, media, and design), for example, are estimated to make up 8% of the national GDP, outstripping auto production, aircraft production, agriculture, electronics, and computer technologies (Siwek, 2002). The annual growth rate of creative industries in OECD countries during the 1990s was twice that of the service industries overall and four times that of manufacturing overall (Howkins, 2001, p. xvi). World exports of visual arts, for example, more than doubled from $10.3 billion in 1996 to $22.1 billion in 2005. Between 2000 and 2005, for example, world trade in creative goods and services reached $424.4 billion in 2005 or 3.4% of the total world trade:

> World exports of creative products were valued at $424.4 billion in 2005 as compared to $227.5 billion in 1996, according to preliminary UNCTAD figures. Creative services in particular enjoyed rapid export growth—8.8. per cent annually between 1996 and 2005. This positive trend occurred in all regions and groups of countries and is expected to continue into the next decade, assuming that the global demand for creative goods and services continues to rise. (UNCTAD, 2008, p. iv)

While developed countries produce and consume the lion's share of the global market in creative products and services, many developing countries, particularly countries in Asia, are beginning to see growing returns. One striking example of this emerging pattern is the increasing dominance of Asia in the area of technology-related creative goods, such as computers, cameras, televisions, and audiovisual equipment. From 1996 to 2005 exports in these key industries grew from $51 billion to $274 billion (UNCTAD, 2008, p. 6). Not surprisingly, China has (since 2005) become the world's leading producer and exporter of value-added creative products.

The major challenge for understanding the creative economy in terms of creative industries, however, lies in defining the scope and breadth of these industries. More problematic than this are their relatively marginal levels of employment. Taken as a whole, the percentage of employment in the creative industries is quite small. In the United States, for example, creative industries accounted for just 2.5 per cent of total employment in 2003. Nederveen Pieterse (this volume) puts it this way,

> If we interpret the cultural economy as a sector (including, e.g., Hollywood, television, the arts, design, fashion) it is vibrant and significant, but not nearly significant enough in job creation to make up for the millions of jobs lost in

manufacturing and through outsourcing . . . The cultural economy, though surely significant, is simply not large and substantial enough to employ enough American workers; just as software, high-tech and back office services in India will never employ enough of India's workforce.

2) Creativity as Axial Principle

The second modality for understanding the creative economy is much broader and more diffuse. It views creativity as vital to the economy in general and fundamental to a technology-driven global economy in particular. Following this line of reasoning, Howkins (2001) and Florida (2002) make creativity the axial principle of postindustrial capitalism. Building out from a "super-creative core" of scientists, engineers, architects, designers, musicians, artists, educators, and entertainers, Florida suggests that the creative economy constitutes 30% of the U.S. workforce (with the supercreative core representing only 12% and a larger contingent of creative professionals in business, finance, health, law, accounting, and related professions representing 18%).

Underlying this version of the creative economy is an argument that creativity is now the key driver of global innovation. This does not mean that creativity is itself an economic activity but that creativity becomes an economic activity "when it produces an idea with economic implications or a tradable product" (Howkins, 2001, p. x).

Critics of this version of the creative economy argue that the celebration of creative workers minimizes class stratification and ignores the systems of exploitation that undergird capitalist economy. From a conventional class analysis, the creative economy does not fundamentally change the nature of exploitation within capitalist production. In this sense, the "Creative Class" is only the newest link in a very long chain of social prophecies extending back through discussions on the evolution of modern Western civilization (Barbrook, 2006). Though differing in emphasis, each of these predictions finds a common root in a Western eschatological approach to history. Oscillating between a "new ruling class" and a "new working class," each prediction has attempted to make sense of the mutations in capitalist economy and society. From Adam Smith's "Philosophers of Industry" (1776) and Karl Marx's "Proletariat" (1848), to Max Weber's "Bureaucrats" (1910/1948), Frederick Taylor's "Scientific Managers" (1911/1967), Joseph Schumpeter's "Entrepreneurs" (1942/1976), Peter Drucker's "Knowledge Workers" (1959), Daniel Bell's "Knowledge Class" (1973), Alvin Toffler's "Prosumers" (1980) and Jean-Francois Lyotard's "Postmodernists" (1984).

As Barbrook observes, there is in fact a deep structural meta-narrative underlying discussions on the creative economy that links it to a much longer history of social prophecy. The basic theme of this eschatological vision is an anticipation of the future by linking the whole of society to the trajectory of a principal segment. In this sense, the contemporary focus on a new revolutionary creative class has deep historical roots:

> Far from rejecting Florida's approach, the most influential thinkers on both the Right and the Left are promoting their own versions of the Creative Class. Just like him, they're also convinced that the new economic paradigm will vindicate their own political stance. According to taste, the growth in the number of information workers can be interpreted as the imminent triumph of either dotcom capitalism or cybernetic communism. Although often bitterly divided in their politics, these gurus still share a common theoretical position. Whether on the Right or the Left, all of them champion the same social prophecy: the new class is prefiguring today how everyone else will work and live tomorrow. (Barbrook, p. 16)

According to Barbrook, contemporary struggles for creative liberation from the stifling limitations of monolithic systems have a significant cultural history in capitalist society, stretching back through the Hippies in the mid-twentieth century and the Bohemians in the early nineteenth century. In the contemporary milieu, however, creativity and innovation remain a privilege of the few. A creative minority can indeed make their living as leaders in the creative economy, but only because of the support afforded them by the mundane labor of everyone else.

> For over two centuries, creativity has been at the centre of the struggle between capital and labour. As the industrial system has evolved, the contending classes have fought not only over the division of the fruits of production, but also over the control of the workplace. In *The Wealth of Nations*, Adam Smith showed how the increasing division of labour allowed capitalists to replace self-governing skilled artisans with more submissive unskilled employees. (Barbrook, 2006, p. 25)

As Barbrook goes on to point out, much of the rhetoric undergirding post-Fordist celebrations of entrepreneurs has simply perpetuated Fordist assumptions that a ruling class is necessary to lead society towards a future "promised land." In the contemporary milieu, however, the focus is increasingly shifting to a new mode of production altogether. While under Fordism, the path to a successful career was found in internalizing the routines and procedures of the corporate machine, today these are exactly opposite to the skills needed to advance contemporary capitalism (Barbrook, 2006). Unlike the rigid hierarchies of industrial capitalism, the dominant model of organization today is not the Fordist bureaucracy but the *network.* This has been occurring in part because of the rise of network capitalism and its capacity to leverage *democratization* in production.

Network Capitalism: Democratizing Innovation

Perhaps the most important strand in understanding the contemporary notion of the creative economy is the recent technology-driven shift from industry to services. Since the onset of the "new economy" in the 1990s, business strategists have been moving beyond efficiency gains in the production of goods and services and become increasingly focused on innovation systems and the exploitation of information. Technology has emerged as the "infostructure" for enterprises competing on a global scale, and IT has provided the platform on top of which knowledge-driven organizations create value (Tapscott, 1997). More recently, networked connectivity has added a new social dimension to business enterprise, transforming IT into ICTs and making multimedia content critical to networked modes of production and consumption.

Much as the assembly line shifted the critical factor of production from labor to capital, today the computer is shifting the critical factor of production from capital to innovation. Beyond the command systems characteristic of industrial production, ICT networks have become infrastructural to new modes of value-driven design and innovation. Underlying this socioeconomic restructuring is the critical importance of information and communications networks (ICNs) to leveraging distributed creativity. As Rycroft & Kash (2004) explain, the capacity of networks to coordinate rapid self-organization is now foundational to global innovation:

> The most valuable and complex technologies are increasingly innovated by networks that self-organize. Networks are those linked organizations (e.g., firms, universities, government agencies) that create, acquire, and integrate the diverse knowledge and skills required to create and bring to the market complex technologies (e.g., aircraft, telecommunications equipment). In other words, innovation networks are organized around constant learning. Self-organization refers to the capacity these networks have for combining and recombining these learning capabilities without centralized, detailed managerial guidance. The proliferation of self-organizing innovation networks may be linked to many factors, but a key one seems to be increasing globalization. Indeed, globalization and self-organizing innovation networks may be coevolving. Changes in the organization of the

> innovation process appear to have facilitated the broadening geographical linkages of products, processes, and markets. At the same time, globalization seems to induce cooperation among innovative organizations. (p. 1)

Moving beyond the simple "one-to-many" linear model of industrial manufacturing, ICNs are facilitating "many-to-many" production. As Eric von Hippel (2005) has pointed out, this new logic is giving rise to a democratization of innovation that is in fact problematizing the divide between producers and consumers:

> When I say that innovation is becoming democratized, I mean that users of products and services—both firms and individual consumers—are increasingly able to innovate for themselves. User-centered innovation processes offer great advantages over manufacturer centric innovation development systems that have been the mainstay of commerce for hundreds of years. Users that innovate can develop exactly what they want, rather than relying on manufacturers to act as their (often very imperfect) agents. Moreover, individual users do not have to develop everything they need on their own: they can benefit from innovations developed and freely shared by others. (p. 1)

This democratization of innovation reflects a larger potential emerging with ICTs in the creative economy. Building out from specialized communities-of-practice, there is a noticeable shift from passive consumption to active cultural production. Fundamental to this shift is an emerging understanding that tools that facilitate design in the context of learning-by-doing are becoming critical to the advancement of both culture and economy (Foray, 2004).

Networks of Prosumer Innovation

Tapscott and Williams (2006) refer to this as *prosumer innovation* (Toffler, 1980). As they suggest, the growing importance of prosumer innovation is directly connected to networks as platforms for mass collaboration. Using examples ranging from software, music, publishing, and pharmaceuticals, Tapscott and Williams link collaboration-driven Web services like Facebook, InnoCentive, Flickr, Second Life, and YouTube to the rising power of prosumer-driven creativity and design. In the online virtual environment of Second Life, for example, prosumers form broad user-communities that create rich value-added products and services. Open business models like Second Life invite customers to add value by offering a platform for creativity. Tapscott and Williams point out that technologies like Apple's iPod and Sony's PSP are now routinely "hacked" to enable creative changes in their design and performance: "Whether it's modifying the casing, installing custom software, or . . . doubling the memory, users are transforming the ubiquitous music and media player[s] into something unique" (p. 133):

The rising influence of prosumer hacking is the result of a convergence of peer-to-peer networks and user-friendly editing tools. While consumers with the skills and inclination to hack commercial products like the iPod remain a minority, they are an expanding consumer segment. Rather than fighting this rising tide, Tapscott and Williams argue that companies should adapt to it by bringing customers into their business webs and giving them lead roles in next-generation products and services:

> Forget about static, immovable products. If your customers are going to treat products as platforms anyway, then you may as well be ahead of the game. Make your products modular, reconfigurable, and editable. Set the context for customer innovation and collaboration. Provide venues. Build user-friendly customer tool kits. Supply the raw materials that customers need to add value to your product. Make it easy to remix and share. We call this designing for prosumption. (p. 148)

As they point out, it may be true that prosumer hacking forces a company to risk losing control of its product platform, but it is also true that "a company that fights its users risks soiling its reputation by shutting out potentially valuable sources of innovation" (pp. 135–136).

Peer-to-Peer Production Ecologies

Prosumer innovation works because it leverages self-organization as a mode of production. Taken as a whole, the Internet represents a global sociotechnological platform in which the knowledge, resources, and computing power of billions of people are coming together into a massive collective force. Bauwens (2006), for example, has outlined a strong case for the rise of peer-to-peer (P2P) systems as a new mode of production. As he points out, what makes peer production systems particularly different from both state and market models is that they are largely independent of monetary incentives and fixed hierarchical organization. In P2P projects like open-source software (OSS), for example, resources are contributed spontaneously. Formal authority is "organic," emerging and receding with the domain-based expertise needed to complete specific tasks. In these democratic production ecologies, authority does not disappear, but neither does it cohere as permanent hierarchical structures. It is literally production that is dependent on the voluntary participation of partners.

According to Bauwens, the Internet as a point-to-point network infrastructure enables "equipotentiality" in the design and development of *commons-based* production regimes. Labor is "permission-less" and bottom-up. P2P is neither hierarchy-less nor structure-less but is shaped by flexible "hierarchicalization," which is entirely dependent on the free cooperation of autonomous agents. In P2P production systems, motivation is intrinsic and passion-based rather than an exchange of labor for financial reward. In the context of OSS, for example, projects are usually led by a core group of founders who head microteams in a patchwork of specialized tasks. Peer production systems are synergistic "hives" in which fluid modes of collaboration support emergent innovation that is collectively *grown*. While hierarchical organizations depend on a *panoptical* logic that steers production from above, networked production systems utilize a *holoptical* logic or "horizontal" intelligence. In P2P projects, all participants have access to the knowledge of what the others are doing, and the vertical knowledge of the project as a whole. As new skill levels evolve, peer contributors move from the periphery to the core without the need for fixed hierarchies or external mediation.

Harnessing the Hive

P2P represents one of the clearest models we have for harnessing complex systems in the production of design and innovation. In his book *The Wealth of Networks*, Benkler (2006) describes this emerging mode of production as "commons-based social production." While traditional systems of production depend on closed proprietary structures, commons-based production utilizes open networks to harness the creative energy of collective intelligence. For Benkler, the key to understanding this democratic cultural practice is that no single entity "owns" the product or manages its direction. While this new mode of production may depend on the technological capacity of networks, it is ultimately configured by an emergent socio-political structure grounded in open systems.

By "importing" energy across permeable boundaries, open systems in nature are continually nourished. It is this capacity for self-creation or *autopoesis* that gives open systems in nature their incredible capacity for growth. When this same boundary permeability is translated into the domain of human socio-economic production, it manifests as a continually evolving collective intelligence. Much as other complex open systems, democratic production systems avoid "creative entropy" by continually absorbing energy and resources from new participants. As free labor is absorbed into shared economic practices, the creative potential for self-organization is continually replenished.

Entering a Creative Age

It has become fairly commonplace to say that creativity and innovation rest on cultural experimentation. In the context of lived reality, however, one can view the world's diverse cultures as "experiments" with innovation. "The more experiments humanity constructs, in other words the greater the cultural diversity, the more knowledgeable and innovative we are likely to be" (Griffin, 2000, p. 193).

We know that creativity flourishes among talented people, but what stokes this process and what sustains it? The answer, according to Florida, lies in geography. That is, a community's cultural capacity for openness or "absorptive capacity." In his view, tolerance for diversity and "low barriers to entry" attract and absorb talent while supporting the rich environments that stoke creative innovation. Zachary (2000) puts it slightly differently. In his view, creativity depends on cultural blending or *hybridity*. While monocultural chauvinism impedes creativity, hybridity renews it. In this sense, creativity emerges with the ability to integrate divergent and even contradictory cultural practices.

The capacity for communities and peoples to work creatively with cultural artifacts in the context of sustained innovation is emerging as a critical challenge today. In contrast to Thomas Friedman's notion of a "flat world," wealth and power are becoming increasingly concentrated in the hands of a highly educated elite living in the world's richest cities.

While the share of the world's population living in urban areas was just 3% in 1800, and 30% in 1950, it is 50% today (and as high as 75% in advanced capitalist countries) (Florida, 2007, p. xviii). Five mega-cities have more than 20 million inhabitants, and another twenty-four cities have 10 million inhabitants. According to Florida, the world's 40 largest mega-regions are now home to some 18 percent of the world's population and produce two-thirds of global economic output (including nearly 9 in 10 new patented innovations).

Even as the world's cities are increasingly absorbing larger and larger numbers of the global population, only a handful of cities make up the dominant share of wealth and power. Whether measured in terms of financial power or commercial innovation, only a very small number of cities in the world today dominate the global economic landscape. Staggering economic peaks like New York, Paris, London, and Tokyo form the major control nodes of the global economy (Porter, 1990; Sassen, 2001). If the world economy were measured for commercial innovation, wealth would in fact be even more concentrated. New York's economy alone is equal in size to that of Russia or Brazil. "Together New York, Los Angeles, Chicago, and Boston have a bigger economy than all of China. If U.S. metropolitan areas were countries, they'd make up forty-seven of the biggest 100 economies in the world" (Florida, 2007, p. xviii).

While theories of a "flat world" (T. Friedman, 2005), for example, accurately register the growing capacities of emerging countries like India and China to compete in the global market, they ignore the growing divide between the super-educated and the vast majority who have little or no access to advanced skills. Beyond the mobile "creative class," whose members are free to migrate between the world's economic peaks, live the vast majority who are left to toil in the world's economic valleys. Put simply, it is not that the world has become "flatter," but that the world's economic peaks have become slightly more dispersed (particularly as industrial and service centers have shifted to Asia).

Education in the Creative Economy

We seem to be entering a new world now, a world in which the major raw materials are no longer coal and steel produced by machines, but creativity and innovation produced by the human imagination. It is certainly true that all human beings are creative—this is a basic capacity of the human species, grounded in its ability to evolve and adapt. Unfortunately, it is only a small minority of people in the world today who are able to tap this creativity. In this sense, Florida is entirely correct when he suggests that the great challenge before us is to develop the systems and policies that harness the creative capacities that lie within all human beings.

If Florida and other advocates of the creative economy are right, then creativity is now fundamental to wealth and prosperity and cultural innovation is critical to its fecundity. Yet it is precisely creativity that is least valued by contemporary institutions. The vast majority of hierarchical organizations today deliberately submerge creativity beneath bureaucratic layers of command-and-control. This is equally true of contemporary systems of education. While it was once true that school systems effectively distributed the necessary skills for an age of industry (numeracy, literacy, symbol manipulation), it is equally true that these same institutions are not equipped to support the skills and capacities for an age of innovation.

Much as Franklin Roosevelt used the New Deal to reform the economic and banking systems in order to construct the infrastructure necessary to emerge from the Depression, so today must we develop the policy framework and infrastructural renewal to reform education for an age of innovation. "Like earlier efforts to build canals, railroads, highways, and other physical infrastructure to power industrial growth, the United States and countries around the world must invest in their *cre-*

ative infrastructure if they want to succeed and prosper in the future" (Florida, 2007, p. 249).

Education is critical to this creative infrastructure. Rather than understanding learning in terms of fixed objects that are transferred from one generation to the next, we need to begin to design educational systems that support knowledge and learning in terms of continuous cultural innovation. Education systems designed for industrial societies do not effectively harness the liquidity of creative innovation because they are too centralized. Transferring a fixed body of knowledge and practices from experts to amateurs is contradictory to an economy increasingly dependent on continuous flows of design and innovation. Allowing students to combine and blend cultural flows as a part of the larger continuum of cultural production is now fundamental to reconfiguring learning and education.

Education and Cultural Production

Cultural innovation is iterative. Contemporary cultural forms are themselves the products of countless prior iterations. In the arts and sciences, past cultural innovations serve as basic resources for future innovations. While established theories of cultural systems have traditionally interpreted cultures as closed systems, the pace of change in a rapidly globalizing world now requires a theory of culture that recognizes the continuous transformation of culture and cultural systems (Kress, 2000).

As Nederveen Pieterse (2004) suggests, our contemporary notion of culture combines two, somewhat contradictory meanings. The first concept of culture (culture 1) assumes that culture stems from a learning process that is geographically fixed: "This is culture in the sense of *a culture*, that is, the culture of a society or social group: a notion that goes back to nineteenth-century romanticism and that has been elaborated in twentieth-century anthropology, in particular cultural relativism—with the notions of culture as a whole, a Gestalt, configuration" (p. 78).

Unlike this self-contained and perpetually colliding notion of culture, however, a second approach to defining culture understands it to be something more akin to a shared and evolving social practice. This broader understanding of culture (culture 2) views it as general human "software," more akin to creative flows than locally bounded knowledge. This second notion of culture is closely linked to translocal learning processes and to theories of evolution and diffusion.

As Nederveen Pieterse points out, these two viewpoints are not incompatible; culture 2 finds expression in culture 1. Nonetheless, they are rooted in shifting ontological and epistemological boundaries. In this sense, culture may be linked to territorial and/or historical contingencies (culture 1), but it is not reduced to them.

If competency in the use of resources within existing cultural systems is the goal of literacy in traditional systems of education (culture 1), then cultural production and the reshaping of cultural systems must be the goal of education today (culture 2). The growing economic challenge for advanced economies is to develop social and economic policies that support sustained cultural innovation. As Venturelli (2005) puts it,

> The challenge for every nation is not how to prescribe an environment of protection for a received body of art and tradition, but how to construct one of creative explosion and innovation in all areas of the arts and sciences (see Venturelli, 2000, 1999, 1998b). Nations that fail to meet this challenge will simply become passive consumers of ideas emanating from societies that are in fact creatively dynamic and able to commercially exploit the new creative forms. (p. 396)

In this sense, legacy approaches to defining cultural policy become deficient to engaging the emerging importance of culture to the economy. As Venturelli observes, the most important question with regard to a given society today is not the cultural legacy of its past but the inventive and creative capacities of its present. This interpretation does not mean that established cultural forms are irrelevant to creativity and innovation in the creative economy. Established cultural forms are themselves the foundations on which new cultural forms are developed. Rather, it is to question the idea that cultural policy is merely a question of protecting cultural traditions:

> In a "museum paradigm," of cultural policy, works of art and artistic traditions are revered and cultural traditions closely guarded and defended. But when these become the predominant measure of cultural resources and the notion of legacy occupies the sole definition of the creative spirit, ultimately the development of that spirit would be undermined. . . . A culture persists in time only to the degree it is inventing, creating, and dynamically evolving in a way that promotes the production of ideas across all social classes and groups. Only in this dynamic context can legacy and tradition have real significance. (Venturelli, p. 395)

Achieving a model of education that supports this radical understanding of innovation is very likely the next major challenge facing countries in the twenty-first century.

Networks, Education, and Communities of Practice

New modes of education are critical to supporting a creative economy. In addition to arguments for investing in the knowledge economy through STEM (science, technology, engineering, and math) disciplines, it is equally critical to invest in art, design, and digital media as interlocking components of the broader creative economy. As Leadbeater (2000, p. 110) points out, contemporary education systems suffer from two lingering traditions that have been combined to severely hinder contemporary school systems: the monastery as knowledge repository and the factory as command production system. In contrast to these fixed hierarchical systems, we now require horizontal networks that allow "student-amateurs" to directly engage with one another in the practice of building and transforming ideas and practices.

In the world of technology, the cultural spotlight is increasingly moving to the "edge of the network," to distributed systems and open platforms in which mass collaboration is used to leverage large-scale projects like Wikipedia, Linux, the World Wide Web, and more recently the Barack Obama presidential campaign. Networked collaboration is highly conducive to learning and innovation because production is grounded in self-organizing systems of collective intelligence. In the increasingly unstable environment of modern education, students must be given access to educational systems that foster collaboration in the context of networked innovation.

John Seely Brown (2005), for example, has suggested that the next generation of education should be more closely linked to apprenticeship models of learning (Lave, 1988). Rather than learning *about* something, Brown argues for a studio-based model that focuses on directly acculturating students into sociocultural practices. Echoing Dewey, Brown's focus on practice emphasizes multimedia literacy (or digital literacy) in the context of the many distributed learning communities found on the Web. Much as Open Source Software production has been catalyzed by open collaboration, Brown argues that education can be catalyzed by social learning communities. With the growing reach and density of global ICTs, social learning networks have proliferated on the Web creating a vast learning platform:

> Note that what has been constructed here, largely as a by-product, is a vast learning platform that is, de facto, training thousands of people about good software practices. A powerful form of distributed cognitive apprenticeship that functions across the world has emerged. Today, there are about one million people engaged in open source projects, and nearly all are improving their practices by

> being part of these networked communities. The key to learning in these environments is that all contributions are subject to scrutiny, comment and improvement by others. There is social pressure to take the feedback from others seriously. (Brown, 2005: 21)

Powered by intrinsic motivation, social learning platforms like Linux and Wikipedia demonstrate the rising tide of participatory learning communities that are rebuilding the education landscape. This new education environment represents a significant shift from Fordist learning systems to passion-based learning systems, in which students are empowered by their own intrinsic motivation to be social agents. No matter how specialized an interest area may be, the Web offers a participatory platform for leveraging mass collaboration. Rather than the *supply-push* mode of learning that undergirds the industrial age, social learning networks enable a *demand-pull* mode of learning that leverages learning through participation. In the context of education, this shifts the focus from building up stocks of knowledge (learning-about something) to enabling participation in flows of cultural production (learning through experience).

The Internet is a rich resource and learning ecosystem that is enabling social learning communities to grow and flourish. There is no doubt that this rising social technology will have a significant impact on education. While in the industrial age, human creativity was divided into distinct activities (art, science, and business enterprise); today technology scaffolds so many varied disciplines that their recombination in new forms is becoming commonplace. In Brown's view, we should focus on shaping education through a kind of "elegant minimalism," in which the core curriculum remains focused on the foundational skills: literacy, numeracy, and critical thinking. However, surrounding this core curriculum is an open curriculum that is largely determined by students themselves as they navigate the proliferation of social learning communities available to them. As Brown points out,

> When new mechanisms for distributing content are combined with new power tools for creating that content, along with social software and recommendations systems for finding the content, we have the beginnings of an infrastructure for enabling the rise of the creative, always learning, class—people who want to create and have others build on, use, critique and, most importantly, acknowledge their creations. This presents a new set of possibilities for unleashing a culture of learning by creating, sharing, and acknowledging the work of others in a way that builds both social capital and intellectual capital simultaneously. (Brown, p. 28)

Educational Policy in the Creative Economy

The capacity of people to work creatively with cultural artifacts in the context of sustained innovation is emerging as a central feature of the global economy today. If we accept the arguments of creative economy theorists like Howkins and Florida, then cultural policy effectively becomes economic policy. This suggests that struggles over resource allocation, competing constituencies, and divergent goals will be even more contentious going forward (Healy, 2002). Educational policy will certainly play a critical role in this.

New policies and planning are vital to making creative work broadly accessible to all and not reserved for an educated elite. As peoples and governments begin to ponder the consequences of the recent collapse of the laissez-faire capitalism in the United States, Britain, and elsewhere, it is becoming obvious that developing coherent policy prescriptions for cultural innovation are now critical to long-term social and economic sustainability. One of the major questions that we must begin to answer today is: "What systems, policies, and structures are most conducive to making it possible for the largest number of people in a society to participate in the creation and development of new cultural forms?"

As the nexus of economic growth increasingly moves from labor-intensive "smokestack" industries to "design work," education is becoming central to both incubating knowledge and harvesting creative innovation. Much as the factory was the core institution of the industrial age, schools and universities may well be the core institutions of the innovation age. In many respects, however, the modern university is now outmoded. Shaped for a different era, the modern university was designed as a state apparatus. Knowledge was perceived as a local commodity and competition between schools mirrored competition in the rest of the marketplace. In a global age, however, the isolated nation-state is being reshaped by global circuits of trade and communication (Toffler, 1990; Castells, 1996). Rather than islands of concentrated knowledge in support of the nation-state, schools and universities must become cultural estuaries in support of creativity and innovation. As students become agents in their own learning trajectories, schools and universities must begin to explore modes of knowledge and learning that facilitate creativity in the context of collective intelligence.

Conclusion

Educational systems today are undergoing an enormously disruptive transformation that is moving them beyond their roots in nineteenth-century industrialization. The interconnected forces of globalization, cultural change, and digital technologies are together democratizing agency and moving authority away from institutions of education. Beyond iterative cultural innovation, national education systems must now explore modes of education that catalyze radical innovation. It is clear that education systems designed for industrial societies do not effectively harness the liquidity of bottom-up innovation because they are embedded in hierarchies of command-and-control. Networks on the other hand, represent a clear model for harnessing radical innovation because they facilitate emergent creativity in the context of mass collaboration. Beyond established arguments for a global knowledge economy, network-driven creativity is critical to revitalizing advanced capitalist countries for an age of innovation.

References

Adorno, T., & Horkheimer, H. (1977). The culture industry: Enlightenment as mass deception. In J. Curran, M. Gurevitch, & J. Wollacott (Eds.), *Mass communication and society* (pp. 349–383. London: Edward Arnold. (Original work published 1944)

Albrow, M. (1997). *The global age.* Stanford, CA: Stanford University Press.

Altman, R. (2009, January-February). The great crash, 2008: A geopolitical setback for the West. *Foreign Affairs,* pp. 2-14.

Amin, A. (1997). Post-Fordism: Models, fantasies and phantoms of transition. In A. Amin (Ed.), *Post-Fordism: A reader.* Cambridge: Blackwell Publishers Ltd.

Aronowitz, S. (2000). *The knowledge factory: Dismantling the corporate university and creating true higher learning.* Boston: Beacon Press.

Barbrook, R. (2006). *The class of the new.* London: OpenMute.

Bartlett, C., Ghoshal, S. & Beamish, P. (2008). *Transnational management.* New York: McGraw-Hill Irwin.

Bauwens, M. (2006). The political economy of peer production. *CTheory.* Retrieved fromhttp://www.ctheory.net/articles.aspx?id=499

Beck, U. (1992). *Risk society.* London: Sage.

Beck, U. (1999). *What is globalization?* Cambridge: Polity Press.

Beck, U. (2002). The cosmopolitanism society and its enemies. *Theory, Culture and Society,* 19, 17–44.

Becker, G. (1964). *Human capital.* Chicago: University of Chicago Press.

Bell, D. (1973). *The coming of post-industrial society.* New York: Basic Books.

Benkler, Y. (2006). *The wealth of networks.* Princeton, NJ: Princeton University Press.

Bereiter, C. (2002). *Education and mind in the knowledge age.* Mahwah, NJ: Lawrence Erlbaum Associates.

Berman, K., & Annexstein, F. (2000). A Future educational tool for the 21st century: Peer-to-peer computing. Retrieved from http://www.ececs.uc.edu/~annexste/Papers/EduP2P.pdf

Bray, M. (1996). *Privatisation of secondary education: Issues and policy implications.* Paris: UNESCO.

Brecher J., & Costello, T. (1994). *Global village or global pillage: Economic restructuring from the bottom up.* Boston: South End Press.

Breen, B. (2003, March). The hard life and restless mind of America's education billionaire. *Fast Company,* 68, 80.

Brown. J. (2005). New learning environments for the 21st century. Retrieved from http://www.johnseelybrown.com/newlearning.pdf

Burbules, N., & Callister, T. (2000). Universities in transition: The promise and the challenge of new technologies. *The Teachers College Record, 102*(2), 271–293.

Buzan, B., & Little, R. (2000). *International systems in world history.* New York: Oxford University Press.

Castells, M. (1996). *The rise of the networked society.* Oxford: Blackwell.

Chomsky, N. (1996). *World orders old and new.* New York: Columbia University Press.

Connexions White Paper. (2004). Rice University.

Cowan, L. G. (1990). *Privatisation in the developing world.* New York: Greenwood Press.

Dewey, J. (1997). *Democracy and education: An introduction to the philosophy of education.* New York: Simon and Schuster.

Drucker, P. (1959). *Landmarks of tomorrow: A report on the new 'post-modern' world.* New York: Harper & Row.

Drucker, P. (1966). *The effective executive.* New York: Harper & Row.

Drucker, P. (1973). *Management: Tasks, responsibilities, practices.* New York: Harper & Row,

Drucker, P. (1985). *Innovation and entrepreneurship.* London: Heinemann.

Drucker, P. (1993). *Post-capitalist society.* Oxford: Butterworth-Heinemann.

Drucker, P. (2001). The next society: A survey of the near future. *The Economist,* November 3–9.

Dyer-Witheford, N. (2000). *Cyber-Marx: Cycles and circuits of struggle in high-technology capitalism.* Chicago: University of Illinois Press.

Emmerij, L. (2000). World economic changes at the threshold of the twenty-first century. In J. Nederveen Pieterse (Ed.), *Global futures: Shaping globalization.* New York: Zed Books.

Falk, R., & Strauss, A. (2001). Bridging the globalization gap: Toward a world parliament. *Foreign Affairs, 80*(1), 212–220.

Florida, R. (2000). America's looming creativity crisis. *Harvard Business Review, 82*(10), 122–136.

Florida, R. (2002). *The rise of the creative class: And how it's transforming work, leisure, community and everyday life.* New York: Basic Books.

Florida, R. (2004, January/February). Creative class war. *Washington Monthly, 36*(1/2), 30.

Florida, R. (2007). *The flight of the creative class: The new global competition for talent.* New York: HarperCollins.

Foray, D. (2004). *Economics of knowledge.* Cambridge: MIT Press.

Frank, A. G. (1998). *ReOrient: Global economy in the Asian age.* Berkeley: University of California Press.

Friedman, B. (2005). *The moral consequences of economic growth.* New York: Knopf. .

Friedman, T. (2005). *The world is flat: A brief history of the twenty-first century.* New York: Farrar, Straus and Giroux.

Friedman, T. (1999). *The lexus and olive tree: Understanding globalization.* New York: Farrar, Straus and Giroux.

Giddens, A. (1990). *The consequences of modernity.* Palo Alto, CA: Stanford University Press.

Giroux, H. A. (2001). Critical education or training: Beyond the commodification of higher education. In H. A. Giroux & K. Myrsiades (Eds.), *Beyond the corporate university culture and pedagogy in the new millennium.* Lanham, MD: Rowman & Littlefield.

Gray, J. (1998). *False dawn: Delusions of global capitalism.* London: Granta Books.

Griffin, K. (2000). Culture and economic growth: The state and globalization. In J. Nederveen Pieterse (Ed.), *Global futures: Shaping globalization.* New York: Zed Books.

Habermas, J. (1992). *Postmetaphysical thinking.* Cambridge: Polity Press.

Habermas, J. (2001). The postnational constellations and the future of democracy. In *The postnational constellation: Political essays* (Max Pensky, Trans.). Cambridge, MA: MIT Press.

Halal, W. (1998). *The new management.* San Francisco: Berrett-Koehler.

Hardt, M., & Negri, A. (2001). *Empire.* Cambridge, MA: Harvard University Press.

Hardt, M., & Negri, A. (2004). *Multitude: War and democracy in the age of empire.* Cambridge, MA: Harvard University Press.

Harris, J. (2008). *The dialectics of globalization: Economic and political conflict in a transnational world.* Newcastle: Cambridge Scholars Publishing.

Head in the clouds. (2005, September 10). *The Economist, 376*(8443), 9–13.

Healy, K. (2002). What's new for culture in the new economy. *The Journal of Arts Management, Law and Society, 32*(2), 86–103.

Hearn, D. R. (n.d.) Education in the workplace: An examination of corporate university models. Retrieved from http://www.newfoundations.com/OrgTheory/Hearn721.html

Held, D. (1995). *Democracy and the global order.* Stanford, CA: Stanford University Press.

Held, D., McGrew, A., Goldblatt, D., & Perraton, J. (1999). *Global transformations—Politics, economics and culture.* Cambridge: Polity Press.

Higher ed., inc. (2005, September 10). *The Economist*, 376 (8443), 19–20.

Hilsenrath, J., & Buckman, R. (2003, October 20). The economy: Factory employment is falling world-wide; study of 20 big economies finds 22 million jobs lost; even China shows decline. *Wall Street Journal*, A2.

Howkins, J. (2001). *The creative economy: How people make money from ideas.* London: Allen Lane.

Huntington, S. (1996). *The clash of civilizations and the remaking of world order.* New York: Simon and Schuster.

Husen, T. (1974). *The learning society.* London: Methuen.

Husen, T. (1986). *The learning society revisited.* Oxford: Pergamon.

Kauffman S. (1996). *At home in the universe.* New York: Oxford University Press.

Kellner, D. (1998). Globalization and the postmodern turn. In R. Axtmann (Ed.), *Globalization and Europe.* London: Cassells. Retrieved from http://www.gseis.ucla.edu/courses/ed253a/dk/GLOBPM.htm

Kirp, D. (2003). *Shakespeare, Einstein and the bottom line: The marketing of higher education.* Cambridge, MA: Harvard University Press.

Klein, N. (2000). *No logo.* London: HarperCollins.

Kolodziej, E. (2005). Plotting an intellectual jailbreak: Rationale for globalizing the campus and university. Retrieved from http://www.cgs.uiuc.edu/resources/papers_and_publications/papers.html

Korten, D. (1995). *When corporations rule the world.* Bloomfield, CT.: Kumarian Press.

Korten, D. (1999). *The post-corporate world: Life after capitalism.* San Francisco, CA: Berrett-Koehler.

Kottke, J. (1999). Review of J. Meister's *Corporate Universities: Lessons in building a world-class work force. Personnel Psychology, 52*(2), 530–533.

Kraidy, M. (2005). *Hybridity: The cultural logic of globalization.* Philadelphia, PA: Temple University Press.

Kreml, W., & Kegley, C. (1996). A global political party: The next step. *Alternatives,* 21: 123–34.

Kress, G. (2000). Design and transformation: New theories of meaning. In B. Cope & M.Kalantzis (Eds.), *Multiliteracies: Literacy learning and the design of social futures* (pp. 153–161). New York: Routledge.

Kukla, A. (2000). *Social constructivism and the philosophy of science.* London: Routledge.

Lave, J. (1988). *Cognition in practice.* Cambridge, UK: Cambridge University Press.

Leadbeater, C. (2000). *Living on thin air: The new economy.* London: Penguin.

Leadbeater, C., & Miller, P. (2004). *The pro-am revolution: How enthusiasts are changing our economy and society.* London: Demos.

Lenski, G., Nolan, P., & Lenski, J. (1995). *Human societies: An introduction to macrosociology* (7th ed.) New York: McGraw-Hill.

Lèvy, P. (1997). *Collective intelligence: Mankind's emerging world in cyberspace.* New York: Plenum Press.

LionShare White Paper. (2004). *Connecting and extending peer-to-peer networks: LionShare White Paper.* Retrieved from http://lionshare.its.psu.edu/main/info/docspresentation/LionShareWP.pdf

Livingstone, D. (1999). *The education-jobs gap: Underemployment or economic democracy.* Toronto: Garamond Press.

Lyotard, J. (1984). *The post-modern condition: A report on knowledge* (G. Bennington & B. Massumi, Trans). Minneapolis: University of Minnesota Press.

Marx, K., & Engels, F. (1848/1955). Manifesto of the communist party. In *Karl Marx and Frederick Engels, Selected Works*. Vol. I. Moscow, Russia: Foreign Languages Publishing House. (Original work published 1848)

Meister, J. C. (1998). *Corporate universities: Lessons in building a world-class work force* (rev. ed.). New York: McGraw-Hill.

Mulgan, G. (1998). *Connexity: Responsibility, freedom, business and power in the new century* (rev. ed.). London: Viking.

Nederveen Pieterse, J. (Ed.). (2000). *Global futures: Shaping globalization*. New York: Zed Books.

Nederveen Pieterse, J. (2003). *Globalization and culture: Global mélange*. Lanham, MD: Rowman and Littlefield.

Peters, M., & Besley, T. (A. C.) (2006). *Building knowledge cultures: Education and development in the age of knowledge capitalism*. Oxford: Rowman and Littlefield.

Pink, D. (2005). *A whole new mind: Moving from the information age to the conceptual age*. New York: Riverhead Books.

Porter, M. (1990). *The competitive advantage of nations*. New York: Free Press.

Readings, B. (1996). *The university in ruins*. Cambridge, MA: Harvard University Press.

Reich, R. (2000). *The future of success: Working and living in the new economy*. New York: Alfred A. Knopf.

Rifkin, J. (1995, October). Vanishing jobs. Mother Jones. Retrievedfrom http://www.motherjones.com/commentary/columns/1995/09/rifkin.html

Rifkin, J. (2000). *The age of access*. New York: Jeremy P. Tarcher/Putnam.

Rycroft, R. and Kash, D. (2004) Self-organizing innovation networks: Implications for globalization. *Technovation*, 24, 187–97.

Said, E. (1994). *Culture and imperialism*. New York: Vintage Books.

Sassen, S. (2001). *The global city*. Princeton, NJ: Princeton University Press.

Schumpeter, J. (1976). *Capitalism, socialism and democracy*. New York: Harper.

Siwek, S. (2002). *Copyright industries in the US economy: The 2002 report*. Washington, DC: International Intellectual Property Alliance.

Slaughter, S., & Leslie, L. (1997). *Academic capitalism: Politics, policies and the entrepreneurial university*. Baltimore, MD: Johns Hopkins University Press.

Slaughter, S., & Rhoades, G. (2004). *Academic capitalism and the new economy: Markets, state and higher education*. Baltimore, MD: Johns Hopkins University Press.

Tapscott, D. (1997). *The digital economy: Promise and peril in the age of networked intelligence*. New York: McGraw-Hill.

Tapscott, D. & Williams, A. (2006). *Wikinomics: How mass collaboration changes everything*. New York: Portfolio.

Taylor, F. (1967). *The principles of scientific management*. New York: W.W. Norton.

Toffler, A. (1970). *Futureshock*. New York: Random House.

Toffler, A. (1980). *The third wave*. New York: Bantam Books.

Toffler, A. (1990). *Powershift: Knowledge, wealth and violence at the edge of the 21st century.* New York: Bantam Books.

United Nations Commission for Trade, Aid and Development. (2008). *Creative economy.* Geneva: UNCTAD.

Venturelli, S. (2005). Culture and the creative economy in the information age. In J. Hartley (Ed.), *Creative industries* (pp. 391–398). Malden, MA: Blackwell.

Von Hippel, E. (2005). *Democratizing innovation.* MIT Press: Cambridge, MA.

Vygotsky, L. S. (1978). *Mind in society: The development of higher psychological processes.* Cambridge, MA: Harvard University Press.

Weber, M. (1947). *The theory of social and economic organization.* London: Collier Macmillan Publishers.

Weber, M. (1948). *Essays in sociology.* London: Routledge.

Wheatley, M. (1998). *What is our work? Insights on leadership: Service, stewardship, spirit, and servant leadership.* New York: John Wiley and Sons.

Womack J., Jones D., & Roos D. (1990). *The machine that changed the world.* New York: Macmillan.

World Bank. (1995). *Priorities and strategies for education: A World Bank review.* Washington, DC.

World Bank. (1999). *World development report 1998/99: Knowledge for development.* Washington, DC:

World Bank. Retrieved from http://www.worldbank.org/wdr/wdr98/contents.htm

World Bank. (2002). *Constructing knowledge societies: New challenges for tertiary education.* Retrieved from http://www1.worldbank.org/education/pdf/Constructing%20Knowledge%20Societies.pdf

Zachary, G. (2000). *The global me: New cosmopolitans and the competitive edge.* New York: Public Affairs.

Zuboff, S. (1988). *In the age of the smart machine: The future of work and power.* New York: Basic Books.

Zukin, S. (1993). *Landscapes of power.* Berkeley: University of California Press.

2

The Policy Journey Toward Education for the Creative Economy

Stuart Cunningham & Luke Jaaniste

The notion of creativity is a Rorschach blot into which a great deal is read. With a range of connotations that sit proudly as democratic ideals—freedom, autonomy, collaboration, renewal, expression, communication—it is little wonder creativity is invoked across numerous trigger points for public debate and policy. We approach the question of the relationship between education and the creative economy, not as education specialists, but from the other side, interested as we are in the evolution of policy agendas around the creative industries and the creative economy. By charting selected public policy milestones, the story of this chapter follows the degree to which education is coming to play a bigger role in creative economy policies and programs. Since creativity can mean different things to different policy actors, some basic distinctions will help clarify the field.

Creative Sector, Creative Attributes

We classify the diverse invocations of creativity into two distinct approaches. On the one hand, talk of creativity can help mark out a particular sector of activity within the economy and society, such as the creative arts or creative industries. On the other hand, creativity is highlighted as generic attributes dispersed across the community, whatever the sector or activities.

There is friction, sometimes productive, between the sectorial and the attributes approaches. Attributes discourses may consider a sectorial focus to be narrow and/or elitist. This perspective manifests itself particularly around the notion of the creative industries and its assumed presumptive divisiveness—if only some economic activity is creative, where does that leave the rest? Conversely, attributes discourses may be considered too all-encompassing to be able to operationalize in policy, and they can fail to identify occupations and activities that draw on creativity in a more focused way built around discipline-specific education and expertise.

At least some of the prestige that accrues to the notion of creativity derives from its historical association with, particularly, romantic models of the artist as an exceptionally gifted individual who articulates and represents human aspiration. When John Howkins (2001, 2009) exemplifies his very broad view of what counts as creative *talent*, he as often as not draws on the creative *sector*. However, there is suspicion on the part of artists, and of arts educators, that celebrating generic creativity threatens to soften the case for arts support and diffuse the specific benefits that may be derived from an education in the arts.

Three Platforms: Culture, Business, Innovation

Discourses of creativity are also underwritten by three co-existing platforms. Creativity is fashioned as a cultural and citizenship issue, as an economic and business issue, and as an innovation and knowledge society issue. Not surprisingly, tensions and intersections abound, with each platform holding its own agendas and assumptions: the cultural platform, whether concerned with the creative arts or generic human attributes of creativity, is preoccupied with questions of identity, meaning, and values; the business platform, whether concerned with the creative industries sector or with generic business and commerce, is preoccupied with questions of economic growth, appropriate regulation (such as issues of intellectual property), and the consumer base; and the innovation platform, whether concerned with the creative sector as a source of innovative ideas or with any form of innovative creativity, is preoccupied with questions of knowledge transformations and transfer, industry, and social change and adaptation.

Two Policy Trajectories

We propose that creativity in public policy has evolved along two major trajectories. One policy trajectory follows the broadening of a sector-based view of creativity. Most narrowly and traditionally, the creative sector has referred to the creative arts, and especially high arts. However, its expansion, under the rubrics of cultural and then creative industries, has seen it come to include a broad range of domains involved in aesthetic, symbolic, and expressive production across the creative arts, design, media, and communications. This approach is exemplified in the widely quoted definition of the creative industries from UK Government Department for Culture, Media, and Sport (DCMS) as "those industries which have their origin in individual creativity, skill and talent and which have a potential for wealth and job creation through the generation and exploitation of intellectual property" (DCMS, 2001, p. 4). But there has been a further expansion of the notion of the creative sector, a movement from creative industries *qua* sector to an understanding of the creative economy; that is, the general economy has been increasingly suffused with creative inputs.

The other policy trajectory follows the focusing and specification of creative attributes in the interests of policy applicability. There are approaches that attempt to mobilize the apprehension of such inherent human attributes for broad development purposes, and these typically are directed at education as a whole. Complementary approaches, though, focus on the mobilization of creativity in commercial applications and in social, household, community, and public sector settings. These methods are often marshalled around a contemporary version of innovation policy.

Cunningham (2002, 2004, 2008) has previously charted the evolution of policies around the creative sector, from culture (creative arts) to business services (creative industries) through to a knowledge and innovation framework (creative economy). To this constellation, we now add that the attributes approach to creativity has similarly moved through the same three platforms, from creativity as related to generic human culture (generic creativity), to all business activity (commercial creativity), and then as it is understood within the scope of the innovation system (knowledge-based creativity). While the former has seen an expansion of its object, the latter has concentrated its focus in the interests of policy applicability. There may be evolution in policy thinking about education and the creative economy, but it is by no means irreversible, and later developments do not cancel out earlier. Positions accrete and overlap, complement and contest one another, and all remain available for appropriation.

A limited number of emblematic policy milestones have been chosen from the United Kingdom and Australia to build the case across an approximate two-decade timeframe: See Figure 1

What is the status of considering a series of policy milestones as markers of the evolution of creativity in public policy? It is to chart a sort of discursive genealogy, noting the interlinked way in which formal, structured knowledge grows, based on and legitimated by highly public processes of expert committee work, public consultation, and testing of ideas with decision makers—in short, it is to treat them as a privileged form of sustained, recursive public communication characteristic of liberal democracies. The term *milestone* is meant to suggest that these documents and the processes that gave rise to them have enjoyed some degree of influence in policy and/or have seeded future exercises in public inquiry. As such, they sit between actual public policy and its legislative and regulatory status, and academic output generated out of various disciplinary imperatives. Cunningham (1992) has previously characterized such discourse as "ideas thick," rather than an "ideas rich" academic form of communication.

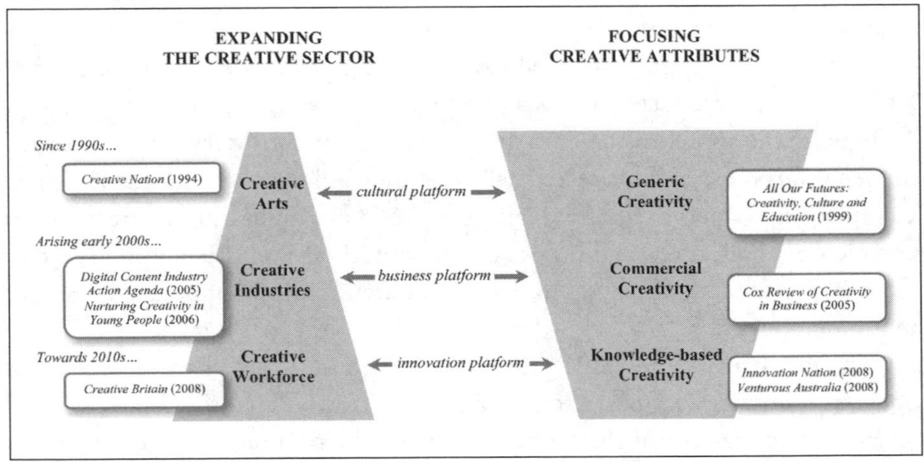

Figure 1: The evolution of creativity in public policy

Education and the Cultural Platform

Creative Nation (Commonwealth of Australia, 1994), Australia's first and only attempt at a national cultural policy statement, was a true milestone. Incorporating a cutting-edge definition of the creative sector for the time, it nevertheless was produced BWE (before the Web era), which also necessarily constrains its articulation of the cultural-commerce dynamic which became the hallmark of the creative industries discourse of some years later. Education is seen in *Creative Nation* as available to perform several roles, for the benefit of future arts practitioners as well as future audiences. On the production side, education can build pools of talent which become the next generation of cultural workers. The sorts of skills seen to be needed include primarily technical and discipline-specific skills. On the consumption side, the education of current and future audiences into particular forms of culture was framed as the province of formal education as well as cultural institutions such as libraries, museums, heritage, and the media. Education should, it is assumed, both disseminate cultural content and events to the wider community and assist the community in understanding that culture to the degree that they would then seek to engage with and consume it. One particular mode of education which gets considerable attention is the partnering of schools with artists and arts organizations, mediating both skills and talent development, as well as participation and audience development. This has a supply-side as much as demand-side rationale. Indeed, it is also proposed as a means of facilitating an extra income stream for the arts sector.

Notwithstanding *Creative Nation*'s focus on sector-specific concerns, themes around generic human attributes occasionally surface: "The development of the skills of artists and workers in arts industries is inadequate," and instead what is needed is "a broadly based education system that focuses on a comprehensive range of educational values stressing imagination and creativity as well as skills." This is a gesture toward what Jeffrey and Craft call the "universalization of creativity" (2001, p. 1).

Such a universalization and its implications for education are exemplified in the UK report *All Our Futures: Creativity, Culture and Education* (NACCCE, 1999) chaired by Sir Ken Robinson. The central premise of *All Our Futures* was that creative and cultural education was essential for a contemporary education and could contribute to raising academic standards just as much as a focus on literacy and numeracy. The initial feat of the report was to grasp the slippery relation of creativity and culture—something *Creative Nation* never attempted. Against a sectorial or elitist view of creativity, *All Our Futures* sought to hold a broad, generic "democratic" view of creativity that looks to all domains and disciplines, all levels of proficiency from beginner to expert, and also to collaborations and group processes rather than a singular focus on individual talent. Here, creativity is defined as "imaginative activity fashioned so as to produce outcomes that are both original and of value" (p. 30). *All Our Futures* also took a broad, generic view of culture, settling on a broad "social definition" of culture as the shared values and behaviors of different groups and communities. Creative and cultural education is thus posited as a cross-cutting issue, relevant for all subject areas and domains, addressing four challenges of education: the economic (developing skills and aptitudes for a changing work environment), the technological (dealing with rapid technological change), the social (engaging with changing social and cultural values) and the personal (building fulfilling lives).

However, when it comes to detailing how to deliver a broad-based creative and cultural education, the report frequently draws on arts programs. While the first of three clusters of recommendations aims at having creative and cultural education recognized and provided for in whole-of-curriculum policies and teaching practices, subsequent recommendations almost entirely focus on the arts concerning teaching, teacher training, partnerships with external organizations and resource provision: "provision for creative and cultural education in early years education should be further developed, in particular through provision for the arts" (p. 196), and "there are many ways in schools of enabling young people to discuss and express their feelings and emotions . . . among the most important are the arts" (p. 37).

Education and the Business Platform

Three sources of public policy can be tracked around the business platform of creativity and its educational imperatives. Policy developments have come out of business-focused policy, business policy directed to skills and training initiatives, and from education-focused initiatives that are informed by business thinking.

The *Cox Review of Creativity in Business*, led by Sir George Cox (Cox, 2005), was commissioned by the UK government in 2005 to examine the ways that the country's creative skills might be exploited more fully for long-run economic success. Cox, then, seeks to apply generic creative attributes to generic business activity, with a particular emphasis on small business. There is inevitably slippage from attributes to sector in the detail of the report, with the term *creative* often reserved for the creative sector, such as creative arts, creative specialists, or creative businesses. However, it is clear that Cox is also including all other business domains in his discussions. The sixth chapter in the *Cox Review* canvassed three main strategies for "preparing future generations of creative specialists and business leaders" (p. 28), focused on linking higher education with business, and promoting multidisciplinary skills. Cox calls for closer links between universities, which train specialists and professionals entering the business workforce, with small-to-medium enterprises, which make up a substantial slice of the economy and are a locus of entrepreneurial behavior. Such links would include workplace learning and graduate placements. Furthermore, higher education is implored to better prepare graduates to work with and understand specialists from other domains. Cox also recommended establishing centers of excellence to deliver postgraduate courses that promote high levels of business creativity through combining management and business skills with technology, science, engineering, and creative sector specialists, building experience and skills in multidisciplinary innovation through linking business, technology, and the creative sector.

The Business End of the Creative Industries

The creative industries idea underwent a rapid evolution such that, by the early 2000s, numerous jurisdictions had appropriated it and commissioned policy reports. In Australia, the Creative Industries Cluster Study (2002–2003) (DCITA

& NOIE, 2002) and ensuing Digital Content Industry Action Agenda (Strategic Industry Leaders Group, 2005) sought to promote the digital end of the creative industries, assessing its needs, and formulating strategies for its promotion and growth. Skills and training, based on industry needs, were critical issues, since a talented and experienced workforce is central to developing the creative industries and to attracting resources, funding, and investment around creative ideas and projects. As is echoed in generic business arguments around skills, the typical skills needed for digital content areas were said to include (in the words of one of these 'cluster' studies): "creativity, a risk taking and innovative mindset, integrative problem solving abilities, high levels of technical knowledge and applications ability, and entrepreneurial business acumen" (Cutler and Company & CIRAC, 2003, p. 47). Besides dealing with this complex range of skills, the Cluster Study also focused on other difficulties in developing appropriate skills, including the embryonic and rapidly changing nature of some parts of the sector, and the fragmentation and disciplinary silos of education beyond the school years.

Entrepreneurship, the focus of so much policy in the UK in recognition of its core status for business, has also been brought to bear on the creative industries. *Creating Entrepreneurship: Higher Education and the Creative Industries* (HEA-ADMSC & NESTA, 2007; summarized in NESTA, 2007) argued that entrepreneurial skills were imperative for creative industries graduates because of the portfolio structure of their career trajectories, high degree of self-employment, and work as sole traders and small businesses. Key recommendations are that entrepreneurship education be explicit and embedded; links between education and industry be improved; funding and quality assurance mechanisms within higher education be revised since they currently act as hindrances; and entrepreneurship education be practice-based and focused on dealing with uncertainty and complexity as opposed to traditional approaches such as writing business plans (NESTA, 2007, pp. 5–6). A parallel DCMS report, *Developing Entrepreneurship for the Creative Industries* (2006), recommended systemic reform, including developing a national coordinated policy framework and overarching program for entrepreneurship training for the creative industries, building the evidence base by conducting research into models of best practice, and introducing incentives and rewards for educational providers to offer entrepreneurial programs within their creative sector courses.

So far at the business end of creative industries, we have looked at implications for the upper levels of education. A UK report, *Nurturing Creativity in Young People* (Roberts, 2006), was charged with increasing creativity at schools and early childhood centers specifically around employment pathways into the creative industries. To bridge the vocational goals with cultural learning goals, a range of creative industries businesses and cultural organizations were asked about the notions of creativity underpinning their workplace, what generates creativity in young people, how it can be assessed and provided for. *Nurturing Creativity* resumes the discussion of key points in *All Our Futures*, such as developing the teacher as a creative professional, partnering with creative practitioners and organizations (now not just the arts, but all the creative industries), and ensuring that regional and national regulations and education system frameworks should be geared to support a focus on creativity.

Other proposals in *Nurturing Creativity* belong to a robust creative industries perspective and push for changes in the way schools operate. For "extended schools" that operate well beyond normal school hours, creative opportunities should not look like normal school activities, or after-hours diversions but instead offer "a space and time to pursue real projects" (p. 35). For schools undergoing renovations, there is opportunity for children to actually help design their spaces and plan for spaces to support creativity activities that could be used within formal schooling and by the community and creative partnerships. All students are to have a "creative portfolio," which could be physical or online, that contains their creative work and links what they do within school (formal learning) to activities out of school (informal learning) such as the many online multi-media creations and co-productions that students participate in, some on a very regular basis. The pathways to creative industries need to be made clearer, easier to navigate, and should lead to a more diverse creative workforce, assisted through career advice, new qualification routes, work-based training, education-business partnerships, mentoring, and demand-led skills provision. *Nurturing Creativity* proposes that the creativity journey begins in early childhood education: collaborating with creative practitioners, beginning their own portfolio of creative work to share with peers and parents, rethinking the learning spaces and environments, training and professional development of early childhood teachers, and thinking about the continuity of the creativity offer into the next level of education.

Education and the Innovation Platform

Tied as it has been historically to narrow, "white lab coat," linear models of commercialization of scientific breakthroughs, innovation thinking has perhaps rightly been viewed with suspicion on the humanities and arts side of the knowledge equation. However, as it has moved across successive stages toward a "fifth-generation" systems approach (Rothwell, 1994), the innovation platform has the potential to bring together many of the strands of our policy journey.

It should be stressed that innovation approaches are a relatively recent public policy framework which has only been in place for a couple of decades. They significantly postdate most educational and cultural policy frameworks. Their recent emergence means that they are still in a state of real contestability—precisely because they undercut the logic of neoliberal rationales for small government, deregulation, and market-only solutions. Innovation policy is being made in a context in which western governments have re-introduced themselves to an active intervening role after a couple of decades of minimalism in the post-stagflation era and the end of the Keynesian settlement. The innovation policy framework is a value-driven orientation to productivity rather than a cost-efficiency driver for intervention, and, in that sense, it is additional to micro-economic rationales for reform which were the mantra of western governments' strategy in the 1980s and early 1990s. Policy makers now exhibit a disposition to focus on emerging industries that exhibit innovation and R&D intensity, improving the skills and the education of the population, and a focus on universalizing the benefits of connectivity through mass ICT literacy upgrades. Governments' roles in innovation systems are to map and help coordinate the system, facilitate linkages where they are inadequate or bring them into existence where they do not exist. On innovation, they must attend to evidence of *system* failure, not only *market* failure.

We see, on the innovation platform, a further focusing of the attributes approach and further fruitful convergence with the sectorial approach. This parity is consistent with organizational psychology's approach to creativity in which scientists, inventors, entrepreneurs, and artists have been comfortably studied side by side and is evidenced in the strategies of the European Year of Creativity and Innovation 2009 (Kaufman & Baer, 2005; see also Haseman & Jaaniste, 2008; Jaaniste, 2009). It is also consistent with the very wide definitions of the creative class advanced by Richard Florida (2002) and with John Howkins's (2001) definition of the creative industries—both included science and technology R&D.

We could not have proceeded to the innovation platform without the antecedent and co-present business platform. The modern business community certainly claimed to subscribe to creativity as a generic attribute but focused on its particularly short-term, bottom-line applicability. It also laid an essential basis for a better articulation of sector and attributes by insisting that innovation in the service sector of the economy was the source of greater, but different, innovation than that derived from science-based commercialization. Today, services innovation is responsible for driving productivity growth in 80 percent of the Australian economy, and 90 percent of Australia's graduates are employed in the services sectors (McCredie, 2009). These figures are largely replicated throughout the OECD. Business has long argued the policy case that such innovation is what policy should be focusing on, and integral to it are the cross-cutting attributes often termed "soft skills" lateral and critical thinking, teamwork, capacity to work across disciplinary knowledge constructs, and deal with complexity and irresolution (see, for instance, Business Council of Australia 2006a, 2006b; and Australian Business Foundation, 2008). At least some of these "skills" have traditionally been seen as the province of the arts and humanities, which thereby become potentially key sources of innovation capacity.

Where, then, is contemporary innovation policy in relation to education? For two milestone reports—the UK's *Innovation Nation* (DIUS 2008) and *Venturous Australia* (Cutler and Company, 2008)—education is critical. As *Venturous Australia* puts it, "high quality human capital is critical to innovation" (Cutler and Company, 2008, p. xi) and *Innovation Nation* states that, "the UK's capacity to unlock and harness the talent, energy, and imagination of all individuals is crucial to making innovation stronger and more sustainable" (DIUS, 2008, p. 8). Governments can assist with this process, and in doing so help create the best possible conditions that maximize the innovation capacity across a society—for the generation and invention of new ideas and knowledge, for the application and incorporation of new ideas from elsewhere, and for adapting to change across society. Several of the innovation-based educational policy recommendations and initiatives carry forward the business platform of skill development, such as identifying and rectifying skill gaps in economic sectors, and building enterprise training networks. On top of technical, management and entrepreneurial skills, innovation policy also looks at inter-disciplinary skills; many complex problems and opportunities emerge from cross-pollination of different sorts of knowledge, drawing on "T-shaped" professionals "with a grounding in their own discipline but with sufficient knowledge and flexibility to embrace the insights of other disciplines" (Cutler and Company, 2008, p. 60).

The innovation platform is also raising a new focus on the overall skills levels of all citizens because it is argued that the general culture of innovation and ability to take on change and new ideas increase as general levels of education increase. There is an innovation imperative then for government to especially assist those parts of the community with low levels of education and to invest in the general expansion of higher education. In other words, education on an innovation platform should not only skill-up future innovation sector workers but should build a culture of innovation in all citizens and consumers, so that a society is more open and available to taking up innovations and adapting to change. The production and consumption sides of the creative economy have been fruitfully articulated to each other.

In principle, *Innovation Nation* and *Venturous Australia* both regard all knowledge sectors as having equal value as inputs to innovation. In this, they have definitively broken with the recent past and its assumed science and technology priorities. However, this is decidedly unevenly applied in practice. *Innovation Nation* does mention creative industries several times, as an important innovation sector within the economy (DIUS, 2008, pp. 4, 7, 14, 20, 32–33, 49) and *Venturous Australia* riffs on the role of cultural institutions and the arts and humanities as inputs into creative human capital (Cutler & Company, 2008, pp. 47, 50–51). However, one is left with the impression that these were place markers—a few parts of the building are in place, but the foundations have not been fully secured.

Stretching the Sector: The "Creative Workforce"

The link between the cultural to innovation platforms was asserted back in the 1990s, in *Creative Nation*'s claim in its introduction that "culture . . . makes an essential contribution to innovation." One of the crucial ways in which such claims have been developed in the intervening years is the move from creative industries as a sector to an understanding of the creative economy as the general economy increasingly suffused with creative inputs. Much of the growth dynamics, the "additionality" in the creative economy, will be found in this move. The creative industries constitute one sector of the economy; the creative economy is formed when we move from sector-specific focus on creative outputs (culture) to creative occupations as inputs into the whole economy, and creative outputs as intermediate inputs into other sectors.

This analysis takes us into territory recently investigated by Richard Florida, whose work on the "creative class" has highlighted the wider economic significance of creative human capital, especially in underpinning high technology industry development. While Florida's work is highly contestable, it is undeniable that his focus on occupation, qualification, and cultural consumption counterbalances the past dependency on "picking winners" in sector-specific industry development debates.

Cunningham and Higgs (2009) and Higgs and Cunningham (2008) at the ARC Centre of Excellence for Creative Industries and Innovation have measured the creative workforce in terms that are significantly wider than a sector focus alone but do not overreach with such an inclusive definition of the creative class that it loses much of its analytical value. In this account, the creative workforce is composed of specialist creatives (people in creative occupations employed within the creative industries); support personnel (people in the creative industries who perform, for example, sales, management, secretarial, accounting, and administrative functions); and embedded creatives (people in creative occupations working outside the creative industries).

The UK's *Creative Britain* White Paper (DCMS, 2008) is arguably the most cognate policy initiative that comprehensively maps pathways from skills and talents to the creative industries, and incorporates insights from the cultural, business, and innovation platforms. *Creative Britain* models the flow of human capital into the creative sector, from early childhood through to workplace education. "Unlocking creative talent" (pp. 7, 12–18), can be achieved with programs for the school years, most notably offering five hours of arts and culture a week, inside and outside of school, to every child. This strategy builds on the national curriculum and the work of the Creative Partnerships program, which links schools with arts organizations and cultural institutions. The creative economy goals are to give all students the possibility of discovering passions and talents that could lead to a future career in the creative sector, although broader cultural goals of enjoyment in participating in arts and culture, thus building self-confidence, are important benefits too. *Creative Britain* then moves to the role of higher education and early career pathways, aimed at "helping creative talent flourish" (pp. 7–8, 19–30) by overcoming barriers of entry into creative sector employment related to geography and low or no pay in the early career years. Key initiatives include an apprenticeship program for the creative industries (with ambitious plans of adding 5,000 new apprenticeship positions by 2013). It addresses a repeated criticism of the creative industries with plans to encourage a more diverse workforce. This goal resonates with a trend toward social equity and innovation. Career and talent pathways schemes are proposed to give better information and guidance for those entering into the creative sector workforce, including talent scouting, mentoring, and national skills camps for young people. The policy will support higher education delivering the right mix of skills for entering creative sector jobs and for better coordination between universities and industry to close skills gaps.

Conclusion: Outline for an Integrated Vision on the Innovation Platform

The journey we have taken has brought us to a human capital model for the importance of creativity in the economy. The human capital model was the brainchild of endogenous growth theory (Romer, 1990), developed by economists seeking to explain the capacity for sustained growth in modern economies, which could not be solely attributed to inputs such as increased capital and exploitation of natural resources. It has greatly influenced contemporary accounts of innovation, and business' focus on getting the best and the most out of its employees. The limited expansion of the notion of the creative sector (rather than Florida's or Howkins's) has also moved from an "exceptional sector" to a human capital model.

This model becomes the basis for articulating the role of education in the creative economy. The theme of human capital in a creative economy allows for an approach to disciplinary training that stresses the distinctive value of each for innovation and moving away from assumed science-based priorities. It also goes to the center of "fifth generation" innovation thinking, where it is dynamic linkages facilitated by personnel transfer or talent mobility that ensure "flows" between the "stock" in the system. This pattern is posited as a better model of knowledge transfer underpinning innovation than a narrow commercialization model; the flow of skilled, knowledgeable people throughout an innovation system is just as important as the flow of ideas and technologies. It is also the domain where government is on surest ground in defining its role in innovation, through education and training and its derivatives. It is also critical for the way in which it addresses both the supply side and the demand side of innovation.

Knowledge transfer via human capital has implications for how disciplines connect to one another. Education must engender a better rapprochement across the arts and science sectors in research and curriculum. It is critical to delay hyper-specialization in the upper years of secondary school and lower years of undergraduate education, not simply by enforcing a broad range of subject choice but by creating some space for problem-based cross-disciplinary approaches. At the post-graduate and research training end, the capacity to bring specializations together in dynamic multidisciplinary formation is equally critical, reconnecting the different knowledge modes. It is not a matter of dissolving disciplinary specificity into a *melange* of fashionable themes and problems (although at the cutting edge of knowledge, we expect to find multiple emergent new fields), but a pedagogical and research funding focus encouraging and enabling multidisciplinary teams to work effectively on the big issues facing us. It is about coordination between disciplines rather than necessarily a subsumption of disciplines. Collaboration recognizes that many, if not most of a country's most important priorities require multiple disciplinary inputs due to their complexity and scale.

Human capital development through education is not only about the supply of expertise into the workforce; it is also about the demand for innovation. As Walt Whitman famously said, "To have great poets, there must be great audiences." The demand side goes to the question of absorptive capacity: critically trained, socially aware, sophisticated consumers, who connect their buying habits with their identity as citizens, who play a critical role in "demanding" innovation and can cope with, respond to, and absorb innovation. They can appropriate and adapt technologies and new knowledge to their own ends in sometimes surprising, unintended, and innovative ways.

The policy journey toward a vision for education for the emergent creative economy is one that now has shape, scope, and dynamism. It shows that the sectorial and attributes approaches, once thought of as distinct and even opposed, are converging. However, the much more challenging task is to take the policy journey into practice, and to track that journey in rigorous and attentive scholarship.

References

URLs provided are current as of 1 September 2009.

Milestone Reports

Commonwealth of Australia. (1994). *Creative nation: Commonwealth cultural policy.* Canberra: Commonwealth of Australia. Retrieved from http://www.nla.gov.au/creative.nation/contents.html

Cox, G. (2005). *Cox review of creativity in business: Building on the UK's strengths.* London: HM Treasury. Retrieved from http://www.hm-treasury.gov.uk/coxreview_index.htm

Cutler and Company & CIRAC (Creative Industries Research and Applications Centre). (2003). *Research and innovation systems in the production of digital content and applications.* Report for stage three of the Creative Industries Cluster Study. Canberra: Commonwealth of Australia. Retrieved from http://cultureandrecreation.gov.au/cics/Research_and_innovation_systems_in_production_of_digital_content.pdf

DCITA (Department of Communications, Information Technology and the Arts) & NOIE (National Office for the Information Economy). (2002). *Outline of findings from stage 1 and stage 2, Creative Industries Cluster Study Reports.* Retrieved from http://www.cultureandrecreation.gov.au/cics/summary1.doc

DCMS (Department for Culture, Media and Sport). (2001). *Creative industries mapping document 2001.* London: DCMS. Retrieved from http://www.culture.gov.uk/reference_library/publications/4632.aspx

DCMS (Department for Culture, Media and Sport). (2006). *Developing entrepreneurship for the creative industries.* London: DCMS. Retrieved from http://www.culture.gov.uk/reference_library/publications/3501.aspx/

DCMS (Department for Culture, Media and Sport). (2008). *Creative Britain: New talents for the new economy.* London: DCMS. Retrieved from http://www.culture.gov.uk/reference_library/publications/3572.aspx/

DIUS (Department for Innovation, Universities and Skills). (2008). *Innovation nation.* London: DIUS. Retrieved from http://www.dius.gov.uk/reports_and_publications%20HIDDEN/~/media/publications/S/ScienceInnovation_web

HEA-ADMSC (Higher Education Academy Art Design Media Subject Centre) & NESTA (National Endowment for Science, Technology and the Arts). (2007). *Creating entrepreneurship: Higher education and the creative industries.* Retrieved from http://www.adm

.heacademy.ac.uk/projects/adm-hea-projects/creatingentrepreneurship-entrepreneurship-education-for-the-creative-industries

NACCCE (National Advisory Committee on Creative and Cultural Education). (1999). *All our futures: Creativity, culture and education.* London: Department for Culture, Media and Sport. Retrieved from http://www.culture.gov.uk/PDF/naccce.PDF

NESTA (National Endowment for Science, Technology and the Arts). (2007). *Entrepreneurship education for the creative industries.* Policy briefing. London: NESTA. Retrieved from http://www.nesta.org.uk/assets/Uploads/pdf/PolicyBriefing/creating_entrepreneurship_policy_briefing_NESTA.pdf

Roberts, P. (2006). *Nurturing creativity in young people.* A report to government to inform future policy. London: Department for Culture, Media and Sport. Retrieved from http://www.culture.gov.uk/reference_library/publications/3524.aspx/

Strategic Industry Leaders Group. (2005). *Unlocking the potential: Digital content industry action agenda.* Canberra: Commonwealth of Australia. Retrieved from http://www.archive.dcita.gov.au/__data/assets/pdf_file/0006/37356/06030055_REPORT.pdf

Other Sources

Australian Business Foundation. (2008). *Inside the innovation matrix: Finding the hidden human dimensions.* Sydney: Australia Business Foundation. Retrieved from http://www.abfoundation.com.au/research_project_files/46/Inside_the_Innovation_Matrix_9_October_2008.pdf

Business Council of Australia. (2006a). *New concepts for innovation: The keys to a growing Australia.* Incorporating the report by Howard Partners Pty Ltd on *Changing paradigms: Rethinking innovation policies, practices and programs.* Melbourne: Business Council of Australia. Retrieved from http://www.bca.com.au/DisplayFile.aspx?FileID=292

Business Council of Australia. (2006b). *New pathways to prosperity: A national innovation framework for Australia.* Produced in conjunction with the Society for Knowledge Economics. Melbourne: Business Council of Australia: Retrieved from http://www.bca.com.au/DisplayFile.aspx?FileID=158

Cunningham, S. (1992). *Framing culture: Criticism and policy in Australia.* Sydney: Allen & Unwin.

Cunningham, S. (2002). Culture, services, knowledge, or, is content king, or are we just drama queens? *International Association for Media and Communication Research,* 23rd Conference, Barcelona, July 2002. Retrieved from http://eprints.qut.edu.au/202/1/cunningham_culture.pdf

Cunningham, S. (2004). The creative industries after cultural policy: A genealogy and some possible preferred futures. *International Journal of Cultural Studies, 7*(1), 105–116.

Cunningham, S. (2008). From creative industries to creative economy. In G. Hearn & D. Rooney (Eds.), *Knowledge policy: Challenges for the 21st century* (pp. 70–82). Cheltenham: Edward Elgar.

Cunningham, S., & Higgs, P. (2009). Measuring creative employment: Implications for innovation policy. *Innovation: Management, Policy and Practice, 11*(2), 190–200.

Florida, R. (2002). *The rise of the creative class: And how it's transforming work, leisure, community and everyday life.* New York: Basic Books.

Haseman, B., & Jaaniste, L. (2008). *The arts and Australia's national innovation system 1994–2008: Arguments, recommendations, challenges.* Occasional Paper No. 7. Canberra: Council for the Humanities, Arts and Social Sciences. Retrieved from http://www.chass.org.au/papers/pdf/PAP20081101BH.pdf (as of 1 September 2009).

Higgs, P., & Cunningham, S. (2008). Creative industries mapping: Where have we come from and where are we going? *Creative Industries Journal, 1*(1), 7–30.

Howkins, J. (2001). *The creative economy: How people make money from ideas.* London: Allen Lane.

Howkins, J. (2009). *Creative ecologies: Where thinking is a proper job.* St. Lucia: University of Queensland Press.

Jaaniste, L. (2009). Placing the creative sector in innovation: The full gamut. *Innovation: Management, Policy and Practice, 11*(2), 215–229.

Jeffrey, B., & Craft, A. (2001). The universalization of creativity. In A. Craft, B. Jeffrey, & M. Leibling (Eds.), *Creativity in Education* (pp. 1–16). London: Continuum.

Kaufman, J. C., & Baer, J. (Eds.). (2005). *Creativity across domains: Faces of the muse.* Mahwah, NJ: Lawrence Erlbaum Associates.

McCredie, A. (2009, 31 August). Nurturing services to spur growth. *Australian Financial Review*, p. 55.

Romer, P. (1990). Endogenous technological change. *Journal of Political Economy, 98*(5), 71–102.

Rothwell, R. (1994). Towards the fifth-generation innovation process. *International Marketing Review, 11*(1), pp. 7–31.

3

The University and the Creative Economy

Richard Florida, Brian Knudsen, and Kevin Stolarick*

Introduction

Most who have commented on the university's role in the economy believe the key lies in increasing its ability to transfer research to industry, generate new inventions and patents, and spin off its technology in the form of startup companies. As such, there has been a movement in the United States and around the world to make universities "engines of innovation" (Feller 1990; David 1997; Gibbons 2000) and to enhance their ability to commercialize their research. Universities have largely bought into this view because it makes their work more economically relevant and as a way to bolster their budgets. Unfortunately, not only does this view oversell the immediately commercial function of the university; it also misses the deeper and more fundamental contributions made by the university to innovation, the larger economy, and society as a whole.

We argue that the university's increasing role in economic growth stems from deeper and more fundamental forces. The university's role in these forces goes beyond technology to both talent and tolerance. To prove this point, our research utilizes Florida's 3Ts theory of economic development, which specifies the role of the 3Ts of technology, talent, and tolerance in economic development. We recognize the ongoing, productive debate over the creative class approach (Kotkin and

Siegel 2004; Malanga 2004; Peck 2005; see also responses by Florida 2004c; and Florida, Mellander, and Stolarick 2007), but note that that debate is outside the scope of his chapter. Our research simply uses the 3Ts theory as a broad and overarching framework to orient our detailed empirical investigation of the university's role in economic development broadly.

This article provides a data-driven, empirical analysis of the university's role in the "3T's" of economic development, looking in detail at the effects of university R&D, technology transfer, students and faculty on regional technology, talent, and tolerance for all 331 U.S. metropolitan regions.

The findings show that the universities plays an important role across all 3Ts. First, as major recipients of both public and private R&D funding, and as important hotbeds of invention and spin-off companies, universities are often at the cutting edge of technological innovation. Second, universities affect talent both directly and indirectly. They directly attract faculty, researchers, and students, while also acting as indirect magnets that encourage other highly educated, talented, and entrepreneurial people and firms to locate nearby, in part to draw on the universities' many resources. Third, research universities help shape a regional environment open to new ideas and diversity. They attract students and faculty from a wide variety of racial and ethnic backgrounds, economic statuses, sexual orientations, and national origins. On the whole, university communities are generally more meritocratic and open to difference and eccentricity; they are places where talented people of all stripes interact in stimulating environments that encourage open thought, self-expression, new ideas, and experimentation.

The findings further suggest that the university's role in the first T, technology, while important, has been overstressed. We find that the university's even more powerful role across the two other axes of economic development—in generating, attracting, and mobilizing talent, and in establishing a tolerant social climate that is open, diverse, meritocratic, and proactively inclusive of new people and new ideas has been neglected.

We conclude that the university comprises a powerful *creative hub* in regional development. On its own, though, the university can be a necessary but insufficient component of successful regional economic development. To harness the university's capability to generate innovation and prosperity, it must be integrated into the region's broader creative ecosystem.

Theory and Concepts

Universities have long played an important role in research, development, and technology generation. Recently, they have been said to support regional development, as well. Any discussion of the university's role in innovation and economic development quickly circles back to the now classic cases of Stanford University and MIT, which played critical roles in the development of Silicon Valley and the greater Boston area and more recently around Austin, Texas, and the North Carolina Research Triangle. (The literature here is vast, but see in particular: Geiger 1986, 1993; Leslie 1990, 1993; Gibbons 2000.) From these cases, many have concluded that the university serves as a catalyst for economic development. Etzkowitz (1989) and Etzkowitz, Webster, Gebhardt, and Terra (2000) argue that the traditional university whose primary missions are research and teaching has been supplanted by an increasingly "entrepreneurial university," which generates revenue and enhances its political viability through technology transfer, the commercial transfer of innovation, the generation of spin-off companies, and direct engagement in regional development. One Silicon Valley entrepreneur, when asked yet again for "the secret of Silicon Valley's success," summed up this perspective by simply responding: "Take one great research university. Add venture capital. Shake vigorously."

There is a broader theoretical underpinning for the view of the university as an "engine of innovation." According to the "linear model of innovation," innovations flow from university science to commercial technology (Smith 1990). This model informs the view that new and better mechanisms can be deployed to make the transfer and commercialization more effective and efficient, increasing the output of university "products" that are of commercial value to the economy.

Solow (1957) argued that productivity growth was only partly attributable to the traditional explanatory factors, gains to capital and labor. The unexplained "residual" productivity growth, he surmised, must have been due to technological change, which he defined broadly. More recent studies suggest that universities have significant effects on both corporate innovation and regional economic development. Mansfield (1991) later found that investments in academic research yield significant returns to the economy and society.

University research has also been found to support private sector innovation. Jaffe (1989) found that businesses located in close proximity to university research generate greater numbers of patents. Anselin, Vargas, and Acs (1997) found that university research tends to attract corporate research labs. A study of MIT by BankBoston (1997) found that MIT-related firms employed over a million people worldwide. However, these firms were highly geographically concentrated. The Cambridge Boston area was home to thirty-six percent of these companies, even though only nine percent of MIT graduates were originally from Massachusetts. Other New England areas were not nearly as successful at hosting MIT firms, even though many of these areas are located within commuting distance of the MIT campus. Silicon Valley was a second major center for MIT-related firms. Included among other regions with significant concentrations of MIT-related firms are Houston, Seattle, Minneapolis, and Dallas, along with several foreign regions.

Goldstein and Drucker (2006) examined the contribution of universities to economic development across U.S. regions, finding that universities tend to increase average annual earnings, with the most substantial effects occurring in small and medium-size regions.

The university as engine of innovation has been criticized as oversimplified for assuming a one-way path from university-based science and R&D, to commercial innovation and also for seeing the steps in the innovation process as discrete (see Florida and Cohen 1999). It has also been criticized for distorting the mission of the university. Robert Merton (1973) long ago contended that academic science should be an open project because it is firmly centered on the efficient creation of knowledge and movement of frontiers. Firms, on the other hand, seek scientific advances in order to increase profits and acquire intellectual property. Dasgupta and David (1994) have argued strongly for keeping academic science separate from industry. Close ties between industry and university might, they argue, draw academic scientists toward research enterprises with immediate short-term benefits to industry, but away from research with broader and long-term impacts to society and the economy. Conversely, Rosenberg and Nelson (1994) argue that university and industry research, basic science and applied science have always been intertwined, and that it is difficult to even discern the divide between science and technology.

Others argue that that regional differences, in addition to university differences, are part of what accounts for differences in commercialization outcomes. Several studies identify the variables that allow firms and regions to better absorb research coming out of the university. Cohen and Levinthal's (1990) concept of absorptive capacity suggests that successful commercialization requires absorptive capability on the part of regional firms. Smilor et al. (2007) identified three success factors across high-tech regions: strong political leadership, a provocative "incendiary" event, and a catalytic organization. Gunasekara (2004) suggests that elements inherent to a strong regional innovation system are likely to improve a region's absorptive capacity.

Fogarty and Sinha (1999) have found a consistent geographical pattern in the flow of patented information from universities. Intellectual property migrates from universities in older industrial regions such as Detroit and Cleveland to high-technology regions such as the greater Boston, San Francisco Bay, and New

York metropolitan areas. Although new knowledge is generated in many places, relatively few actually absorb and apply those ideas.

At the firm level, Kirchhoff et al. (2007) identify a positive relationship between a company's R&D budget and its absorptive capacity. The R&D level trails only market size and the size of the foreign born population in terms of its influence on firm formation. Agrawal and Cockburn (2003) find a significant relationship between innovative firms and their proximity to so-called "anchor tenants."

A substantial amount of literature seeks to reformulate and move beyond "university as engine" metaphor (Wolfe 2004; Huggins et al. 2008). The "triple helix school" (i.e., Etzkowitz and Leydesdorff, 2000; Etzkowitz and Klofsten, 2005) suggests that ad hoc alliances between the public, private, and educational sectors preclude the discussion of a discrete university that is autonomous from industry, government, and institutions.

We contribute to these literatures by examining whether the university plays additional roles in regional economies—a role beyond technological development. On its own, a university may be a substantial regional resource, but its mere presence is not enough. The region must have the will and capacity to transform and capitalize on what the university produces. It requires a geographically defined ecosystem that can mobilize and harness creative energy. In order to be an effective contributor to regional creativity, innovation, and economic growth, the university must be seamlessly integrated into that broader creative ecosystem.

As noted earlier, we argue that the university's increasing role in the innovation process and in economic growth stems from deeper and more fundamental forces. The changing role of the university is bound up with the broader shift from an older industrial economy to an emerging creative economy, which harnesses knowledge and creativity as sources of innovation and productivity growth (see Florida 2002, 2003, 2004a, 2004b, 2005). We argue that the university plays a role not just in technology, but in all three Ts of economic development: technology, talent, and tolerance.

It is important to note that Florida's creativity theories have stimulated controversy and debate on the drivers and determinants of economic development (Kotkin and Siegel 2004; Malanga 2004; Peck 2005); Florida has responded in detail to these criticisms (2004c), clarifying and refining his theory, and providing additional empirical support for the 3Ts framework (Florida, Mellander, and Stolarick 2007). We recognize this debate and consider it to be useful and important, but its parameters are outside the scope of this study. In our research, we use the 3Ts as a logical guiding framework for an empirical investigation of the role of the broad, multidimensional impact of the university on economic development.

Furthermore, the 3Ts enable us to contextualize the technological contributions of the university which are noted in the literature within the broader context of talent (human capital) and tolerance (or an open social and cultural climate) thus facilitating a more holistic approach to understanding the university's role in economic development.

There is wide consensus among economists and other students of economic development about the primary factors that drive economic development. Solow (1957) found that technology is critically important. Today, drawing primarily upon the work of Lucas (1988), who in turn drew upon Jacobs (1961, 1969), the primary factor is seen to be human capital or what Florida refers to as talent. Drawing on Jacobs's insights, Lucas declared the multiplier effects that stem from talent clustering to be the primary determinant of growth, and he dubbed this multiplier effect "human capital externalities." Places that bring together diverse talent accelerate the local rate of economic development.

Florida's 3T's model is in line with the human capital theory of economic development. It agrees that human capital is the driving force in economic development, but it seeks to amend or supplement it. Foremost, it offers an alternative measure of human capital or talent, which has advantages both conceptually and practically. Most studies of human capital measure it as educational attainment. Florida instead substitutes an occupational measure for the traditional attainment measure, for two primary reasons. First, attainment measures omit people who have been incredibly important to the economy, but who for one reason or another did not go to or finish college. Second, attainment measures do not allow regions to identify, quantify, or build strategy around specific types of human capital or talent. It is clear that nations and regions are specializing in particular kinds of economic activity, and occupational measures highlight this trend.

Additionally, Florida seeks an answer to the question of why some places are better able to develop, attract, and retain human capital/skills/creative capabilities. Recent work by Florida, Mellander, and Stolarick (2007) determined that the distribution of human capital or talent across regions is influenced by a university presence, available consumer service amenities, and regional tolerance. They also found that the creative class outperforms conventional educational attainment measures in accounting for regional labor productivity measured as wages and that tolerance is significantly associated with both human capital and the creative class as well as with wages and income.

Since Schumpeter (1962, 1982), economists have noted the role of the first T, technology, in economic growth (Romer 1986, 1990). More recently, there has been increased interest in the role of the second T, talent or human capital in economic growth (Lucas 1988). However, technology and talent have been mainly seen as *stocks* that accumulate in regions or nations. In reality, these stocks are accumulations of flows between these regions. The ability to capture these flows requires understanding the third T, tolerance, the openness of a place to new ideas and new people. Places increase their ability to capture these flows by being open to the widest range of people across categories of ethnicity, race, national origin, age, social class, and sexual orientation. The places that can attract the widest pool of creative talent—harnessing the creative contributions of the most diverse range of people—gain considerable economic advantage emerging as creativity magnets. They simultaneously catalyze talent from within and attract talent from the outside environment. With the rise of the Creative Economy, the university—as a center for research and technology generation, a hub for talent production and attraction, and a catalyst for establishing an open and tolerant regional milieu—becomes increasingly essential to both innovation and economic growth.

Data and Methods

To explore these issues, we conducted an empirical analysis of the university's role in the 3T's of economic development for all 331 U.S. metropolitan regions. Our university indicators include measures of students, faculty, research and development, technological innovation, and commercialization. The measures of students and faculty are from Integrated Post-Secondary Education Dataset (IPEDs) from the Department of Education; measures of research are from the National Science Foundation's Science and Engineering data series; and measures of technology transfer (such as license income and startups) are from the annual survey of the Association of University Technology Managers (AUTM) and indicators.

The technology measures include indicators of high-tech industry from the Milken Institute and from the patent database of the U.S. Patent and Trademark Office. Talent measures include conventional measures of human capital based on educational attainment and measures of the creative class based on Florida (2002) and from the Bureau of Labor Statistics occupational data files. Tolerance measures are from the U.S. Census and include specific measures of integration (Integration Index), foreign-born people (Melting Pot Index), artistic communities (Bohemian Index), and the gay and lesbian population (Gay/Lesbian Index). (See the Appendix for a full description of all variables and data sources.)

We introduce a new measure of talent, the *Brain Drain/Gain Index*—a measure of the extent to which a region is gaining or losing college-educated talent. We also introduce a new comparative measure of the university in the Creative

Economy, the *University-Creativity Index*, a combined ranking of a region's university strength *and* its creative class. We employ a variety of statistical methods and tests to shed additional light on the university's role in the 3T's of economic development.

Chapple et al. (2004) point out some of the limitations of using specific rankings to evaluate regions and find that older, more diversified economies are often penalized by specific technology-ranking approaches. Their study takes issue with the Milken Institute measure of regional high technology. We utilize the Milken Index measure in light of this critique. We note however, that criticism is primarily directed at understanding regional economies from an output-oriented, industry-based perspective. Both our measures and approach are much more holistic in nature and many of our measures are themselves composites of individual factors. Our diversified measures include industry, human capital, occupational, and numerous perspectives. We also appreciate the complex nature of these relationships and have created multi-dimensional measures to more fully reflect that complexity.

Technology

Technology is the first T. As noted above, various studies have found that universities play a significant role in regional technology. We begin with a listing of the top 25 regions in R&D intensity (measured as R&D spending per capita). One can already see a limit to the university as engine of innovation perspective. The top five regions are State College, PA (Penn State); Bryan-College Station, TX (Texas A&M); Iowa City, IA (University of Iowa); Rochester, MN (Mayo Clinic); and Lawrence, KS (University of Kansas). Rounding out the top 10 are Champaign-Urbana, IL (University of Illinois); Corvallis, OR (Oregon State University); Athens, GA (University of Georgia); and Lafayette, IN (Purdue University). In fact, the entire list is dominated by regions home to large state universities. Of the leading high-tech centers, only Raleigh-Durham-Chapel Hill (15th) and Boston (19th) are represented in the top twenty. Silicon Valley is conspicuously absent from the list.

Table 2 ranks the top 25 regions across the country in terms of licensing income per faculty and university-generated spin-off companies. Two regions generate more than $40,000 per faculty in licensing income—Rochester, MN, and Tallahassee, FL. These are also not regions that top the popular lists of high-tech industrial centers. Two others, Santa Cruz and Santa Barbara, CA, generate more than $20,000, while seven others generate more than $10,000 in licensing income.

San Jose, Boston, and Seattle, three noted high-tech industry centers, make this list, though a wide variety of other types of regions are on it, including a lot of classic college towns.

The ability of universities to generate new startup companies has frequently been noted as a key spur to regional growth of high-tech industry. The roles played by Stanford University in the Silicon Valley and of MIT in the growth of the greater Boston-Route 128 corridor are legendary. When considering the number of start-up companies per faculty member Rochester, MN, ranks first. This set of cities is followed by Galveston, TX, Charlottesville, VA, Birmingham, AL, and Salt Lake City. None of these cities is known as being a hotbed of entrepreneurial activity. However, the top ten is rounded out with Boston, the Research Triangle area, Madison, WI, Athens, GA, and Mobile, AL. Again, major state university centers also do rather well.

Table 1
University R&D, Inventions and Patent Applications

Rank	Regions	R&D per Capita	Invention Disclosures per Faculty	Patent Applications per Faculty
1	State College	$3242.97	0.104	0.149
2	Bryan-College Station	2606.49	0.085	0.057
3	Iowa City	2259.52	0.081	0.081
4	Rochester MN	2146.82	1.434	0.717
5	Lawrence KS	1932.31	0.054	0.012
6	Champaign-Urbana	1913.54	0.062	0.031
7	Bloomington IN	1858.77	0.047	0.042
8	Corvallis	1775.23	0.044	0.034
9	Athens	1684.50	0.041	0.041
10	Lafayette IN	1440.97	0.076	0.046
11	Gainesville	1352.11	0.099	0.080
12	Charlottesville	1312.92	0.116	0.137
13	Madison	1299.71	0.194	0.109
14	Ann Arbor	863.43	0.054	0.045
15	Raleigh-Durham	805.49	0.143	0.093
16	Auburn	769.90	0.020	0.019
17	Columbia MO	695.05	0.024	0.012
18	Fort Collins	609.12	0.049	0.029
19	Boston	591.68	0.103	0.098
20	Bangor	583.29	0.005	0.003
21	Santa Barbara	582.07	0.079	0.069
22	Lincoln	543.46	0.013	0.016
23	Santa Cruz	510.79	0.095	0.083
24	Lansing	508.64	0.044	0.032
25	Baltimore	489.39	0.096	0.089

N = 107 MSAs for which AUTM data are available

Table 2
University Licensing Income and Startups

Rank	Regions	Licensing Income per Faculty	Total Licensing Income ($ M)	Startups per 1000 Faculty	Total Startups (still in business)
1	Rochester MN	47,460	5.36	17.699	5
2	Tallahassee	43,603	67.50	1.292	6
3	Santa Cruz	29,318	16.77	2.847	0
4	Santa Barbara	24,514	29.86	2.380	0
5	Madison	16,028	22.94	4.193	32
6	Gainesville	15,621	26.27	3.567	33
7	Orange County	13,133	28.07	1.275	0
8	Sacramento	13,084	38.70	1.270	0
9	Oakland	11,982	47.75	1.163	0
10	Lansing	11,864	25.72	0.461	15
11	San Jose	11,516	36.94	2.494	88
12	New York	9,977	164.09	0.934	54
13	Los Angeles	9,078	108.52	2.212	82
14	Seattle	7,914	30.30	1.567	127
15	Boston	7,558	73.33	5.154	271
16	San Diego	7,223	29.51	1.188	5
17	Rochester NY	5,879	14.63	0.923	5
18	Birmingham	5,421	3.72	7.278	28
19	Iowa City	4,915	5.07	0.000	17
20	Galveston	4,446	0.96	13.953	4
21	Houston	4,344	18.45	2.119	33
22	Minneapolis	4,291	23.14	2.039	50
23	Springfield MA	3,911	9.05	0.864	8
24	Riverside	3,754	15.60	0.364	0
25	Charlottesville	3,752	4.02	9.346	29

N = 107 MSAs for which AUTM data are available

We conducted a variety of statistical analyses to better gauge the relationship between university research and regional high technology. In particular, we looked at the relationship between university technology outputs and the Milken Institute's commonly used measures of high-technology industry. The main findings are as follows. There is a considerable overall relationship between university technology and regional high-technology industry. The correlations between university technology outcomes (invention disclosures, patent applications, licensing income, startups), and regional innovation and high-tech industry are consistently positive and significant. It should be noted that license income correlations are considerably stronger for the 49 large regions (those with populations of more than one million) than for all 107 regions for which data are available, but the rest of the correlations show no such large city bias. This finding confounds research by Matthiessen and Schwarz (1999), which suggests that successful commercialization is associated with large urban agglomerations.

Table 3
Correlations between University and Regional Technology Measures

	Invention Disclosures	Patent Applications	License Income	Startups
Regional Patents	0.344	0.390	0.376	0.291
	0.376	0.342	0.687	0.288
Tech-Pole	0.312	0.409	0.485	0.287
	All insignificant			

All correlations are significant at the 0.05 level (2-tailed).

Note: First row for each indicator is for the 49 regions over 1 million; the second row is for all 107 regions for which university data is available

The relationship between university technology and regional innovation is complex, however. There are some regions where university technology has a strong effect on regional innovation and high-tech industry, and others where it does not. Table 1 is a two-by-two matrix that we use to illustrate the pattern of relationships between university technology to regional innovation. It compares regions with high and low scores on the Milken Institute's Tech-Pole Index (a measure of high-tech industry concentration) to the level of university innovation (measured as university patenting in the region). Its quadrants identity four types of regions.

Table 1
University Patenting versus Regional High-Technology

	Low Tech-Pole Index	**High Tech-Pole Index**
High University Patenting	Galveston Charlottesville Athens Bryan-College Station State College N=8	Los Angeles Houston Atlanta Boston San Jose N=8
Low University Patenting	Detroit Baton Rouge Springfield MA Mobile Lexington N=13	New York Washington DC Nassau Newark Portland OR N=6

Strong university innovation does not necessarily translate into strong local high-tech industry. An apt, if oversimplified, metaphor for this dynamic is the university as the transmitter and the region as the receiver. In a few, highly selective cases the university sends out a strong signal which is picked up well by the region.

However, this is far from the norm. In a large number of cases, the university may be sending out a strong signal—it is carrying out a lot of technical R&D and producing patents—but the region's receiver is switched off and unable to take in the signal the university sends out. As numerous studies suggest, these signals can be and are frequently picked up by other regions outside the local region (Bathelt et al. 2004; Saxenian 2002; BankBoston 1997). This allows regions with weak local university signals to capitalize on the technology signals they absorb from outside regions. The extent to which regions exhibit the capacity to absorb ideas and knowledge into their economies is indicative of the presence of a local ecosystem of creativity, places that, with their universities, create an environment amenable to the attraction of both new ideas and creative and knowledgeable people.

As Jane Jacobs (1961, also see Ellerman 2004) pointed out, it might be best to see the university in biological terms, where the talent and technology being produced by a university are "seeds." These seeds can land close to the parent plant; they can be carried by animals to other (generally nearby) locations; or they can be carried by the winds around the globe. However, like all seeds, just landing somewhere is not enough—if the soil is not fertile, if there is not enough water or light, or if there is too much, the seed will not sprout. Further, the seed might sprout but then not grow very much or be stunted. If the conditions are not right, many seeds will not sprout and will instead be carried on the next breeze or passing animal to better locations.

Additionally, like several plants, the university can change its surrounding ecosystem to make conditions more favorable for its seeds to take root. It can also create an environment in which more and different types of species—ideas and people—can combine, compete, reproduce, and evolve. However, if the ecosystem is not receptive, those seeds will only grow in more amenable regions.

Talent

Talent is the second T. Lucas (1988) long ago argued that economic growth stems from clusters of talented people and high human capital. Glaeser (2000a, 2000b; Berry and Glaeser 2005) finds a close association between human capital and economic growth. He shows that firms locate not to gain advantages from linked networks of customers and suppliers, as many economists have argued, but to take advantage of common labor pools of talented workers. Glendon (1998) found that human capital levels in cities in the early twentieth century provided a strong predictor for city growth over the course of the entire century. Wolfe (2004) notes the university's role in talent generation and attraction. In their study of the economic effects of universities, Goldstein and Drucker (2006) found that universities effect economic growth more through the production of human capital than from research and development. Universities are themselves generators of human capital. They attract and produce two primary types of talent—students and faculty. Regions that can retain these locally produced resources gain competitive advantage. Students represent the core production of universities. However, faculty members are important talent in their own right. In addition to teaching students and doing research, star faculty are magnets for faculty and staff from abroad. Star faculty can and often do have a magnetic effect in the attraction of people and even companies.

Table 4
Student and Faculty Concentration: Top 25 Regions

Rank	Region	College Students per 10,000	Total College Students	Faculty per 10,000	Total Faculty
1	Bryan-College Station	3,086	47,039	108.3	1,651
2	Bloomington IN	2,896	34,916	116.9	1,409
3	State College	2,678	36,356	144.4	1,961
4	Lawrence KS	2,565	25,640	104.5	1,045
5	Gainesville	2,449	53,371	77.2	1,682
6	Iowa City	2,422	26,885	92.9	1,031
7	Champaign-Urbana	2,377	42,713	104.2	1,873
8	Corvallis	2,153	16,823	119.3	932
9	Auburn	2,123	24,433	98.9	1,138
10	Athens	2,047	31,409	115.4	1,771
11	Lafayette IN	2,018	36,888	84.6	1,547
12	Tallahassee	1,845	52,485	54.4	1,548
13	Columbia MO	1,833	24,827	66.2	897
14	Yolo	1,785	30,104	n/a	n/a
15	Bloomington IL	1,633	24,570	67.7	1,019
16	Provo	1,547	57,002	43.1	1,589
17	Greenville NC	1,506	20,154	5.8	78
18	Charlottesville	1,391	22,199	67.1	1,070
19	Muncie	1,366	16,227	77.1	916
20	Grand Forks	1,339	13,051	48.6	474
21	Lansing	1,302	58,283	48.4	2,168
22	Tuscaloosa	1,282	21,141	60.8	1,003
23	Lubbock	1,271	30,844	40.4	981
24	San Luis Obispo	1,270	31,338	14.5	358
25	Chico	1,269	25,780	6.7	137

Table 4 lists the top 25 regions by student and faculty concentration. The list here is dominated by college towns. The top five large (high population) regions in terms of student concentration are Austin, the Research Triangle, San Francisco, San Diego, and San Jose, but none of these regions ranks higher than 50th in terms of overall student concentration. Production of students is only a small part of the overall regional talent story. It is important to examine the larger role of the university in the region's overall talent or human capital system. To get a first glimpse of this lay-out, we look at the correlations between the talent produced by the university and the region's overall talent base. Table 5 shows the correlations between university strength and talent.

Table 5
Correlation of University and Talent Measures (N=331)

	Students per Capita	Faculty per Capita
Human Capital (BA and above)	0.572	0.429
Super-Creative	0.251	0.134
Creative Class	0.208	0.150

All correlations are significant at the 0.05 level (2-tailed).

There is a positive and significant correlation between both students and faculty and regional talent, measured by the percentage of the working age population with a college degree. A positive but less strong relationship is also found between students and faculty and the creative and super-creative classes. Here, it is important to note that university faculty are members of both the creative and super-creative class and when faculty are removed from those categories the correlation disappears. While there is a strong tie between regional talent and technology outcomes, the relationships between university talent and regional technology outcomes are mixed. The relationship is much stronger for students than for faculty. Students are significantly associated with the regions' patents per capita (0.490), patent growth (0.473), and high-technology industry (using the Milken Institute Tech-Pole Measure, 0.431). The correlation coefficient is not a sensitive enough to isolate "star faculty." Subsequent research should examine the degree to which an elite group of faculty members might serve as magnets for talent. Our research does not mean to suggest a firm boundary between the university's technology generating and talent attraction roles. In fact, when universities attract talent to the region they are assisting in the commercialization of new discoveries. Technological knowledge is not completely codifiable. Many crucial forms of knowledge (skills, practices, memories) are embedded within individuals and their social networks (Wolfe 2004; Pavitt 1991). Attraction and retention of talent can also be seen as attraction and retention of technological knowledge. Furthermore, the ability of regions to retain human capital from local universities can be seen as a key indicator of absorptive capacity. We now look specifically at the issue of talent retention and attraction, using a new indicator developed for this purpose.

Brain Drain or Gain

There has been mounting concern in the United States and elsewhere over the so-called "brain drain," the movement of talented, high human capital people from one region to another, as seen from the losing region's perspective. Low retention rates of local graduates is troubling to parents and economic developers alike, and many regions are trying to figure out ways to keep graduates from leaving or to lure them back when they get older.

However, focusing only on retention misses a crucial part of the picture. A region that retains many of its own graduates but fails to attract degree-holders from other regions will most likely fall behind. The availability of a strong pool of local talent can trump both physical resources and cost in attracting corporations and growing regional economies. Talented people are a very mobile means of production. Students often leave regions after their four years are up; and young, highly educated people are the most mobile of virtually any demographic group. Some regions produce talent and export it, while others are talent importers.

To get at this issue, we developed an index that quantifies the combined retention and attraction rates of university-educated talent. We call it the *Brain Drain/Gain Index* (BDGI). This measure makes no distinction between graduates retained and those drawn from other regions. It just computes the net result: the relative gain (or drain) of people progressing from students to degree-holding workers.

The BDGI for a region is calculated as the percent of the population age 25 and over with bachelor's degree or above, divided by the percent of the population ages 18 to 34 currently in college or university (postsecondary school). A region with a BDGI above 1.0 is a *brain gain* region, a net recipient of highly educated talent. A region with a BDGI below 1.0 is a *brain drain* region, a net *breeder* or *donor* of university talent. It retains proportionately fewer degree-holders than degree-earners.[1] We consider the BDGI to be the best available simple and easily tractable indicator of a region's combined talent attraction and retention capability. Table 6 shows the 25 regions on the BDGI along with the percentage of the total population in college, percentage of 18 to 34 year-olds in college and percentage of those 25 and above with a college degree or above.

The Brain Drain/Gain Index is not a perfect measure. It does not capture the actual "flows" of college-educated persons, whether recently graduated or in mid-career, to/from a region, nor does it measure whether previous generations of college graduates have been retained. Instead, it measures the current "state" of education utilization and production across the region. It is designed to determine if a region is producing people with college degrees at the same rate in which it is using them. It could be considered a "temporal quotient" in that it compares percentages across two different life-stages for a region in the same way that a location quotient compares a regional-special concentration to a national one. The numerator for the BDGI is the standard and widely accepted measure for human capital, but it does include retired individuals. While retired people can have a college degree that they essentially are no longer putting to effective use, there is no clear cut-off age. In addition, many retired individuals continue to make significant regional social and economic contributions. The measure as currently constructed is straightforward and can be easily calculated at almost any geographic level using readily available Census data.

Table 6
Leading Brain Drain/Gain Index Regions

Rank	Region	BDGI	% of Entire Population in College	%18-34 in College	% 25 and above with Degree
1	Stamford	2.04	4.3%	24.2%	49.4%
2	Naples	1.67	2.7%	16.7%	27.9%
3	Danbury	1.50	4.7%	26.3%	39.4%
4	Atlanta	1.45	5.4%	22.0%	32.1%
5	Rochester MN	1.41	5.1%	24.7%	34.7%
6	Denver	1.38	5.8%	24.8%	34.2%
6	Dallas	1.38	5.4%	21.7%	30.0%
8	Washington DC	1.31	7.1%	31.9%	41.8%
9	Barnstable	1.25	3.4%	26.8%	33.5%
9	San Francisco	1.25	8.7%	35.0%	43.6%
11	Seattle	1.24	6.6%	28.9%	35.9%
12	Nashua	1.23	5.0%	26.9%	33.2%
13	Middlesex	1.22	6.4%	30.8%	37.4%
13	Charlotte	1.22	5.1%	21.8%	26.5%
15	Indianapolis	1.21	4.6%	21.4%	25.8%
16	Minneapolis	1.19	6.1%	28.0%	33.3%
16	Houston	1.19	5.3%	22.9%	27.2%
18	San Jose	1.18	8.4%	34.4%	40.5%
19	Kansas City	1.17	5.0%	24.4%	28.5%
20	Portland ME	1.16	5.7%	28.9%	33.6%
21	Des Moines	1.15	5.5%	24.9%	28.7%
21	Richland	1.15	3.9%	20.3%	23.3%
21	Santa Fe	1.15	6.3%	34.8%	39.9%
24	Elkhart	1.14	2.9%	13.6%	15.5%
24	Newark	1.14	5.5%	27.7%	31.5%

The most striking finding of our geographic data is that just 10 percent of all 331 U.S. metro regions are net attractors of talent. Of all regions, only 10 boast BDGI scores of 1.25 or above. Another 5 score over 1.20, and 8 more over 1.15. Only 23 regions nationwide do better than 1.15. Especially notable here are San

Francisco, San Jose, Washington, DC, and Santa Fe, in that a large part of the population is college educated (more than 30 percent) and many employees have a college degree (more than 40 percent). We should also note that six regions score high on both the BDGI and our overall measure of university strength: Austin, Boston, Raleigh-Durham, San Francisco, San Jose, and Portland, ME. Our findings support the work of Stephan et al. (2004), who have previously commented on the geographic dimensions of the "brain drain." Their study of PhD students revealed significant hemorrhaging from the Midwest toward the Pacific and Northeast.

To get at the relationship between talent and regional growth, we estimated correlations between the BDGI and a variety of regional outcome measures: patent growth, high-tech industry, population growth, job growth, and income growth (see Table 7). The correlations are uniformly high. The BDGI is related to key regional outcomes, especially employment growth and high-technology industry but also regional innovation, population growth, and income growth.

Table 7
Correlations among BDGI, Regional Innovation and Growth

Outcome	BDGI
Patent Growth	0.395
Tech-Pole	0.361
Tech Share	0.434
Tech Share Growth	0.432
Population Growth	0.443
Job Growth	0.520
Per Capita Income Growth	0.320

In our view, the relationship between the BDGI and regional growth is multifaceted. High BDGI regions have thick and thriving labor markets that are able to capture and absorb growth. However, high BDGI regions also have higher talent levels, which in turn are associated with higher technology levels. In effect, the correlation results for the BDGI reflect a "virtuous circle" where higher levels of talent lead to more technology generation, innovation and entrepreneurship, leading over time to higher rates of economic growth, more job generation and in turn to higher rates of talent production, retention, and attraction.

Tolerance

Tolerance is the third T. Major research universities can do much to "seed" tolerance and diversity in a region. Nationwide, university towns tend to be among the most diverse regions. Tolerance means being open to different kinds of people and ideas—ideally being *proactively inclusive*—not just "tolerating" their presence but welcoming diverse people as neighbors and entertaining their views as valid and worthwhile.

A key mechanism by which universities—both singularly and in partnership with communities—help build ecosystems of innovation and contribute to talent retention and attraction is through the promotion of tolerance and diversity, which have been shown to be important factors in individuals' location decisions.

Scholars such as Joel Mokyr (1990) and Simonton (1999) have found that societies through history tend to flourish when they are open and eclectic but stagnate during periods of insularity and orthodoxy. Florida and Gates (2001) find that openness and tolerance are associated with differential rates of regional innovation and high-tech industry in the United States. Florida (2002) has found that talented and creative people favor diversity and a wide variety of social and cultural options. Openness to ideas—to *creativity*—is paramount to both talent attraction and economic success. Talented and creative people vote with their feet—and they tend to move away from communities where their ideas and identities are not accepted. Indeed, regions with large numbers of high-tech engineers and entrepreneurs also tend to be havens for artists, musicians, and culturally creative people. Seattle, Austin, and Boston are cases in point. Some scholars (Gunasekara 2004, 2008; Cooke 2002; Cooke and Morgan 2000) have pointed out that openness to learning is a key feature of successful regional innovation system. Kirchhoff et al. (2007) find the size of the foreign-born population is the second largest influence on the creation of new firms in a location.

The university has long functioned as a hub for diversity and tolerance. Universities have been called "Ellis Islands" of our time, noting their ability to attract large numbers of foreign-born students. The Silicon Valley venture capitalist, John Doerr, has frequently remarked that the United States should "staple a green card" to the diplomas of foreign-born engineering and science students who contribute significantly to the nation's innovative capability (Miller 2008).

Indeed, universities can serve as an incredibly productive refuge for minorities seeking education as a hedge against discrimination. Gay men and lesbians show higher than average education levels and are often disproportionately represented on college campuses and in college towns (Black, Gates, Sanders, and Taylor 2000). Lifelong learning provides older citizens with a way to actively engage in a community. In general, the universities and university communities have long been places that are open to free speech, self-expression, political activism, and a broad diversity of ideas.

The university itself becomes an "island" of tolerance or at least a "spike." By its very nature, the university is more diverse, both faculty and students—generally more diverse than the surrounding community. The university's diversity can be contagious with the university's diversity extending beyond the campus to the surrounding community. Moreover, if conditions are right, it can spread well beyond the immediate neighborhood and even have a multiplicative impact on broader regional diversity. This extra-university diversity changes the nature of the surrounding community so that it can attract and retain more of the talent and technology that the university is producing.

Cooke (2005) sees the university as a knowledge "transceiver," which links local actors with global knowledge sources, in addition to transmitting knowledge across a global pipeline. We would emphasize the role of foreign human capital in the "transceiving" process. Foreign students and faculty members do not simply augment the diversity of their regional environments; they act as links to knowledge and financial networks that are crucial in the dissemination of technological knowledge. Saxenian's (2002) construct of "brain circulation" highlights the key role of the university in fostering links between local economies and offshore networks.

Until relatively recently, though, the university had been a very insular environment, often purposely and intentionally separating itself from the broader society. In a way, university communities provided a function sort of like the old bohemian communities of Greenwich Village where eccentricity and difference were readily accepted, even encouraged. With the rise of creativity as the primary driver of economic growth, the norms and values of these once limited and isolated "creative communities" become more widely generalized and diffused throughout greater segments of society.

We conducted statistical analyses to gauge the relationship between the university and regional tolerance. We employ various measures of tolerance including an overall Tolerance Index, which is composed in turn of separate measures of racial integration (Integration Index), foreign born population (Melting Pot Index), artistic and bohemian communities (Bohemian Index), and the gay and lesbian population (Gay/Lesbian Index).

Between the original and paperback editions of *The Rise of the Creative Class,* the tolerance measures were revised to reflect the need to have and engage minority populations. The addition was not simply adding the percentage of the population that is African American to the metrics. Instead, a much more nuanced argument is made. An Integration Index was added to the measures (for a complete discussion, see Appendix B of the paperback edition). The important point is that regional growth and talent attraction are not impacted simply because a region is diverse (racially, ethnically, other ways). As Jane Jacobs (1969) pointed out, the interactions generate innovation. Separate "islands" of diverse populations do not have the impact that interactions among those groups can create. Integration provides a much better measure of regional openness and acceptance than simple percentages.

We found a considerable correlation between tolerance and the log of students and faculty, as Table 8 shows. Tolerance increases with both overall population and number of faculty, but the strongest relationships are almost always with the number of students. This is true in all but one case, the Melting Pot Index, which is roughly the same for total population and number of students. While integration does decline as population increases, this relationship is not as strong in communities with larger university-based populations.

There is also a significant, negative correlation between the integration index and logged populations of students and faculty. This finding is in line with those of Thomas and Darnton (2006). We suspect several causes. First, there may be a relationship between racial integration in a jurisdiction and the emphasis it puts on university funding. Student and faculty numbers are key indicators of a jurisdiction's resource investment. Perhaps both of these variables can be associated with a more "liberal" or "progressive" political environment. Homogenous regions may also experience high integration scores because the index compares neighborhood diversity to total regional diversity. In other words, the negative integration correlation may also suggest a positive relationship between regional heterogeneity and economic growth.

Each of the tolerance measures was regressed against the logs of total population, total students and total faculty for all 331 metro regions. As Table 9 shows, students appear to play the key role here. The correlations for the total number of students are positive and highly significant for the overall Tolerance Index and the separate Melting Pot, Gay, and Bohemian Indexes. The correlations for both population and faculty are generally negative and significant. The negative coefficients for population suggest that the impact that the total number of students has on diversity declines with increasing population. In other words, the universities have a bigger and more pronounced effect on tolerance when they are located in smaller regions.

Table 8
Correlations between University Strength and Tolerance

	N	Tolerance Index	Melting Pot Index	Gay/Lesbian Index	Bohemian Index	Integration Index
Log Total Students	331	0.510	0.463	0.502	0.548	-0.480
Log Total Faculty	324	0.427	0.322	0.420	0.478	-0.351
Log Total Population	331	0.386	0.467	0.415	0.440	-0.538

All correlations are significant at the 0.05 level (2-tailed).

Universities are institutions that value diversity and whose effects on diversity and tolerance extend far beyond their classrooms and laboratories. This scenario is especially true in smaller regions where the universities play larger and more significant roles in shaping regional norms and values. As with the dimensions we have examined up until now, a university's tolerance "signals" are subject to differential rates of absorption. We suspect that much of this has to do with, among other things, the level of cooperation/animosity between a university and its surroundings. In addition the quality of the "signal" itself can vary, as many universities must overcome hurdles before they are beacons of diversity and meritocracy. However, universities, in general, do foster social environments of openness, self-expression and meritocratic norms and help to establish the regional milieu required to attract and retain talent and spur growth in the Creative Economy.

Table 9
Regression Results for Diversity

	Dependent Variable			
	Tolerance Index	Melting Pot Index	Gay/Lesbian Index	Bohemian Index
Intercept	-0.004 n/s	-0.384	-0.389 n/s	-0.176 n/s
Log Students	0.541	0.136	0.757	0.834
Log Faculty	-0.123	-0.075	-0.169	-0.151
Log Population	-0.272	0.012 n/s	-0.290	-0.382
Adjusted R^2	0.33	0.28	0.273	0.341

n/s = not significant
All other correlations are significant at the 0.05 level (2-tailed).

The University-Creativity Index

In order to get at the broader relationship between the university and regional creativity, we constructed a *University-Creativity Index* or UCI. The index combines a measure of student concentration with the percent of a region's work force in the creative class. In keeping with Chapple et al. (2004), this is a diversified measure, which is more likely to capture more of the big picture. We view this not as a measure of actual creative performance but rather as a measure of how a region's absorptive capacity is capitalizing on its university capabilities and how it combines them with other creative assets. In our view, a ranking in the top 50 means a region has considerable assets to work with and is well positioned to leverage those assets for improved innovative and economic performance. Table 10 shows regions on the University-Creativity Index for four regional size classes.

The top five large regions are all noted high-tech regions: San Jose, San Francisco, San Diego, Austin, and Boston. Rounding out the top 10 are Sacramento and Oakland (both in the San Francisco Bay Area), Seattle, Denver, Los Angeles, and Chicago. The rankings for small and medium-size regions, not surprisingly, are dominated by major state university centers, such as Lansing, MI (Michigan State); Ann Arbor, MI (University of Michigan); Madison, WI (University of Wisconsin); Provo, UT (University of Utah); Gainesville, FL (University of Florida); Bryan-College Station, TX (Texas A&M); and Corvallis, OR (University of Oregon), among many others. These findings suggest there is tremendous potential for harnessing university assets for regional economic growth in these communities. This trend is already occurring in some of these places, notably Madison's recent ascendance as a center for high-technology industry and spin-off companies.

A wide variety of regions that are not usually seen as topping the lists of high-technology centers also do well on the UCI. These include: Albany and Syracuse, NY; Omaha and Lincoln, NE; Dayton, OH; Trenton, NJ; Des Moines, IA; Spokane, WA; Muncie, IN; and Portland, ME. Our sense is that there is considerable unrealized creative potential in these regions. Of older industrial regions, only Chicago places in the top 50. Other older industrial regions with superb universities and colleges—like St. Louis, Baltimore, Philadelphia, and Pittsburgh—rank only between 50 and 100. It is our view that these regions suffer from a significant absorptive capacity deficit. Alongside efforts to improve university research and technology transfer, these regions need to work on their ability to absorb the significant signals their universities are sending out.

Our findings suggest that there are many considerable advantages for developing inter-regional partnerships between older industrial regions and their surrounding university centers. Two places that jump out from the data are Central Indiana and Greater Detroit. Indianapolis, for example, which ranks 239th on the UCI, is flanked by Bloomington and Lafayette, which rank 3rd and 10th, respectively. Detroit, which ranks 140th on the UCI, is flanked by Lansing and Ann Arbor, which rank 4th and 21st, respectively. In our view, the economic future of these regions lies less in their older commercial centers and downtowns (which are in part legacies of the industrial age) and much more in the major university centers that are on their peripheries. These places would benefit from broad interregional partnerships—and the development of "superregional" strategies that combine the size and scale of their older centers with the considerable 3T capabilities of their major research university communities.

Table 10
University-Creativity Index

Rank	Region	Overall Rank	University/Creativity Interaction
Regions with population 1 million and above			
1	San Jose	6	0.924
2	San Francisco	11	0.896
3	San Diego	19	0.856
3	Austin	19	0.856
5	Boston	24	0.841
6	Sacramento	26	0.837
7	Oakland	29	0.814
8	Seattle	34	0.801
9	Denver	35	0.795
10	Los Angeles	42	0.772
10	Chicago	42	0.772
Regions with population between 500,000 and 1,000,000			
Rank	**Region**	**Overall Rank**	**University/Creativity Interaction**
1	Albany NY	15	0.876
2	Ann Arbor	21	0.855
3	Columbia SC	37	0.789
4	Omaha	42	0.772
5	Albuquerque	48	0.761
6	Springfield MA	51	0.754
7	Dayton	54	0.748
8	New Haven	59	0.745
9	Syracuse	61	0.737
10	Baton Rouge	68	0.710
Regions with population between 250,000 and 500,000			
Rank	**Region**	**Overall Rank**	**University/Creativity Interaction**
1	Lansing	4	0.926
2	Madison	8	0.917
3	Montgomery	9	0.914
4	Provo	11	0.896
5	Trenton	13	0.893
6	Tallahassee	14	0.891
7	Huntsville	22	0.853
8	Lincoln	28	0.828
9	Des Moines	36	0.790
10	Spokane	38	0.787
Regions with population below 250,000			
Rank	**Region**	**Overall Rank**	**University/Creativity Interaction**
1	Gainesville	1	0.980
2	Bryan-College Station	2	0.976
3	Bloomington IL	3	0.965
4	Corvallis	4	0.926
5	Missoula	7	0.923
6	Lafayette IN	10	0.899
7	Charlottesville	15	0.876
8	Muncie	17	0.869
9	Santa Fe	18	0.861
10	Portland ME	23	0.849

Conclusion

This study has examined the role of the university in the 3Ts of economic growth—technology, talent, and tolerance—suggesting that the role of the university encompasses much more than the simple generation of technology. We examined these issues for all 331 metropolitan regions in the United States, analyzing the performance of universities in producing technology and talent and in shaping the tolerance of their regions. We introduced a new indicator for talent flows, the *Brain Drain/Gain Index* (BDGI), a measure of the extent to which a region is attracting and retaining college-educated talent. We also introduced a new comparative measure of the university in the Creative Economy, the *University-Creativity Index*, a combined ranking of a region's university *and* its overall strength in the Creative Economy. We have used statistical methods to further illuminate the university's role in the 3Ts and hope to shed new light on its broad role in economic growth and development.

Our findings suggest that the role of the university goes far beyond the "engine of innovation" perspective. Universities contribute much more than simply pumping out commercial technology or generating startup companies. In fact, we believe that the university's role in the first T, technology, while important, has been overemphasized to date, and that experts and policy-makers have somewhat neglected the university's even more powerful roles in the two other Ts—in generating, attracting, and mobilizing talent and in establishing a tolerant and diverse social climate.

Future research should attempt to parse out how much of the statistical relationships are due solely to the presence of the university, whose employment of highly educated and often non-white individuals factors directly into the dependent variables in question. This scenario is particularly problematic for "college towns" where the university represents the bulk of the local economy. Testing separately for the presence (or absence) of knowledge-based activity and talent *outside* the university can be completed to investigate the extent to which the university generates a "spillover" effect into the regional economy. This more specific investigation was beyond the scope of this chapter, which is focused more broadly on the regional impact of the university across all regions, both those with and without a major university presence.

In short, the university comprises a potential—and, in some places, actual—*creative hub* that sits at the center of regional development. It is a catalyst for stimulating the spillover of technology, talent, and tolerance into the community.

First, in terms of technology: as major recipients of both public and private research and development funding and as sources of innovations and spin-off companies, universities are often at the cutting edge of technological innovation. However, university invention does not necessarily translate into regional high-tech industry and economic growth. In fact, we found that many regions have universities at the cutting edge of technology, but this does not develop into local regional growth. While universities comprise an important precondition for regional innovations, to be effective, they must be embedded in a broader regional ecosystem that can absorb their research and inventions and turn them into commercial innovations, industrial development, and long-term growth.

Second, universities play a powerful role in generating, attracting, and retaining talent. On the one hand, they directly attract top faculty, researchers, and students. On the other hand, they can also act as magnets for other talent, attracting talented people, research laboratories, and even companies to locate near them to access their research and amenities.

Third, universities and colleges have a significant effect on the third T, tolerance, shaping regional environments that are open to new ideas and diversity. Universities are the "Ellis Islands" of the creative age, attracting students and faculty from a wide variety of racial and ethnic backgrounds, income levels, sexual orientations, and national origins. University communities and college towns are places that are open to new ideas, cultivate freedom of expression, and are accepting of differences, eccentricity, and diversity. These norms and values play an increasingly important role in attracting talent and in generating the new ideas, innovations and entrepreneurial enterprises that lead to economic growth.

Our findings also indicate the simultaneity of university-economy relationships. Studies of technology note that a region's "absorptive capacity" affects its ability to capitalize on technological research. We suggest that a region's ability to absorb human capital is also important to regional retention of non-codified knowledge. We likewise suggest that a more open social and cultural climate also works to bolster greater regional absorptive capacity.

In order to be an effective contributor to regional creativity, innovation, and economic growth, the university must be integrated into the region's broader creative ecosystem. On its own, a university's actions are limited. In this sense, universities are necessary but insufficient conditions for regional innovation and growth. To be successful and prosperous, regions need absorptive capacity—the ability to absorb the science, innovation, and technologies that universities create. Universities and regions need to work together to build greater connective tissue across all 3Ts of economic development. The regions and universities that are able to synergistically and simultaneously bolster their capabilities in technology, talent, and tolerance will realize considerable advantage in generating innovations, attracting and retaining talent, and in creating sustained prosperity and rising living standards for all their people. Most of all, we encourage future research that probes the non-technological dimensions of the university in economy and society.

*Gary Gates contributed to an earlier version of this paper. Thanks to Patrick Adler and Andrew Bell for research assistance.

References

Agrawal, A., and Cockburn, I. 2003. "The Anchor Tenant Hypothesis: Exploring the Role of Large, Local, R&D-Intensive Firms in Regional Innovation Systems." *International Journal of Industrial Organization* 21: 1227–1253.

Anselin, L., Vargas, A., and Acs, Z. 1997. "Local Geographic Spillovers between University Research and High Technology Innovations." *Journal of Urban Economics* 42:3, 422–448.

BankBoston. 1997. "MIT: The Impact of Innovation." Boston: BankBoston.

Bathelt, H., Malmberg, M., and Maskell, P. 2004. "Clusters and Knowledge: Local Buzz, Global Pipelines and the Process of Knowledge Creation." *Progress in Human Geography* 28:31, 31–56.

Berry, C., and Glaeser, E. 2005. "The Divergence of Human Capital across Cities." *Papers in Regional Science* 84:3, 407–444.

Black, D., Gates, G., Sanders, S., and Taylor, L. 2000. "Demographics of the Gay and Lesbian Population in the United States: Evidence from Available Systematic Data Sources." *Demography* 37:2, 139–154.

Chapple, K., Markusen, A., Schrock, G., Yamamoto, D., and Pingkang, Y. 2004. "Gauging Metropolitan "High-Tech" and "I-Tech" Activity." *Economic Development Quarterly* 18:10–29.

Cohen, W., and Levinthal, D. 1990. "Absorptive Capacity: A New Perspective on Learning and Innovation." *Administrative Science Quarterly* 35:1, 128–152.

Cooke, P. 2002. *Knowledge Economies: Clusters, Learning and Cooperative Advantage.* New York: Routledge.

Cooke, P. 2005. "Regionally Asymmetric Knowledge Capabilities and Open Innovation: Exploring 'Globalisation 2'—A New Model of Industry Organisation." *Research Policy* 34, 1128–1149.

Cooke, P. and Morgan K. 2000. *The Associational Economy: Firms, Regions, and Innovation.* Oxford: Oxford University Press.

Dasgupta, P., and David, P. 1994. "Toward a New Economics of Science." *Research Policy* 23:3, 487–521.

David, P. 1997. "The Knowledge Factor: A Survey of Universities." *The Economist,* 4 October.

Ellerman, D. 2004. "Jane Jacobs on Development." *Oxford Development Studies,* 32:4 December 2004: 507–521.

Etzkowitz, H. 1989. "Entrepreneurial Science in the Academy: A Case for the Transformation of Norms." *Social Problems* 36:1, 14–29.

Etzkowitz, H., Webster, A., Gebhardt, C., and Terra, B. 2000. "The Future of the University and the University of the Future: Evolution of Ivory Tower to Entrepreneurial Paradigm." *Research Policy* 29, 313–330.

Etzkowitz, H., and Leydesdorff, L. 2000. "The Dynamics of Innovation: From National Systems and 'Mode 2' to a Triple Helix of University-Industry-Government Relations." *Research Policy* 29, 109–125.

Etzkowitz, H., and Klofsten, M. 2005. "The Innovating Region: Toward a Theory of Knowledge-Based Regional Development." *R&D Management* 35:3, 243–255.

Feller, Irwin. 1990. "Universities as Engines of R&D-based Economic Growth: They Think They Can." *Research Policy* 19:4, 335–348.

Florida, R. 1999. "The Role of the University: Leveraging Talent, Not Technology." *Issues in Science and Technology* 15:4, 67–73.

Florida, R. 2002. *The Rise of the Creative Class.* New York: Basic Books.

Florida, R. 2003. "Cities and the Creative Class." *City and Community* 2.1, 3–19.

Florida, R. 2004a. *The Rise of the Creative Class* (updated paperback edition). New York: Basic Books.

Florida, R. 2004b. *Cities and the Creative Class.* New York: Routledge.

Florida, R. 2004c. "Revenge of the Squelchers." *Next American City,* 5, July.

Florida, R. 2005. *The Flight of the Creative Class.* New York: HarperBusiness.

Florida, R., and Cohen, W. 1999. "Engine or Infrastructure? The University Role in Economic Development." In *Industrializing Knowledge: University—Industry Linkages in Japan and the United States,* edited by Lewis M. Branscomb, Fumio Kodama, and Richard Florida. Cambridge, MA: MIT Press.

Florida, R., and Gates, G. 2001. "Technology and Tolerance: The Importance of Diversity to High-Technology Growth." Washington, DC: Brookings Institute, Center on Urban and Metropolitan Policy.

Florida, R., Mellander, C., and Stolarick, K. 2007. "Inside the Black Box of Regional Development—Human Capital, the Creative Class, and Tolerance." *Journal of Economic Geography* 8.5: 615–649.

Fogarty, M., and Sinha, A. 1999. "University-Industry Relationships and Regional Innovation Systems—Why Older Industrial Regions Can't Generalize from Route 128 and Silicon Valley." In *Industrializing Knowledge: University-Industry Linkages in Japan and the United States,* edited by Lewis M. Branscomb, Fumio Kodama, and Richard Florida. Cambridge, MA: MIT Press.

Geiger, R. 1986. *To Advance Knowledge: The Growth of American Research Universities, 1900–1940.* New York: Oxford University Press.

Geiger, R. 1993. *Research and Relevant Knowledge.* New York: Oxford University Press.

Gibbons, J. 2000. "The Role of Stanford University: A Dean's Reflections." In *The Silicon Valley Edge: A Habitat for Innovation and Entrepreneurship,* edited by Chong-Moon Lee, William F. Miller, Marguerite Gong Hancock, and Henry S. Rowen. Stanford, CA: Stanford University Press.

Glaeser, E. 2000a. The New Economics of Urban and Regional Growth. In *The Oxford Handbook of Economic Geography,* edited by Gordon Clark, Meric Gertler, and Maryann Feldman. New York: Oxford University Press.

Glaeser, E., 2000b. "The Future of Urban Research: Non-Market Interactions." Pp. 101–149 in *Brookings—Wharton Papers on Urban Affairs,* edited by William G. Gale and Janet Rothenberg Pack. Washington, DC: Brookings Institution Press.

Glendon, S. 1998. "Urban Life Cycles." Harvard University, Department of Economics, unpublished working paper.

Goldstein, H., and Drucker, J. 2006. "The Economic Development Impacts of Universities on Regions: Do Size and Distance Matter?" *Economic Development Quarterly* 20:1, 22–43.

Gunasekara, C. 2004. "The Regional Role of Universities in Technology Transfer and Economic Development." *British Academy of Management Conference.* St. Andrews, Scotland.

Hall, B., Jaffe, A., and Tratjenberg, M. 2000. "The NBER Patent Citation Data File: Lessons, Insights and Methodological Tools." NBER Working Paper 8498.

Huggins, R., Johnston, A., and Steffenson, R. 2008. "Universities, Knowledge Networks and Regional Policy." *Cambridge Journal of Regions, Economy, and Society* 1, 321–340.

Jacobs, J. 1961. *The Death and Life of Great American Cities.* New York: Random House.

Jacobs, J. 1969. *The Economy of Cities.* New York: Random House.

Jaffe, A. 1989. "Real Effects of Academic Research." *American Economic Review* 76:5, 984–1001.

Kirchhoff , B. A., Newbert, S. L., and Hasan I., et al. 2007. "The Influence of University R&D Expenditures on New Business Formations and Employment Growth." *Entrepreneurship Theory and Practice,* 31: 543–559.

Kotkin, J., and Siegel, F. 2004. "Too Much Froth." *Blueprint Magazine.*

Leslie, S. 1990. "Profit and Loss: The Military and MIT in the Postwar Era." *Historical Studies in the Physical and Biological Sciences* 21:1, 59–86.

Leslie, S. 1993. *The Cold War and American Science.* New York: Columbia University Press.

Lucas, R. 1988. "The Mechanics of Economic Development." *Journal of Monetary Economics* 22:1, 3–42.

Malanga, S. 2004. "The Curse of the Creative Class." *City Journal,* Winter. 36–45.

Mansfield, E. 1991. "Academic Research and Industrial Innovation." *Research Policy* 20:1, 1–12.

Matthiessen, C., and Schwarz, A. 1999. "Scientific Centres in Europe: An Analysis of Research Strength and Patterns of Specialization Based on Bibliometric Indicators." *Urban Studies* 36:3, 453–477.

Miller, C. 2008. "John Doerr's Advice for Barack Obama: Hire Bill Joy." *New York Times.com* November 5.

Merton, R. 1973. *The Sociology of Science.* Chicago: University of Chicago Press.

Mokyr, J. 1990. *The Lever of Riches: Technological Creativity and Economic Progress.* New York: Oxford University Press.

Pavitt, K. 1991. "What Makes Basic Research Economically Useful?" *Research Policy* 20:109–119.

Peck, J. 2005. "Struggling with the Creative Class." *International Journal of Urban and Regional Research* 29:4, 740–770.

Romer, P. 1986. "Increasing Returns and Long-Run Growth." *Journal of Political Economy* 94:5, 1002–1037.

Romer, P. 1990. "Endogenous Technological Change." *Journal of Political Economy* 98:5, S72–S102.

Rosenberg, N., and Nelson, R. 1994. "American Universities and Technical Advance in Industry." *Research Policy* 23:3, 323–348.

Saxenian, A. 2002. "The Silicon Valley Connection: Transnational Networks and Regional Development in Taiwan, China, and India." *Science Technology Society* 2:117, 118–149.

Schumpeter, J. 1962. *Capitalism Socialism and Democracy.* New York: Harper Perennial.

Schumpeter, J. 1982. *The Theory of Economic Development.* Somerset, NJ: Transaction.

Simonton, D. 1999. *Origins of Genius: Darwinian Perspectives on Creativity.* New York: Oxford University Press.

Smilor, R., O'Donnell, N., Stein, G., and Welborn, R. 2007. "The Research University and the Development of High-Technology Centers in the United States." *Economic Development Quarterly.* 21:3, 203–222.

Smith, B. L. R. 1990. "American Science Policy Since World War II." Washington, DC: The Brookings Institution.

Solow, R. 1957. "Technical Change and the Aggregate Production Function." *Review of Economics and Statistics* 39:3, 312–320.

Stephan, P., Sumell, A., Black, G., and Adams, J. 2004. "Doctoral Education and Economic Development: The Flow of New Ph.D.s to Industry." *Economic Development Quarterly* 18, 151–167.

Thomas, J. M., and Darnton, J. 2006. "Social Diversity and Economic Development in the Metropolis." *Journal of Planning Literature* 21, 153–168.

Wolfe, D. A. 2004. *The Role of Universities in Regional Development and Cluster Formation.* University of Toronto: Centre of International Studies.

Appendix
Indicators and Data Sources

This appendix provides a brief description of the major variables and data sources used. The unit of analysis is the region or Metropolitan Statistical Area (MSA).

University Measures

University Technology: Data for university technology outputs, including research and development, invention disclosures, patent applications, licensing income, and startups are from the Association of University Technology Managers annual survey. The data are for the year 2000 and cover 107 metropolitan areas.

University Strength: This measure is the sum of inverse rankings of college students per capita and faculty members per capita, and it covers all 331 MSAs. The faculty data are from the Integrated Postsecondary Education (IPEDS) dataset and are for the year 2000. Students per capita come from the 2000 Census which counts students in the metropolitan region. IPEDS also has student numbers, but they are based on the number of students who attend institutions within the metropolitan area, so those who attend the school and commute from outside the MSA are counted. The IPEDS and Census student counts are closely correlated (0.98 correlation).

University-Creativity Index: This measure is the sum of inverse (or reverse) rankings of students per capita and percent Creative Class (see below), with that quantity divided by 662. In this system the highest score corresponds with the highest rank.

Technology Measures

Tech-Pole Index: The tech-pole index measures the prevalence or spatial concentration of high-tech industry in a metropolitan area and is based on two factors: (1) high-tech location quotient and (2) the metro area proportion of national high-tech output (referred to in the text as "tech share"). It is based on data provided by Ross De Vol and colleagues at the Milken Institute.

Patents: There are two measures of patents: patents per capita and patent growth. This variable measures innovation by using simple utility patent count data available from the NBER Patent Citations Data File (Hall, Jaffe, and Tratjenberg 2000).

Talent Measures

Human Capital: This is the standard human capital index which measures the percentage of residents 25 years of age and older with a bachelor's degree and above.

Creative Class: Percentage of the region's employees in the following categories:

- Super-Creative Core: Computer and mathematical occupations, architecture and engineering occupations; life, physical, and social science occupations; education (not including education support), training and library occupations; arts, design, entertainment, and media occupations
- Management occupations
- Business and financial operations occupations
- Legal occupations
- Healthcare practitioners and technical occupations (not including Healthcare support)
- High-end sales and sales management

These definitions are based on Florida, *The Rise of the Creative Class* and are from the 2000 Bureau of Labor Statistics Occupational Employment Statistics Survey (Florida 2002; 2005).

Tolerance Measures

Bohemian Index: A location quotient of the number of those working in bohemian occupations in an MSA. It includes authors, designers, musicians, composers, actors, directors, painters, sculptors, craft-artists, artist printmakers, photographers, dancers, artists, and performers.

Gay/Lesbian Index: Originally calculated by Black et al. (2000) for gay men only, it is a location quotient measuring the over- or underrepresentation of coupled gays and lesbians in an MSA.

Melting Pot Index: This variable measures the percentage of foreign-born residents in an MSA. It is based on the 2000 Census.

Integration Index: The Integration Index measures how closely the racial percentages within each Census tract within a metropolitan area compare to the racial composition of the region as a whole. This measure takes into account six racial/ethnic groups: white, non-Hispanic; black, non-Hispanic; Asian/Pacific Islander, non-Hispanic; other races (including mixed races), non-Hispanic; white Hispanic; and nonwhite Hispanic.

Tolerance Index: The Tolerance Index is a composite of four separate measures, each of which captures a different dimension of tolerance or diversity: the Integration Index, Melting Pot Index, the Bohemian Index, and the Gay/Lesbian Index.[2]

Notes

1. It is important to point out that the numerator does not count people under 25 who already have a degree and are working, while it does count those who have a degree but are not working. Although this data limitation is regrettable, we do not expect for there to be significant inter-regional differences in under the age of 25 human capital levels. Another caveat is that small regions with universities students actually tend to score lower on the BDGI because the denominator (percent of younger people currently in school) is so large.
2. See the paperback edition of *The Rise of the Creative Class* (New York: Basic Books, 2004) for further definitions of the Integration and Tolerance indices.

4

Creative Clusters and Universities: The Cluster Concept in Economics and Geography

Terry Flew

For much of its history, economics as a discipline has tended to work with a limited understanding of the significance of space. Models of economic equilibrium have very often assumed that markets operate, as the geographer Doreen Massey put it, "like angels dancing on the head of a pin" (Massey, 1984, p. 52). Whenever the question of where economic activity takes place emerges, the focus of the answer has commonly been on the economic development of nations as the relevant spatial unit, most famously articulated by Adam Smith in *The Wealth of Nations*. The macroeconomic revolution that followed the publication of John Maynard Keynes' *General Theory* in 1936 focused on the flow of goods, services, people, and money within and between nations, in line with the orientation toward the nation-state that came to characterize the social sciences from the late nineteenth century onwards (Taylor, 1996).

One of the few major economists to have given explicit attention to questions of where economic activity takes place was Alfred Marshall. In his *Principles of Economics*, first published in 1890, Marshall addressed this question with particular reference to the changing industrial geography of nineteenth-century Britain. In the first instance, industries set up business close to where the physical raw materials are most available (such as steel mills near coal mines), but the patronage of wealthy individuals as well as government could also attract skilled people to a city or region, as with artisans and tailors moving to be near particular courts. He observed that the concentration of a particular region on a single industry had advantages and disadvantages. The advantages are that labor markets develop in such places, and what we today term tacit knowledge is fostered by the clustering of a particular group of workers in a region. As Marshall put it:

> When an industry has chosen a locality for itself, it is likely to stay there long: so great are the advantages which people following the same skilled trade get from near neighborhood to one another. The mysteries of the trade become no mysteries; but are as it were in the air, and children learn many of them unconsciously. Good work is rightly appreciated, inventions and improvements in machinery, in processes and the general organization of the business have their merits promptly discussed: if one man starts a new idea, it is taken up by others and combined with suggestions of their own; and thus it becomes the source of new ideas. And presently subsidiary trades grow up in the neighborhood, supplying it with implements and materials, organizing its traffic, and in many ways conducing to the economy of its material. (Marshall, 1890/1990, p. 225).

The obvious disadvantage of industrial concentration is that the economic fortunes of the region are very much hostage to developments in that industry. Over time, cities tend to develop a more diverse range of activities, and improvements in transport and communication further this trend because, even though they make it easier to move goods from one place to another (thereby promoting regional specialization), they also make it easier for people to relocate. What Marshall observed in nineteenth-century Britain was not so much the movement of the population from agriculture to manufacturing, but rather that the use of large-scale machinery meant that growth in output was steadily less dependent on additional supplies of labor, but instead on the growth of service occupations and industries that cluster around growth centers. The rise of services, for Marshall, "tended to increase the specialization and localization of industries" (Marshall, 1890/1990, p. 230), as they can make a region less vulnerable to the cyclical fluctuations and the rise and fall of particular manufacturing industries.

Marshall's observations on regional specialization were not widely taken up by economists, partly because they opened up the thorny question of what happens to equilibrium economic models if we allow for falling costs and increasing returns to scale, which would make monopolies more prevalent and challenge assumptions that the price mechanism operates primarily to ration scarce goods and services (Warsh, 2006). The French economist Francois Perroux developed the concept of *growth poles* to assist policy-makers in understanding how particular regions developed economic dynamism based on industrial specialization, and Swedish economist Gunnar Myrdal developed the concept of *cumulative causation* to explain why particular regions could experience ongoing growth based on the agglomeration of industries and skilled labor, which could occur at the expense of other regions. For the most part, however, these insights were not taken up in the Anglo-American economic mainstream, and they were also at the margins of economic geography until the 1970s, which became more interested in developing a "science of the spatial" that could make the sorts of universal claims that characterized economics, rather than studies of regional differentiation, which came to be seen as limited and parochial (Barnes, 2003).

The rise of *clusters* as a stand-alone concept emerges out of the business management literature, and particularly with the work of Michael Porter from the Harvard Business School (Porter, 1990). In extending his *competitive advantage* model from firms to nations, Porter observed that understanding the dynamic and sustainable sources of competitive advantage required a shift of thinking away from costs and production efficiencies toward those elements that promote productivity growth over time and innovation, and in particular the *spillover* benefits that can emerge from being in particular locations, including the presence of related and supporting industries. Porter argued that one's location within particular clusters provides three sources of competitive advantage to the firms that are a part of them:

1. *Productivity gains,* deriving from access to specialist inputs and skilled labor, access to specialized information and industry knowledge, the development of complementary relationships among firms (e.g., hotels or restaurants based around tourism centers), and access to institutions providing public or quasi-public goods, such as universities and training institutions;

2. *Innovation opportunities,* derived from proximity to buyers and suppliers, ongoing face-to-face contact with others in the industry, and the presence of competitors that generates pressure to innovate in circumstances where cost factors are similar;

3. *New business formation,* as there is better information about opportunities, better access to resources required by business start-ups (e.g., venture capitalists, skilled workforce), and reduced barriers to exit from existing businesses as takeovers and mergers are more readily facilitated due to shared informational resources.

Localization, Urbanization, and Creative Clusters

Cluster theories bring together two dynamic trends in economic geography. The first is the tendency toward *localization*, or the clustering of firms in similar or related industries in a particular city or region, and the positive externalities that can arise from such co-location. Marshall's pioneering analysis of such externalities pointed to the benefits in terms of labor market specialization, tacit knowledge, and institutional specialization, and was developed in three directions in the 1980s and 1990s. First, there was a growing interest in the significance of *industrial districts*, or those cities and regions that appeared to defy trends toward de-industrialization and the shifting of manufacturing industry toward lower cost centers in the developing world. Work on manufacturing districts in the "Third Italy," to mention, one example, pointed to an evolving historical nexus between clusters of small and medium-sized enterprises (SMEs), embedded trust relations that acted as a positive stimulus to innovation, and the production of quality goods that retained global market demand even in the face of lower-cost alternatives from countries such as China (Piore & Sabel, 1984; Asheim, 2000). Second, there were those regions where conspicuous value adding to a primary product had occurred through cluster developments that had a global impact, such as in wine-making regions such as Northern California and South Australia. Finally, there was the focus on developing new high-technology districts that could become the "next Silicon Valley" (Castells & Hall, 1994; Kenney & Von Burg, 2000). The costliness and lack of results associated with many of the ventures, combined with the realization that the lessons of Silicon Valley were hard to generalize to other locations (Leslie & Kargon, 1996), have generated skepticism among economic geographers about the cluster concept, with Martin and Sunley observing that "it is being applied so widely that its explanation of causality and determination becomes overly stretched, thin and fractured" (Martin & Sunley, 2003, p. 29).

Urbanization is generally understood as involving the large-scale movement of people to cities, whether through migration from the countryside or from other parts of the world. Amin (2003) observes that, historically, "the Western city was the factory and the center of commercial life, in short, the engine of capital accumulation . . . the city became the source of 'immobile' resources and agglomeration economies for competitive advantage" (Amin, 2003, p. 115). While the disadvantages of cities—such as pollution, overcrowding, and high land rents—have seen large parts of industry leave cities to take advantage of locational advantages elsewhere, cities remain central to post-industrial or knowledge-based economies on the basis of factors such as the benefits of proximity for diverse businesses, concentrated consumer demand for services, culture and entertainment, diversity of populations, the concentration of business, professional and legal services in cities, and the location of corporate headquarters in "global" cities. Lorenzen and Frederiksen (2008) differentiate urbanization economics from those associated with localization on the basis of the large city attracting a diverse range of industries and types of employment, in contrast to the concentration of a particular industry coming to define specialized industrial districts. The positive externalities that cities develop include their diversity of industries, the sharing of knowledge among unrelated firms and industries, the diversity of labor, skills, knowledge, and ideas that act as stimuli to innovation and entrepreneurship, and the range and diversity of institutions and infrastructures (Lorenzen & Frederiksen, 2008, pp. 159–160). Surveying the literature from economics, geography, and sociology, Amin concludes:

> There appears little evidence to support the claim that cities are becoming less important in an economy marked by increasing geographical dispersal. . . . [They] assert, in one way or another, the powers of agglomeration, proximity, and density, now perhaps less significant for the production of mass manufactures than for the production of knowledge, information and innovation, as well as specialized inputs (Amin, 2003, p. 120).

It is insufficient, however, to simply understand the continuing growth of cities as the result of economic forces. In his epic *Cities in Civilization*, Peter Hall (1998) observes that because the city "continues to attract the talented and the ambitious . . . it remains a unique crucible of creativity" (Hall, 1998, p. 7). Through his historical account of great cities, Hall argues that "while no one kind of city, or any one size of city, has a monopoly on creativity or the good life . . . the biggest and most cosmopolitan cities, for all their evident disadvantages and obvious problems, have throughout history been the places that ignited the sacred flame of human intelligence and the human imagination" (Hall, 1998, p. 7). The need to think about cultural and economic factors together has, for Hall, become even more imperative in the advanced industrial or postindustrial nations, as their cities "have become more and more preoccupied by the notion that cultural industries . . . may provide the basis for economic regeneration, filling the gap left by vanishing factories and warehouses, and creating an urban image that would make them more attractive to mobile capital and mobile professional workers" (Hall, 1998, p. 8).

The Rise of Creative Clusters

The notion that city cultures could constitute a key source of location-based competitive advantage became one of the big ideas of urban economic geography in the 2000s. Charles Landry (2000) drew attention to the role played by *creative cities* in catalyzing economic and social innovation, particularly through the formation of a *creative milieu*, whose participants form what he terms a *soft infrastructure* of "social networks, connections and human interactions, that underpins and encourages the flow of ideas between individuals and institutions" (Landry, 2000, p. 133). Richard Florida (2002, 2008) has widely proclaimed that cities with a reputation for tolerance, diversity, and openness to new ideas and cultural "buzz" act as talent magnets for what he terms the *creative class* of idea-generators, who are central to the knowledge-based and creative industries. This creative class is, for Florida, the fastest growing segment of the U.S. economy, as creativity becomes "the decisive source of competitive advantage in 21st century global knowledge economies" (Florida, 2002, p. 5).

The growing interest in creative cities has arisen in part out of the awareness that, in the twenty-first century, cities have become more important. This development occurred despite forces emerging since the 1970s, such as economic globalization, the suburbanization of major cities, the movement of large-scale manufacturing to the developing world, and the rise of the Internet and globally networked information, and communication technologies (ICTs), which could have promoted population dispersal and the decline of cities. Scott (2008) links the resurgence of cities to the rise of what he terms the *cognitive-cultural economy*, and others term the rise of the *creative industries* (Hartley, 2005) or the *creative economy* (UNCTAD, 2008). Scott links the centrality of cities, or what have elsewhere been termed *global city-regions* (Scott *et al.*, 2001), to three core elements of this "new" economy:

1. The contractual and transactional nature of production in knowledge-intensive and creative industries, which involve ongoing relationships between shifting networks of specialized but complementary firms. Geographical proximity reduces the transaction costs of joining and maintaining such networks across projects and over time.

2. Specialist workers engaged in these industries are drawn to such urban agglomerations as the centre of activity, thereby reducing job search costs, and as "talent magnets" for those aspiring to work in such industries.

3. The resulting local system of production, employment and social life in turn generates learning and innovation, and "a 'creative field' or a structured set of interrelationships that stimulate and channel various kinds of creative energies" (Scott, 2008, p. 313). This dynamic is further promoted by the existence of complementary forms of "social overhead capital" that include the role played by universities, research centers, design centers, and other sites that generate specialist knowledge capital that can be applied in these sectors.

Alongside the resurgence of cities has been a rethinking of the role of culture, from a set of activities defined by their distance from the economy (the noncommercial arts), toward *culture as a resource.* Landry argued that "cultural resources are the raw materials of a city and its value base. . . . Culture, therefore, should shape the technicalities of urban planning rather than be seen as a marginal add-on to be considered once the important planning questions like housing, transport and land-use have been dealt with" (Landry, 2000, p. 7). In a similar vein, Venturelli identified culture as the "gold" of the global information economy:

> Culture can be seen as the key to success in the Information Economy, because for the very first time in the modern age, the ability to create new ideas and new forms of expression forms a valuable resource base of a society and not merely mineral, agricultural and manufacturing assets. Cultural wealth can no longer be regarded in the legacy and industrial terms of our common understanding, as something fixed, inherited, and mass-distributed, but as a measure of the vitality, knowledge, energy, and dynamism in the production of ideas that pervades a given community . . . the greater cultural concern should be for forging the right environment (policy, legal, institutional, educational, infrastructure, access, etc.) that contributes to this dynamism and not solely for the defence of a cultural legacy or industrial base. (Venturelli, 2005, p. 396)

In terms of urban policy, thinking about culture as an economic resource and as an asset-generating competitive advantage has given rise to what Stevenson refers to as a *new civic gold rush* in urban planning and cultural policy alike, promoting strategies aimed at "fostering strategically the cultures of cities and regions . . . [where] culture and creativity have become forms of 'capital' . . . traded in an international marketplace comprised of cities eager to compete with each other on the basis of imagery, amenity, liveability and visitability" (Stevenson, 2004, pp. 119–120).

The creative cities debate can be understood at two levels (Stevenson, 2004; Mommaas, 2004; Cooke, 2008; Costa, 2008). First, there are debates about whether whole cities are creative, and whether some cities are more creative than others. Such claims have been made about cities such as London (Landry, 2005), New York (Currid, 2007), Los Angeles and Paris (Scott, 2000). "Creative city" indices inspired by the work of Florida and Landry have generated "league tables" designed to address such questions. Is San Francisco more creative than Los Angeles? Is Dublin more creative than Glasgow? Is Barcelona more creative than Madrid? Is Melbourne more creative than Sydney? Storper and Scott (2009) observe that aside from problems arising from the metrics used for such exercises, they are premised upon assumptions that urban growth and the capacity to attract creative and knowledge-intensive industries are primarily driven by "supply" factors, or the ability of local authorities or cultural elites to generate the right "settings" to attract creative workers; they systematically downplay the role played by global macro-economic forces in driving the location of such industries. It is not surprising, then, that cities such as New York, London, Los Angeles, and Paris feature in such discussions, as these are global cities and centers of global information and service industries more generally.

A second approach focuses on *creative clusters* and the capacity of local authorities to incubate creative industries growth in particular parts of major cities, sometimes referred to as cultural quarters (Bassett et al., 2005; Cooke, 2008). Such strategies are closer to the Marshall-Porter tradition of cluster development, as they are premised upon the spatial agglomeration of related activities more than a hard-to-define creative ethos residing in some sections of an urban population. In an evaluation of creative cluster initiatives in four cities in The Netherlands (Amsterdam, Rotterdam, Tilburg, and Utrecht), Mommaas (2004) observed that these strategies have been driven by a heterogeneous mix of policy priorities including:

- Attracting globally mobile capital and skilled labor to particular locations;
- Stimulating a more entrepreneurial and demand-oriented approach to arts and cultural policy;
- Promoting innovation and creativity in the society more generally, through opening up possibilities for greater interaction between culturally vibrant locales and innovation in other sectors of the economy;
- Finding new uses for derelict industrial-era sites such as warehouses, power plants, etc. as sites for postindustrial activities, such as residential apartments, arts centers, and business incubators;
- Promoting cultural diversity and cultural democratization, and being more inclusive of the cultural practices of hitherto marginalized social groups and communities.

Given such an eclectic mix of motivations, it is not surprising that the scorecard for the new "creative" urban cultural policies is mixed. In an overview of such developments in European cities, Bassett et al. (2005, pp. 150–153) argue that some of the benefits have included:

- Moving questions of culture from the margins to the centre of urban development strategies;
- Broadening understandings of culture from elite arts and formally defined arts centers to the wider spectrum of informal arts practices, popular culture, and cultural consumption in urban spaces;
- More integrated approaches to urban planning and zoning that recognize the significance of lifestyle and consumption activities as well as production;
- Development of new cultural infrastructures that have acted as catalysts for urban regeneration and given cities more of a cultural image that also acts as an attractor for tourism and possibly investment.

Problems with these policies have included:

- Blurring of the distinctiveness of arts and culture, and absorption into civic "boosterism" and strategies primarily focused upon real estate development;
- Possibly contradictory aims and policy agendas, particularly between economic development and social inclusion;
- The danger that the drive to harness the cultural and creative assets of locations acts as a homogenizing force, promoting minor variations on the same thematic elements (Peck, 2005);
- The tendency of creative cities to bifurcate between urban "creative" elites and a large supporting army of low-wage service industry workers (Peck, 2005; Scott, 2008; Storper & Scott, 2009).

Locating the University in Creative Clusters

While there is an extensive literature on the relationship between cities and globalization, and the role of universities in knowledge-based and creative economies, there has been surprisingly little work undertaken on the relationship between cities and universities. The result is that "the role of cities in the globalizing environment are studied independently of the institutional place of their universities" (Perry & Wiewel, 2008, p. 4). We find no consistent theme over time or across nations around the question of where universities should be located, and while almost all major cities have universities, there are many major universities located outside of major cities. Indeed, there is an influential tradition of the modern research university that emphasized campuses being located away from major population and industry centers in order to promote independent scholarship, whose most visible manifestation can be found with many of the major U.S. public universities established in the late nineteenth century. Indeed, the word *campus*, first used to describe the grounds of Princeton University in New Jersey, refers to a field, and has implied a space located outside of the city grounds (Haila, 2008, p. 31).

The literature on universities and clusters has been overwhelmingly focused on the high-technology sectors, with the relationship between the Massachusetts Institute of Technology (MIT) and the Route 128 high-tech cluster outside of Boston, and the relationship between Stanford University and Silicon Valley south of San Francisco, providing key case studies. Both cases have proved difficult to replicate in other contexts, despite various attempts worldwide to do so. In the case of MIT, it is important to understand not only its relationship to proximate ICT companies, but also to nearby Harvard University, and a range of prestigious universities also located around Boston (Northeastern University, Boston University, Brandeis, Tufts University, Brown University, and University of Massachusetts), which make the Boston area one of the most research-intensive regions in the world. Hulsink et al. (2007) find that the ICT companies around Route 128 have not been particularly strong on knowledge sharing, and that much of the research intensity of the region derives from relations among the universities and colleges themselves, rather than knowledge transfer with industry. The Stanford University/Silicon Valley link is a more successful example of knowledge transfer through clustering of a research university with knowledge-intensive industries, although Leslie and Kargon (1996) note that most attempts to replicate the "Silicon Valley model" have failed, and suggest that one distinctive feature of this model is that both the university and the start-up businesses emerged together. Interestingly, they observe that attempts to replicate the model in countries such as South Korea, with the Korean Advanced Institute of Science and Technology (KAIST), have been more successful than the various attempts to reproduce it in the United States.

Seeking to align universities more closely to industry and policy agendas is consistent with what Gibbons et al. (1994) referred to as "Mode 2" of knowledge production, differentiated from 'Mode 1," or the traditional university model, on the basis of the following criteria:

Although this debate has largely occurred in the science and technology areas, Ang (2004) has noted its relevance to the humanities in general, and cultural studies in particular, where "knowledge production has become much more widely distributed, taking place in many more types of social settings, and involving many different types of individuals and organizations . . . [and] to the extent that universities continue to provide quality graduates, they undermine their monopoly as knowledge producers" (Ang, 2004, p. 479). Hartley (2005) also identifies its particular significance in the context of the rise of creative industries, and for universities based in cities where "there is also a large number of people who are trend-conscious, early adopters, curious about the new, and relatively unencumbered by family commitments. . . . Universities are not just destinations, but hubs, and young people with time on their hands who are just hanging around are just as important to the creative sector as more traditional forms of investment" (Hartley, 2005, pp. 24–25).

	MODE 1	MODE 2
Conditions of knowledge production	Grounded within rules and practices of an academic discipline	Grounded in context of application and expectations of external clients
Conditions of knowledge valorization	Academic discipline as a "single collective stakeholder"	Multiple stakeholders, both within and outside the academy
Purpose of knowledge	Advancement of disciplinary knowledge	Solving of practical problems as they arise in social context
Mode of knowledge production	Individuals or discipline-based groups	Trans-disciplinary, project-based teams
Where knowledge is produced	Traditional sites: universities and research centers	Multiple sites: universitics, corporations, government agencies, "think tanks," activist organizations, consultants, and so on
Quality control mechanisms	Internal mechanisms (e.g., academic peer review)	Multiple criteria (contribution to economic productivity, social cohesion etc.)

Source: Derived from Gibbons et al., 1994.

In discussing the possible relationship of universities to creative clusters, we need to be aware of three endemic questions that arise with the clusters concept itself:

1. Is it primarily about mapping existing centers of cultural development and leadership, or about policy-driven strategies to create such sites? Storper and Scott (2009) observe a sleight of hand in existing "creative cities" literature, which downplays the role of macro-trends in the global creative economy in promoting certain sites as creative cities vis-à-vis the enabling role of supply-driven or "atmospheric" factors such as a thriving arts scene or a tolerant and

diverse culture. The result is that every city is presented as having the potential to become a creative city, even though in practice there are strong correlations between those cities that are leaders in global financial, service, and entertainment industries and those deemed to have a strong creative infrastructure.

2. Is the focus upon bottom-up, grassroots initiatives to cultivate the "soft infrastructure" of cultural development, or on the top-down initiatives of government authorities to bring together cultural and educational activities in designated cultural quarters? There is considerable evidence that the two can be in conflict, particularly insofar as creative cluster initiatives come to be more associated with urban "branding" and real estate development than with questions of cultural access and cultural diversity.

3. Are creative clusters seen primarily as sites of cultural production or cultural consumption? Pratt (2009) has observed that considerably more attention has been given to the latter than the former, and there tends to be an implicit assumption that consumption-led urban cultural regeneration will in itself provide the basis for attracting cultural producers and sustaining cultural infrastructures. There are also major issues that arise from the characteristically "hourglass" structure of the creative industries, with a small number of large employers and a very large number of individual providers and small-medium enterprises (SMEs), which means that employment structures in these industries can be highly volatile. (Cunningham, 2005)

When we bring universities into the mix, we need to note a further range of questions that arise:

- Does the university have a range of teaching activities and associated student recruitment strategies that link to the activities associated with a creative cluster?
- Does the university prioritize research that links in with the firms, industries, and activities associated with a creative cluster?
- Does the university see itself as having a role in developing the local cultural infrastructure and enabling its graduates to pursue careers linked to this creative cluster?
- Does the university see its graduates as being primarily employed in and around its local catchment area, or are they expected to move elsewhere upon graduation?

This set of questions indicates that the relationship of universities to creative clusters is likely to be very contingent. For those universities that have been located outside of major urban centers, there would not appear to be much point in seeking to re-badge local cultural activities as part of a cultural quarter or creative cluster in the hope that this activity would be part of redefining the local area as a creative city. Universities located in parts of cities that are hubs of cultural activity will need to make some strategic decisions. First, there are arguments against going down the path of being a more applied "Mode 2" university. Marginson (2006) has argued that globalization and the rise of global "league tables" for universities mean that those institutions aspiring to global research university status should not go down the path of applied, locality-based, and industry-focused research, as global research indicators remain largely driven by what can be termed "Mode 1" priorities.

Second, there is a great deal of fluidity within urban spaces for the emergence of creative clusters, and policy-makers and university administrators find it difficult to respond to them. If we take one of the better known recent collaborations between a university and a creative movement—the role played by Goldsmiths, University of London in the rise of the "Young British Artists" of the 1990s and 2000s (Damian Hirst, Tracey Emin, and so on)—it is not apparent that this collaboration played much of a role in the development of the South London area of which Goldsmiths is a part, as the dynamics of developing cultural sites in London were far more contingent in their nature (Pratt, 2009).

Finally, universities that see their future development as being linked to creative clusters will need to make serious commitments to the individuals and sectors involved. There is a need to think about curriculum, resourcing, student recruitment, research activities, cultural development, and community engagement, and graduate destinations as a package, and a will to make genuine changes to institutional practice as required. It will not simply be enough to point to evidence of co-location as proof of a cluster, since the clusters literature points to real and substantive differences between simple co-location of activities and the development of dynamic synergies.

References

Amin, A. (2003) The Economic Base of Contemporary Cities. In G. Bridge and S. Watson (eds.), *A Companion to the City* (pp. 115-129). Oxford: Blackwell.

Ang, I. (2004). Who needs cultural research? In P. Leistyna (Ed.), *Cultural studies: From theory to action* (pp. 478–483). Malden, MA: Blackwell.

Asheim, B. (2000). Industrial districts: The contributions of Marshall and beyond. In G. Clark, M. Feldman, & M. Gertler (Eds.), *The Oxford handbook of economic geography* (pp. 413–431). Oxford: Oxford University Press.

Barnes, T. (2003). Inventing Anglo-American economic geography, 1889–1960. In E. Sheppard & T. Barnes (Eds.), *A Companion to economic geography* (pp. 11–26). Oxford: Blackwell.

Bassett, K, Smith, I., Banks, M., & O'Connor, J. (2005). Urban dilemmas of competition and cohesion in cultural policy. In N. Buck, I. Gordon, A. Harding & I. Turok (eds.), *Changing cities: Rethinking urban competitiveness, cohesion and governance* (pp. 132–153). Basingstoke: Palgrave Macmillan.

Castells, M., & Hall, P. (1994). *Technopoles of the world.* London: Routledge.

Cooke, P. (2008). Culture, clusters, districts and quarters: Some reflections on the scale question. In P. Cooke &L. Lazzeretti (Eds.), *Creative cities, cultural clusters and local economic development* (pp. 25–47). Cheltenham: Edward Elgar.

Costa, P. (2008) Creativity, innovation and territorial agglomeration in cultural activities: The roots of the creative city. In P. Cooke and L. Lazzeretti (eds.), *Creative cities, cultural clusters and local economic development* (pp. 183–210). Cheltenham, UK: Edward Elgar.

Cunningham, S. (2005). Creative enterprises. In J. Hartley (Ed.), *Creative industries* (pp. 282–298). Malden, MA: Blackwell.

Currid, E. (2007). *The Warhol economy: How fashion, art and music drive New York City.* Princeton, NJ: Princeton University Press.

Florida, R. (2002). *The rise of the creative class.* New York: Basic Books

Florida, R. (2008). *Who's your city?.* New York: Basic Books.

Gibbons. M., Limoges, C., Nowotny, H., Schwartzman, S., Scott, P., & Trow, M. (1994). *The new production of knowledge: The dynamics of science and research in contemporary societies.* London: Sage.

Haila, A. (2008). The university of Helsinki as developer. In D. Perry & W. Wiewel (Eds.), *Global universities and urban development* (pp. 27–39). Armonk, NY: M. E. Sharpe.

Hall, P. (1998) *Cities in civilization: Culture, innovation, and urban order.* London: Weidenfeld & Nicolson.

Hartley, J. (2005). Creative industries. In J. Hartley (Ed.), *Creative industries* (pp. 1–43). Malden, MA: Blackwell.

Hulsink, W., Manuel, D., & Bouwman, H. (2007). Clustering in ICT: From Route 128 to Silicon Valley, from DEC to Google, from hardware to content. Erasmus Research Institute of Management Research Paper ERS-2007–064-ORG.URL: http://repub.eur.nl/publications/index/147990084/. Accessed 10 August, 2009.

Kenney, M., & Von Burg, U. (2000). Institutions and economies: Creating Silicon Valley. In M. Kenney (Ed.), *Understanding Silicon Valley: The anatomy of an entrepreneurial region* (pp. 218–240). Stanford, CA: Stanford University Press.

Landry, C. (2000). *The creative city: A toolkit for urban innovators.* London: Earthscan.

Landry, C. (2005). London as a creative city. In J. Hartley (Ed.), *Creative industries* (pp. 233–243). Malden, MA: Blackwell.

Leslie, S., & Kargon, R. (1996) Selling Silicon Valley: Frederick Terman's model for regional advantage. *Business History Review, 70*(4), 435–472.

Lorenzen, M., & Frederiksen, L. (2008). Why do cultural industries cluster? Localization, urbanization, products and projects. In P. Cooke & L. Lazzeretti (Eds.), *Creative cities, cultural clusters and local economic development* (pp. 155–179). Cheltenham: Edward Elgar.

Marginson, S. (2006). Dynamics of national and global competition in higher education, *Higher Education 52*(1), 1–39.

Marshall. A. (1990 [first published 1890]). *Principles of economics* (8th ed.). London: Macmillan.

Martin, R., & Sunley, P. (2003). Deconstructing clusters: Chaotic concept or policy panacea? *Journal of Economic Geography, 3*(1), 3–35.

Massey, D. (1984). *Spatial divisions of labour: Social relations and the geography of production.* London: Macmillan.

Mommaas, H. (2004). Creative clusters and the post-industrial city: Towards the remapping of urban cultural policy. *Urban Studies, 41*(3), 507–532.

Peck, J. (2005). Struggling with the creative class. *International Journal of Urban and Regional Research, 29*(4), 740–770.

Perry, D., & Wiewel, W. (2008). The university, the city, and land: Context and introduction. In D. Perry & W. Wiewel (Eds.), *Global universities and urban development* (pp. 3–24). Armonk, NY: M. E. Sharpe.

Piore, M., & Sabel, C. (1984). *The second industrial divide.* New York: Basic Books.

Porter, M. (1990). *The competitive advantage of nations.* New York: Free Press.

Pratt, A. C. (2009). Urban regeneration: From the arts "feel good" factor to the cultural economy—A case study of Hoxton, London. *Urban Studies, 46*(5), 1041–1061.

Scott, A. J. (2000). *The cultural economy of cities.* Thousand Oaks: Sage.

Scott, A. J, J. Agnew, E. J. Soja & M. Storper (2001). Global city-regions. In A. J. Scott (ed.), *Global city-regions: Trends, theory, policy* (pp. 11–30). Oxford: Oxford University Press.

Scott, A. J. (2008). Cultural economy: Retrospect and prospect. In H. Anheier & Y. R. Isar (Eds.), *The cultural economy* (pp. 307–323). Los Angeles: Sage.

Stevenson, D. (2004). "Civic gold" rush: Cultural planning and the politics of the "third Way." *International Journal of Cultural Policy, 10*(1), 119–131.

Storper, M., & Scott, A. J. (2009). Rethinking human capital, creativity and urban growth. *Journal of Economic Geography, 9*(1), 147–167.

Taylor, P. (1996). Embedded statism and the social sciences: Opening up to new spaces. *Environment and Planning A, 28*(1), 1917–1928.

United Nations Commission for Trade, Aid and Development. (2008). *Creative economy.* Geneva: UNCTAD.

Venturelli, S. (2005). Culture and the creative economy in the information age. In J. Hartley (Ed.), *Creative industries* (pp. 391–398). Malden, MA: Blackwell.

Warsh, D. (2006). *Knowledge and the wealth of nations: A story of economic discovery.* New York: W. W. Norton & Co.

5

Education for the Creative Economy: Innovation, Transdisciplinarity and Networks

Greg Hearn and Ruth Bridgstock

The question posed in this chapter is: To what extent does current education theory and practice prepare graduates for the creative economy? We first define what we mean by the term creative economy, explain why we think it is a significant point of focus, derive its key features, describe the human capital requirements of these features, and then discuss whether current education theory and practice are producing these human capital requirements.

The term *creative economy* can be critiqued as a shibboleth, but as a high level metaphor, it nevertheless has value in directing us away from certain sorts of economic activity and toward other kinds. Much economic activity is in no way creative. If I have a monopoly on some valued resource, I do not need to be creative. Other forms of economic activity are intensely creative. If I have no valued resources, I must create something that is valued. At its simplest and yet most profound, the idea of a creative economy suggests a capacity to compete based on engaging in a gainful activity that is different from everyone else's, rather than pursuing the same endeavor more competitively than everyone else. The ability to differentiate on novelty is key to the concept of creative economy and key to our analysis of education for this economy.

Therefore, we follow Potts and Cunningham (2008, p. 18) and Potts, Cunningham, Hartley, and Ormerod (2008) in their discussion of the economic significance of the creative industries and see the creative economy not as a sector but as a set of economic processes that act on the economy as a whole to invigorate innovation-based growth. We see the creative economy as suffused with all industry rather than as a sector in its own right. These economic processes are essentially concerned with the production of new ideas that ultimately become new products, service, industry sectors, or, in some cases, process or product innovations in older sectors. Therefore, our starting point is that modern economies depend on innovation, and we see the core of innovation as new knowledge of some kind. We commence with some observations about innovation.

Innovation

Many economies have a base in primary industries such as resources and agriculture. These industries are continuing to grow in terms of revenues, in some cases rapidly. However, the developed economies overall grow faster than either of these specific sectors because whole new categories of economic activity are constantly being invented. Economist Brian Arthur suggests that "the underlying mechanisms that determine economic behaviour have shifted from ones of diminishing to ones of increasing returns" (Arthur, 1996, p. 100). Investments in primary resources run down over time as the resource is exhausted, whereas investments in new knowledge (e.g., Google) ramp up as the new knowledge is monetised and consumers are captured. Increasing returns eventuate because the cost of product development is up-front (the overall unit cost of a product falls as sales increase), due to network effects, the likelihood of a product emerging as standard increases with greater use. Furthermore, because expansion into future markets becomes easier as more market is captured ("customer groove-in") (Arthur, 1996). The new high tech industries—computers, aircraft, and telecommunications, for example—clearly illustrate this dynamic. Service industries, Arthur suggests, are characterized by a hybrid of both the old and the new. Although demand for services is limited geographically and met by generally low-tech processing models, increasing returns can accrue to give market leaders an advantage, for example, via brand loyalty. In this case, it is the intangible resource of brand equity that accrues value.

Increasingly, economic growth can be argued to depend on continuous waves of innovation and entrepreneurship. It is the ability to generate wholly new products and services rather than deriving greater efficiency and economies of scale from existing production processes that has been a defining factor in the transition from an industrial to a creative economy (Flew, 2008). Much of the creative economy is built around knowledge and products that, in some cases, did not exist a few decades ago. For example, 75% of Siemens AG revenue is based on products that were invented in the last five years. However, the difference between new products and new efficiencies is not necessarily as pronounced as it might first appear. Many increases in efficiency are actually developed by new companies supplying service innovation to an industry's supply chain. Innovation may, therefore, also involve generating new efficiencies in the basic modes of production or grafting new features onto existing products or services. For example, the motorcar sees innovation necessarily trickle down to entry level models (twin airbags, electric windows, power steering, CD and MP3 for $14,990 drive-away). Innovation can, in fact, occur right across the supply chain from production to consumption. In each case, the heart of innovation is new knowledge. How then is this new knowledge created?

Why Is Creative/Cultural New Knowledge Important to Innovation?

There are two broad fields of human endeavor concerned with the production of new knowledge, namely the sciences and the arts.[1] However, far more attention has been paid to science as a source of innovation than to arts, and so we intend to spend some time teasing out what the specific contribution to the creative economy might be from the arts side. Part of this process requires us to think through how scientific and artistic creativity coalesce and differ.

Potts and Cunningham (2008, p. 239) consider this question explicitly. One possibility they consider is that the creative industries may be a driver of innovation by introducing novel ideas that then percolate through to other sectors—design-led innovation is a case in point. They also suggest that creative industries may facilitate the adoption and retention of new ideas and technologies—the creative digital sector is an example of this. Furthermore, they propose that the creative industries may be better thought of as part of the innovation system of the economy as a whole. In particular, the creative industries originate and coordinate change in the knowledge base of the economy and can be understood "as a kind of industrial entrepreneurship operating on the consumer side of the economy."

In fact, most economic activity is driven by consumption (60% to 70%), and increasingly this consumption is directed toward the pursuit of cultural goods or goods with cultural components—the so-called culturization of the economy (Lash & Urry, 1994). Fuelled by their ability to modify and process the building blocks of identity (images, visual codes, phrases, and ideas), our current mass media, via identity construction, have expanded consumption in advanced industrial societies. This style of consumption has increased with the emergence of digital media. The ability to provide individualized identity construction commodities has made this self-construction process even more compelling. It can be expected, therefore, under this trajectory that media that allow for individualization will spawn increased commodification, although with less reliance on mass images. The ability of a switched broadband network to support delivery and billing for diverse, informationally or culturally significant commodities further increases the momentum toward increased cultural commodification.

In a sense then, a techno/cultural-economic paradigm has replaced the techno-economic paradigm. It is defined by an amalgam of technology and culture that creates new market spaces. Sectors that derive in large part from the applied social and creative disciplines (e.g., business, media, entertainment, and education) represent 25% of exemplary economies while the new science sector (e.g., agricultural biotech, fibre, construction materials, energy, and pharmaceuticals) accounts for only about 15% of these economies (Rifkin, 2000, p. 52). That is, science, technology, and engineering are essential for economic growth; they are no longer a sufficient condition for future economic success. Technology + culture is the formula for twenty-first century problem solving, and, hence, for growing the creative economy.

Put another way, all scientific innovations must eventually feed into markets, and the disciplines that govern speed and access to, and exploitation of, markets all derive from the applied social and creative disciplines. Then, as affluence increases, functionality and price cease to be sufficient for market dominance. Consumers are increasingly influenced by the aesthetic and experiential components of products. Current forms of innovation are therefore based on intimate knowledge of, and facility in, creating consumer culture. Lucrative "blue ocean" markets—where there are no initial competitors—are only created by radical innovation in consumer spaces, not by technology innovation alone. This kind of innovation requires technology, plus design, plus cultural innovation.

We suggest then, that innovation occurs primarily at the intersection of three knowledge regimes: Scientific/Technical, Creative/Cultural, and Business. The first two produce new knowledge and the third translates it into valued and therefore consumed commodities.[2] New knowledge in the form of technical innovation might manifest in the product itself, in its manufacture, or distribution. New knowledge in the form of artistic innovation might again constitute the new product, its design, or marketing. It is the interplay of all these three knowledge regimes that is the second defining feature of the creative economy. This is what we term the transdisciplinary imperative for the creative economy.

Transdisciplinarity

Innovations that reach the market are rarely the products of single disciplines, but rather they involve compound multidisciplinary knowledge regimes. Indeed, modern corporations, for example, may be most distinguished by their ability to bring together composite knowledge (e.g., technical, marketing, and legal knowledge). Commercialisation depends on "whole product value propositions" not just basic research in one or two disciplines. Creativity is found across the scientific, technological, economic, and cultural domains, in diverse forms such as patents and designs, entrepreneurship, and artistic product: "no intellectual domain or economic sector has a monopoly on creativity" (Mitchell, Inouye, & Blumenthal, 2003, p. 18).

The rapidly growing knowledge intensive business services (KIBS) sector illustrates this principle well. It combines generic knowledge from a broad range of domains with information from clients to diagnose problems, provide advice, and prescribe or implement solutions (Miles, 2008). The domains of knowledge on which KIBS may draw include those associated with social systems and institutions, especially administrative rules and regulations; supply chain management; educational and clinical psychology and psychiatry; engineering; and IT services. As problem-solvers, KIBS are involved in generating new solutions and new knowledge, and their client can be understood as co-producer of this innovation. Technology-oriented KIBS assist in diffusing new techniques and systems throughout the economy, and R&D services are, of course, intimately involved with innovation as they undertake knowledge-creation for their clients. Many KIBS are hybrid technology-oriented and professional services. For example, lawyers specialize in IT or patent law, and financial advisors and market analysts provide expertise in high-tech or consumer innovation fields. The KIBS sector is among the fastest growing service sectors (Miles, 2008).

Exactly how do these knowledge regimes interplay? They interplay via agents capable of learning and communicating new knowledge. These agents are organized via complex social networks. Why networks rather than other forms of social organization? We point here to the requirement to generate new knowledge rather than replicate knowledge. Networks are flexible multi-path structures that accommodate rather than eliminate diversity. A process of economic evolution is generated in network economies by the development of new connections between tasks, technologies, firms, industries, and markets (Potts, 2000). As more of the economy becomes connected to previously unconnected parts, the scope and depth of innovation processes increase significantly (Morrison & Potts, 2007). Moreover, far from being a local process, new connections and networks form between regions, nations, and entire industries. Thus, the creative economy exhibits the prevalence of networks, and networks are the third defining feature. Indeed, Potts et al. define creative industries as "the set of agents and agencies in a market characterized by the adoption of novel ideas within social networks for production and consumption" (2008, p. 172).

Networks

From an information science perspective, networks are ideal mechanisms of information resource allocation and flow; these properties may be part of their growing importance. Structurally, they put people in direct contact via the provision of horizontal links across institutional boundaries, thus facilitating rapid information transfer. In addition to transmitting information, networks also help create it. New ideas may develop as each person in the network receives and synthesizes information; information easily builds on information. Thus, new ideas are both shared and created via networks. For example, Ahuja (2000) indicates that strong ties facilitate resource sharing and knowledge spill-over benefits. He also has pointed out that the network benefits of strong and weak ties are dependent on a number of other features and are limited to specific contexts.

The structural dynamics of networks are different from some of the other patterning mechanisms that educators take for granted: for example, hierarchies and grids. Barabási (2002) and Watts (2003) show that the basic structure of what they term scale free networks applies to many phenomena, ranging from cellular metabolism to the physical structure of the Internet and from protein regulatory networks to social relationships, as manifested by research collaborations, actors' appearances in different movies, or sexual relationship networks. Scale free networks are composed of connected nodes. Most of these nodes are connected by a small number of links, whereas some—called hubs—may have hundreds, thousands, or even millions of links while retaining the basic distributive characteristics of the network, hence the term "scale free." The distribution of connections between nodes is not even or random but obeys a power curve. This property makes scale free networks very robust against failure; only coordinated attacks against a number of hubs will break down such a network. The consistent features of scale free networks are evidence of the self-organizing processes at work; they work via an internal "logic" that requires no external guidance. Modern economies are characterized by the proliferation of these scale-free networks.

One of the defining features of many "new" products and services is that they exhibit network effects; their functionality depends not just on the functionality of one material artefact but on a total network of functional connections. This assessment is true in a technical sense (e.g., mobile phone networks), a service sense (e.g., credit cards), a software sense (e.g., operating systems), and a cultural sense (e.g., English language MBAs). A key component of the value of new products is the ability to connect, no matter whether the issue is software, communications, or more generally, participation in a taste culture via enjoyment of a movie or the purchase of a lifestyle. This approach is in contrast to the economic transaction of material goods, which are mostly valuable regardless of their precise relation to others. The term externality has been used by economists to describe situations where the value (or cost) of a product derives from anything outside the product itself. The best example of network externalities is the telephone—its value increases with the number of connections it allows.

The growth in credit card issuing and associated branding is a good example of externalities. Visa Card and American Express are vast international companies whose value is not in the physical plastic, but the network they open up and how this is communicated to prospective and current cardholders (rewards, clubs, affiliated company offers, the Amex and Diners brand). Both companies, while examples of expansive networks, are nevertheless closed and offer no interoperability, making entry "a crucial issue" (Economides, 1995) and coercion a valuable tool in signing up new members (Low interest balance transfers, anyone?). As with the telephone, the more persons join the network, the more the value of the network increases for members ("American Express Cards Accepted Here").

To reiterate, we commenced our discussion of the creative economy by arguing that at its core it is about innovation. Innovation requires new knowledge combined with the capacity to turn this new knowledge into valued commodities. The combination of knowledge regimes, or transdisciplinarity, is a second defining feature of the creative economy. Finally, the generation and combination of knowledge occur in networks, and hence the prevalence of networks becomes the final feature of the creative economy. So having defined what we mean by the creative economy and described what we see as its defining imperatives, we now turn to the second part of our question and ask: why is education important to the creative economy?

The simple answer is that knowledge—both new and combined—is carried by people. More specifically, human capital is a major consideration as is, therefore, how human capital is attained. Florida, amongst many, argues that human capital is central to success in the creative economy; "studies of national growth find a clear connection between the economic success of nations and their human capital, as measured by the level of education" (Florida, 2003, p. 222). He suggests the same principle is true for regions and cities. The centrality of human capital to the creative economy provides a point of departure in thinking about the design of education systems for the creative economy.

However, a point of clarification is required first. Human capital cannot be understood simply as individuals with skills and competencies. It is also concerned with the embedding of human resources in social and cultural capital networks. It is this embeddedness that is the defining characteristic of the creative economy.

Granovetter (1985; 2005, p. 36), for example, provides a useful model that explains why a socialized conception of human capital is needed. This model explains the importance of social networks underlying economic activity. Economic activity is embedded in the social sphere. That is, economic transactions depend on actions or institutions that are non-economic in content goals or processes (Granovetter, 2005, p. 36). For example, a culture of corruption may impose high economic costs that dampen efficiencies. Alternatively, in terms of cost savings, recruitment of personnel through social networks can reduce costs, and these networks contain social resources of trust and obligation that are built up from the social sphere. Social networks are important in reinforcing norms that govern the proper way to behave in economic activities and also play a key role in labor markets in terms of transmitting information about employers, employees, and job flows.[3] In Granovetter's view, social structure also affects productivity. Many tasks in any value chain require cooperation from others. Tacit knowledge is also a requirement, and tacit knowledge can only be transmitted via interaction with knowledgeable others. Group norms also shape skills and, hence, productivity. Finally, networks affect innovation capacity. Networks assist in transmitting unique and non-redundant (potentially innovative) information; they determine how different knowledge regimes may be combined in new products and services. "Innovation means breaking away from established routines. Development of resources outside their usual spheres may often be a source of profit and new institutional forms can facilitate such deployment" (Granovetter, 2005, p. 46). The operation of venture capital is a case in point.

There is strong economic evidence then that human capital is central to success in modern economies where knowledge, creativity, and innovation are particularly valued (e.g., De la Fuente & Ciccone, 2002; Florida, 2003; Machin & Vignoles, 2005). These sentiments have translated to "more education for all" at the level of education policy, providing part of the impetus for the widening participation and massification agendas in higher education in many parts of the world.

There is widespread agreement that more education makes for more innovation and, therefore, more economic growth. However, in light of the distinctive characteristics of creative economies as just outlined, we argue that more education is not a sufficient answer to the question of how best to leverage human capital for development in this economic context. Rather, we ask: what are the kinds of capabilities required to drive the four key features of the creative economy: first, domain specific creativity (a concept analogous to Florida's (2003) "creative class," i.e., professional domain-specific creativity of scientific/technical and creative/cultural kinds in fields like science, performing arts, architecture, engineering, and film making); second, innovation; third, transdisciplinarity; and fourth, networks?

The remainder of this chapter is devoted to a broad discussion regarding the extent to which, and also the ways in which, our education systems meet these specific human capital challenges of the twenty-first century, and how they might be better positioned to do so in the future. We focus our discussion on higher education and other professional-level education and training provision, as purveyors of the specialist and advanced skills likely to be vital to innovation processes.

Domain-Specific Creative Professions

Of the four defining characteristics of the creative economy, post-compulsory education can be argued to address domain-specific skills and knowledge the most adequately. Domain-specific skills and knowledge are acknowledged to be a central function of higher and professional education, with a significant and increasing proportion of courses devoted to the development of occupational, or at least sector-specific, competence. The dominant view at present is that the primary role of post-compulsory education is to provide training for employment and employability (Masjuan & Troiano, 2003; Prokou et al., 2008). While non-occupationally specific courses in the humanities, sciences, and the social sciences are the norm in undergraduate education in the United States and elsewhere, the "market driven" approach means that increasingly often the terminal program in an individual's initial sequence of education or training is designed to prepare him or her for a particular profession or occupation (although these courses also contain generic skill content designed to enhance employability more generally). Courses that deal with domain-specific knowledge and skills devoted to scientific, technological, and cultural/creative occupations are ubiquitous in post-compulsory education.

However, much has also been written about perceived shortcomings in the domain-specific content of such courses. These criticisms relate primarily to educational responsiveness to labor market requirements, in terms of skills mismatches—the oversupply or undersupply of graduate workers in certain fields and sectors, and the oversupply or undersupply of specific types of skills and knowledge that graduates possess. For instance, literature documents a tendency toward oversupply of graduate level creative workers in arts fields in a number of countries (Blackwell & Harvey, 1999; Bridgstock, 2005), and undersupply of science, technology, and engineering graduates in certain specialties (Teitelbaum, 2004; Wilson, 2009). In respect of the problem of graduate undersupply, governments can do much to attract students to desired fields. Augmentation of educational funding tied specifically to science, engineering, technology, and digital creative courses, with specific strategies such as scholarships, recruitment drives, enrolment targets, and expansion of teaching facilities, are well established approaches. However, care should be taken in implementing recruitment strategies so that relaxation of course entry requirements does not occur.

In respect of the problem of undersupply of skill sets demanded by the labor market, it is very difficult to plan educational provision to match current or projected skill needs. Historically, attempts to do this have been fairly unsuccessful (Boswell, Stiller, & Straubhaar, 2004; Gordon, 1986). Rapid change in workforce requirements, particularly in technology-based skills, results in forecast inaccuracy. In addition, the process of renovating course content can be unwieldy, time-consuming, and expensive, and by the time the process is completed, courses are often once again out of date. Although periodic course updating and creation of new courses as subdisciplines emerge are both desirable, pursuing a perfect skills match is probably an erroneous approach. We suggest that a far better strategy is to focus on stable, enduring domain-specific knowledge and skills. In addition, we believe that post-compulsory education has an important role in engendering flexibility and ongoing educational self-management in students so that they can propel their own skill development in response to continually changing workforce demands. The need for labor market-responsive skilling of this type may be best addressed by fast-turnaround, targeted workplace-based professional and vocational training programs.

Innovation

Earlier in this chapter, we argued that the core of the creative economy is innovation, which is defined as: first, the formation of new knowledge, which is then, second, transmuted into valued products, services, or processes. Thus, the skills involved in innovation can be divided into two main subgroups: those creative skills concerned with formation of new knowledge more generally, and business/enterprise skills concerned with turning new knowledge into commodities. We discuss each category of education in turn.

Creativity Education

The development of creativity has not traditionally been a priority for higher and professional education, except in the creative arts (Fleming, 2008; Jackson, 2003). In fact, it has often been argued that formal education has a tendency to militate against such creative qualities as curiosity, imaginativeness, and intuition (Robinson, 2007). Until relatively recently, many educators in science, technology, and engineering-related fields viewed creativity not just as superfluous, but as undesirable and in opposition to the rigorous scientific thinking required by professionals in these fields (DeHaan, 2005; McWilliam, Poronnik, & Taylor, 2008).

However, the last decade has seen increasing policy acknowledgment of the importance of creativity education to the economy and society. *All Our Futures: Creativity, Culture and Education* (Robinson, 1999, 5), published by the UK National Advisory Committee on Creative and Cultural Education, was influential in promoting universal creative and cultural education for the "development of capacities for original ideas and action" as essential precursors to national economic development. Its publication stimulated widespread discussion internationally about possible types of creativity that might be more or less useful and whether creativity can be "taught" at all.

A significant movement in creativity education has subsequently shifted away from what McWilliam (2008) terms "first generation," also known as "Big C" (Craft, 2001) thinking, about the nature of creativity as being extraordinary, paradigm-shifting, and associated with giftedness. This type of creativity is unlikely to be able to be developed through education. Rather, these thinkers argue for a "second generation" of thinking about creativity as being "little c" capabilities involved in problem-identification and solving, such as the adaptation and synthesis of existing knowledge (McWilliam, 2008; Robinson, 2007). There is some evidence that this second type of creativity is likely to have more impact on innovation (Weisberg, 2006), and it can be cultivated in most people through educational experiences (Runco, 2007).

Two recent studies conducted in Australia and the UK (McWilliam & Dawson, 2008b; The Creativity Centre, 2006) suggest that many university teachers now aim to incorporate creative capacity building into their teaching. McWilliam and Dawson's (2008a) analysis of higher education policy documents indicated that 75% of Australian universities at that time were committed to "creative" learning outcomes of some sort. However, significant conjecture remains about what creative curricula should look like, and assessment of creative capacity development remains a thorny issue. Ongoing attempts to render creativity observable, developable, and evaluable in higher and professional education are very important to the growth of innovation capacity through post-compulsory education, and require continued policy and institutional support.

Before moving on to the second aspect of innovation capacity, enterprise education, we make two observations regarding creative learning. First, when undertaking creative capacity building with a view to enhancing innovation in work contexts later on, it is preferable to maximize the domain-specificity and authenticity in learning and assessment tasks. Creative tasks that as far as possible emulate "real world" scenarios and are perceived by the learner to be personally relevant will stimulate interest and motivation (Herrington & Herrington, 2005) and will encourage transferability of creative thinking and behaviors to workplace situations. Second, creative capacity building should also recognize and make use of the social nature of creativity: the idea that co-creative capacity can be far greater than individual creativity if conditions are conducive to it (Csikszentmihalyi, 1997). McWilliam and Dawson (2008a, 2008b) describe several guiding principles for optimizing co-creativity, including fostering connectedness while maintaining group diversity and a degree of individual autonomy, collective responsibility for leadership, and providing appropriate scaffolding to allow individuals to act in ways that optimize group performance.

Entrepreneurship and Business Education

In this section, we discuss education for the second component of innovation: business and entrepreneurship skills concerned with turning new knowledge into commodities. Entrepreneurship is defined by Curran and Stanworth as the process of creating "a new economic entity centred on a novel product or service or, at the very least, one which differs significantly from products or services offered elsewhere in the market" (1989, p.12). Entrepreneurship programs tend to focus on new business creation, which we suggest is a key mechanism by which new knowledge can be commodified in the creative economy. However, we recognize that skills and knowledge associated with product manufacturing and commercialization more broadly are also likely to be important to innovation.

The last decade has seen a rapid acceleration of support for education in business and entrepreneurship (Matlay & Carey, 2007), both in terms of the volume of programs offered and the range of content within those programs. This rapid acceleration has emerged from widespread acceptance of the idea that entrepreneurship and business education play an integral role in economic development across all sectors, and that these types of education will produce an increase in the number and quality of emerging entrepreneurs and business people throughout the economy (Brockhaus, Hills, Klandt, & Welch, 2001). However, significant definitional, conceptual, and pedagogical issues remain, with no consensus about outcomes or teaching methods, or much consistency in course provision (Fiet, 2000; Henry, Hill, & Leitch, 2004; Matlay, 2008). Commentators have also reported pervasive difficulties in evaluating the "real world" impact of business and entrepreneurship programs (Matlay, 2008).

In fact, debate continues about the extent to which entrepreneurship can be taught, and how much is dispositional (Fiet, 2000; Henry, Hill, & Leitch, 2005). Jamieson (1984) considered the essential components of entrepreneurship training to be: education *about* enterprise, comprising awareness and theoretical knowledge to do with setting up and running a business; education *for* enterprise, comprising the practical skills for business set-up and management; and education *in* enterprise, relating to business growth through product development, marketing, and management. It seems likely that these categories of entrepreneurial knowledge and skills can indeed be taught.

However, qualities such as entrepreneurial drive, competitiveness, persistent optimism, propensity for risk-taking, comfort with change, and visionary business leadership have also been observed repeatedly to be important to entrepreneurial behaviour and enterprise success (e.g., Jacobowitz & Vidler, 1982). It is the importance of these more trait-like characteristics that has led theorists to argue that entrepreneurs are born, not made.

While being entrepreneurial probably comes more easily to some, entrepreneurship education will be useful even to those in scientific/technical and creative fields who have little natural propensity for, or inclination toward, entrepreneurialism. These students may never need to start a business or commodify products themselves, but it is likely that they will need to be conscious of the commercial sphere and have an appreciation of core business concepts so they can deal effectively with the business specialists who will be performing these tasks. Thus, rather than attempting to develop in students knowledge and skills that they are unlikely to use, may have minimal interest in and motivation for (Richards, 2005), the focus of entrepreneurship education becomes growth of commercial awareness, and the provision of underpinning knowledge for productive partnerships between domain specialists and business specialists. As such, it would also be advantageous for programs aimed at training business and entrepreneurship specialists to include elements of the scientific/technical and creative/cultural.

In terms of entrepreneurship course delivery, Matlay (2008) has noted that very often programs are developed and delivered by business schools rather than the home faculty. This approach can be problematic in that there is a tendency for such courses to be tacked on to the core domain program, with insufficient regard for, or integration with, discipline or industry-based specificities. A second issue with business and entrepreneurship courses is that although a wide range of classroom-based teaching approaches have been adopted (e.g., case studies, theoretical didactic approaches, role plays and problem solving), they may not emulate the demands of "business in the real world" sufficiently, including coping with challenges like limited or incomplete information, multiple stakeholders with differing agendas, judgement-based decision making with no right answer, and an element of risk (Gibb, 1987). To maximize the efficacy of entrepreneurship coursework, learning should, at least in part, comprise experiences that are similar to situations students are likely to experience. As with creativity education, an authentic learning approach seems appropriate to entrepreneurship education, wherein students with different domain-specific and business backgrounds engage in project work to turn new knowledge into new products, processes or services.

Transdisciplinarity

The third defining characteristic of the creative economy we suggested is transdisciplinarity: the interaction, translation, and synthesis of knowledge between and among scientific/technical, creative/cultural, and business/entrepreneurial disciplines, and also between different subdisciplines within each (e.g., between aerospace engineering and astrophysics in the design and construction of new types of radio telescope). Unlike the creative economy skills and knowledge types we have discussed so far, that is, domain-specific expertise and skills for innovation (comprised of creativity and business/entrepreneurship), transdisciplinarity has yet to find general recognition and acceptance in higher and professional education. Theorists have argued for transdisciplinarity in post-compulsory education since the early 1970s (Jantsch, 1972) on the basis that most professional scenarios involve players from different disciplines, and also that many problems require solutions that draw upon knowledge from multiple areas. However, significant deep-seated obstacles to do with institutional structures and differing epistemologies mean that the transdisciplinarity agenda remains underdeveloped. Kezar (2005) reported that half of attempted interdisciplinary teaching and learning collaborations fail.

Unfortunately, the very structures that permit universities to be such strong purveyors of discipline-specific knowledge and skills tend to make transdisciplinary offerings difficult to obtain institutional mandate for, let alone plan, deliver, and assess. The traditional discipline-based department/faculty structures of most institutions continue to be relatively rigid, stable, and cumbersome, with funding for teaching and learning initiatives similarly devolved on a departmental/disciplinary basis. This type of institutional structure does not lend itself easily to cross-disciplinary communication or collaboration ("The university of the future," 2007). Encouragingly, some institutions have started to overlay multidisciplinary "centres" or "institutes" on traditional faculties/departments in an endeavor to encourage dialogue between fields, often with a specific research focus. The success of these centres as crucibles for transdisciplinarity varies and depends on more than staff co-location or shared financial support.

The different perspectives maintained by different disciplines represent a basic challenge to transdisciplinary education. Not only does use of language and terminology within disciplines create specialized dialects, making communication potentially problematic (Hammer & Soderqvist, 2001), but there may also be enormous divergence between disciplines in terms of fundamental belief systems, worldviews, and approaches to professional problems. The more theoretical divergence between the disciplines involved, the more difficult it is for educators (who of course have strong disciplinary backgrounds) to find enough common ground conceptually to make transdisciplinary courses workable, except at a superficial level (e.g., bringing in occasional guest lecturers from other disciplines). Unfortunately, unidisciplinarity in educational institutions and professions can easily become a self-perpetuating cycle, with ever more entrenched theoretical positions and specialized dialects resulting in ever-greater divisions between disciplines.

Courses for transdisciplinarity also need to do more than supply information from different disciplines in parallel, as can often be the case with dual degree programs (Michael & Balraj, 2003). Multidisciplinary programs, which do not attempt the difficult task of disciplinary dialogue and rapprochement in addition to providing solid disciplinary grounding, seem to be less effective than integrated programs using strategies like mixing students from different programs for joint projects (Hammer & Soderqvist, 2001).

To engage with the transdisciplinarity imperative effectively, policy and institutional support to enhance creativity education needs to occur. Some basic institutional reorganization and restructuring may be necessary (McWilliam, Hearn, & Haseman, 2008; Srikanthan & Dalrymple, 2005), including the use of strategies such as joint staff appointments and joint funding arrangements. Staff and students must also be willing and able to enter into research and teaching/learning partnerships with others who do not have the same professional vocabularies, and in some cases, fundamental assumptions.

Social Networks and Embeddedness

Much has been written about the network mechanisms that are responsible for the generation of social capital and also the strength and groupings of connections in networks (e.g., Burt, 2000). However, much less is known about these connections at the level of the individual and small groups. Further research is required to better characterize the nature and development of individual relationships in the social network, particularly the skill and resource flows involved in innovation. Also, despite a plethora of popular literature in the area, we know a surprisingly modest amount about how people can go about creating and maintaining optimal social networks for innovation and career success (c.f., Seibert, Kraimer, & Liden, 2001). In light of this lack of knowledge, it is to be expected that relatively little has been documented about the skills and abilities needed to grow and manage social networks.

At the individual level, social networking capability involves the capacity to build and maintain personal and professional relationships with others for mutual benefit in work or career (Forret & Dougherty, 2001). It is through these relationships that information and resource transmission for generation and commodification of new knowledge occurs. Being "social network capable" also involves what McWilliam and Dawson (2008a) refer to as "network agility"—the ability to develop and navigate social networks in a strategic and enterprising manner.

Thus, social networking capability depends on an understanding of the nature and distinctive characteristics of social networks. It also includes the ability to use this understanding to position oneself optimally in networks; to develop and maintain weak and strong ties as appropriate with individuals who possess complementary skill sets, knowledge and resources, and to manage the exchange of these for mutual benefit. Social networking capability also relies on an awareness of the characteristics of the creative economy, and within this the "ingredients" for successful innovation: creative/cultural and scientific/technical domain-specific knowledge, transdisciplinarity, creativity, and entrepreneurship. While all of these ingredients are important to the ongoing generation, combination, and dissemination of knowledge that drives the creative economy, one person need not possess all of them individually. They can rely on their social networks to provide the remainder. However, it seems clear that all individuals will require some degree of social networking capability in order to make the most of the capabilities of others. We turn now to examine the extent to which post-compulsory education takes account of social network capability development. Skills such as "communication," "teamwork," and "interpersonal skills" are commonplace in generic skill lists, and are routinely included in higher and professional education programs (e.g., Bennett, Dunne, & Carré, 1999). These socially oriented generic skills are certainly essential to social network capability. However, they form a small subset of the skills required to form, develop, and maintain networks.

The emergence of Web 2.0 technologies and online social networks provides another source for social network capability development via post-compulsory education, particularly because of the related upsurge of interest in "online learning communities" (Barab & Duffy, 2000; Kaplan, 2002). In an online learning community, people use computer-mediated communication to work together to achieve shared learning objectives.

The online learning community movement is important to social network capability development in more ways than just helping people to learn how to work in groups using technology. It represents a widespread shift in educational perspective from an individualistic focus to one that recognizes the contribution of others to individual learning (Vygotsky, 1978). In short, for the last few years we have been undergoing a gradual shift from the "age of the Individual to the Era of Community" (Feldman, 2000, p. xiii) in education. We are now beginning to understand social networks offer far more than just interacting with friends online. Rather, social networks can be "object-oriented communities" of collective intelligence (Araya, 2009); networked people can form powerful mechanisms by which educational, social, and economic ends may be met. Alas, while recent theoretical and pedagogic advances in the ICT field have brought social networks into the educational arena, policy lags far behind on this issue. At present, universities are prepared to support pedagogic "experiments" such as collaborative learning tools embedded in online course management systems, yet in general maintain an unswerving emphasis on educating (and assessing) the individual to the exclusion of his/her network.

It should also be noted that social network capability is a larger concept than that implied by the current educational emphasis on ICTs, and online learning communities are not sufficient to engender social networking capability for the creative economy. Social networking involves face-to-face as well as online relationships. It involves developing and managing those relationships in considerate ways, a feature that is often not addressed by online learning communities where students working together already know one another, or at least are enrolled in the same course. Education for social network capability requires elements of self-directed relationship building, and as such lends itself to project-based work-integrated learning scenarios, in conjunction with theoretical learning about the nature and use of social networks.

Related to the issue of social network capability is that of embeddedness: how can graduates be embedded in the social and cultural capital networks for success after course completion? This question is concerned with how education systems can maximize their responsiveness to labor market demands (i.e., graduate employability) and represents another underexplored area within higher and professional education.

Some of the issues of graduate embeddedness can be addressed by the development of network capability, as discussed above. If graduates are able to develop effective relationships and interact socially, they should be well on the way to being able to embed themselves professionally. As discussed by Lin (2001) and Granovetter (Granovetter, 2005), valuable information about employers, employees, jobs, and industries be transmitted by effective social networks. People with well developed social capital can rely on informal networks (such as personal contacts) in addition to more formal networks (such as employment agencies) when looking for work. Indeed, informal networks may have a stronger positive influence on individual employability than formal networks (Fugate, Kinicki, & Ashforth, 2004).

Education systems have an excellent opportunity to take a more responsive role in assisting students with the formation of job related networks through work integrated learning programs and other opportunities for industry and employer contact. Further, it is highly advantageous for education programs to aim to enhance student awareness of the 'rules of the game' in the industry they have chosen to work in, the workings of the labour market (Bridgstock, 2009), and the broader characteristics of the creative economy.

Conclusion

Our aim in this chapter has been to draw attention to three defining features of the creative economy and to conduct a preliminary review how professional education is, in broad terms, addressing them. Some aspects of these features (for example, domain creativity and entrepreneurship) are understood and catered for while other aspects (for example, transdisciplinarity) remain underdeveloped in general, with real institutional barriers in place that are not easy to resolve. Moreover, an additional assessment that we might make is to ask to what extent education that addresses these three imperatives operates holistically. That is, education for innovation, transdisciplinarity, and networks should not be exclusive of one another. The features of the creative economy are inextricably linked—in the creative economy, innovation takes place under transdisciplinary, networked conditions—education interventions should emulate this dynamic and treat these three holistically. Our review did not find many examples of this interconnectedness, and future work might usefully look for examples of professional programs that do embody a holistic approach. Where education systems are inadequate, we suggest that the way forward for education systems is not, for example, simply to add creativity modules to the curriculum. Rather, education systems themselves need to embody the same dynamics in their own evolution. This may mean redesigning policy, digital infrastructure, resourcing models, and processes for the professional development of educators as much as curriculum.

References

Ahuja, G. (2000). Collaboration networks, structural holes and innovation: A longitudinal study. *Administrative Science Quarterly, 45*(3), 425–455.

Araya, D. (2009). personal communication.

Arthur, W. B. (1996). Increasing returns and the new world of business. *Harvard Business Review 74,* 100–109.

Barab, S. A., & Duffy, T. (2000). From practice fields to communities of practice. In D. J. S. Land (Ed.), *Theoretical Foundations of Learning Environments* (pp. 25-56). Mahwah, NJ: Erlbaum.

Barabási, A. L. (2002). *Linked: The new science of networks.* New York: Perseus.

Bennett, N., Dunne, E., & Carré, C. (1999). Patterns of core and generic skill provision in higher education. *Higher Education,* 37(1), 71–93.

Blackwell, A., & Harvey, L. (1999). *Destinations and Reflections: Careers of British art, craft and design graduates.* Birmingham: Centre for Research into Quality, University of Central England.

Boswell, C., Stiller, S., & Straubhaar, T. (2004). *Forecasting labour and skills shortages: How can projections better inform labour migration policies?* Hamburg: European Commission, DG Employment and Social Affairs.

Bridgstock, R. (2005). Australian artists, starving and well-nourished: What can we learn from the prototypical protean career? *Australian Journal of Career Development,* 14(3), 40–48.

Bridgstock, R. (2009). The graduate attributes we've overlooked: Enhancing graduate employability through career management skills. *Higher Education Research and Development,* 28(1), 27–39.

Brockhaus, R., Hills, G., Klandt, H., & Welch, H. (2001). *Entrepreneurship education: A global view.* Aldershot: Ashgate Publishing.

Burt, R. S. (2000). The network structure of social capital. *Research in Organizational Behavior,* 22, 345-423.

Craft, A. (2001). Little c creativity. In A. Craft, R. Jeffrey & M. Leibling (Eds.), *Creativity in Education* (pp. 45-61). London and New York: Continuum.

Csikszentmihalyi, M. (1997). *Creativity: Flow and the psychology of discovery and invention.* New York: Harper Collins.

Curran, J., & Stanworth, J. (1989). Education and training for enterprise: some problems of classification, evaluation, policy and research. *International Small Business Journal, 7*(2), 11-23.

De la Fuente, A., & Ciccone, A. (2002). *Human capital in a global and knowledge-based economy.* Brussels: European Commission Directorate-General for Employment and Social Affairs.

DeHaan, R. L. (2005). The impending revolution in undergraduate science education. *Journal of Science Education and Technology, 14*(2), 253–269.

Economides, N. (1995). 'Commentary—Credit Card Networks,' *Federal Reserve Bank of St. Louis Review,* Nov–Dec 1995. from http://www.findarticles.com/p/articles/mi_m3367/is_n6_v77/ai_18305126.from http://www.findarticles.com/p/articles/mi_m3367/is_n6_v77/ai_18305126.

Feldman, D. H. (2000). Foreword. In V. John-Steiner (Ed.), *Creative collaboration* (pp. ix–xiii). New York: Oxford University Press.

Fiet, J. (2000). The pedagogical side of entrepreneurship theory. *Journal of Business Venturing,* 16(2), 101–117.

Fleming, M. P. (2008). *Arts Education and Creativity.* London: Arts Council: Creative Partnerships.

Flew, T. (2008). Cultural and creative industries. In G. Hearn & D. Rooney (Eds.), *Knowledge Policy: Challenges for the 21st Century.* Cheltenham, UK & Northampton, MA, USA: Edward Elgar.

Florida, R. (2003). *The rise of the creative class: And how it's transforming work, leisure, community and everyday life.* North Melbourne: Pluto Press.

Forret, M. L., & Dougherty, T. W. (2001). Correlates of networking behavior for managerial and professional employees. *Group & Organization Management, 26*(3), 283.

Fugate, M., Kinicki, A. J., & Ashforth, B. E. (2004). Employability: A psycho-social construct, its dimensions, and applications. *Journal of Vocational Behavior, 65*(1), 14–38.

Gibb, A. (1987). Enterprise culture—its meaning and implications for education and training. *Journal of European Industrial Training, 18*(8), 3–12.

Gordon, A. (1986). Education and training for information technology. *Studies in Higher Education, 11*(2), 189–198.

Granovetter, M. (1985). Economic action and social structure: The problem of embeddedness *American Journal of Sociology, 91*(3), 481–510.

Granovetter, M. (2005). The impact of social structure on economic outcomes. *The Journal of Economic Perspectives, 19*(1), 33–50.

Hammer, M., & Soderqvist, T. (2001). Enhancing trandsciplinary dialogue in curricula development. *Ecological Economics, 38,* 1–5.

Henry, C., Hill, F., & Leitch, C. (2004). The effectiveness of training for new business creation. *International Small Business Journal, 22*(3), 249–269.

Henry, C., Hill, F., & Leitch, C. (2005). Entrepreneurship education and training: Can entrepreneurship be taught? *Education + Training, 47*(2), 98–111.

Herrington, J., & Herrington, T. (2005). *Authentic learning environments in higher education.* London: Information Science Publishers.

Jackson, N. (2003). *Creativity in higher education.* Heslington, York: Higher Education Academy.

Jacobowitz, A., & Vidler, D. C. (1982). Characteristics of entrepreneurs: Implications for vocational guidance. *Vocational Guidance Quarterly, 30*(3), 252–257.

Jamieson, I. (1984). Education for enterprise. In A. Watts & P. Moran (Eds.), *CRAC* (pp. 19–27). Cambridge: Ballinger.

Jantsch, E. (1972). Inter- and transdisciplinary university: A systems approach to education and innovation. *Higher Education, 1*(1), 7–37.

Kaplan, S. (2002). Building communities: Strategies for collaborative learning. *Learning Circuits* Retrieved May 1, 2009, from http://www.astd.org/LC/2002/0802_kaplan.htm

Kezar, A. (2005). Redesigning for collaboration within higher education institutions: An exploration of the developmental process. *Research in Higher Education, 46*(7), 831–860.

Lash, S., & Urry, J. (1994). *Economies of signs and space.* London: Sage Publications.

Lin, N. (2001). Building a network theory of social capital. In N. Lin, K. Cook & R. Burt (Eds.), *Social capital: Theory and research* (pp. 3–31). New York: Aldine de Gruyter.

Machin, S., & Vignoles, A. (2005). *What's the good of education? The economics of education in the UK.* Princeton: Princeton University Press.

Masjuan, J. M., & Troiano, H. (2003). University, employability and employment. In C. Pritchard & P. Trowler (Eds.), *Realizing qualitative research into higher education* (pp. 61–87). Aldershot: Ashgate.

Matlay, H. (2008). The impact of entrepreneurship education on entrepreneurial outcomes. *Journal of Small Business and Enterprise Development, 15*(2), 382–396.

Matlay, H., & Carey, C. (2007). Entrepreneurship education and the UK: A longitudinal perspective. *Journal of Small Business and Enterprise Development, 14*(2), 252–263.

McWilliam, E. (2008). *The creative workforce: How to launch young people into high-flying futures.* Sydney: UNSW Press.

McWilliam, E., & Dawson, S. (2008a). Teaching for creativity: Towards sustainable and replicable pedagogical practice. *Higher Education, 56*(6), 633–643.

McWilliam, E., & Dawson, S. (2008b). *Understanding creativity: A survey of 'creative' academic teachers.* Canberra: Carrick Institute of Teaching and Learning.

McWilliam, E., Hearn, G., & Haseman, B. (2008). Transdisciplinarity for creative futures: What barriers and opportunities? *Innovations in Education and Teaching International, 45*(3), 247–253.

McWilliam, E., Poronnik, P., & Taylor, P. G. (2008). Re-designing science pedagogy: Reversing the flight from science. *Journal of Science Education and Technology, 17*(3), 226–235.

Michael, S., & Balraj, L. (2003). Higher education institutional collaborations: An analysis of models of joint degree programs. *Journal of Higher Education Policy and Management, 25*(2), 131–145.

Miles, I. (2008). Knowledge services. In G. Hearn & D. Rooney (Eds.), *Knowledge Policy: Challenges for the 21st Century.* Cheltenham, UK & Northampton, MA, USA: Edward Elgar.

Mitchell, W., Inouye, A., & Blumenthal, M. (2003). *Beyond productivity: Information technology, innovation, and creativity.* Washington, DC: National Research Council of the National Academies, National Academies Press.

Morrison, K., & Potts, J. (2007). Industry policy as innovation policy. In G. Hearn & D. Rooney (Eds.), *Knowledge policy: Challenges for the 21st century.* Cheltenham, UK & Northampton, MA, USA: Edward Elgar.

Ninan, A. (2005). *What are the roles of networks and clusters in the operation of an industry? The case of Queensland music.* Unpublished PhD Thesis, Queensland University of Technology.

Potts, J. (2000). *The new evolutionary microeconomics: Complexity, competence, and adaptive behaviour.* Cheltenham, UK & Northampton, MA, USA: Edward Elgar.

Potts, J., & Cunningham, S. (2008). Four models of the creative industries. *International Journal of Cultural Policy, 14*(3), 233–247.

Potts, J., S. Cunningham, Hartley, J., & Ormerod, P. (2008). Social network markets: A new definition of the creative industries. *Journal of Cultural Economics, 32*(3), 167–185.

Prokou, E., Guth, J., Andrews, J., Higson, H., Omerzel, D. G., Sirca, N. T., et al. (2008). The emphasis on employability and the changing role of the university in Europe. *Higher Education in Europe, 33*(4), 387–394.

Richards, G. (2005). *Development for the creative industries: The role of higher and further education media and sport.* London: Skills and Entrepreneurship Task Group.

Rifkin, J. (2000). *The age of access: The new culture of hypercapitalism where all of life is a paid-for experience.* New York: J. P. Tarcher.

Robinson, K. (1999). *All our futures: Creativity, culture and education.* London: National Advisory Committee on Creative and Cultural Education.

Robinson, K. (2007). *Out of our minds: Learning to be creative.* New York: Wiley.

Runco, M. A. (2007). *Creativity: Theories and themes, research, development and practice.* Amsterdam: Elsevier Academic Press.

Seibert, S. E., Kraimer, M. L., & Liden, R. C. (2001). A social capital theory of career success. *Academy of Management Journal, 44*(2), 219–237.

Srikanthan, G., & Dalrymple, J. (2005). Implementation of a holistic model for quality in higher education. *Quality in Higher Education, 11*(1), 69–81.

Teitelbaum, M. S. (2004). Do we need more scientists? *The Public Interest, 153,* 40–53.

The Creativity Centre. (2006). *Facilitating creativity in higher education: The views of national teaching fellows.* London: Higher Education Academy.

The university of the future. (2007). *Nature, 446*(7139), 949.

Vygotsky, L. S. (1978). *Mind in society.* Cambridge, MA: Harvard University Press.

Watts, D. (2003). *Six degrees: The science of a connected age.* New York: W.W. Norton.

Weisberg, R. W. (2006). *Creativity: Understanding innovation in problem solving, science, invention, and the arts.* Hoboken, NJ: Wiley.

Wilson, R. (2009). *The demand for STEM graduates: Some benchmark projections.* London: Warwick Institute for Employment Research.

Notes

1. Much of the history of creative industries discourse can be interpreted as a rhetorical move by arts-based disciplines to gain equal status with science in national innovation agendas. This approach is understandable, but we are completely agnostic about an allegiance to arts advocacy in the form of the creative industries, and so we intend to incorporate scientific creativity as much as artistic creativity in the purview of the creative economy (as indeed, for example, does Richard Florida).
2. All of these forms of knowledge are socially inscribed, and, therefore, a social knowledge regime undergirds them all. More specifically, four broad knowledge regimes are in operation in today's society, namely scientific and technical; creative and cultural; and business and social. Science and engineering; creative arts and humanities; economics business; and the social sciences are all involved in the creative economy.
3. Granovetter's ideas have been applied in detail to a study of the creative industries specifically Queensland's music industry (Ninan, 2005).

6

The Knowledge Economy, Knowledge Capitalism, Creativity and Globalization

Phillip Brown & Hugh Lauder

Introduction

In Western economies, there is an assumption that the knowledge economy has provided impetus to the idea that progress can be seen as a technocratic model of evolutionary social change that progressively demands higher skilled workers who increasingly have "permission to think": the autonomy to exercise creativity. This view is elegantly captured in Peter Drucker's (1993) declaration that we are on the threshold of a new form of capitalism in which knowledge workers would replace the owners of capital as the locus of power. He argued that we were in a new stage of post-capitalist development that would lead to a fundamental shift in power from the owners and managers of capital to knowledge workers. Not only would they assume power, but with it would come greater autonomy, creativity, and rewards. While this assessment is perhaps the most radical account of the knowledge economy, the thinking behind it has a long history in the social sciences. In recent years several theories with shared assumptions have been used to buttress the views of politicians and policy makers that if individuals invest in education, they will be able to achieve the freedom in the knowledge economy to gain significant rewards and to be creative.

In this chapter we argue that these technocratic approaches are in Marx's terms one-sided; they provide an ideological account of economic change that serves prevailing political interests because they focus on the technological story: the forces of production, if you will, rather than the social relations of production—the way the technology, human skill, and creativity are exploited in the service of profit. It is remarkable how through the disciplines of economics, sociology, and management the guiding principle of the knowledge economy has enabled a coherent technicist account of the knowledge economy to be developed.

In contrast to the concept of a knowledge economy, we examine the key global drivers of knowledge capitalism. Global knowledge capitalism is characterized by supra-national emergent properties or institutions, of which transnational companies (TNCs) are of particular interest in this chapter. It is the way that they have exploited the digital revolution to organize global production that is key to understanding recent economic developments. However, while there are new elements that characterize global capitalism, there are others germane to an understanding of work in global knowledge capitalism that have been recognizable throughout the history of capitalism: here we focus on the way that technological innovations, rapidly become translated into routine work. As Jay Tate (2001) has noted, "Industrial revolutions are revolutions in standardization" (442). The logic behind standardization is that it reduces the cost of skilled or knowledgeable labor and hence enhances competitiveness and profitability. It is when this form of routinization is applied to the digital revolution that we confront one of the central realities of knowledge capitalism: digital Taylorism.

A Report from the Field

What is surprising about policy and academic debates about the impact of globalisation is the lack of detailed empirical evidence. Much of the evidence is derived from consultancy companies that invariably conflate prognosis with prescription in order to profit from their knowledge. It has also been dominated by American writers on management and business issues that have tended to see the world from a position of assumed political and economic pre-eminence.

Our argument is based on research with leading transnational companies at the vanguard of global economic change. The United Nations estimates that there are around 64,000 transnational companies, a rise from 37,000 in the early 1990s. These transnational companies comprise parent enterprises and foreign affiliates that vary in size and influence. The foreign affiliates of these companies generated circa 53 million jobs around the world (UNCTAD, 2005).[1] General Electric had the largest foreign assets in 2003 with 330 enterprises in the United States and over 1,000 foreign affiliates (UNCTAD, 2005; these figures exclude TNCs in the final sector).[2]

The key role that these firms play in shaping the global economy is reflected in the fact that a third of global trade is due to intra-firm activities where components, products, services, and software are sold between affiliates within the same company. Equally, it is estimated that over 60 percent of the goods exported from China in 2005 came from foreign-owned firms that had moved manufacturing plants to increase profit margins.[3]

Over the last three years, we have interviewed 180 senior managers and executives in twenty leading transnational companies in financial services, telecoms, electronics, and the automotive sector to achieve a better understanding of their global corporate strategies and the future of skills. We investigated how transnational companies were globalising their human resources and whether high skilled jobs were concentrated in the developed economies as predicted within the official discourse. We interviewed the same companies in different countries, often including the "home" country where the head office is typically found and in two other countries including Britain, China, Germany, India, Korea, Singapore, and the United States. We also interviewed government policy-makers in each of these countries to understand their competition strategies in respect to high value inward investment from foreign transnational companies.

To understand why the knowledge economy discourse has failed to grasp the nature and implications of the economic transformation which is now in train, we will focus on a number of interrelated issues to explain why this vision of a high skills, high wage economy is illusory.

Theories of the Knowledge Economy

The concept of the knowledge economy is deeply implicated in a tradition of technicist theory that sees technological change and education as the key drivers of progress. In the social sciences, Clark Kerr and his colleagues (1973) highlighted the progressive nature of industrialization based on science and technological innovation, which in turn demanded high levels of education and meritocratic opportunity.

Economists have also bought into the idea that the knowledge economy offers great opportunities for workers with the relevant knowledge and skills. Gary Becker, the doyen of human capital theorists, has argued that: Human capital and skill bias theorists have also played a significant part in developing the misrepresentation of the knowledge economy. The doyen of human capital theorists Gary Becker has argued that "Human capital refers to the knowledge, information, ideas, skills, and health of individuals. This is the 'age of human capital' . . . the economic success of individuals, and also whole economies, depends on how extensively and effectively people invest in themselves."

It is clear here that Becker makes the classic human capital assumption that high skilled workers will be in demand because employers will respond to their greater productive capacity by investing in the appropriate technologies and systems to utilize their skills.

More recently David Baker (2009) has provided additional support for this view in a theory of how education produces both minds and characters that are productive in the labor market. He argues that the schooled society creates "thinking and choosing actors, embodying professional expertise and capable of rational and creative behaviour." This is part of what he calls the rise of the culture of academic intelligence, which he claims has influenced the construction of formal organizations that are global and are similar "regardless of their mission." Educated workers of the kind described above facilitate intensive rationalization in the workplace through the rise of accounting and auditing, elaborate legal contracts, corporate social responsibility, human relations, and strategic planning amongst other traits.

These theories, and there are more we could identify in a similar vein (Brown, Lauder, & Ashton, 2010), tend to assume that there is an infinite demand for highly educated knowledge workers irrespective of development in the global economy. Indeed, one of the most striking features of these theories is how insular they are: how little they take into account the emerging nature of global capitalism.

However, there are some theorists who have linked the assumptions about the significance of education and knowledge explicitly to the global economy. For example, Robert Reich (1991) explained the growth in income polarization in the United States in the 1980s in terms of the relative ability of workers to sell their skills, knowledge, and insights in the global job market. He argues that the incomes of the top 20 percent have pulled away from the rest because of their ability to break free of the constraints of local and national labor markets. The global labor market offers far greater rewards to "symbolic analysts" or "knowledge workers" precisely because the market for their services has grown, whereas those workers who remain locked into national or local markets have experienced stagnation or a decline in income.

Reich, among others, interprets rising wage inequalities as proof of both the realities of the global labour market and as evidence of the failure of the existing education system (Brown & Lauder, 2006). The reason why income inequalities have grown is not explained as a structural problem—that the proportion of high skilled, high waged jobs is limited by the occupational structure—but due to the failure of the education system to make a larger proportion of the workforce employable in the global competition for high skilled, high waged work.

Thomas Friedman (2005) is also upbeat about what can be achieved by investing in the knowledge and skills of the workforce: "America, as a whole, will do fine in a flat world with free trade—provided it continues to churn out knowledge workers who are able to produce idea-based goods that can be sold globally and who are able to fill the knowledge jobs that will be created as we not only expand the global economy but connect all the knowledge pools in the world. There may be a limit to the number of good factory jobs in the world, but there is no limit to the number of idea-generating jobs in the world" (p. 230).

It is believed, therefore, that there is now a global auction for jobs. Low skilled jobs will be auctioned on price and will tend to migrate to low waged economies such as those in Asia or Eastern Europe, while high skilled jobs will continue to attract higher wages. These jobs will be auctioned on "quality" rather than price, including the skills, knowledge, and insights of employees. The main bidders for "quality" jobs are assumed to be today's advanced economies. This offers the potential for countries such as Britain, France, and the United States to become *magnet* economies, attracting a disproportionate share of high skilled, high wages jobs (Brown & Lauder, 2001). The idea that knowledge work will be retained in the west due to superior education and innovation systems is an example of a form of neocolonial thinking, which has failed to understand the speed with which countries like China and India are mobilizing for high skilled work.

Quality and Price

Much of the literature has suggested that the comparative advantage of nations depends on their ability to compete on quality or price. We have described how developing economies are assumed to be restricted to price competition for low skilled, low-value goods and services because they lack the skilled labor and hi-tech capabilities of OECD countries, such as the United States, Germany, or the United Kingdom. In turn, to maintain their prosperity, workers and businesses in developed economies must move up the value chain towards the "quality" end of the market, based on the assumption that the value of knowledge will continue to rise.

However, at a time when human knowledge is being taught, certified, and applied on a scale unprecedented in human history, the overall value of human knowledge is likely to decline rather than increase. We are witnessing an increasing polarisation in the market value of different kinds of qualifications, knowledge, and occupational roles. If knowledge is the key asset of the new economy, the task of business is not to pay more for it but less. There are two aspects to the strategies that companies adopt to pay less for more. The first is by accessing the increasing supply of graduates from across the globe, many of whom will work for far lower incomes than those in the West, either by offshoring or by locating their high skills work, such as research and development, in developing nations including China and India. The second is by standardizing knowledge work through processes that we call Digital Taylorism (see below).

Companies will continue to pay a premium for outstanding "talent" (however it is defined) as part of the hierarchical segmentation of "knowledge" work. This reality has long been a feature of capitalism, but today it has greater significance because the incomes of so many workers in Western economies depend on maintaining if not increasing the market value of what they know. It has also become more significant because the global economy offers employers new ways of reducing costs and raising productivity that were not available until now. A high profile political example is the growth in offshoring in key sectors such as financial services and information technologies. The cost of employing a chip design engineer in the United States is over four times more than a designer in Korea and 10 times or over the costs associated with the same workers in India and China (Brown et al., 2006). In financial services, relocations increasingly involve "front" and well as "back" office functions, including financial analysis, research, regulatory reporting, accounting, human resources, and graphic design.[4] Quality has become price sensitive, and labor arbitrage (profiting from differences in labor costs around the world) no longer stops with factory workers and call-centre operatives.

The new competition is based on quality *and* price that are enabling companies to raise their game and lower their costs at the same time. While national governments in the developed economies may see the knowledge economy as a way of increasing prosperity, and while there is a tendency in the policy literature to understand competitiveness and productivity as a question of competing for knowledge and skills rather than profits, it is far removed from the way companies understand the new competition which involves getting smart things done at a lower price.

High Skills: A Declining Advantage

The argument that a knowledge-driven economy demands a larger proportion of the workforce with a university education and with access to lifelong learning opportunities has had a major impact on participation rates in tertiary education. In OECD countries, university is no longer the preserve of an elite, whatever the merits of the economic case for expanding higher education. There has been a significant expansion in all OECD countries with the exception of Germany. Canada was the first country to achieve the target of over 50 percent of people aged 25 and 34 entering the job market with a tertiary level qualification. Korea is not far behind having engineered a massive growth in tertiary provision since 1991. Germany is the exception due to its continued commitment to the dual system of workplace and off-the-job training.[5]

This expansionary phase is unlikely to end in the near future as most countries benchmark themselves against those with the highest participation rates, although its relationship to employment, productivity and economic growth remain unclear (Ashton & Green, 1996). This expansion is consistent with the Western view that low skilled jobs will be auctioned on price and will tend to migrate to low waged economies such as those in Asia or Eastern Europe, while high skilled jobs will continue to attract higher wages. However, this analysis fails to recognize the mass production of well-qualified candidates from developing economies who will enable transnational companies to export some of their "brain" work as well as their "body" work to low cost economies.

The collapse of communism, economic integration, and advances in information technologies have brought China, India, and Russia along with a number of smaller nations into the global competition for education, knowledge, and high skilled employment. As one commentator has noted: "the composition of China's exports has begun to change rapidly, away from reliance on cheap low-margin goods to more value-added manufacturers offering much higher profits."[6] China and India want to move their cost advantage further up the value chain. As we were told by a government official in Beijing, "today China is the world's factory, tomorrow the world's competitor."

In an interview with a senior Indian government official in New Delhi, we discussed India's expansion into manufacturing. This was his response, "the Chinese have a great advantage when it's mass production. We will not be able to compete with them there . . . but increasingly every item is requiring new inputs like design inputs, it's requiring innovation and embedded software. That is our skills advantage, we are moving up the value chain in manufacturing." It is this attempt to move up the value chain that will transform the global auction for jobs as knowledge workers, and the developed economies are no longer immune from price competition with highly qualified workers in low cost locations.

China had over six times as many students in higher education than the UK and almost as many as the United States in 2002, including 600,000 engaged in postgraduate studies. The latest figures suggest that China has now overtaken the

United States with around 20 million students enrolled in higher education.[7] In India, there has also been a major expansion of higher education with the aim of increasing the participation rate of 18- to 23-year-olds in higher education from 6 percent in 2002 to 10 percent in 2007.[8] India's Prime Minister, Mr. Manmohan Singh, recently observed that: "In the next one or two years, the knowledge sector will receive our attention to the extent that it deserves. I do recognise that India has to be the centre, the hub of activity as far as the knowledge economy is concerned. We don't want to miss the chance."[9] There is little sense for countries such as China, India, Malaysian, Poland, or the Czech Republic to be content with doing the body work within the global economy while the brain work is left to the developed economies such as the United States, Japan, Germany, and Britain.

Although the quality of education is likely to vary in countries experiencing rapid expansion of educational provision, it is nevertheless the case that Asia is producing more engineers than Europe and North America combined. In the natural and agricultural sciences (including physical, biological, earth, atmospheric, and ocean sciences) Asia is also ahead, although this lead is not the case for mathematics and computer sciences.

In the United States, close to half of those gaining a doctoral degree in engineering, mathematics, and computer science are foreign students. Some of these remain within the developed economies, but others return to their indigenous countries, adding to the stock of highly skilled workers (Saxenian, 2006). South Korea alone graduates as many engineers as the United States, and, according to recent evidence from a U.S. Business Roundtable report, more than 90 percent of all scientists and engineers in the world will soon be living in Asia.[10] The World Bank also estimates that Russia has the third-highest numbers of scientists and engineers per capita in the world, and other Eastern European countries also have a growing proportion of well educated scientists and IT specialists.[11]

Critics have argued that the mass higher education systems in India and China have limited potential for high level research and hence innovation and that, therefore, the excellence for innovation in America and the West will remain unrivalled. However, this criticism fails to understand the position of universities in America in relation to economic globalisation. Once universities have been freed to play the market, they no longer need to be or are attached to national loyalties: research breakthroughs in California can be capitalized on in Beijing.

On this evidence, the view that it will take decades for developing economies to compete in the global market for high skilled jobs has grossly underestimated the speed of educational reform and business innovation in emerging economies including China and India.

Where to Think?

Innovation remains a crucial source of competitive advantage as mass customization has assumed greater importance in virtually all industrial sectors. The demand for constant innovation has also been fuelled by rapid technological advancement and consumer tastes. Over 80 percent of BMW Minis produced in Britain for the global market are built to customer order, offering a range of over 250 factory-fit options and dealer-fit accessories making every Mini uniquely similar. In the United States the Toyota Tundra sports has 22,000 possible configurations and the Chrysler Dodge Ram is available in 1.2 million variations.[12] The use of build-to-order where products are only made to the specific requirements of customers is not restricted to the auto industry. Dell computers has established a sophisticated made-to-order business that gives customers the opportunity to build a computer based on a choice of the twenty or so product features that go into a computer including memory (RAM), disk space, modem, processor, screen, and software. The same processes are being applied to clothes, watches, sneakers, cosmetics, window-frames, and houses. Nike offers customized sports shoes where customers can choose between a range of "uppers" and "soles" and have their names embroidered on the back of each sneaker, while "Customatix," an Internet company, allows customers to design their own shoes based on an almost limitless combinations of colors, graphics, logos, and materials.[13] These trends not only highlight the importance of accelerating the development of new ideas and improving on existing ones, but also on reducing the time and cost to get them into the market place. To reduce the time from "innovation to invoice" some companies use twenty-four hour design teams that work around the clock moving through time zones across Asia, Europe, and North America. This dynamic is not only intended to reduce the time between invention, application, and market launch, but also to reduce costs, due to lower salary levels in much of Asia. As a senior executive in a German multinational told us, "we have to drive innovation, we have to be at the leading edge at reasonable cost . . . we have to try to get higher skills at reasonable cost and high flexibility."

This approach is leading companies to give more thought to "where to think." Typically, this new mode has led them to question the role of the appropriately named "head" office as the primary source of corporate brain power. However, where to think is more than a question of finding the cheapest locations, as it reflects a number of other considerations such as the need for a critical mass of people that understand the organization, or share the collective intelligence necessary for advanced R & D. It is also assumed to reflect the importance of embedded capabilities as innovation rarely depends on the skills of individuals working in isolation but on a culture of mutual collaboration and purpose. However, companies are increasingly experimenting with research, design, market, and product development activities in the emerging economies.[14]

Such trends reflect a quality revolution within emerging economies that challenges much of the existing literature on the social foundations of economic performance. It is, for instance, assumed that quality depends on particular "regimes of production," such as the dual system of workplace and college training in Germany or high trust relations in the "third Italy," that are difficult if not impossible to duplicate (Hall & Soskice, 2001). However, companies have discovered as they experiment with higher end activities such as research and design in lower cost countries that quality may not be impaired and may even be improved, although there are also companies who retreat because they struggle to achieve the standards they require or due to fears about intellectual property rights.

Our studies show that the assumption that hi-tech depends on social sophistication in the form of democratic politics, welfare provision, and high GDP per capita, fails to capture the extreme forms of uneven development where the preindustrial and the post-industrial share the same postcode. There is a tendency to study economic activity from the outside looking in, based on an assumed correspondence between society and economy, but business is being turned "inside, out." While companies need a decent infrastructure (roads, communications) and supply of well educated and motivated workers, they are able to set up "oasis operations" (high-tech factories, offices, and research facilities in low-tech societies). Moreover, it is a mistake to assume that the rapid development, especially in China, is at the price of quality. One does not need to spend much time in Beijing, Shanghai, or Guangzhou to understand that they are building to compete with America, Japan, and Germany rather than other developing economies.

The rise in quality standards around the world is making it more difficult for highly qualified workers in developed economies to shelter from the global competition for jobs. Equally, as the performance gap rapidly narrows, differences in labor costs between developed and developing economies are narrowing slowly except for in a few hot spots in China and India, and even here, it is still a long way before the price advantage is seriously eroded. Consequently, companies have greater scope to extract value from international webs of people, processes, and suppliers, based on a Dutch or reverse auction where quality is maintained while labor costs go down.

In the late 1990s, when we asked a leading German car manufacturer whether they could make their executive range anywhere in the world, the answer was an emphatic "no." Today it is an equally emphatic "yes." Another car maker, this time from the United States, added, "If you had asked me 5 years ago I would have said that the skill sets probably are still in the advanced economies but I think that is changing very, very quickly. . . . The advantage from our perspective is that you are paying those guys anywhere from sort of 12 to 15 thousand dollars a year versus say a European or a US engineer at anywhere from 75 to 95 thousand dollars a year with a whole bunch of benefits as well."

A leading engineering corporation also told us there has been a significant narrowing in the performance of operations and factories around the world, "those in emerging countries are catching up fast and this is making it more difficult for plants in the West. It's really a bit of a rat race." Research in China revealed that many enterprises had adopted the latest high performance management practices, which flourish in the context of a highly educated labor force, enabling them to produce high value-added goods at much lower costs (Venter, Ashton, & Sung, 2002). Moreover, a United Nations survey of transnational companies also found that China was the most attractive prospective R&D location between 2005 and 2009, followed by the United States, India, Japan, United Kingdom, and the Russian Federation.[15]

As differences in quality and productivity narrow between operations in different parts of the world, the cost and working conditions of Western employees are no longer the global benchmark. This assessment has been true for various kinds of low-skilled activities in the manufacturing sector for thirty years. However, the same rule may now be true for high skilled workers in the developed economies as a growing proportion of high skilled, high value activities can be undertaken in low-cost locations. In moving inward investment up the value-chain of products and services, transnational companies are not only "following the business" into rapidly expanding emerging markets, but they are also adopting a deliberate strategy to establish leading edge operations in parallel to those in the developed economies. This assessment not only gives them global flexibility and continuity if there are industrial relations problems or problems of underperformance in a specific regional center, but it also enables companies to point to their lower cost operations in the emerging economies when negotiating with employees in the West.[16]

Digital Taylorism

While the policy spotlight has focused on the creation of new ideas, products and services, the ability of companies to leverage new technologies to globally align and coordinate business activities has also brought to the fore a different agenda involving the standardization of functions and jobs within the service sector, including an increasing proportion of technical, managerial, and professional roles. As Jay Tate (2001) has observed "industrial revolutions are revolutions in standardization."

Standardization is well understood in manufacturing where the same standard components such as wheels, brake linings, and windscreens, can be made in different factories around the world and shipped for final assembly at one location in the knowledge that all the components meet international quality standards and will fit together. This method not only gives companies flexibility but enables them to reduce costs. The same logic is now being applied to service sector occupations that were previously difficult to standardize because there were no digital equivalents to mechanical drills, jigs, presses and ships, all required to create global supply chains in manufacturing.

The potential to transform work in the service sector, that is work that does not involve physical proximity to the customer, client, or patient (although our understanding of what can be done "remotely" is being transformed by new communication technologies), is inevitably limited so long as knowledge remains in the heads of individuals working in idiosyncratic ways using different computer systems and application software. However, the communication technologies that we have today, including the capacity for digital processing, Internet capability; and increasing bandwidth (that determines the volume and speed that data, information, or live video can be transferred across a network) have created the realistic possibility of developing global standards that reduce technical complexity and diversity (Davenport, 2005).[17]

Through building modular applications, business processes including ordering, marketing, selling, delivering, invoicing, auditing, and hiring, can be broken down into their component parts, which include the unbundling of occupational roles so that job tasks can be simplified and sourced in different ways. In other words, an increasing proportion of managerial and professional jobs that were previously sheltered because they were not tradable are being redesigned, although it is difficult to predict how far this process can transform technical, managerial, and professional occupations (Bryant, 2006).

Terms such as "financial services factory" and "industrialization" are being applied by leading consultancy companies to describe the transformation of the service sector. Accenture Consulting (2007, p. 1) is a proponent of "the concept of industrialization—breaking down processes and products into constituent components that can be recombined in a tailored, automated fashion—to non-manufacturing settings." Likewise, Gupta (2006) states that "by componentizing their business processes, the Financial Services firms have begun to look at each component independently of the other components while selecting the best sourcing option (i.e., insourced or outsourced, onshore and/or offshore, etc.). Should the trend continue tomorrow's banks would look and behave no differently to a factory" (p. 43).

It is this form of organisational innovation in the way companies hire, order, market, sell, deliver, distribute, invoice, and account, driven by new information technologies and greater choices in terms of where to produce, partner, or purchase goods and services that define today's knowledge capitalism. These trends remain in their "craft" stage resembling manufacturing in the early twentieth century. However, while it took decades for manufactures to "lift and shift" through standardization, the process is likely to be much quicker when applied to service sector employment because the only hardware you need can fit on the average office desk.[18]

This part of our analysis suggests that if the twentieth century brought what can be described as *mechanical Taylorism* characterized by the Fordist production line, where the knowledge of craft workers was captured by management, codified and re-engineered in the shape of the moving assembly line, the twenty-first century is the age of *digital Taylorism.* This method of operation involves translating *knowledge work* into *working knowledge* through the extraction, codification, and digitalization of knowledge into software prescripts that can be transmitted and manipulated by others regardless of location.

Anell and Wilson (2002) argue that "the question of how to extract and distribute knowledge efficiently will not be answered by recommendations about how to build and use human and structural capital. The solution resides in the ability of knowledge firms to extract and translate more or less tacit, personal knowledge into explicit, codified knowledge, into what we call prescripts. Prescripts constitute a form of capital, to be regarded in the same vein as the company's human, structural, social and financial capital" (pp. 7–8).

While there seems little doubt that the extent to which companies can capture the knowledge of those who think for a living is often exaggerated, the problem for knowledge workers was recognized by Harold Wilensky in 1960 when he envisaged a time when the distinction between conception and execution would move further up the occupational hierarchy as new technologies would give senior managers and executives much greater control of the white collar workforce:

> Top executives, surrounded by programmers, research and development men [and women], and other staff experts, would be more sharply separated from everybody else. The line between those who decide, 'What is to be done and how' and those who do it—that dividing line would move up. The men who once applied Taylor to the proletariat would themselves by Taylorized (1960, p. 557).

Whereas the distinction between conception and execution in a period of mechanical Taylorism transformed the relationship between the working and middle classes, digital Taylorism also takes the form of a power struggle within the middle classes, as these processes depend on reducing the autonomy and discretion of the majority of managers and professionals. It encourages the segmentation of talent in ways that reserve the "permission to think" to a small proportion of employees responsible for driving the business forward.[19] However, the loss of autonomy for managers and professionals remains significantly different from the era of mechanical Taylorism because its digital variety eliminates the need for close, over-the-shoulder, supervision. Control is remote because it is built into the software, so that the monitoring of activities is at a distance. Equally, it does not eliminate the importance of employee motivation or the need for good customer-facing skills as the standardization required to achieve mass customization still needs customers to feel that they are receiving a personalized service. This fact may contribute to a continuing demand for university graduates, but their occupational roles are far removed from the archetypal graduate jobs of the past.

Creating a "War for Talent"

While the official account of the knowledge economy assumes a linear relationship between education, jobs, and rewards, where mass higher education is predicted to reduce income inequalities as people gain access to high skilled, high waged jobs, the reality is more complex. In America and Britain the expansion of higher education has been associated with an increase in wage differentials (Mishel, Bernstein, & Allegretto, 2007). This gap exists not only between university graduates and non-graduates but within the graduate workforce. Frank and Cook (1996) argue that income inequalities are not the result of changes in the distribution of human capital—that some have invested more in their education and training that others—but due to the changing structure of the job market (Brown, 2006). Even within "graduate" occupations those at the top of the occupational pyramid receive a disproportionate share of rewards, in what Frank and Cook call "winner-takes-all markets." They argue that changes in domestic and global competition make "the most productive individuals more valuable, and at the same time have led to more open bidding for their services" (p. 6).

This argument is consistent with that of consultants from McKinsey's who popularized the idea of a "war for talent" (Michaels et al., 2001). They argue that reliance on talent has increased dramatically over the last century. "In the 1900s, only 17 percent of all jobs required knowledge workers; now over 60 percent do. More knowledge workers means it's important to get great talent, since the differential value created by the most talented knowledge workers is enormous" (Michaels et al., 2001, p. 2). Whatever the merits of this argument, virtually all those we spoke to in China, Korea, India, and Singapore, as well as the United States, Germany, and Britain believed that they were in a war for talent, which was increasingly global.

Therefore, is the war for talent essential to higher productivity and competitiveness, or can it be explained in terms of positional conflict (i.e., bosses taking a larger share of the profits)? It seems clear that there is a more intense positional conflict within organizations, especially when the emphasis is on shareholder value (Lazonick & O'Sullivan, 2000). When the focus is on maximizing the returns to shareholders, senior managers, and executives need to be aligned to short-term profit maximization often through share options that require a consistent attempt to reduce costs. Workers who are not defined as top talent will constantly come under pressure to "prove their worth" within an increasingly global context. We know that in many transnational companies a larger share of the profits is also going to shareholders rather than the workforce predicted by pundits of the knowledge economy (Roach, 2006). There is also evidence of corporate executives in the United States and Britain gaining massive wage hikes that often bear little relationship to business performance (Bebchuk & Grinstein, 2005). However, this story is still incomplete because the war for talent also reflects the changing nature of economic competition. The value of a company is not simply determined by the "value" of what it produces, but on its "reputational" capital (Brown & Hesketh, 2004), or what is commonly referred to as "branding." As Samsung, a leading global electronics firm, has observed, "in the digital era, a product will be distinguished by its brand more than by its functions or by its quality."[20]

This emphasis on the "social" rather than the "technical" facets of business success is also highlighted in the nature of services that include management consultancy and the creative industries. As Alvesson (2001) has suggested, "the ambiguity of knowledge and the work of knowledge-intensive companies means that 'knowledge,' 'expertise' and 'solving problems' to a large degree become matters of belief, impressions and negotiations of meaning. Institutionalized assumptions, expectations, reputations, images, etc. feature strongly in the perception of the products of knowledge-intensive organizations and workers" (p. 863).

Value added in knowledge intensive industries, e.g., consultancy or financial services, stems from branding the company in order to maximize the price of its professional knowledge. However, the value of corporate branding is not restricted to the image of the goods or services sold to consumers around the world. It also relates to the workforce. The more corporate value is "embodied" in the people who work for it, the more companies want to be seen to recruit "the best" (Brown & Hesketh, 2004).

It is assumed that the best graduates gravitate toward the elite universities. This view is actively promoted by leading universities as higher education has become a global business. The branding of universities and faculty members is integral to the organization of academic inquiry. Claims to world-class standards depend on attracting "the best" academics and forming alliances with elite universities elsewhere in the world, while recruiting the "right" kinds of students. Universities play the same reputational games as companies because it is a logical consequence of market competition.

We can also see how a new global hierarchy is being created that transforms national hierarchies; this dynamic has been exemplified by recent reforms in German higher education. Until recently it has been based on "parity of esteem" between universities. To date there has been little difference in the market value of a degree from one German university rather than another. However, the introduction of excellence reforms is leading more resources to be targeted at a small number of universities. In short, this policy will create an elite in an attempt to lift the profile of German higher education within global rankings of leading universities.

In ripping up the level playing field, it will transform the positional relationship between students from different universities. In an attempt to recruit the best and to be seen to do so, leading companies will target this elite group, based on the assumption that the most talented students will go to these universities because they are the most difficult to get into. Hence, the idea of a war for talent in Germany is real in its consequences, as a likely outcome will be growing income inequalities between German graduates.

As it becomes impossible for employers to have firsthand knowledge of universities or the quality of their students, reputation (like branding) becomes key. All companies benchmark leading universities around the world based on their own formulations often in conjunction with public rankings of top universities. Despite much talk of greater diversity, the ranking of universities by reputation has made it more important to study at a leading national university with an international reputation. Notions of diversity are being transformed from a concern to recruit from a broad range of social backgrounds within a given national context, toward viewing diversity as the recruitment of foreign nationals as part of the internationalization of human resource management. In reality, this form of diversity is about recruiting elites from different countries in the global war for talent. To qualify, individuals have to go to the *best* universities whatever country they live in.

These issues have profound implications for understanding the relationship between education, jobs, and rewards, as human capital theory, with its emphasis on technical knowledge, fails to account for positional conflict surrounding shareholder models of corporate governance, or the increasing importance of "reputational" capital in assessing the differential value of individual credentials and knowledge. Although the relationship between reputation and performance is hazy, its consequences are stark, as reputation and performance are woven together through the exercise of symbolic power to define which employees are to be truly valued as exhibiting high potential or outstanding performance. Employees defined as "top talent" are able to draw on this reputational capital to leverage a better remuneration package, whereas other equally well qualified employees find themselves in a reverse bidding war as companies try to reduce the cost of knowledge.

In short, almost without exception, companies were not only "segmenting" their educated workforce based on occupational function but also on "performance" driven by an attempt to reduce the cost of knowledge work, while retaining what they perceived at top talent. Within a context of increasing globalisation, digital Taylorism and the expansion of high skilled, low cost workers from developing economies, companies are developing new ways to compete for the best ideas at the same time as delivering them at the lower cost. Within this new economy of knowledge, employees are caught in a pincer movement where those defined as "top talent" are judged to have high market value, while others in the same occupations increasingly find themselves in a cost-driven competition whether domestic or global.

Conclusions

This chapter challenges the dominant theories on education in a global knowledge-based economy. Rather we argue that we need to understand the key developments in global knowledge capitalism if we are to understand the possibilities for creativity for graduate workers. We argue that Britain and the United States are not knowledge economies, where the value of knowledge continues to rise, but they are characterized by an economy of knowledge that is transforming the relationship between education, jobs, and rewards. This dynamic will inevitably lead to claims that education is failing to meet the needs of industry, but the overriding problem is a failure to lift the demand for knowledge workers to meet the increasing numbers entering the job market with a bachelor's degree (Keep, 2004).

The disjunction between education, jobs, and rewards has profound implications for our understanding of educational opportunity, justice and social mobility. Ernest Gellner (1983) observed that "modern society is not mobile because it is egalitarian; it is egalitarian because it is mobile" (pp. 24–25). This statement suggests that the growing evidence of declining social mobility in both the United

States and Britain is not simply due to increasing inequalities in opportunity, but this lack of social mobility reflects the transformation of work that we are beginning to capture in this chapter.

The technocratic model of skills upgrading and rising value of investments in human capital is subject to the laws of diminishing returns. Human capital theory does not offer a universal theory of the relationship between education, job, and rewards but represents a "transitional" case in the second half of the twentieth century characterized by educational expansion and a rising middle class.

Today, the "positional" advantage of many with university credentials is not only declining domestically (as higher education is expanded) but also globally as access to tertiary education becomes more widespread both within and across countries. We predict that the global expansion of tertiary education will lead to downward pressure on the incomes of skilled workers in the developed economies, along with some upward pressure on those in emerging economies. At the same time, there are trends towards "winner-takes-all" markets, which reveal that people with similar qualifications in the same occupations, organizations and countries will experience increasing polarization in future career prospects (Frank & Cook, 1995).

The trends identified in this chapter raise doubts about the efficacy of economic policies based on educational reform. In much the same way that we have misunderstood the source of social mobility in the developed economies to stem from an extension of meritocratic opportunity rather than changes in the occupational structure, we have also misunderstood the role of high skilled workers to economic development and national prosperity. There has been a tendency to understand the supply of high skilled workers as a direct cause of national prosperity rather than a consequence of innovative enterprise (Lazonick, 2003) that lifts the demand for skilled workers.

It could be argued that what we are witnessing is a process of creative destruction, which can be seen in the decline of key industries such as automobiles in America or through corporate restructuring as a result of the current economic recession. Out of the "destruction" fresh forms of creativity will be released that will enable well educated workers to capitalize on new opportunities. This interpretation reflects the myth that economies are always open to new ideas and practices. However, one of the key arguments of this chapter has been that innovation in capitalist economies follows a path whereby creative destruction becomes the destruction of the creative. In mass modern economies key innovations have to become embedded in routines in order for profits to be made. This assessment does not mean that at the fringes individual entrepreneurs cannot make a good living out of their creativity, but it does mean that the mass employment of educated workers implied by the rhetoric of the knowledge economy is a myth.

References

Accenture. (2007). Automation for the people: Industrializing Europe's insurance industry. Retrieved from http://www.accenture.com/Global/Services/By_Industry/Financial_Services/Insurance/The_Point/Y2007/fsi_thepoint47a.htm

Alvesson, M. (2001). Knowledge work: Ambiguity, image and identity. *Human Relations, 54*(7), 863–886.

Anell, B., & Wilson, T. (2002). Prescripts: Creating competitive advantage in the knowledge economy. *Competitiveness Review, 12*(1), 26–37.

Ashton, D., Brown, P. and Lauder, H. (forthcoming). Skill webs and international human resource management: Lessons from a study of the global skill strategies of Transnational companies, *International Journal of Human Resource Management.*

Ashton, D., & Green, F. (1996). *Education, training and the global economy.* Aldershot: Edward Elgar.

ATKearny Consultants. (n. d.). Retrieved from http://www.atkearney.com/main.taf?p=1,5,1,130

Baker, D. (2009). The educational transformation of work: Towards a new synthesis. *Journal of Education and Work, 22*(3), 163–191.

Bebchuk, L., & Grinstein, Y. (2005). The growth of executive pay. *Oxford Review of Economic Policy, 21*(2), 283–303.

Becker, G. (2006). The age of human capital. In H. Lauder et al. (Eds.), *Education, globalization and social change.* Oxford: Oxford University Press.

Brown, G. (2004, November 9). Full text: Gordon Brown's confederation of British industry. Speech. Retrieved from http://news.ft.com/cms/s/eb4dc42a3239–11d9–8498–00000e2511c8.html

Brown, P. (2006). The opportunity trap. In H. Lauder, P. Brown, J. A. Dillabough, & A. H. Halsey (Eds.), *Education, globalization and social change.* Oxford: Oxford University Press.

Brown, P., Green, A. and Lauder, H. (2001). *High skills: Globalization, competitiveness and skill formation.* Oxford: Oxford University Press

Brown, P., & Lauder, H. (2001). *Capitalism and social progress.* Basingstoke: Palgrave.

Brown, P., & Hesketh, A. (2004). *The mismanagement of talent.* Oxford: Oxford University Press.

Brown, P., & Lauder, H. (2006). Globalisation, knowledge and the myth of the magnet economy. *Globalisation, Societies and Education, 4*(1), 25–57.

Brown, P., Lauder, H., Ashton, D., & Tholen, G. (2006). Towards a high-skilled, low-waged economy? A review of global trends in education, employment and the labour market. In S. Porter & M. Campbell, (Eds.), *Skills and economic performance* (pp. 55–90). London: Caspian Publishing.

Challenge of Customization: Bringing Operations and Marketing Together, The. (n. d.)Retrieved from http://www.strategy-business.com/sbkwarticle/sbkw040616?pg=all&tid=230

Chanover, M. Mass customize—Who? (n. d.) What Dell, Nike and others have in store for you. Retrieved from http://www.core77.com/reactor/mass_customization.html

Davenport, T. H. (2005, June). The coming commoditization of processes. *Harvard Business Review*, pp. 101–108.

Drucker, P. (1993). *Post-capitalist society.* Oxford: Butterworth-Heinemann.

Fevre, R. (2003). *The sociology of economic behaviour.* London: Sage.

Frank, R. H., & Cook, P. J. (1996). *The winner-take-all society: Why the few at the top get so much more than the rest of us.* New York: Penguin.

Friedman, T. (2005). *The world is flat.* London: Allen Lane.

Gellner, E. (1983). *Nations and nationalism.* Oxford: Blackwell.

Grubb, W. N., & Lazerson, M. (2006). The globalization of rhetoric and practice: The education gospel and vocationalism. In H. Lauder et al. (Eds.), *Education, globalization and social change.* Oxford: Oxford University Press.

Gupta, R. K. (2005). India's economic agenda: An interview with Manmohan Singh. *McKinsey Quarterly, Special Edition: Fulfilling India's Promise.* Retrieved from http://www.mckinseyquarterly.com/

Gupta, S. (2006). Financial services factory. *Journal of Financial Transformation, 18,* 43–50.

Hall, P. A., & Soskice, D. (2001). *Varieties of capitalism.* Oxford: Oxford University Press.

Halsey, A. H. (1961). "Introduction" in A. H. Halsey, J. Floud, & J. Anderson (Eds.), *Education, economy and society.* Oxford: Oxford University Press.

Jones, B. J. (1999). *Knowledge capitalism.* Oxford: Oxford University Press.

Keep, E. (2004). After access: Researching labour market issues. In J. Gallacher (Ed.), *Researching access to higher education.* London: Routledge.

Kerr, C., Dunlop, J., Harbison, F., & Myer, C. (1973). *Industrialism and industrial man.* Harmondsworth: Penguin.

Lazonick, W. (2003). The theory of the market economy and the social foundations of innovative enterprise. *Economic and Industrial Democracy, 24*(1), 9–44.

Lazonick, W., & O'Sullivan, M. (2000). Maximizing shareholder value: A new ideology for corporate governance. *Economy and Society, 29*(1), 13–35.

Leitch Review of Skills. (2006). *Prosperity for all in the global economy—World class skills.* Final Report. Norwich: HMSO.

Marginson, S. (2006). National and global competition in higher education. In H. Lauder et al. (Eds.), *Education, globalization and social change.* Oxford: Oxford University Press.

McGregor, R. (2006, July 4). Report by Richard McGregor in Beijing. *Financial Times*, p. 10.

Michaels, E., Jones, H. H., & Axelrod, B. (2001). *The war for talent.* Boston: Harvard Business School Press.

Mishel, L., Bernstein, J., & Allegretto, S. (2007). *The state of working America 2006/2007.* Economic Policy Institute. Ithaca, NY: Cornell University Press.

Reich, R. (1991). *The work of nations.* New York: Simon and Schuster.

Rosecrance, R. (1999). *The rise of the virtual state.* New York: Basic Books.

Roach, S. (2006, December 14). From globalization to localization. Morgan Stanley, Global Economic Forum, Retrieved from http://www.morganstanley.com/views/gef/archive/2006/20061214-Thu.html

Samsung Company Report. (n. d.). Retrieved from http://www.samsung.com/AboutSAMSUNG/ValuesPhilosophy/DigitalVision/index.html

Saxenian, A. (2006). *The new Argonauts: Regional advantage in a global economy.* Cambridge, MA: Harvard University Press.

Stewart, H. (2005, June 12). The West sees red. *The Observer.*

Stewart, T. A. (2001). *The wealth of knowledge.* London: Nicholas Brealey.

Streeck, W. (1997). German capitalism: Does it exist? Can it survive? *New Political Economy, 2*(2), 237–256.

Tapping America's Potential: The Education for Innovation Initiative. (2005). Retrieved from http://www.businessroundtable.org/publications/publication.aspx?qs=2AF6BF807822B0F1AD1478E

Tate, J. (2001). National varieties of standardization. In P.A. Hall & D. Soskice (Eds.), *Varieties of capitalism.* Oxford: Oxford University Press.

Trombly, M. (2003, September 15). There's a treasure trove of scientific talent—And lots of government bureaucracy. Retrieved from http://www.computerworld.com/managementtopics/outsourcing/story/0,10801,84874,00.html

United Nations Conference on Trade and Development (UNCTAD). (2005). *World Investment Report, 2005, Transnational Corporations and the Internationalization of R&D.* Retrieved from http://www.unctad.org/wir.

Venter, K., Ashton, D. N., & Sung, J. (2002). *Education and skills in the People's Republic of China: Employers' perceptions,* ILO/CLMS (p. 64). Leicester: University of Leicester.

Wilensky, H. (1960). Work, careers, and social integration. *International Social Science Journal, 12,* 543–60.

Acknowledgment

We would like to acknowledge the support of the Economic and Social Research Council of Great Britain (ESRC), who funded this three-year comparative project on "Global Corporate Strategies and the Future of Skills." We would also like to thank Susan Wright and Ian Jones for their comments on an earlier draft of the chapter.

Notes

Elements of this paper were first published in the Web-based *European Educational Research Journal, 7*(2) in 2008.

1. See United Nations Conference on Trade and Development (UNCTAD) World Investment Report. (2005). Transnational Corporations and the Internationalization of R&D. Retrieved from www.unctad.org/wir.
2. Ibid. Appendix A, pp. 267–268. These figures exclude TNCs in the financial sector.
3. See Heather Stewart (2005).
4. See ATKearny Consultants. (n. d.). Retrieved from http://www.atkearney.com/main.taf?p=1,5,1,130
5. The relative merits of the German dual system and its future have been widely debated. See Brown, Green, & Lauder (2001); Streeck, (1997).
6. Report by Richard McGregor (2006).
7. Private communication with the Department of Education in Beijing (based on figures for October 2005).
8. India's "Tenth Plan" for education is focused on increasing access; quality; adoption of state specific strategies; liberalization of the higher education system; relevance including curriculum, vocationalization, networking and information technology; distance education; convergence of formal, non-formal, distance and IT education institutions; increased private participation in establishing and running of colleges and deemed to be universities; research in frontier areas of knowledge and meeting challenges in the area of Internationalisation of Indian Education. http://www.education.nic.in/htmlweb/approach_paper_on_education.htm
9. See Rajat K. Gupta (2005).
10. See Tapping America's Potential: The Education for Innovation Initiative (2005). Retrieved from http://www.businessroundtable.org/publications/publication.aspx?qs=2AF6BF807822B0F1AD1478E
11. See Maria Trombly (2003).
12. See The Challenge of Customization: Bringing Operations and Marketing Together(n. d.) Retrieved from http://www.strategy-business.com/sbkwarticle/sbkw040616?pg=all&tid=230
13. Michael Chanover (n. d.) Mass customize—Who? What Dell, Nike and others have in store for you. Retrieved from http://www.core77.com/reactor/mass_customization.html
14. However, while companies may want to offshore some of their R&D activities, there is a constant concern about "reverse" engineering and technology transfer. The opportunity to extend into new markets of the size of China and India also raises the threat of low-cost competitors able to create competing products or services. This possibility makes MNCs reluctant to share their state of the art knowledge, technologies, and know how, but at the same, they need access to emerging markets and to reduce devel-

opment costs. The problem is illustrated in the electronics sector. We were told by a leading MNC that the Chinese were capable of copying the latest mobile phones in two months. This had led this company to re-trench its R&D activities within the home base to protect its product developments for as long as possible. They have also started launching the same product simultaneously in different countries to gain a lead on the competition, even if they can catch up very fast.

15. See United Nations Conference on Trade and Development (UNCTAD) World Investment Report (2005). Transnational Corporations and the Internationalization of R&D, Figure IV.11, p.153. Retrieved from http://www.unctad.org/wir
16. Germany is an obvious example.
17. Davenport identifies various initiatives that have been introduced to standardize and commodify business processes such as the Supply-Chain Operations Reference (SCOR) model that outlines five key steps of plan, source, make, deliver, and return. Another is the Software Engineering Institute's Capability Maturity Model (CMM), and ISO 9000 for quality standards for product development. ISO 9000 is based on the design, development, production, installation, and servicing of products. ISO 9000-9003 were created by the International Organisation for Standardization, which is a global consortium of national standard bodies. Six Sigma focuses less on management process and more on the output of the process, especially defect reduction.
18. Combined with offshoring the potential is huge as Suresh Gupta (2006) notes: "Our research indicates that when used in conjunction with offshoring, componentization can deliver massive benefits. This model assumes three important capabilities: disaggregating (and digitizing) a process into self-contained components and using broadband to ship them offshore; processing each component using best mix of offshore resources and shipping them back to the original location; and reassembling the 'processed' components into a coherent whole" (p. 45, "Financial Services Factory," *Journal of Financial Transformation,* 18, 45).
19. We are grateful to Ian Jones, Innovation and Engagement Officer, Cardiff School of Social Sciences, for the term "permission to think," which he has used in discussion with Phil Brown.
20. Samsung Company Report. Retrieved from http://www.samsung.com/AboutSAMSUNG/ValuesPhilosophy/DigitalVision/index.htm

7

The National Knowledge Commission: Education and the Future of India

Sam Pitroda

It is widely recognized that the economies of the new generation will be powered by knowledge systems and innovative solutions in the education field. To empower future generations, we will need to herald a paradigm shift in the knowledge structures, with a focus on education and skill development. This change will provide the impetus for new growth opportunities and lift many individuals to a better quality of life. The need for such a systemic change in the knowledge sector was addressed at the National Knowledge Commission (NKC) in India, a high-level advisory body that was set up by the Prime Minister of India in 2005 to provide a blueprint for reform of the knowledge institutions and infrastructure in the country. To outline this roadmap for reform, the Commission focused on five core aspects of knowledge: enhancing access to knowledge, reinvigorating institutions where knowledge concepts are imparted, building a world class environment for creation of knowledge, promoting applications of knowledge for sustained and inclusive growth, and using knowledge applications in efficient delivery of public services.

Under these five focus areas, the Commission has submitted around 300 recommendations on various subjects related to the Right to Education, Languages, Translation, Libraries, National Knowledge Network, Portals, Health Information Network, School Education, Vocational Education and Training, Higher

Education, More Students in Science and Mathematics, Professional Education, More Quality Ph.D.s, Open and Distance Education, Open Education Resources, Intellectual Property Rights, Legal Framework for Publicly Funded Research, National Science and Social Science Foundation, Innovation, Entrepreneurship, Traditional Health Systems, Agriculture, Enhancing Quality of Life and E-governance. These recommendations have been formulated after consultations with numerous experts and stakeholders, both within and outside the Government.

The National Knowledge Commission's recommendations have been crafted to achieve the objective of tapping into India's enormous reservoir of knowledge, to mobilize national talent, and create an empowered generation with access to tremendous possibilities. Building a vibrant and innovative education system is crucial in a country such as India in order to face the daunting challenges posed by demography, disparity, and development. In the past two decades while our economy has made significant strides, our education system has not kept pace with the aspirations of the 550 million below the age of 25. Consequently we have not been able to harness our greatest asset—our human capital. By recommending reforms in the education and associated sectors, our aim has been to provide a platform to harness this demographic dividend, which has the ability to change the course of development in the country. India is a country of glaring disparities, with a significant number of people on the periphery of the development process. To assimilate those on the margins, we need a massive expansion in educational opportunities, with a special emphasis on inclusion to ensure that nobody is left out of the system. Finally, to set the country on the path of sustainable development, we need to invest in an educational system that nourishes innovation and entrepreneurship and addresses the skill requirements of an economy on the move.

Access to quality school education is imperative to create the foundation of a vibrant knowledge society. To address the issues of low levels of enrolment and high drop-out rates in India, NKC has recommended a central legislation affirming the Right to Education. To become effective, this legislation must entail a financial provision requiring the central government to provide the bulk of the additional funds needed for realizing the Right to Education. Unless steps are taken to include children who are left out of the system, we as a nation will never be able to move toward equitable growth. However, apart from enhancing access we need to take steps toward improving the kind of education being imparted, the teaching strategies being employed, and the overall direction of our schooling system. To this end, NKC has recommended a massive expansion at the elementary and secondary levels as well as generational changes in the school system with special emphasis on local autonomy in management of schools, decentralization, and flexibility in the disbursal of funds. Further, we have also suggested changes in the curricula and examination systems to encourage conceptual understanding of subjects and a move away from an assessment system that rewards rote learning. With English being seen as an important determinant of access to higher education and employment possibilities, NKC has recommended that English language teaching should be introduced, along with the first language, from Class I. We at NKC strongly feel that a strong foundation in the pure sciences is crucial for India's transformation into a knowledge economy. Consequently, NKC has underlined the need for due emphasis on teaching subjects such as math and science, which are usually perceived as difficult, in ways that encourage students to take them up. Most importantly, ICT should be leveraged in a big way by students and teachers to supplement learning, and by administrators to facilitate transparency in the system.

In order to transform a nation into a knowledge society with the capability to compete in the global market, a focused agenda for skill development is imperative. There is a significant gap between what the education system is producing and the skill requirements of the market place. Lack of vocational skills is a critical problem in India. NSS data (61st round 2004–2005) indicate that only 2% of the individuals in the labor force between the ages 15 and 29 have received formal vocational training and that another 8% are reported to have received non-formal vocational training. This figure is far higher in developed countries: 96% in South Korea, 80% in Japan, 75% in Germany, 68% in the UK and even in developing countries, 28% in Mexico, 22% in Botswana. To avail the fruits of our demographic dividend and to strengthen the link between education and employability, we need to overhaul the system of Vocational Education and Training (VET) in the country. NKC has recommended to increase the flexibility of VET within the mainstream education system by providing due linkages with school and higher education. Further, to effectively provide quality skill development, NKC has suggested steps to expand capacity through innovative delivery models, including robust public private partnerships. Given that only 7% of the country's labor force is in the organized sector, training options available for the unorganized and informal sector need to be enhanced for improving the productivity of the bulk of our working population. We have also underlined the need for a robust regulatory and accreditation framework, along with proper certification of vocational education and training. This formalization of the education process will allow easier mobility into higher education streams, thus enhancing the value of such training.

To further encourage mobility and flexibility in VET, NKC has also recommended creating models for community colleges that provide credit and non-credit courses leading to two-year associate degrees. These would include general education programs as well as employment oriented programs, creating the flexibility for students to pursue higher education later in life.

At the other end of the spectrum, in Higher Education, the Commission has recommended reforms to address the concerns of Expansion, Excellence, and Inclusion in the system. Our current Gross Enrolment Ratio (GER) for higher education (percentage of the 18 to 24 age group enrolled in a higher education institution) is around 10% to 11%, whereas it is 25% for many other developing countries. The quality of education in the higher education sector is highly uneven, both in the government-financed and private unaided sector, showing very poor standards. Further, the higher education system is ill equipped to face the challenge of inclusion. There are large disparities in enrolment rates across states, urban and rural areas, sex, caste, and poor-non-poor.

These are significant challenges that require a comprehensive and committed response. NKC has recommended increasing the GER in higher education to 15% by 2015 through a massive expansion of the system. This expansion can be brought about by setting up new universities and by restructuring existing ones. In addition to increased public spending, this endeavor would require diversifying sources of financing to encourage private participation. Further, to reduce the current barriers to entry, NKC has recommended an Independent Regulatory Authority for Higher Education. This office would create a new paradigm of governance which will encourage openness and transparency and remove the cumbersome entry barriers to new institutions. To enhance quality in the system, NKC has recommended reforms in existing universities to ensure frequent curriculum revisions, the introduction of the course credit system, and reliance on internal assessment. Reforms have also been suggested to improve the quality of faculty as well as the structures of governance.

To ensure that all deserving students have access to higher education, irrespective of their socio-economic background, NKC has also recommended strategies for inclusion. While the government heavily subsidizes university education by keeping fees low, there is better value created for this subsidization by ensuring well funded scholarships and affirmative action that takes into account the multidimensionality of deprivation. Qualitative and quantitative reforms have also been suggested in professional education streams, which are also plagued by similar problems. Reforming the higher education landscape would go a long way in ensuring that we get the demographic dividends from our growing young population.

Further, more than one-fifth of the students enrolled in higher education are in the ODE stream. Keeping this fact in mind, NKC recommendations have also focused on developing and strengthening the Open and Distance education stream and Open Educational Resources.

The quality of original research in a country is a crucial component of a competitive knowledge and skills economy. To invigorate research in the country,

NKC has recommended steps to improve the quality of doctoral programs. It has suggested massive investment in education and research at all levels, together with the reform of the university system, and the fostering of a global outlook in research. Further, NKC has recommended revitalizing the synergies between teaching and research to enrich both the streams. Stand alone research institutions have moved research out of universities, and we need to foster a change in mindsets that would reverse this trend and make universities the hub of research once again and to integrate and enhance teaching and research.

The Commission has also recommended setting up a National Knowledge Network (NKN) linking all research and education institutions in the country through a high-width broadband network that would allow live data and resource sharing. This network is now being implemented by the Government and will be a revolutionary step toward altering the research landscape in the country.

This development agenda will not be complete unless associated sectors that impact education and learning are also revitalized, including libraries, translation activities, the IPR regime, knowledge networking, and the innovation and entrepreneurship eco-systems in the country. These fundamental reforms will require strong commitment and political will, as well as an effective implementation of strategies at the central and state levels. Education will be the key to future development, and, fortunately, we as a nation are realizing the need for changing our mindsets and opening up to experimentation, new delivery systems, new processes and organizational innovation in the knowledge sector. Only if we begin the process of reform now will we see real dividends for our future generations.

8

Education in the Learning Economy: A European Perspective

Bengt-Åke Lundvall
Palle Rasmussen
Edward Lorenz

Introduction

Over the last decades the discourse about international competitiveness and wealth creation has changed, pointing to *knowledge* as the strategic factor (OECD, 2000). In this chapter we argue that different perspectives on how knowledge drives economic growth point to different challenges for Europe's education systems. It makes a major difference whether economic growth is seen as being fuelled by investments in codified scientific and technological knowledge, or whether it is seen as being driven by learning processes resulting in a combination of codified and tacit knowledge. The former perspective—often captured by the term *the knowledge-based economy*—can most clearly be seen in international benchmarking exercises where lags in performance between Europe and the United States and Japan are accounted for mainly in terms of such science and technology indicators as expenditures on research and development (R&D) and the number of third-level science and technology graduates. The latter perspective—captured by the term *the learning economy*—can be seen in work focusing on the way informal networking relations, practical problem-solving on the job, and investments in life-long learning contribute to competence building.

If the most important role of schools were to implant codified knowledge (embraining knowledge), the emphasis would be on teaching skills in mathematics and science. If the role of schools were seen as preparing students for lifelong learning, the emphasis would be on linking theory to practice, and more attention would be given to "relational learning" and formation of other "personal skills" (embodying knowledge).

The first perspective evokes the image of a traditional school isolated from society, where teaching is organized according to traditional disciplines with use of traditional educational methods, while the second perspective evokes a picture of a school that is open to the rest of society, with room for interdisciplinary activities and learning based on practical problems. These two ways of organizing learning may be called mode 1 and mode 2 learning. We will argue that the current context of a "learning economy" requires a movement toward mode 2 learning in most European countries. However, we recognize that a mix of the two modes is necessary and that diversity among institutions will often be better than standardization.

The knowledge-based and the learning economy perspectives have in common that they point to an inherent tendency toward a polarization in labor markets between people with strong and those with weak educational foundations. Against this background we discuss how a movement towards mode 2 learning may be combined with strategies that aim at reducing inequality. We argue that the reform strategy that is the most adequate will be different in European education systems with more elitist characteristics than in those with fewer, and with different emphases on formal third-level education versus vocational education and training.

In the next section we make distinctions between different categories of knowledge and discuss how they may be used to specify different perspectives on education. The section following that, "The Learning Economy," introduces the learning economy and discusses what role schools may play in this context. The section entitled "How Europe's Economies Learn" shows dramatic differences in how European economies work and learn in the workplace and links such differences to differences in educational patterns. "Inequality, Learning and Economic Development" presents data on the inequality of access to workplace learning in different parts of Europe.

Then "Mainstream Education and Types of Knowledge" discusses the characteristics of mainstream institutionalized education, national differences in European education, and the consequences for knowledge and learning. "The Problem of Equality in Education" discusses the tension between the principles of mode 2 learning and educational equality. Finally, "What Role for the Education System in the Learning Economy?" reflects on the dangers of market regulation and emphasizes the necessity to combine mode 1 and mode 2 perspectives on knowledge, learning, and education.

What Kind of Knowledge Matters for the Economy?

It has become commonplace among policy-makers to refer to the current period as a knowledge-based economy, and increasingly it is emphasized that the most promising strategy for economic growth is one aiming at strengthening the knowledge base of the economy.[1] This discourse raises questions about what kinds of knowledge matters for the economy. In this section we try to answer this question by introducing some basic concepts related to the role of knowledge and learning in the economy.

Is Knowledge a Public Good?

Is knowledge easy to transfer? This issue is at the core of two different strands of economic debate, one of them emphasizing transferability and another emphasizing that knowledge may be local and difficult to move from one context to another.

On the one hand, it has been argued that knowledge is a public good that can flow freely in time and space (Nelson, 1959; Arrow, 1962a, 1962b). When this is the case, there is a need for public intervention that supports the production of knowledge through subsidies, through public production, or through legal protection of intellectual property rights.

On the other hand, Marshall (1923), who was concerned to explain the real-world phenomenon of *industrial districts*, emphasized the local character of knowledge. He found that specific specialized industries were concentrated in certain regions and that such industrial districts remained competitive for long historical periods. The modern Silicon Valley phenomenon has resulted in a renaissance for this perspective among industrial and regional economists over the last decades.

In order to understand the apparent contradiction between the two perspectives, it is necessary to make distinctions between different kinds of knowledge.

Tacit and Explicit Knowledge

There is a link between the debate on knowledge as public good and the *tacitness* of knowledge (Cowan et al., 2000; Johnson et al., 2002). It has been assumed that the more knowledge is tacit, the more difficult it is to share it between people, firms, and regions. The reason why industrial districts remain successful in specific fields is that they flourish on the basis of tacit knowledge (according to Marshall, "the secrets of industry are in the air"; we would add that they are also implanted in the backbone of the people living in the region).

Tacit knowledge is knowledge that has not been documented and made explicit by the persons who use and control it (Polanyi, 1958/1978, 1966). The fact that a certain piece of knowledge is tacit does not rule out the possibility of making it explicit—to codify the knowledge—if incentives to do so are strong enough. To make this clear, it is useful to distinguish between tacit knowledge that can be made explicit—tacit for lack of incentives—and knowledge that cannot be made explicit—tacit by principle (Cowan et al., 2000).

Sectors where the knowledge base is dominated by non-codified knowledge may be sectors where systematic progress toward more efficient practices is slow and difficult. Economists have used *education* as a typical example of a production process characterized by tacit techniques (Murnane & Nelson, 1984). The OECD (2000) presents a unique attempt to compare the production, diffusion and use of knowledge across some important sectors—with health and education as the most prominent.

Behind the OECD (2000) study there is an assumption that the education sector could learn from other sectors and especially from the health sector in developing a more codified, scientific, and evidence-based knowledge background. One of the interesting outcomes of the study was that the education sector was not the only field where the degree of codification of knowledge was weak. Disciplines related to management were in a similar situation. The failure of IBM, Microsoft, and Hewlett-Packard to develop management information systems that could substitute for "the art of managing" is quite striking since the economic resources as well as the incentives to do so are enormous in these cases. This failure helps to understand that, while increased efforts to establish a formal, evidence-based knowledge may be useful, some aspects of education (and management) referring to human interaction may be inherently difficult to codify.

Four Different Kinds of Knowledge

Tacit and codified knowledge is an important distinction, but it is somewhat abstract, and to understand this distinction more concretely, it is helpful to introduce categories that are more concrete. We propose a taxonomy of knowledge where it is divided into four categories (Lundvall & Johnson, 1994)[2]:

- *Know-what* refers to knowledge about "facts." Here, knowledge is close to what is normally called information—it can be broken down into bits and communicated as data.

- *Know-why* refers to knowledge about the nature of causality in the human mind and in society. This kind of knowledge is important for technological development in science-based industries.
- *Know-how* refers to the ability to do something. It may be related to the skills of artisans and workers. However, it actually plays a role in all economic activities, including science and management.
- *Know-who* involves information about who knows what and who knows what to do as well as the social ability to cooperate and communicate with different kinds of people and experts.

There are important differences in the degree to which these four categories of knowledge can be codified and in how education systems are affected by the degree of codification. Databases can bring together *know-what* in a more or less user-friendly form. The need to learn this kind of knowledge at school may become less important while it becomes more important to learn to use information technology including search machines (Shapiro & Varian, 1999). Another reason for giving less weight to the details of factual information is that rapid change makes this kind of knowledge obsolete, while the capacity to locate, assess, and use factual knowledge becomes increasingly important.

Scientific work aims at producing theoretical models of the *know-why* type, and some of this work is highly codified and globally accessible—but often only to experts operating in a specific field who can read the specialized codes and make use of it. Furthermore, scientific work is based not only on codified knowledge. The personal knowledge of the scientist is crucial for research outcomes (Polanyi, 1958/1978). The role of *know-how* is even more important in engineering. Often technology is used to solve problems or perform functions without engineers having a clear model or scientific understanding of why certain solutions work (Vincenti, 1990).

The explosive growth of productivity within the sciences makes it a major challenge to mediate such knowledge in the education system. Rather than trying to cover all new theoretical developments, it is necessary to define and teach "basic tools" and "basic perspectives." The way to link the theoretical universe to practical problems may be to let students conduct experiments and find technical solutions to specific problems. Letting students go into some depth in a specific real-world problem and giving them a chance to make use of theoretical and engineering knowledge is a way of fostering a more coherent understanding of theory and practice.

Know-how always has tacit elements. It is impossible completely to separate the competence from the person or organization that acts. The outstanding expert—cook, violinist, physician, manager—may write a book explaining how to do things, but what is done by the new beginner on the basis of that explanation is, of course, less perfect than what the expert would produce. Attempts to use information technology to develop expert systems show that it is difficult and costly to transform expert skills into information that can be used by others (Dreyfus & Dreyfus, 1986). It can be done only in reasonably stable contexts. The transformation always involves changes in the content of the expert knowledge (Hatchuel & Weil, 1995).

It is obvious that know-how is crucial for economic performance. Employers hire people not because they know something but because they can do something. *Know-what* may be crucial for many professionals (doctors learn Latin terms for all parts of the body and lawyers learn references to a series of critical legal cases), but it is the use of this knowledge as supporting know-how that counts for the performance of the organization. The same is true for know-why knowledge (at a minimum, the knower needs the capability to express his or her knowledge and communicate it to others). In spite of the importance of know-how, attempts to evaluate learning effects or performance of education programs tend to focus more on know-what and know-why than they do on know-how. The kind of know-how (solving mathematical problems or reading literary texts) that is tested is often quite different from the kind of know-how required in real-world situations, and so is the test situation.

Know-how is often implicitly regarded in Western society as a kind of knowledge of secondary importance and often it is associated with the practical skills of blue-collar workers. This is a serious mistake since it is at the core of the knowledge that outstanding managers and scientists use when they manage or do research. Much of it has to be learned in practice, but the educational methods and programs that are used may make the transition from school to real-world practice more or less difficult. The elements of linking theory to practice through problem-based learning and through integrating periods of external activities into study programs are important in making the transition less cumbersome.

Know-who refers to a combination of information and social relationships. Telephone directories that list professions, as well as databases that list producers of certain goods and services, are in the public domain and can, in principle, be accessed by anyone. In the economic sphere, however, it is extremely important to obtain quite specialized competencies and to find the most reliable experts—hence the enormous importance of good personal relationships with key persons one can trust.

Education systems differ radically in terms of how they develop social skills and how far they promote "relational learning" where students learn to collaborate and communicate. If the system is based on individual competition it may undermine know-who competence. Elite systems where the elite get clearly separated from the rest promote a certain type of know-who knowledge.

On the one hand, the elite graduate who has passed through the "Grandes Ecoles" in France will be part of a closely knit community, but the social distance to ordinary workers will be correspondingly large. The education systems in the Nordic countries where there are several ways to join the upper class and where the separation between theory and practice is smaller produce less social distance and more "generalised trust." As we will see below, the mode of organization and learning in different European countries will reflect such differences.

The Learning Economy

In various contexts we have introduced an interpretation of what has actually taken place in the economy over the last few decades under the heading "the learning economy" (Lundvall & Johnson, 1994). The intention is to mark a distinction from the more generally used term "the knowledge-based economy." The learning-economy concept signals that the most important trend shift is not the more intensive use of knowledge in the economy but rather that *knowledge becomes obsolete more rapidly than before*; therefore, it is imperative that firms engage in organizational learning and that workers constantly attain new competences.

The speed-up of change can be illustrated by the fact that it is claimed that half of the skills that a computer engineer has obtained during his or her education will have become obsolete one year after the exam has been passed, while the "halving period" for other wage earners with higher education is estimated to be eight years (Danish Ministry of Education, 1997, p. 56).

The transition to a learning economy confronts individuals and organizations with new demands and it has important *implications for education.* The most obvious is that the education system needs to give attention to *enhancing the learning capacity* of students. This objective does not conflict with teaching basic tools and complex bodies of theory. However, it implies that the way teachers teach and the way students learn become crucial since the methods used affect the future learning capability of the student.

A second major implication is that education institutions need to be ready to support *continuous and lifelong learning.* Especially in fast-moving fields of knowledge, there is a need to give regular and frequent opportunities for experts to renew their professional knowledge. The current boom in MBA and MPA programs like Master of Business Administration and Master of Public Administration may be seen as indicating the growing insight among individuals and in management that continuously renewing competences is of great importance. So far, though, they tend to operate mainly in relation to management functions. Similar programs are needed in many other areas where "effective demand" is less strong.

Finally, rapid change in science and technology and the need to move quickly from invention to innovation present a strong argument for keeping a reasonably *close connection between education and research*, especially in higher education. Teachers who have little or obsolete knowledge about what is going on in current research are not helpful when it comes to giving students useful insights in dynamic knowledge fields.

Skill Requirements and Organisational Change—New Challenges for Education

Since one major role of schools is to educate and provide qualified labor, it is important to capture new tendencies in skill and competence requirements. In this section we take a closer look at how management in a selection of Danish firms refer to changes in the content of work in the 1990s. It is of particular interest to focus on changes in the competences demanded within firms that have engaged in organizational change, since change is oriented towards establishing "learning organization" and thus may be seen as responding to the emergence of a learning economy.

A series of surveys of the Danish national innovation system (the DISKO surveys) showed, among other results, that organizational change gives rise to new demands for qualifications. Table 1 reveals substantial differences in the pattern of answers between the firms that have introduced new forms of organization and those that have not (percentage of firms with numbers in parentheses). The importance of general skills reflected in growing demands for independence in the work situation, for cooperation with external partners, especially customers, and for cooperation with management and colleagues, has grown remarkably in firms that have pursued organizational change and much less so in firms that have not changed their organization. There are correspondingly large differences between the two types of organization in the rate of occurrence of a reduction in routine work.

Table 1. Changes in task content for employees in the period 1993–1995 for firms that have made organizational changes, compared with firms that have not made organizational changes (in parentheses).*

	More	Less	Unchanged	No answer
a. Independence of work	72.6 (37.1)	4.2 (2.7)	21.2 (56.3)	2.0 (3.8)
b. Professional qualifications	56.4 (36.3)	7.5 (5.3)	33.3 (53.8)	2.8 (4.4)
c. Degree of specialisation	33.9 (26.2)	20.8 (7.8)	39.3 (58.4)	6.0 (7.5)
d. Routine character of tasks	5.6 (8.2)	41.8 (15.5)	45.0 (67.1)	7.7 (9.1)
e. Customer contact	51.6 (29.3)	5.1 (3.1)	37.2 (59.9)	6.1 (7.6)
f. Contact with suppliers	34.9 (18.0)	7.1 (4.3)	46.4 (62.0)	11.6 (15.6)
g. Contact with other firms	24.7 (14.0)	5.5 (4.3)	56.8 (68.9)	13.0 (13.7)
h. Cooperation with colleagues	59.1 (27.1)	5.8 (4.5)	31.8 (63.3)	3.2 (5.0)
i. Cooperation with management	64.9 (28.6)	5.9 (4.2)	26.1 (62.2)	3.1 (4.9)

Source: Voxted 1999, DISKO-Survey, *n* = 952 (981).

*Management representatives in 4000 Danish private firms, excluding agriculture, were asked: "Did the firm introduce a non-trivial change in the organisation in the period 1993–95?" The response rate was close to 50%. For more detailed information, see Lundvall (2002a).

All firms, and especially those that engage in organizational change, require that employees can communicate and collaborate internally and externally. If education is to respond to that, more efforts should be made to prepare students for communicating and cooperating with experts from other disciplines. Firms that are change oriented are at the same time asking for both more and less specialization. This may be seen as an argument in favor of a differentiation in teaching methods and study programs among the institutions in the educational system. The ideal team for a development task is probably a combination of workers trained at institutions that are more discipline-oriented with workers from schools where education makes more use of interdisciplinary problem-based learning.

In general, it might be a good idea to allow for institutional diversity in the education system. It can be argued that this kind of differentiation may be difficult to cope with for employers who hire graduates. On the other hand, the differentiation helps the society to evolve in a complex and rapidly changing global context. It gives the system an ability to readjust and respond to the complex and differentiated emerging social and economic needs.[3]

These are implications for fast-changing and wealthy societies with strong emphasis on innovation and learning. In the next session we will introduce some European-wide data to show that workplace learning takes place differently in different European countries with different patterns of education and that, therefore, the challenges for educational reform may be very different in different national systems.

How Europe's Economies Learn

Lorenz & Valeyre (2006) developed an original and informative EU-wide mapping of how employees work and learn in the private sector. This template has wider implications for the pattern of economic growth in different parts of Europe. In Arundel et al. (2007) international comparisons show that there is a positive correlation between the national share of private sector workers engaged in advanced forms of learning and the percentage of private sector enterprises doing more radical forms of innovation involving the introduction of new products or processes into international markets. In this section, we will focus upon international differences relevant for national education systems.

The Third European Survey of Working Conditions on which the mapping is based was directed to approximately 1,500 active persons in each country, with the exception of Luxembourg which had only 500 respondents. The analysis presented here is based on the responses of the 8081 salaried employees working in establishments with at least 10 persons in both industry and services.

Cluster analysis is used to identify four different systems of work organization:

- Discretionary learning (DL)
- Lean
- Taylorist
- Traditional forms

Two of these, the discretionary learning and lean forms, are characterized by high levels of learning and problem-solving in work. The principal difference between the discretionary learning and the lean clusters is the relatively high level of discretion or autonomy in work exercised by employees grouped in the former. Task complexity is also higher in the discretionary learning cluster than in the lean cluster.

Discretionary learning thus refers to work settings where a lot of responsibility is allocated to the employee who is expected to solve problems on his or her own. Employees operating in these modes are constantly confronted with "disequilibria," and as they cope with those, they learn and become more competent. However, in this process they also experience the fact that some of their earlier insights and skills become obsolete.

Lean production also involves problem-solving and learning, but here the problems are more narrowly defined and the set of possible solutions less wide and less diverse. The work is highly constrained, and this points to a more structured or bureaucratic style of organizational learning that corresponds rather closely to the characteristics of the Japanese or "lean production" model.

The other two clusters are characterized by relatively low levels of learning and problem-solving. The Taylorist form leaves very little autonomy to the employee in making decisions. In the traditional cluster there is more autonomy, but learning and task complexity is the lowest among the four types of work organization. This cluster includes employees working in small-scale establishments in personal services and transport where methods are for the most part informal and non-codified.

Table 2. National differences in organizational models (percentage of employees by organizational class).

	Discretionary learning	Lean production learning	Taylorist organisation	Simple organisation
North				
Netherlands	64.0	17.2	5.3	13.5
Denmark	60.0	21.9	6.8	11.3
Sweden	52.6	18.5	7.1	21.7
Finland	47.8	27.6	12.5	12.1
Centre				
Austria	47.5	21.5	13.1	18.0
Germany	44.3	19.6	14.3	21.9
Luxemb.	42.8	25.4	11.9	20.0
Belgium	38.9	25.1	13.9	22.1
France	38.0	33.3	11.1	17.7
West				
United Kingdom	34.8	40.6	10.9	13.7
Ireland	24.0	37.8	20.7	17.6
South				
Italy	30.0	23.6	20.9	25.4
Portugal	26.1	28.1	23.0	22.8
Spain	20.1	38.8	18.5	22.5
Greece	18.7	25.6	28.0	27.7
EU-15	39.1	28.2	13.6	19.1

Source: Adapted version based on Lorenz & Valeyre (2006)

Table 2 shows that people working in different national systems of innovation and competence work and learn differently. Discretionary learning is most widely diffused in the Netherlands, the Nordic countries, and to a lesser extent in Austria and Germany. The lean model is most in evidence in the UK, Ireland, and Spain.

The Taylorist forms are more present in Portugal, Spain, Greece, and Italy, while the traditional forms are similarly more in evidence in these four southern European countries.[4] Within the Nordic group, Denmark is extreme in terms of its high share of discretionary learning and low share of Taylorist workplaces. The share of discretionary learning is higher in Germany than it is in the UK and France.

Table 2 indicates unequal access to learning in different parts of Europe. The fact that there is broader access to learning environments in the wealthier countries may contribute to increasing inequality within Europe. It is interesting to see how the national learning modes relate to national patterns of education. In this respect, we will draw upon recent work by Lorenz (2006).

Education and Training for Learning Organizations

Since discretionary learning depends on the capacity of employees to undertake complex problem-solving tasks, it can be expected that nations with a high frequency of these forms will have made substantial investments in education and training. In what follows, we compare tertiary education in universities and other institutions of higher education with the continuing vocational training offered by enterprises.

Tertiary education develops both problem-solving skills and formal and transferable technical and scientific skills. While most of the qualifications acquired through third-level education will be relatively general and hence transferable on the labor market, the qualifications an employee acquires through continuing vocational training will be more firm specific. Some of this training will be designed to renew employees' technical skills and knowledge in order to respond to the firm's requirements in terms of ongoing product and process innovation.

Figure 1 shows the correlations between the frequency of the discretionary learning forms and two of the four measures of human resources for innovation used in Trendchart's innovation benchmarking exercise: the proportion of the population with third-level education; and the number of science and engineering graduates since 1993 as a percentage of the population aged 20 to 29 years in 2000. The results show a modest positive correlation (R-squared = .26) between the discretionary learning forms and the percentage of the population with third-level education, and no discernible correlation between the discretionary learning forms and the measure of the importance of new science and engineering graduates.

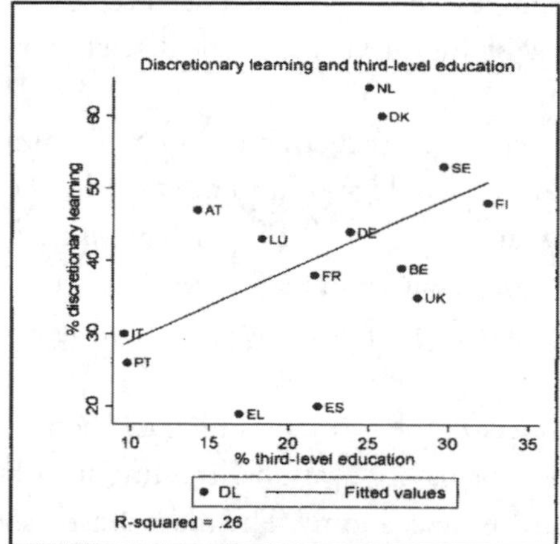

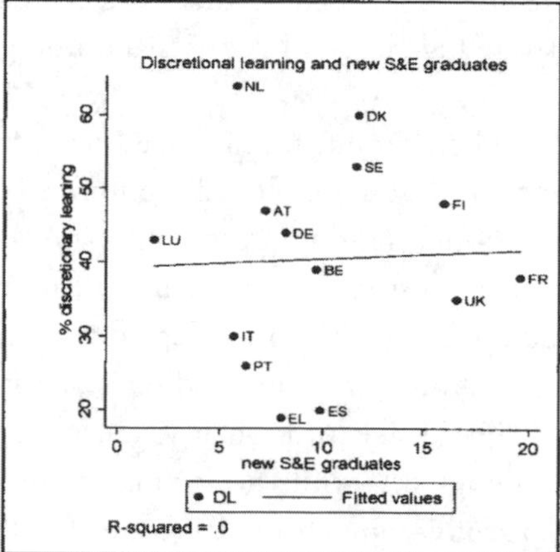

Figure 1. Discretionary learning and tertiary education

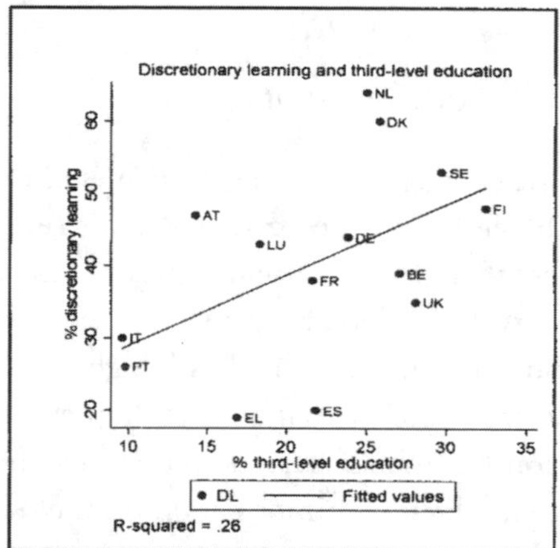

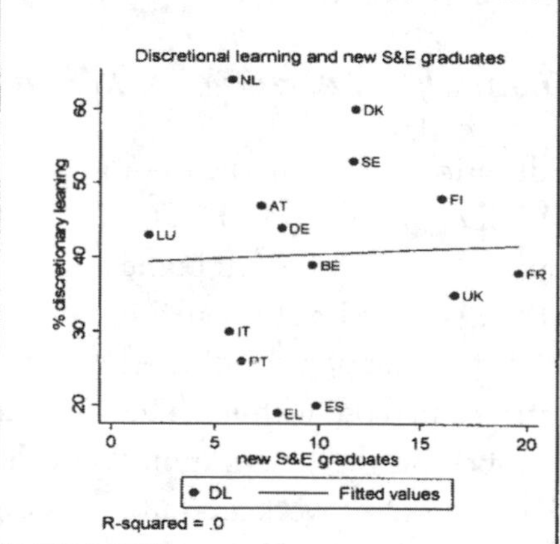

Figure 2. Discretionary learning and employee vocational training

Figure 2 shows that there are fairly strong positive correlations (R-squared = .75 and .52, respectively) between the frequency of the discretionary learning forms and two measures of firms' investments in continuing vocational training: the percentage of all firms offering such training, and the participants in continuing vocational education as a percentage of employees in all enterprises. The results suggest that these forms of firm-specific training are key complementary resources in the development of the firm's capacity for knowledge exploration and innovation. The figure also points to a strong north/south divide within Europe. The four less technologically developed southern nations are characterized both by low levels of continuing vocational training and by low use of discretionary learning, while the more developed northern and central European nations are characterised by relatively high levels of vocational training and by high-level use of the discretionary learning forms.

These results indicate that one bottleneck for constructing learning organizations in the less developed economies of Europe would appear to be at the level of vocational training. Portugal, Spain, Italy, and Greece, all of which have made important strides in increasing the number of science and engineering graduates, stand out for their low levels of investment in continuing vocational training and for ranking lowest on the discretionary learning scale.

National educational systems that emphasize the formal training of scientists and engineers while neglecting the broader forms of vocational training may be especially vulnerable in the context of the learning economy. The more drastic the status difference and distinction between theory and practice in education programs, the more difficult it will be to install participatory learning in the private sector.

Inequality, Learning, and Economic Development

Education matters in the labor market. One of the most striking outcomes of the OECD's Jobs Study (OECD, 1994) was that it documented that people with education became better off in the labor market than people without education in *all* OECD-countries in the period from 1985 to 1992. Either income differences or employment opportunities became more unequal between those with higher education and those without. The Jobs Study referred to trade and technology as factors behind the polarization. Globalization brought increased competition for many unskilled workers, and it was especially the use of information technology that tended to benefit educated employees.

We have proposed an alternative model where globalization and new technologies, together with other factors such as deregulation and privatization, speed up the rate of change and give rise to "a learning divide." The learning divide indicates both that people with education are better prepared to engage in learning and that they get more opportunities to do so. We thus propose that the learning economy perspective captures how these and other factors are combined in a new model of economic growth characterized by growing inequality (Lundvall, 1996).

In Lundvall (2002) we find a strong "Matthews's effect" in the distribution of training opportunities in Danish firms. Employees with higher education were offered opportunities to attend courses much more frequently than were workers with less formal education. As we will see in the next section, access to "discretionary learning" is easier for managers than for workers. A different way to capture the learning divide where the emphasis is on the productivity of educated workers has been proposed by Nelson and Phelps (1966) and Schultz (1975).

Nelson and Phelps (1966)assume that people with higher education contribute to economic growth through two mechanisms. First, they are able to pursue regular activities more efficiently than the average worker. Second, and here is the new insight brought by the article, they are more competent when it comes to exploiting new technical opportunities in the economy. Schultz (1975) follows a similar line of thought but takes the reasoning some steps further. Schultz assumes that education makes individuals better prepared to "deal with disequilibria."

With reference to Nelson and Phelps, we would assume that the relative demand for higher education would increase as the rate of change accelerates in the learning economy. As mentioned, this was the case in *all* OECD countries in the period 1985–1995.

Access to Organizational Learning in Europe

In order to develop a measure of the social distribution of workplace learning opportunities, we distinguish in Table 3, for the group of employees having access to discretionary learning, between the share of "managers" and "workers" having such access.[5]

Table 3 shows that employees at the high end of the professional hierarchy have more easy access to jobs involving discretionary learning. This rule of thumb applies to all the countries listed.

However, it is also noteworthy that the data indicate different learning modes in different countries. In the more developed economies (with the exception of France and the UK), we find that the inequality in the distribution of learning opportunities is moderate, while it is very substantial in the less developed south. For instance, the proportion of the management category engaged in discretionary learning in Portugal is almost as high as in Finland (62% in Finland and 59% in Portugal), but the proportion of workers engaged in discretionary learning is much lower in Portugal (18.2% versus 38.2%).

Table 3. National differences in organizational models (percentage of employees by organizational class)

	Discretionary learning	Share of managers in discretionary learning	Share of workers in discretionary learning	Learning Inequality index*
North				
Netherlands	64.0	81.6	51.1	37.3
Denmark	60.0	85.0	56.2	35.9
Sweden	52.6	76.4	38.2	50.3
Finland	47.8	62.0	38.5	37.9
Austria	47.5	74.1	44.6	39.9
Centre				
Germany	44.3	65.4	36.8	43.8
Luxemb.	42.8	70.3	33.1	52.9
Belgium	38.9	65.7	30.8	53.1
France	38.0	66.5	25.4	61.9
West				
UK	34.8	58.9	20.1	65.9
Ireland	24.0	46.7	16.4	64.9
South				
Italy	30.0	63.7	20.8	67.3
Portugal	26.1	59.0	18.2	69.2
Spain	20.1	52.4	19.1	63.5
Greece	18.7	40.4	17.0	57.9

Source: Calculations made by Edward Lorenz specifically for this chapter on the basis of the data referred to in connection with Table 1.

*The index is constructed by dividing the share of workers engaged in discretionary learning by the share of managers engaged in discretionary learning and subtracting the resulting percentage from 100. If the share of workers and managers was the same, the index would equal 0, and if the share of workers was 0, the index would equal 100.

This pattern indicates that a movement toward a learning economy is not necessarily accompanied by growing inequality. At least in the present situation, it is rather associated with reduced inequality in learning opportunities and with more equal access both to formal vocational training and to on-the-job learning. The countries at the top of the table are countries where income inequality is low, and they are highly successful in adapting to the changes imposed upon them by new technologies and new forms of more intense and global competition. Thus, while it might be true that higher education fosters people who are successful as equilibrators and innovators, it is when those people interact with a broader segment of the workforce in promoting or coping with change that the innovation system as a whole turns out to be most efficient.

For the design of education systems in the south of Europe, these observations raise an issue of how education programs can be designed in such a way that the social distance in working life between "management" and "workers" does not become too great. The experience of the Nordic countries, always appearing at the top of global competitiveness assessments, indicates that public policies reducing income inequality may actually promote innovation and growth through their positive effect on the participation in learning and change.

Mainstream Education and Types of Knowledge

Schools, colleges, and universities are institutions organized for the main purpose of transmitting knowledge and facilitating learning. The early history of organized education was closely connected to religious institutions, but the modern concept of schools and schooling was shaped by the Enlightenment and the emergence of nation-states. The states that evolved during the nineteenth century generally shared two features: a steadily growing commitment to national policies for mass schooling and a steady increase in the share of children and young people participating in basic schooling.

In explaining this development, scholars such as Ramirez and Boli (1987) argue that the nation-states were influenced by a common set of factors and followed generally similar doctrines. Nation-states were only to some extent independent units; they functioned within a common cultural framework. In the second half of the nineteenth century, education, citizenship, and social development became closely interwoven themes in a political discourse, which connected mass schooling to political progress. The aim of educational policy-makers was to create culturally homogeneous, loyal, and productive mass citizens.

The educational institutions that emerged had characteristics that are still dominant today. Schools are more or less full-time "work places" for pupils and students in different age groups. Internally they are organized along three main principles: disciplines/subjects; age progression; and classes. The knowledge to be transmitted is organized in school subjects that reflect and draw on some main scientific disciplines, such as mathematics, history, geography, and languages. In primary school all areas of knowledge are supposed to be represented; in secondary and higher education studies become more specialized, students in universities often concentrating on one discipline. Most schools organize students into age groups, on the assumption that the development of learning ability increases with age, so that pupils progress from first grade to second grade to third grade and so on. Moreover, most schools organize the teaching activity in classes where a group of, for instance 20 children, is taught by one teacher. All these elements are tied together by national regulation (more or less detailed) of educational structures and by bureaucratic management at the level of the school. If one looks inside a classroom, one will generally find students in one age group being taught in one subject by one teacher.

To some degree, primary and secondary schooling, vocational education and higher education constitute a common system, tied together by a number of institutional structures and processes. The educational system is not closed; it fulfils certain functions in other areas of social life, and its actors are involved in communication with many other types of actors and institutions. Still, the educational system is characterized by an administrative logic, and by deeply rooted traditions, upheld not only by states but also, for instance, by the teachers as a social group.

Archer (1982) has drawn attention to the significance of what she calls "the principle of sequence"—that is, the principle that completion of the initial levels of schooling gives the right, and often also the impulse, to continue at higher levels within the educational system. The precondition for the operation of the principle of sequence is the establishment of a certain degree of coherence and continuity in the educational system. For example, in Denmark an important step in this direction was taken in 1903, when the "middle school" was introduced as a link between the public primary school and higher secondary education. Archer contends that in an educational system of this kind, growth will be partly self-reinforcing, because the primary actors in the daily life of the school (especially the teachers, who are motivated by professional interests, but also parents and students) will most often work for continued participation in education. As the educational system continues to grow, its "inner life" gets more difficult to control, and the environment for the principle of sequence becomes more favorable.

The type of educational institution and practice described here constitutes the mainstream of the educational system in Europe and indeed in most of the world. In primary education it dominates completely, in secondary education, higher education, and adult education less so. In both secondary and higher education vocational and professional programs and institutions play an important role, and their interaction with businesses and workplaces challenges the academic logic and institutional isolation of schools. Adult education is generally less institutionalized and includes a wide variety of programs, some academic ("second chance" courses), some vocational and some aimed at leisure activities. Adult education and training reflect more distinctly the structures in some main areas of social life: work and the labor market, political institutions, and "civil society" (Rasmussen, 1996; Rasmussen & Rasmussen, 2008).

This description of mainstream education may seem commonplace, but if we are to consider what kind of knowledge is communicated and learned in schools, these fundamentals are important to consider.

The predominant types of knowledge communicated in mainstream education are undoubtedly know-what and know-why. Educational curricula draw their logic and content from academic disciplines, and the ability of students to perform well enough in a given subject to progress to a higher level of education in that subject is probably the main criterion for assessment. Most primary and secondary schools also have the mission of developing more general cultural and social skills, but this has little impact on the organization of knowledge. Basic facts and explanations from scientific disciplines constitute the bulk of knowledge transmitted in schools, and much of this is quickly forgotten (after exams) unless the learners encounter contexts where this knowledge is needed. This temporary acquisition of knowledge is what some researchers (for instance, Entwistle, 1988) have called surface learning, in contrast to deep learning. Of course, deep learning, which may be seen as linked to know-why, does take place in schools, not least in secondary education; but the emphasis on factual knowledge in many curricula restricts the time and opportunities available for deep learning. Moreover, the institutional closure of schools means that type of deep learning cultivated is mostly abstract academic reasoning. For the same reason, the type of know-who available in schools consists mostly of knowledge of names and texts, not of experience of interacting with knowledgeable persons.

Know-how is not absent from schools. Most curricula include a fair amount of skills training, for instance, in calculation, reading, writing, composition, and foreign language use. Through this regiment, learners may acquire valuable basic skills for both work and other areas of life. The limitation is, however, that the skills are trained in the school environment, and beyond the basic level, it is often difficult for learners to transfer the skills for use in other contexts.

The training of such skills in schools is generally not tacit because it is accompanied by explication of rules and principles. However, other types of skills are also taught in schools, resulting in tacit knowledge. This form of knowledge involves hierarchical power relations, of distinctions between academically relevant knowledge and personal experience, of managing homework and classroom participation—in short, the elements of what some researchers have called the hidden curriculum. Children and young people spend a major part of their life in schools, and the social organization and practices in school are a strong socializing force. This dynamic is generally recognized by social scientists and educational researchers, but the assessment of this socialization varies. For instance, Parsons (1967) argued that schooling distances the person from the intimate and personalized environment of the family and thus prepares him or her for life in wider adult society, while Bowles and Gintis (1976) argued that the hierarchical and standardized organization of schooling mirrored industrial workplaces and in fact socialized young people for passive participation in these.

While the basic institutional features of mainstream education are fairly general, there are important differences between national school systems (and also within national systems). Such dimensions have many differences, of which only some are focused on in current transnational debate and policymaking. Much attention is given to national levels of achievement in mathematics and other school subjects (measured in surveys like the Programme for International Student Assessment [PISA] and the Trends in International Mathematics and Science Study [TIMSS]), to national completion rates for different levels of education, and to recruitment for higher education programs in mathematics and science. These dimensions are not unimportant, but they relate mainly to the efficiency of mainstream educational institutions and to the transmission of codified knowledge in certain fields. We argue that in assessing the contribution of education to the learning economy, other dimensions are also vital. We have in mind especially the tacit knowledge present in patterns of culture and communication in schools; the relationship between codified and tacit, academic and practical elements in vocational education; and the provision for and participation in adult education and training. In the next section we will give examples of how education systems in Europe differ in respect to these dimensions.

European Differences

Some important differences between European school systems are highlighted in a comparative study of schools, pupils and learning in the UK, France and Denmark (Osborn et al., 2003). The study used observation, interviews, and surveys to investigate both the experiences and identities of pupils and the cultural and educational contexts for these.

There are many similarities between the school systems of these countries. They have a core of mandatory schooling lasting approximately ten years and curricula mainly consisting of school subjects with much the same content across borders. All three countries have objectives for schooling including the acquisition of knowledge and skills necessary for further study and for employment but also transmission of national culture and advancing personal and social development. However, the balance between and the implementation of these objectives are often determined by the historical and cultural roots of the individual school system.

An example of this dynamic is found in the different principles for organizing pupils in classes and groups. In Denmark, pupils generally remain in the same class most of the years they spend in school; classes are academically heterogeneous but provide a stable social environment. In France, most subjects are also studied together with the same class, but only for one year at a time, because classes are organized on the basis of academic standards. In the UK the organization of pupils differentiates them according both to subjects and to performance levels, so the concept of *class* is weak.

The English school system has its roots in a liberal laissez-faire tradition. It draws on a humanist approach to learning emphasizing spiritual and moral values as well as on traditions of individualism and early specialization. This led to a segregated school system where different types of schooling catered for the needs of different social groups. Later—not least in the 1960s—liberal and egalitarian views took hold, and initiatives have been taken to lessen the hold of grammar schools and universities on the culture of schooling. Osborn et al. (2003) observe that English students are not so positive about schooling and learning and often express a wish to leave school as soon as they can. They acknowledge, however, that their teachers concerned themselves with pupils as people and encouraged them to express their own ideas. They feel that they receive good feedback from teachers about their work and that teachers make them work hard. Such responses reflect the traditional English educational concern with the development of the whole person as well as the current emphasis on individualization and differentiation.

Schooling in France has been organized on the basis of a republican ideal emphasizing values such as universalism, rationalism, and utilitarianism. The state has the responsibility of providing a general school system with equal access for all. Schooling is expected to contribute to the promotion of national values and social solidarity. The French state traditionally exercises detailed control in all matters of primary schooling, including teaching materials. There is a distinct professional distance between teachers on the one side, and pupils and parents on the other. French students in the study found few personal and social elements in their school experience and thought that teachers almost exclusively emphasized cognitive goals. Despite this focus, they lacked guidance from teachers on how to improve their work, and they also had strong concerns about teachers who did not respect pupils. There was a distinct difference between boys and girls, with French boys being less positive about schools and teachers than girls in the study.

In Denmark, basic schooling has embodied strong communitarian values (also present in a cultural emphasis on local democracy and social partnerships), with emphasis on inter-student collaboration, but with less importance attached

to the professional autonomy of teachers. Public primary and secondary schooling emerged in response to the need to educate the rural population, but it was also informed by the ideas of the Enlightenment and later by the tradition of popular education. An important element in the school organization in Denmark has been the role of the "class teacher," a teacher with many teaching hours in the individual class and with a responsibility for social contact with pupils and parents. The Danish students were in general the most positive toward schooling, toward learning, and toward teachers. They saw school as helping them to fit into a group situation rather than emphasizing the development of the individual. On the other hand, they often felt that the teachers did not give the detailed feedback that they needed to improve their skills, and some pupils felt they had to "play down" intellectual effort and achievement in the class community.
Some main findings of this study are summarized in Table 4.

	Denmark	England	France
National policy discourse	Collaboration/ consensus	Differentiation Quasi-market	Universalism Equal entitlement
Teachers	Low distance Expressive	Medium distance Expressive Instrumental	High distance Instrumental
Pupil culture: peer groups	Community within the class 'Organic' solidarity	Differentiation into sub-sets	Solidarity/lack of difference

Adapted from Osborn et al. (2003, p. 215).

Table 4. National differences in policy discourse, teacher roles, and pupil culture

Another area where important differences are found between national education systems in Europe is the organization of vocational education and training (VET) (Münck, 1997; Green et al., 1999). A crucial dimension here is to what degree and how educational programs combine theoretical schooling with practical work training. At one end of this scale we find national systems like those of France and Sweden, dominated by school-based learning; at the other end we find the UK and several countries in southern Europe, where vocational training is largely a question of learning on the job, with little or no theoretical schooling. In between these cases are countries like Denmark and Germany, which each in its own way combines schooling and practical training. These are examples of so-called dual systems that combine school education with workplace learning and sequentially organized programs. In other systems, like that of Norway, a phase of school education is followed by a phase of workplace learning. Historically most of the

European VET systems are descendants of three "classical" training models that emerged in the wake of the first Industrial Revolution: the liberal market economy training model in England; the state bureaucratic model in France; and the corporatist dual vocational training system in Germany (Greinert, 2005).

The relative merits of the different ways of organizing VET have often been debated. It can be argued that vocational schooling in publicly regulated schools has the advantage of providing a broader general and theoretically underpinned approach to the vocation, an approach that also facilitates links to general education. In principle, the competences acquired through this type of education should also be adaptable to changing demands in work, and thus constitute a basis for lifelong learning. However, school-based VET programs are often dominated by a "grammar of schooling" well known from primary education, and this nomenclature undermines the motivation of vocational students and the relevance of the curriculum to real-life work. Training based mainly on workplace learning, on the other hand, provides skills closely linked to the demands and cultures of work, which also facilitates student motivation. The potential disadvantages are low individual flexibility and the risk that acquired skills may soon be outdated.

The different national systems and cultures of schooling and learning cultures imply that different types of competence and different resources for learning are available in the adult population and in the workplaces of these three European countries. The perceptions of learning, of the balance between academic disciplines and other objectives, e.g., collaboration, and of relationships between students and teachers constitute much of the know-how learned in school and taken along into later life situations. The emphasis on collaboration in a trustful environment and on students being recognized as persons in Danish school culture, supplemented by the organized combination of schooling and practical training in Danish VET, fits well with the high level of participation in adult education and the emphasis on discretionary learning in Danish organizations.

The virtual absence of a VET system in the UK probably means that the culture of schools impacts more directly on the relationship to work; and English schools seem to prepare for individual careerism, less for collaboration and trust. This structure may be one of the reasons for the predominance of the lean model of work organization, characterized by relatively low levels of employee autonomy, in UK workplaces (see Tables 2 and 3).

The formal organization and distanced relationship between teachers and students in French schools, combined with the school-based VET system, does little to prepare individuals for participation and initiative in work contexts. This assessment is supported by the findings of the classic comparative study of education and work organization in France and Germany by Maurice et al. (1986), where it was observed that the low social status attached to VET as well as the poor integration of formal training and practical on-the-job training contributed to a more hierarchical organization with a narrower and more specialized employee skill-base. This low-level skill base may be indicative of a lack of interest on the part of the students and thus may account for the limited involvement of French employees in discretionary learning forms of work organization as shown in Table 3.

It should be noted that neither Denmark, France, nor the UK were able to engage all pupils positively in learning, but the discouragement with education seems more prominent in the UK and France, which may have an impact on participation in both vocational and higher education.

The national differences in European educational systems are complex and cannot be reduced to simple polarities. Nevertheless, as stated in the introduction, it is our view that two contrasting principles of learning are generally present, often with one dominating and often distributed unequally among the different levels and sectors of educational systems. The principle we call mode 1 learning (which is often in a dominant position) situates learning in traditional school environments, separate and professionalized institutions where curricula are modelled on the traditional disciplines, and teaching follows traditional methods. The other principle, which we call mode 2 learning,[6] situates learning in more open contexts—for instance, in schools which systematically include practical training or experience in their activities, which reorganize the curriculum interdisciplinary activities and problem-based learning, and which encourage students to work collaboratively, or even demand that they do. There is no doubt that both modes of learning are necessary in all education systems, and that the optimal mix will depend on the national context. However, we find it obvious that a movement towards the learning economy makes it urgent for most European states to strengthen the provision for mode 2 learning throughout their education systems (Guile, 2003).

The Problem of Equality in Education

Our main line of argument in this article is that learning and innovation are crucial to the competitiveness of the economies of Europe; that learning involves different types of knowledge of which the less formalized (knowing how and knowing who) are often no less important than the formalized; and that educational principles and cultures focusing on collaboration, interdisciplinarity, and engagement with real-life problems will prepare people better for flexible and innovative participation in work organizations. However, this line of argument runs counter to the way progressive educational principles have sometimes been assessed in sociological research on education. Here it has been argued that progressive educational principles like group work and project methods favor middle-class students and families.

One of the best known examples of this is Bernstein's analysis of visible and invisible pedagogies and their class basis (Bernstein, 1997). For Bernstein, visible pedagogy is an organization of teaching where the distinction between different knowledge areas or school subjects is clearly drawn, and where the teacher exercises clear (though not necessarily authoritarian) control over what happens in the class, while invisible pedagogy does uphold these clear distinctions. Bernstein sees invisible pedagogy as a continuation of the child-centered and play-related methods of much pre-school pedagogy, and he argues that these educational principles correspond to and give preference to the culture in modern middle-class families. In this culture children learn to navigate flexible relationships between child and adult, between work and leisure, between relevant and irrelevant, and this back-and-forth dynamic helps them manage the role of learner in progressive classrooms. To the children and parents from working-class families, however, the strong distinctions of visible pedagogy and the explicit criteria for assessment that it usually employs are more readily understandable and help them and make school a more transparent context for learning strategies. Educational strategies that aim at promoting learning and innovation through progressive educational principles may thus end up reproducing or even deepening social inequalities.

There is undoubtedly some truth in this argument, and some studies do in fact indicate that children from working-class backgrounds may focus on and achieve better results in science and other "exact" subjects than in more "soft" areas of knowledge (see, for instance, Campbell et al., 1996; Hirata et al., 2006; Iannelli, 2007). However, we find Bernstein's argument too narrow, and that a consideration of other aspects changes the picture.

First, it is necessary to look at the relationship between knowledge in the school curriculum and in the world outside school. If the perspective is confined to the institutionalized setting of the school, it makes sense that a student unacquainted with the culture of the school will look to clear distinctions between school subjects and clear assessment criteria as stable points of reference for his or her learning strategies. However, organizing learning around problems and including practical experience and training in educational programs is not only a reorganization within the school, it also opens the curriculum and the learning processes to the organization of knowledge in other contexts, including, of course, work contexts (Kolb, 1984). While know-how in the school context is often a question of how to manage the student role, know-how in problem-solving projects and practical training becomes professionally relevant knowledge. The invisible pedagogy of the school loses some of its grip on the learning process, and learning becomes less of a foreign territory for students from outside the elites and the middle classes.

Second, equality in schooling is multidimensional. In their discussion of education and equality, Lynch and Baker (2005) distinguish between equality of resources, equality of respect and recognition, equality of power, and equality of love, care, and solidarity. Educational institutions and programs must promote equality in all these dimensions if they are to make real contributions to social equality. Educational achievement, for instance, in examination results or completion rates, is an insufficient measure for levels of equality. If young people are to be prepared and equipped for an active and satisfactory adult life, they also have to experience caring learning environments, recognition, and respect from teachers, managers and co-students, and chances to participate in decision-making. The dominant paradigms of schooling and pedagogy may make it easier for students to achieve results and academic recognition if they neglect their needs in other dimensions, but in a wider perspective this compromise is unhealthy for the individuals and for society. Educational strategies for learning economies are fundamentally a question of empowering employees and citizens to manage their own lives in a modern society.

In sum, we find that the problem of inequality in education must be seen in a wider perspective including both the relationship between knowledge in and outside school and the non-cognitive dimensions of equality. In this perspective, we continue to maintain that educational principles like problem-based learning, group work, and the linking of theoretical and practical learning have the potential to contribute to economic growth through learning and innovation as well as to social equality.

What Role for the Education System in the Learning Economy?

If the regulation of the learning economy is left to market forces, these will tend to open up learning divides in Europe both between regions and within national systems. Growing inequality is problematic not only from a normative perspective where social cohesion is seen as a positive characteristic of society. As we have demonstrated, the most advanced economies that perform well in terms of growth and innovation are those where there is wide participation in organizational learning. Increasing inequality undermines the learning capability of organizations, regions, and countries.

How the national education systems respond to this challenge has to do both with the overall structure of the system and with the micro-processes that take place in schools. The overall structure of the education system will have a major impact on the reproduction or reduction of inequality. One aspect of this is that systems where students at an early stage have to choose between (and be selected for) "practical" and "academic" branches of the school system will tend to increase stratification and reduce mobility and flexibility both in education and in the labor market. Another point is that systems with little provision for vocational education and training programs will not be able to counterbalance the hierarchical bias and short-term perspective evident in much workplace training. A further aspect is that systems where there are alternative ways to reach the highest level of education and where adults who have originally have had little education or training can join education programs and in principle reach the highest level will result in more egalitarian culture than systems aiming at picking and isolating the elite at an early stage.

National differences in the cultural assessment of theoretical and practical knowledge, respectively, constitute an important issue in relation to inequality. In systems where know-why and codified knowledge are held in much higher regard than know-how and experience-based tacit knowledge, a big gap between those with higher education and the rest will tend to open up.

The organization and culture of the basic school system is a crucial element. This is where all members of society acquire not only basic skills and knowledge, but also basic experience with systematic learning and handling of different types of knowledge. The traditional model of schooling—mode 1 learning, as we have called it here—too often fails to give young people skills in and motivation for collaboration, problem-solving, and learning. To many, the experience of basic schooling discourages them from later engagement in education and learning activities. Schools in all European countries need to embrace—in different ways, depending on the national conditions—elements of mode 2 learning in order to prepare young people for participation in the learning economy.

It is obvious that the education system has a key role to play in the learning economy. However, it is important to remember that education systems embrace more (potentially much more) than educational institutions with strongly codified curricula, and that formal education is only one, albeit important, aspect of the learning economy. The concept signals the fact that learning in connection with everyday economic activities has become much more important than ever before. Economic success reflects the capability of regions and organizations to mobilize many different institutions (the organization of the firm, the network position of firms, knowledge infrastructure organizations, incentive systems, and so on) in support of learning. Again, this trend toward interaction between organizations in the education of children points to a need for operating with a broad definition of the education system and especially to the importance of the interface between the educational system and the broader socioeconomic system. The interaction between everyday learning and formal education is a key to tackling some of the most fundamental contradictions in the learning economy.

Competence building can be the outcome of formal training in specific institutions (schools and universities), of learning-by-doing in ordinary work situations, and of a mix of the two. Rethinking the relationships between the two spheres of competence building is necessary in order to take on the most important challenge: simultaneously to speed up learning and to compensate weak learners. What really matters in the future is the capacity to learn, and the distribution of this capacity will determine the economic fate of individuals, regions, and organizations. This observation has far-reaching implications for education and training systems and for what will matter most in its knowledge base. New pedagogical methods and new psychological and sociological insights into the processes of learning need to be linked to insights about socio-economic developments and organization theory in order to confront these issues.

Much of the training in formal institutions puts a (too) strong emphasis on codified knowledge and on "know-what" and "know-why." It is a great challenge for research on education and learning to analyze to what degree training can exploit more efficiently the spiral movement between tacit and codified knowledge. Problem-oriented and "practice-related" methods that focus on creating know-how may reduce the gap between formal training and what is going on in the regular labor market. Group work and practice in communicating insights and results to others are methods that may be used to strengthen competence in relation to the 'know-who' kind of knowledge.

One of the most fundamental characteristics of the learning economy is the high rate of change in the market sector. Education and training systems will always tend to lag behind the most dynamic parts of the private sector regarding the skills and the competences they can implant in their pupils. In the learning economy this lag tends to grow as the rate of change accelerates. Again, one way to narrow the gap is to strengthen the practice-oriented elements in the formal education and training system and of course to develop new forms of cooperation between knowledge institutions and the work organizations in the private and public sectors.

References

Archer, Margaret (1982). Introduction: Theorizing about the expansion of educational systems. In M. Archer (Ed.), *The sociology of educational expansion.* Beverly Hills: Sage.

Arrow, K. J. (1962a). The economic implications of learning by doing. *Review of Economic Studies, 29*(80).

Arrow, K. J. (1962b). Economic welfare and the allocation of resources for invention. In R. R. Nelson (Ed.), *The rate and direction of inventive activity: Economic and social factors.* Princeton: Princeton University Press.

Arundel, A., Lorenz, E., Lundvall, B.-Å., & Valeyre, A. (2007). How Europe's economies learn: A comparison of work organization and innovation mode for the EU-15. *Industrial and Corporate Change, 16*(6), 1175–1210. http://dx.doi.org/10.1093/icc/dtm035

Bernstein, B. (1997). Class and pedagogies: Visible and invisible. In A. H. Halsey, H. Lauder, P. . Brown & A. S. Wells (Eds.), *Education: Culture, economy and society.* Oxford: Oxford University Press.

Bowles, S. ,& Gintis, H. (1976). *Schooling in capitalist America.* London: Routledge & Kegan Paul.

Campbell, J. R., Voelkl, K. E., & Donahue, P. L. (1996). *NAEP 1996 Trends in academic progress: Achievement of U.S. students in science, 1969 to 1996; mathematics, 1973 to 1996; reading, 1971 to 1996; writing, 1984 to 1996.* Princeton, NJ: US Department of Education.

Cowan, M., David, P., & Foray, D. (2000). The explicit economics of knowledge codification and tacitness. *Industrial and Corporate Change, 9,* 211–253. /icc/9.2.211

Danish Ministry of Education. (1997). *National kompetenceudvikling.* Copenhagen: Ministry of Education.

Dreyfus, H. L.,& Dreyfus, S. E. (1986).*Mind over machine.* Oxford: Basil Blackwell.

Entwistle, N. (1988). *Styles of learning and teaching.* London: David Fulton.

Foray, D., & Lundvall, B.-Å. (1996). The knowledge-based economy: From the economics of knowledge to the learning economy. In D. Foray & B.-Å. Lundvall (Eds.), *Employment and growth in the knowledge-based economy.* OECD Documents. Paris: Organisation for Economic Cooperation and Development.

Gibbons, M., Limoges, C., Nowotny, H. et al. (1994.) *The new production of knowledge: The dynamics of science and research in contemporary societies.* London: Sage.

Green, A., Wolf, A., & Leney, T. (1999). *Convergence and divergence in European education and training Systems.* London: Institute of Education, University of London.

Greinert, W. (2005). *Mass vocational education and training in Europe.* Cedefop Panorama Series 118. Luxemburg: Office for Official Publications of the European Communities.

Guile, D. (2003). From 'credentialism' to the 'practice of learning': Reconceptualising learning for the knowledge economy. *Policy Futures in Education, 1*(1), 83–105. http://dx.doi.org/10.2304/pfie.2003.1.1.10

Hatchuel, A., & Weil, B. (1995). *Experts in organisations.* Berlin: Walter de Gruyter.

Hirata, J., Nishimura, K., Urasaka, J., &Yagi, T. (2006). Parents' educational background, subjects 'good-at' in school and income: An empirical study. *Japanese Economic Review*, 57(4), 533–546. http://dx.doi.org/10.1111/j.1468–5876.2006.00343.x

Iannelli, C. (2007). Inequalities in entry to higher education: A comparison over time between Scotland and England and Wales, *Higher Education Quarterly*, *61*(3), 306–333. http://dx.doi.org/10.1111/j.1468–73.2007.00357.x

Johnson, B., Lorenz, E., & Lundvall, B.-Å. (2002). Why all this fuss about codified and tacit knowledge? *Industrial and Corporate Change*, 11, 245–262. /icc/11.2.245

Kolb, D. A. (1984). *Experiential learning.* Englewood Cliffs, NJ: Prentice Hall.

Lorenz, E. (2006). The organisation of work, education and training and innovation. Keynote presentation prepared for the Conference on Education, Innovation and Development, Calouste Gulbenkian Foundation, 27–28 November, Lisbon, Portugal.

Lorenz, E., & Valeyre, A. (2006). Organizational forms and innovative performance: A comparison of the EU-15, in E. Lorenz & B.-Å. Lundvall (Eds.). *How Europe's economies learn: Coordinating competing models.* Oxford: Oxford University Press.

Lundvall, B.-A. (1996). The social dimension of the learning economy. *DRUID Working Papers*, no. 96–1. http://www.druid.dk/wp/pdf_files/96–1.pdf

Lundvall, B.-Å. (2002a). *Innovation, growth and social cohesion: The Danish model.* Cheltenham: Edward Elgar.

Lundvall, B.-Å. (2002b). The university in the learning economy. DRUID Working Papers, no. 02–06. http://www.druid.dk/wp/pdf_files/02–06.pdf

Lundvall, B.-Å., & Johnson, B. (1994). The learning economy. *Journal of Industry Studies*, *1*(2), 23–42.

Lynch, K., & Baker, J. (2005). Equality in education: The importance of equality of condition. *Theory and Research in Education*, *3*(2), 131–164. /1477878505053298

Marshall, A. P. (1923). *Industry and trade.* London: Macmillan.

Maurice, M., Sellier, F., & Silvestre, J. J. (1986). *The social foundation of industrial power: A comparison of France and Germany.* Cambridge, MA: MIT Press.

Münck, D. (1997). Berufsbildung in der EU zwischen Dualität und 'Monalität'—eine Alternative ohne Alternativen? *Berufsbildung*, *45*(6), 5–8.

Murnane, R. J., & Nelson, R. R. (1984). Production and innovation: When techniques are tacit. *Journal of Economic Behaviour and Organization*, 5, 353–373.

Nelson, R. R. (1959). The simple economics of basic economic research. *Journal of Political Economy*, 67, 323–348. /258177

Nelson, R. R., & Phelps, E. S. (1966). Investment in humans, technological diffusion, and economic growth. *American Economic Review*, *56*(1/2), 69–75.

Nowotny, H., Scott, P., & Gibbons, M. (2001). *Re-thinking science.* Cambridge: Polity Press.

Organisation for Economic Co-operation and Development (OECD). (1994). *The OECD jobs study.* Paris: OECD.

Organisation for Economic Co-operation and Development (OECD). (2000). *Knowledge management in the learning society.* Paris: OECD.

Osborn, M., Broadfoot, P., McNess, E., Planel, C., Ravn, B. & Triggs, P. (2003). *A world of difference: Comparing learners across Europe.* Maidenhead: Open University Press.

Parsons, T. (1967). The school class as a social system. In T. Parsons, *Sociological theory and modern society.* New York: Free Press.

Polanyi, M. (1958/1978).*Personal knowledge.* London: Routledge & Kegan Paul.

Polanyi, M. (1966) *The tacit dimension.* London: Routledge & Kegan Paul.

Ramirez, F. O., & Boli, J. (1987). Global patterns of educational institutionalization. In G. M. Thomas, J. W. Meyer, F. Ramirez & J. Boli (Eds.), *Institutional Structure.* London: Sage.

Rasmussen, P. (1996). Adult education, gender and social change. In H. S. Olesen & P. Rasmussen (Eds.), *Theoretical issues in adult education: Danish research and experiences* (pp. 155–170). Frederiksberg: Roskilde University Press.

Rasmussen, A., & Rasmussen, P. (2008). Educational knowledge at work. In V. Aarkrog & C. H. Joergensen (Eds.), *Divergence and convergence in education and work.* Pieterlen: Peter Lang.

Schultz, T. W. (1975). The value of the ability to deal with disequilibria. *Journal of Economic Literature, 13*(3), 827–846.

Shapiro, C., & Varian, H. R. (1999). *Information rules: A strategic guide to the network economy.* Boston: Harvard Business School Press.

Vincenti, W.G. (1990). *What engineers know and how they know it: Analytical studies from aeronautical history.* Baltimore: Johns Hopkins University Press.

Notes

1. The OECD has pursued several analytical activities along these lines (Foray & Lundvall, 1996). The Portuguese chairmanship for the EU ministerial council for the first half of 2000 was pursued under the theme of "a Europe based on knowledge and innovation."
2. These distinctions draw on different inspirations, among them Aristotle's distinction between *epistêmê* (knowledge that is universal and theoretical) and *technê* (knowledge that is instrumental, context specific, and practice related); the analytic philosophy of Ryle, Moore, and others; and Polanyi's analysis of personal knowledge.
3. In the European Commission there has been a tendency to see the strength of the U.S. knowledge system as connected to big scale, and the Framework programs may be seen as attempts to emulate the United States in this respect. This idea is contradicted by leading innovation experts such as Richard Nelson and David Mowery, who point to the fact that the most important asset of the U.S. innovation system is its high degree of institutional diversity.

4. Lorenz and Valeyre (2006) use logit regression analysis in order to control for differences in sector, occupation and establishment size when estimating the impact of nation on the likelihood of employees being grouped in the various forms of work organization. The results show a statistically significant "national effect" also when controlling for the structural variables, thus pointing to considerable latitude in how work is organized for the same occupation or within the same industrial sector.
5. The class of managers includes not only top and middle management but also professionals and technicians (International Standard Classification of Occupations [ISCO] major groups 1, 2 and 3). The worker category includes clerks, service, and sales workers, as well as craft, plant and machine operators and unskilled occupations (ISCO major groups 4–9).
6. These concepts are of course inspired by the recent theories of knowledge production (Gibbons et al., 1994; Nowotny et al., 2001).

9

Creatively Wise Education in a Knowledge Economy

David Rooney

Contemporary policy debates are increasingly focused on creativity, innovation, and knowledge. The European Union's Lisbon Strategy is one notable outcome of this focus. Assumptions driving such debates are that using knowledge for innovation is what energizes modern economies (Arora & Gambardella, 1994; Conceição, Gibson, Heitor, & Shariq, 2000; Corno, Reinmoeller, & Nonaka, 2000; Reich, 1991; Rooney & Mandeville, 1998). Furthermore, creativity is also being seen as economically important because it underpins the creative industries. However, what is frequently embedded in policymakers' views about creativity, innovation, and knowledge are narrowly instrumental assumptions about the value of knowledge and creativity (Graham & Rooney, 2001). I argue that a lack of adequate conceptual frameworks keeps policy unnecessarily and unhelpfully anchored in this instrumental discourse (Rooney, 2005). To better guide development of the next generation of education policy and practice, this chapter explores what contemporary creativity, wisdom, and knowledge theory suggest about how creativity fits in the larger context of knowledge systems. By taking this approach, a more nuanced view of how to situate creativity education in a knowledge economy can be developed. In particular, I expand the range of applications for creativity beyond simple links to industrial innovation and the creative arts or creative industries (important though they are) to more general social roles for creativity education. To this end, I examine what creativity is; explain what knowledge systems are; discuss the politics of creative production; explain how creativity, wisdom, and values are linked; and, finally, explore how creatively wise education can be formulated in light of these ideas.

Creativity

Creativity does not simply happen in the heads of innately exceptional people, and if that were all there was to it there would be little point in education policymakers and curriculum specialists bringing a creativity focus to their work. Creativity is better seen as occurring in the interaction of a person's thoughts with other people in social spaces (Csikszentmihalyi, 1996). Creativity is therefore very much a social process. What is also important is that disparate ideas are pulled together in novel and socially meaningful and valuable ways. Importantly, it is not the creator who decides his or her work is meaningful and valuable and therefore creative. The decision about the creative value of something is made in communities of practice, or markets, or in the formation of popular opinion, depending on the fields of practice in which the creator works and the created artifacts are being applied. Negus and Pickering (2004) call this creative or communicative value. In other words, "creativity" has to communicate its value. One must be able to imagine a novel state of affairs, but creativity requires such imaginings be turned into an artifact, say a text, that has communicative or creative value.

An understanding of the discursive context of creativity is helpful when considering education. Because creativity involves making novel artifacts, it is important to consider the cultural and institutional characteristics of creativity in the context of a politics of creativity. Creative products that challenge normative assumptions create tensions and introduce issues related to power and persuasion into the process of accepting new creative artifacts (Ranciére, 2004). Therefore, students/citizens need to be skilled at ethically navigating this political context.

A useful way to start an analysis of how creativity fits within education is to look at the links between knowledge and creativity. Most creativity research either explicitly or tacitly assumes that knowledge is linked to creativity. The assumptions are that knowledge and creativity interact and are either in an uneasy tension with each other, or that knowledge is the foundation upon which creativity is built (Weisberg, 1999). In the tension theory one requires knowledge, but not too much of it, to be creative. One must know some basic facts, but if one has too much knowledge, then one is caught in the tram tracks of the certainties possession of a body of knowledge elicits. One becomes stuck and resistant to new ideas because too much knowledge is a conservative force. In the more recent foundation approach, researchers frequently invoke the so called "ten-year-rule," which says that long immersion (for a decade or more) in a field of practice, developing deep knowledge and experience is necessary for creativity. The difference between very creative and less creative people is likely to be found in the depth of knowledge they bring, their attitudes to knowledge, and how nimbly they use knowledge (Weisberg, 1999). More creative people have more cognitive resources to work with. Ericsson, Krampe, and Clemens (1993) have added to this view in their Deliberate Practice Theory. Through developing experience in a deeply engaged or deliberate practice, one gains more and more knowledge and therefore a richer palette of ideas and heuristics to pull together in different and creative ways. Much of the deliberate practice research has studied successful musicians. Deliberate Practice Theory acknowledges many aspects of education, including that formal learning and coaches or teachers play central roles in developing learners' creativity. Importantly, immersion also involves the creator having been immersed in the works of those successful exponents of the relevant art or skill who went before them (Weisberg, 1999). This foundation approach is highly relevant for creativity education and appears to match most closely with reality.

Studies of jazz improvisation have proved very useful for eliciting the discursive dynamics of creativity, and they illustrate a range of processes creativity education professionals need to consider. In his examination of jazz improvisation, Weick (1998) shows how systems of institutions and knowledgeable and creative actors merge. Importantly, Weick uses this approach to demonstrate important ideas for managers to grasp if they are to improve their organizations' use of knowledge and creativity. Weick's insights are just as instructive for education professionals. The process of improvisation reflects the deliberate practice and foundation approaches to creativity because it is based on experience and knowledge gained through practice and learning from and about other practitioners to create a palette of heuristics, knowledge, and skills that can be creatively deployed. "[J]azz improvisation involves conversation between an emerging pattern and such things as formal features of the underlying composition, previous interpretations, the player's own logic, responsiveness of the instrument, other musicians, and the audience" (p. 549). Significantly, Weick characterizes improvisation as processes of communication or conversation in which patterns of creative activity emerge from individuals and groups based on past experience, knowledge, retrospection, imagination, and formal and informal institutions. Thus, knowledge of music theory, riffs, standard progressions, and styles, as well as the traits and habits of fellow band members, and the tools of the trade (instruments) are key parts of the improvisation process. Creative acts like improvisation, therefore, rely on complex processes that use a wide range of resources. More broadly, economic, political, and social events have long been shown to influence creativity (Simonton, 1990), and so these external aspects of the context of creative work are worth dwelling on.

Given the preceding discussion, it is reasonable to say that creativity presupposes the fact that groups of people who share more or less compatible ways of thinking and acting are linked in the creative process. These shared ways of thinking and behaving are the over-arching patterns of thought and behavior in a society that constitute its culture. Creativity therefore is not simply a social phenomenon, it is also a cultural phenomenon: thus culture is often seen as the domain within which creativity happens (Csikszentmihalyi, 1999). With this discursive process view comes the possibility of identifying mechanisms, structures, resources, and organizing principles surrounding creativity and the idea that creativity education practice can be anchored in them. In this view, creativity is constrained but ongoing, indeed, endless. There are inexhaustible possibilities for creativity through the course of time, even if not, perhaps, at one particular time. Creativity therefore happens as part of a constrained stochastic system in which creativity is not totally random, because constraints apply (e.g., lack of resources, cultural inhibitions, path dependency, limits to knowledge, and so on), but is, nevertheless, still unpredictable and embedded with endless potential (Simonton, 2003).

Creativity can therefore be thought of as a system of processes and institutions. Process and institutions are necessary conditions for the ongoing existence of all human systems. Furthermore, human systems are by definition adaptive (self-organizing or auto-poietic) (Luhmann, 1995). The fact that they are adaptive also implies the existence of imaginative, innovative, and creative processes within them. In educational terms, this concept means the important unit of analysis is the system and its broad processes of creative production, rather than only the minds of individual creators. The goal of this chapter, therefore, is to see creativity as part of a much larger system and not to understand it as a stand-alone phenomenon. For the sake of convenience, I will call this larger system a knowledge system or knowledge economy. This nomenclature reflects my views about creativity and knowledge as a "foundationalist"; but knowledge economy, creative economy, and conceptual economy can be used as largely interchangeable terms.

Knowledge Systems and Intentionality

Because knowledge is important to creativity as an economic, social, and cultural phenomenon, it is important to understand what knowledge is. Knowledge is difficult to define, but the challenges of defining it have yielded interesting and useful results that reinforce, amplify, and add important details that can be overlaid on creativity research. In social practice, what counts as knowledge means different things to different people in different situations. One person's knowledge is another's speculation or guess. One reason for this phenomenon is that we all have different epistemic values. We all regard the truthfulness, veracity, creditability, or usefulness of different ideas in different ways depending on our social, cultural, economic, and political positions. Foucault (1970, 1972) acknowledges these debates, or epistemic contests, as largely happening in contests between elites (the Church, scientists, political leaders) through power discourses. Consensus has emerged that knowledge is not understandable in the context of work practices simply as a list of facts, but, rather, as social and systemic, and comprising of actors and various resources (Chia, 1998; Stacey, 2001; Tsoukas, 2005). Knowledge systems are complex and adaptive systems of facts, information, logics, insights, imaginations, intuitions, assumptions, judgments, and so on that are turbulent but, nevertheless, form coherent and creative systems of social understandings, and social practices (Hearn, Rooney, & Mandeville, 2003). As with creativity theory, knowledge theory points to constrained stochastic systems that create, hold, and diffuse ideas.

Sociologists have done very useful research that illuminates the patterning of knowledge in terms of the links between knowledge systems and culture. By acknowledging patterning, researchers point to the existence of phenomenological structures in knowledge systems that underpin creativity. This patterning or organization is very much related to patterns and structures of culture and language. Indeed, knowledge is often viewed as a primary manifestation of culture (including language). Going further, Durkheim, Lévi-Strauss, and Saussure saw the essential roles of language, culture, and knowledge in making society possible, and, reciprocally, in how collective consciousness is also derived from specific forms of social organization (McCarthy, 1996). The idea that knowledge has (social) structure is also linked to Kant's idea that people are active, purposive participants in the production of knowledge, and the idea that the structure of knowledge has a temporal shape comes from the Hegelian view that knowledge is an historical product (McCarthy, 1996). Culture, language, social organization, purposiveness, and history, then, are structures and structuring forces that can be incorporated into creativity-oriented education.

Extending the idea of structure, I argue that knowledge systems are multilevel systems comprised of ideas (cognitions, knowledge), people, history, and also physical entities such as technology and other aspects of built and natural environments (Rooney & Schneider, 2005). In this view, at one level, ideas are connected to other ideas to create meaning. These networks of ideas are integrated with networks of people. At another level, sentient people make sense of particular arrangements or constellations of ideas and then enact these meaningful ideas as work or some other kind of activity. These sentient and enacting people are, at a third level, situated in places or spaces at particular times that influence how they think and act. A stark, almost empty lecture theater provides a very different intellectual environment than a full and engaged one does. Technologies, including information technologies, can also influence how and what we think. If a person only has a hammer, a lot of problems will appear to be solvable by using the hammer. However, if a person has more resources than a hammer, the person has more options and can be more creative. Education policy and practice need to identify all these resources and understand how they work together. Moreover, education professionals need to understand how the parts can be integrated properly within their systems. To find workable educational processes to work with these dynamics, it is important to ask some further and basic social questions about education.

Given the organizing, structuring, and patterning activity in knowledge systems, we can say that knowledge guides us in how we live. More specifically, knowledge is an instrument of collective action with which groups seek authority and power and to satisfy intentions and purposes. A function of knowledge is creating and maintaining social order and coherence, thus rendering social relations meaningful. Particular bodies of knowledge operate within culture as systems of meanings that provide categories and conceptions that enable their users to understand their worlds as something (Percy, 1958), and to assert values, power, and necessity. Within this process, knowledge as ideology hides or mystifies reality, social orders, and so on, and authenticates particular truths. Thus, when actors put forward challenging new and creative ideas, political tensions emerge. Politics and policy are, therefore, not automatically incompatible with knowledge and creativity, they are part of it. Thus, I suggest getting the politics of knowledge and creativity right is important.

To begin exploring the politics of creativity, I emphasize that for knowledge and creativity social and cultural imperatives are more fundamental than commercial imperatives. Importantly, culture, insofar as it is constituted by shared conceptual elements such as beliefs, assumptions, norms, theories, ideologies, myths, and language, has, by definition, considerable internal logical consistency (Archer, 1996). These consistencies hold societies and cultures together because they create sense and purpose (polity). The ideas that societies have about themselves matter, and their cultures are produced, reproduced, and transformed in large part by their ideas about themselves. These ideas include a society's ideas about what it is and should be. However, societies need to evolve and change by actively developing new and creative ideas about themselves. An important and difficult challenge for societies, though, is to entertain new and challenging ideas about themselves, and, going further, entertain creative ideas about themselves generated in nondominant groups.

For equity reasons, education professionals must be aware that knowledge, and therefore creativity, is unevenly distributed in society because of the ability of elites to influence and occlude (see Jacques, 1996 on semantic occlusion) meanings. One potential effect of uneven distribution is a lack of diversity in cultural and intellectual life and a tendency for elites to resist transformational ideas. Creativity depends on diversity of ideas so that disparate ideas can be drawn together, and so too does social renewal and transformation. More diversity of knowledge, though, means more worldviews and the possibility of more ideological conflict. Education needs to deal with this dilemma because ideology is part of what constitutes knowledge systems. Thus, a knowledge economy is likely to be a place in which competing ideas and ideologies clash. Sociology of creativity and knowledge therefore highlight political challenges for a knowledge economy and for creativity in education. Education systems need to respond wisely to this dynamic by understanding and working with these political processes.

Creativity, Wisdom, and Values

To cope with the challenges and uncertainties just described, it is useful to consider wisdom. More knowledge, creativity, and innovation do not necessarily make the world a better place. In his book on intelligence, creativity, and wisdom, Sternberg (2003) begins with the observation that over time average IQ is increasing. However, Sternberg asks, what is the use of all this intelligence if we are actually using it to dig ourselves into a very large hole in the form of global climate change, conflict, and economic chaos? I conjecture that not only is intelligence not automatically a preventative against degradation of life, but that intelligence, innovation, and knowledge in the absence of wisdom actively contribute to it.

Much of the rhetoric about knowledge economies is based on the assumption that more knowledge, creativity, and innovation necessarily trigger positive effects, but it is a dangerous assumption. Furthermore, knowledge and innovation treated only in utilitarian, technocratic terms are likely to lead to savant-like economies rather than wise ones (Rooney & McKenna, 2005). Wisdom is about the creative balancing and integration of the full range of human mental capacities and using that integrative ability to form sound, creative judgments, and to act in the best interests of oneself and society. In this view, values, facts, rationality, reflexivity, intuition, imagination, insight, judgment, and creativity are integrated to promote and guide positive action in society.

Baltes and Staudinger (2000) see wisdom as a way of knowing, understanding, and behaving that is characterized by good judgment and advice; is deeply concerned with being ethical, prudent, and practical; uses reason but is able to transcend the limits of method, data, and information with elegance, insight, foresight, and imagination; and uses and displays knowledge with great scope, depth, and balance (Baltes & Staudinger, 2000). Thus, wisdom, because of its wide cognitive, intuitive, and ethical scope, unifies rationality and intellectual transcendence, reduces fragmentation and lack of decisiveness, and bridges technical expertise with non-technical forms of insight. Similarly, the balance theory of wisdom defines wisdom as "the application of tacit as well as explicit knowledge as mediated by values towards achievement of a common good through a balance among (a) intrapersonal, (b) interpersonal, and (c) extrapersonal interests over the (a) short term and (b) long term to achieve a balance among (a) adaptation to existing environments, (b) shaping of existing environments, and (c) selection of new environments" (Sternberg, 2001, p. 231). The elegant and insightful integrating, weighing, and balancing of wisdom are creative acts. In fact, being wise can be seen as an art. Aristotle saw wisdom as embracing aesthetics. Aristotle explained aesthetics as the art of creative, imaginative, speculative composition and expression producing creative or "poetic" artifacts. In this sense, Aristotle says aesthetics is the work of creating narrative and cultural expression to bring about and communicate important and complex meaning and knowledge, stimulate inspiration and imagination, and attend to individual and group emotional development, thus facilitating social well-being. Vico (1948) draws on the same ideas in his elaboration of poetic wisdom. The essence of Aristotle's position is that aesthetics enhance the communication and use of knowledge (including using it creatively and for it to become a social good). Aristotelian aesthetics, is different to modern notions of aesthetics that focus on taste, and sophisticated consumption of art and beauty, it is more a social practice aesthetics in everyday social life (Rooney, McKenna, & Liesch, 2009). However, aesthetics, imagination, and creativity are also part of the practices of an ethical knower. Thus, Aristotle also considered the role of creativity in intellectual or epistemic virtue. "The central epistemic virtues Aristotle considers are ingenuity (which includes intellectual creativity), perceptual creativity, acuity of inference, a sound sense of relevance, and an active ability to determine the relative importance of heterogeneous and sometimes incommensurable ends" (Eflin, 2003, p. 61). Much of the focus of creativity discourse is linked, not surprisingly, as I said earlier, to the creative arts/industries and even to industrial innovation. There is nothing wrong with this reality; in fact, it makes sense. However, as I have also said, I want to add another dimension to the value of creativity in education. I want to link it to wisdom, values, ethics, and also to social renewal and transformation through the reflexivity and ethical imagination and creativity embedded in wisdom. Somerville (2006), like Aristotle, argues that an ethical imagination is necessary for human flourishing and maintaining a decent society. She says myths, stories, poetry, examined emotions, intuition, and the human spirit are critical to being ethical because they are the resources for the moral or ethical imagination. Ethical imagination is necessary for ethical action. Ethical imagination is needed to see consequences creatively as new possibilities, and to navigate ethical dilemmas and tensions found in everyday life. Ethicality, then, is based on creative processes and requires the same resources needed for creative knowledge systems to work.

Going further, Somerville says ethical practice needs to be based in authentic ethical commitments made by individuals; they must be heart-felt commitments rather than pseudo-commitments. To accomplish this goal, one must have a deep respect for human dignity and spirit. This attitude transcends the individual and brings one in touch with a larger purpose and humanity. One has to see (imagine) oneself as part of, as belonging to, a larger community, web of life, and universe. Common good and values have to be defined in terms larger than oneself or family; they must be defined in larger human and global terms (Sternberg, 2008). This concept extends to having a mindful empathy (imagining others' experiences), and a sense of the need to care (Code, 1987). While this more metaphysical connection to humanity and life is important, it is also instructive for educators that Eflin (2003) says virtues are learned and therefore develop over time and with practice. They become part of one's identity. We are not just conscious deciders but also conscientious deciders using creative cognitive processes rather than just habitual patterns of thought.

Eflin (2003) says: "The greater an enquirer's ability to move fluidly between producing alternatives and evaluating them and to operate at both levels simultaneously, the better she will be at finding valuable discoveries" (p. 62). She also notes that "Higher-order virtues are metacognitive, and they must be cultivated and practiced if deep insights are to be generated" (p. 63). In these two statements, Eflin links knowledge, creativity, deliberate practice, values, and virtue. These core processes in wisdom are very important social resources, and they should be taken seriously in a knowledge economy, in education, and as aspects of creativity. The creativity of wisdom helps not just to integrate but to evaluate. Therefore, I will discuss values as part of knowledge systems.

Values include assumptions and evaluations about what is good, bad, necessary, and unnecessary. In this way, values are the components of knowledge systems that we use to evaluate the desirability or otherwise of what happens or could happen around us. Values exist as important, indeed, fundamental knowledge in a knowledge economy and are central to identifying what is valuable enough to be considered creative. Values are produced and deployed in systems of ideas and can positively or negatively influence society. Because values are involved in social consciousness and in social transformation, they should greatly concern education professionals. Therefore, an issue education professionals need to face is how to foster the production of positive and transforming values.

Vico's poetic wisdom uses creativity and imagination and combines them with reason and logic, and with ethics: it combines sense and intellect (Vico, 1948, p. 110). Poetic wisdom is needed to deal with social complexity and to build strong and healthy societies and civil institutions. Importantly, Vico is also explicit about the centrality of communication and language in creativity and wisdom. Specifically, the development of wisdom depends on civic discourse to create a collective sense or *sensus communis* (Vico, 1948).

Values and the ability to influence discourse are not uniformly distributed. Different people possess different values, and not all values will be considered as virtuous as others because there are political forces operating that promote certain values over others. In this context, it is necessary to also acknowledge that powerful elites promote certain evaluative knowledge over others in the service of their particular interests. Hence, esteem for economic modes of evaluation that are prized by powerful industrial elites are readily adopted by governments and citizens. Elites are successful at embedding and privileging certain evaluations and evaluative processes in discourse. Nevertheless, there are tensions between competing value systems within social systems, and games are played between competing actors with different values. Language and communication dynamics are important components of these contests.

In discourse, values are formulated as "texts," which actors may seek to embed or institutionalize in a community. Texts are taken to be linguistic or visual images that convey ideas (knowledge, beliefs, and ideologies) and representations and have an interpersonal function. Importantly, if people's texts are convincingly and persuasively presented the texts' ideas (i.e., values) are more likely to become accepted or institutionalized community values. Phillips, Lawrence, and Hardy (2004) have modeled how such discursive games are played (Figure 1).

If new ideas are to be embedded and institutionalized, those ideas must be presented as compelling texts that people want to open-mindedly engage with to consider new alternatives. Compelling novel texts are essential in social renewal and transformation but are often threatening to vested interests whose first impulse may be to summarily reject them. An inherently compelling text needs, therefore, to do more than command attention to its message: if it is to bring new ideas, values, and behaviors to a group; it needs also to compel people to fairly consider them, to adopt them as their own beliefs, and to act on them. Positive new texts, therefore, need to be skillfully and creatively embedded in discourse. For this to happen, they have to make sense, have legitimacy, be coherent, and be presented in an appropriate and effective genre.

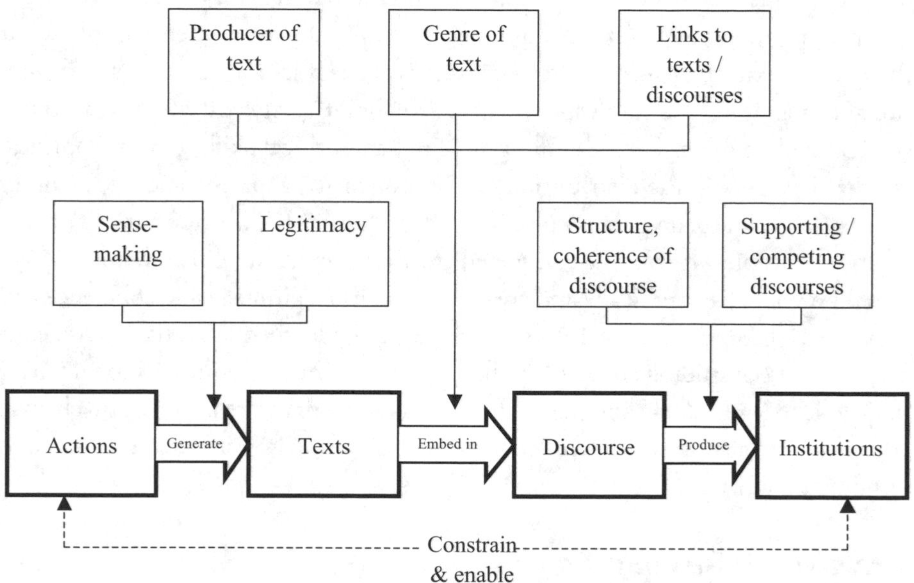

Figure 1. A Discursive Model of Institutionalization

Source: Phillips, Lawrence, and Hardy (2004, p. 641)

Genre, structure, and coherence of a new idea must creatively coalesce to appear relevant and usable and to be shown to have some meaningful link to existing assumptions, values, goals, and knowledge. Moreover, the producer of the idea must be seen to have legitimacy for it to be embedded in discourse to produce socially reconstructed institutions. The idea's producer must demonstrate how he or she can work around existing discursive structures to compete with them as a provider of understandable alternatives. For this reason new ideas that are supported by broader macro-discourses appear more persuasive and more legitimate (Phillips et al., 2004). Clearly, a goal of teaching and learning is to develop skills for championing and evaluating creative artifacts. For education, an explicit acknowledgment of the creative communication skills and communication ethics that engender the intellectual and emotional stimulus for change to the established order is needed.

In this section, then, I am less interested in what the novel ideas or insights are, and more interested in how their possessors effectively present their ideas to bring about positive change, despite any threats adopting and acting on those ideas might bring to the established order. Text producers who are able to embed creative new ideas in discourse do so by attending to genre, the way of presenting and organizing their ideas, and equally, to "ways of acting and interacting in the course of social events" (Fairclough, 2003, p. 65). This interaction is about persuasiveness, communication strategy, and communication skill. Discourse theory implies that the creatively aesthetic component of wisdom is essential for positive transformational change and renewal. A specific objective for creativity education is to understand the place of aesthetics, communication, and ethical skills within a political context influencing the distribution of the propensity, capacity, and opportunity to present, listen to, and act on new texts. For educators, it is important to understand that people who may be considered creatively wise, but who do not have positional power, need to ethically and creatively establish legitimacy and coherence and that the skills needed to reach this objective can be learned. Creatively wise education can assist students to develop the skills for citizens without positional power to be able to legitimately and persuasively speak in favor of the new. Creatively wise education can also assist students to become open and wise listeners or readers and writers of novel texts.

Creatively Wise Education

Creativity serves many social and economic purposes, and education systems need to acknowledge this fact. However, although an education system that seeks to enhance a society's capacity for creativity is starting out in the right way, the power to realise this goal will not be easy to acquire without a framework to guide practice.

Creativity must be understood as part of a larger set of phenomena that includes knowledge, values, and wisdom, and must be considered as valuable in artistic endeavor and the creative industries. In addition, creativity must also be seen as valuable in a more general social or civic sense. Creativity and knowledge can contribute to making the world a better place, but this effect does not come about automatically. It is the wise use of these valuable human capacities that makes the difference. I therefore argue that an overriding reason for focusing on creativity in education is to engender in students the capacity for wise communication in bringing about transformational change. Doing so will both enhance creativity and facilitate its better application. I am also arguing that creativity, as an aspect of a larger knowledge system, is dependent on a range of discursive processes to make an impact. In line with this observation, I recommend that classroom practices seek to engender in students skills in reflexive understanding and awareness of discursive processes. This teaching objective is about preparing students as creative citizens who are conscientious deciders.

Reflexive Learning for Wisdom

In this final section, I want to discuss Sternberg's work on wisdom and education. I deal with Sternberg in two parts. First, I discuss why intelligent people can be foolish because understanding this fact can inform our attitudes to knowledge and its role in creativity education to show how openness and creative virtue are more likely to induce wisdom. Second, I outline Sternberg's 16 principles of teaching for wisdom and build links from them to the discursive process approach to form an initial structure for educational practices that facilitate creative wisdom.

Not only stupid people act foolishly: smart people can act foolishly by being too smart and falling victim to one or more of five cognitive fallacies (Sternberg, 2004):

1. Unrealistic optimism, whereby they believe they are so smart that they can do whatever they want and not worry about it;

2. Egocentrism, whereby they focus on themselves and what benefits them while discounting or even totally ignoring their responsibilities to others;

3. Omniscience, whereby they believe they know everything, instead of knowing what they do not know;

4. Omnipotence, whereby they believe they can do whatever they want because they are all-powerful; and

5. Invulnerability, whereby they believe they will get away with whatever they do, no matter how inappropriate or irresponsible it may be.

Underneath all this foolishness is a lack of humility, a complacent attitude, and narcissism, which conspire against both creativity and wisdom. Against this framework of foolishness, we can say humility, alertness or vigilance, and unselfishness are creative and intellectual virtues and should be central parts of education for creative wisdom. To reiterate, knowledge systems need to be creative to evolve and adapt in ways that facilitate social and economic transformations. Lack of intellectual humility, alertness, and selflessness stifle creativity; they foster a loss of interest in creating and accepting new and better ideas. These five fallacies presented by Sternberg can be countered through education in a way that also promotes the communication and discursive skills discussed earlier.

Sternberg's (2001, p. 238) educational antidote to the kind of intellectual dysfunction or unbalance described above is contained in his 16 principles for teaching for wisdom. For the purposes of this chapter, I have broken Sternberg's principles into six groups and I discuss each of those groups in turn. The first group:

1. Explore with students the notion that conventional abilities and achievements are not enough for a satisfying life. Many people become trapped in their lives and, despite feeling conventionally successful, feel that their lives lack fulfillment. Fulfillment is not an alternative to success, but rather, is an aspect of it that, for most people, goes beyond money, promotions, large houses, and so forth.

2. Demonstrate how wisdom is critical for a satisfying life. In the long run, wise decisions benefit people in ways that foolish decisions never do.

3. Teach students the usefulness of interdependence—a rising tide raises all ships; a falling tide can sink them.

Wisdom has a higher purpose and therefore has a metaphysical and empathetic dimension requiring imagination. Therefore, a satisfying life is more than making money, or becoming important, or knowing clever things; satisfaction requires more than material success. This point is also about developing a holistic longer term vision in students. Schools might teach "skills," but they may fail in fundamental ways at being strategic or in providing guidance to students about the benefits of acting wisely. Citizenship, co-operation, and collaboration are part of this dynamic, but they depend on students being able to imagine and to understand intangible or metaphysical rewards. More important in this first set of principles is for students to understand their own lives and how they fit into the world. To reach this goal, students need to acknowledge their interdependence, that being actively reciprocal in this interdependence is worthwhile and a mark of success, and how they could be more satisfied and more socially useful by focusing on the "other." Clearly, Sternberg is suggesting education needs to acknowledge the metaphysics of everyday life to guard against narcissism and the misery it brings. To this train of thought I add the observation that understanding how one is positioned socially through conventions and interdependence is essential if effective communication strategies are to be developed and applied.

4. Role-model wisdom because what you do is more important than what you say. Wisdom is action dependent, and wise actions need to be demonstrated.

5. Have students read about wise judgments and decision making so that students understand that such means of judging and decision making exist.

How one thinks is important for wisdom, but wisdom theory also demands an emphasis on what people do (Rooney et al., 2010). Wisdom, after all, is a way of being rather than only a way of thinking. An important part of education, then, can be learning about what and how wise people do. Learning from history, teachers, and other role models is important in wisdom development. An important quality role models have is their own legitimacy and a record of using successful approaches to establishing legitimacy of new ideas. Learning this way includes but goes beyond book learning to include modeling within a school's walls. Schools and their teachers can foster wisdom by modeling wise action themselves, that is, creating bodily texts. There are benefits to wise role modeling in schools and teachers developing clearer and deeper views of who they are, what their purposes are, and what practices they should use. Having decided about these things teachers and schools should then be confident in communicating in wisdom-developing ways to students.

Teachers should strive for legitimacy in the eyes of students. In fact, the cases they teach about can be about people who have established widely accepted legitimacy for their transformational activities. Students benefit if they can explore case studies of remarkable and difficult decisions made and actions taken. Many of these could be famous cases, or examples from literature, but unsung community workers who have not been given hero status despite the great things they achieve are also important (Hart, 2001). Furthermore, one can teach about "particular courses of action that would be considered wise" (Sternberg, 2001, p. 230) and that exhibit the appropriate communication skills needed to be creatively wise.

It is not experience but reflexive experience that is important to wisdom because through reflexive experiential learning students and teachers can build a coherent and legitimate vocabulary and library of stories about wise social change. This form of instruction is more than facilitative teaching; it expressly respects the special role of teachers in developing wisdom. Good teachers matter; they make a difference. All the ancient wisdom traditions have recognized how important teachers are (Armstrong, 2006).

6. Help students to learn to recognize their own interests, those of other people, and those of institutions.

7. Help students learn to balance their own interests, those of other people, and those of institutions.

8. Teach students that the "means" by which the end is obtained matters, not just the end.

9. Help students learn the roles of adaptation, shaping, and selection, and how to balance them. Wise judgments are dependent in part on selecting among these environmental responses.

These points are about recognizing, evaluating, and selecting what needs to be balanced and about thinking how best to integrate ideas, actions, and purposes. Part of this is to balance one's own needs and those of others. These are, arguably, the most complex tasks in Sternberg's educational vision. What is suggested here is learning the evaluative and judgmental foundations of creative ethical political action. The idea of selecting among environmental responses also suggests adaptability in an evolving context is important and thus rejects rigid ideology. A fluid and dynamic sense of proportion or perspective is linked to ethical imagination (Somerville, 2006) and creativity. Sternberg also points to the fact that wise people bring benefits to society because they select and shape aspects of society's renewal. We can think of this as improvisation and part of the creative art of living. Learning to be comfortable with and good at improvisation can be part of a curriculum. As in improvised music, there is nothing like actually doing it for developing it. Reflection on such matters can be aided by "biography, freestyle writing, probing questioning with a colleague or coach, free associating and analytical interpretation of drawings and dreams" (Gosling & Mintzberg, 2006, p. 422).

10. Encourage students to form, critique, and integrate their own values in their thinking.

11. Encourage students to think dialectically, realizing that both questions and their answers evolve over time, and that the answers to an important life question can differ at different times in one's life.

12. Show students the importance of dialogical thinking, whereby they understand interests and ideas from multiple points of view.

13. Teach students to search for and then try to reach the common good—a good position where everyone wins, not only those with whom one identifies.

Different and evolving ways of communicating, searching, and thinking are emphasized here. Going beyond the agility or improvisational components in the previous set of principles, intellectual flexibility using a range of resources is emphasized in principles ten to thirteen because the world always changes. In particular, subjective, interpretive, and insightful thinking is foregrounded, and importantly, so too are social learning and communication or dialogical thinking. We are reminded, therefore, that there are many good ways of thinking, and we should be able to use all of them when needed. In particular, we must recognize that good thinking is part of a social process, and students need to understand that good thinking is worthwhile (Kuhn & Udell, 2001) and that reaching a state where the common good is found is not likely to occur without considerable thought. Looked at through the lens of the Discursive Model (Figure 1), it is important that Sternberg reminds us about the common good being more widely defined than as what is good for those groups with whom we easily identify.

14. Encourage and reward wisdom.

15. The reward structures for teachers and other education professionals need also to be reconsidered. Where are the rewards for creatively wise teaching?

Maxwell (1984) argues universities encourage and reward knowledge rather than wisdom. This approach may also be true of schools. Different forms of knowledge and practice need to be integrated, and different encouragement and reward processes need to be integrated, too. To this end, assessment should move beyond its narrow focus on passing assignments and exams with traditional grades. These elements of the education process do not have to be discarded, they still have a role to play, but grades are not enough as encouragement and reward for wisdom. Some aspects of wisdom are not sensibly gradable, as indeed are some examples of creativity. It is easy to think of alternative rewards and encouragements, but it is not always easy to change attitudes that prevent alternatives from being adopted. "[M]any people will not see the value of teaching something that shows no promise of raising conventional test scores" (Sternberg, 2001, p. 242). However, test scores do not have to be discarded, they still have an important function; the point is to balance them against other methods of assessment, encouragement, and reward.

16. Teach students to monitor events in their lives and their own thought processes about these events. One way to learn to recognize others' interests is to begin to identify one's own.

Help students understand the importance of inoculating oneself against the pressures of unbalanced self-interest and small-group interest.

Wisdom involves the capacity to be reflexive. Reflexive, self-critical processes have a place in many of the processes already discussed; however, what is emphasized here is the development of processes that help develop self-knowledge. Teachers need to become skilful at providing the space for reflection so students do not, for example, ignore seemingly redundant thoughts, questions, and memories (Gosling & Mintzberg, 2006). Contemplation is important to wisdom's creative, reflexive, and self-critical aspects. To this end, schools should ask students to stop and do nothing, except notice and feel for a while. I refer to the benefits of contemplative or meditative practices in education here. In teaching for wisdom, schools need to teach students to notice and feel in general, and specifically to notice and feel what is going on within them. Contemplative noticing and feeling can also be a social process. An important part of these reflexive processes is remembering. Remembering is best done creatively as part of a conversation with other people sharing similar contexts or concerns, collective reflection (Gosling & Mintzberg, 2006).

In closing, I suggest education should promote creativity for more reasons than are usually given. Creativity is necessary in all aspects of life. Creativity provides an important element of humanity in a knowledge economy and contributes to wisdom. What is needed are creatively wise societies rather than savant economies. Thus, education has to provide more than good knowledge workers. Moreover, education has to do more than, for example, develop critical thinking in students. As Sternberg says:

> Some of the world's worst leaders have been notably well educated. Many of the Nazi leaders had doctorates. Alberto Fujimori, the former president of Peru, has a doctorate. Presumably, these individuals have not lacked in critical thinking skills. Yet, I suggest and many agree, they all have been atrocious leaders, at least from the standpoint of the well-being of their countries and the people in them. They used their critical-thinking skills to figure out how they could gain and maintain power, and they did so in quite rational ways. They knew exactly what they were doing and why. They lacked not in critical or rational thinking skills, but, rather, in wisdom. They placed their own interests and those of a small band of cronies around them above the interests of almost everyone else. Dictators and ruthless leaders continue to plague the world. They stay in power not despite their critical thinking, but because of it. We need to encourage students not only to learn to think, but to take responsibility for the products of their thinking. (Sternberg, 2008, p. 272)

Critical thinking has an important place in education, but it is insufficient on its own. Creative wisdom is a better objective for education. Imagination, ethics, and empathy are clearly needed in people, and so too are open cultures that defend against narcissism, narrow-mindedness, and antisocial attitudes generally. It is also important to note that wisdom does not do away with politics and power, or is not somehow apolitical. All social systems involve purposive political dynamics, and creative wisdom is also decisively purposive. Ethical politics and exercise of power have been brought into focus in this chapter as important aspects of creativity. Creativity education, therefore, needs to do more than teach students how to have creative ideas; it has to assist students in dealing with the politics and discursivity of being creative and in evaluating the creativity of others.

References

Archer, M. S. (1996). *Culture and agency: The place of culture in social theory* (rev. ed.). Cambridge: Cambridge University Press.

Armstrong, K. (2006). *The great transformation: The beginning of our religious traditions.* New York: Alfred A. Knopf.

Arora, A., & Gambardella, A. (1994). The changing technology of technological change: General and abstract knowledge and the division of innovative labour. *Research Policy,* 23(5), 523–532.

Baltes, P. B., & Staudinger, U. M. (2000). A metaheuristic (pragmatic) to orchestrate mind and virtue towards excellence. *American Psychologist,* 55(1), 122–136.

Chia, R. (1998). From complexity science to complex thinking: Organization as simple location. *Organization,* 5(3), 341–369.

Code, L. (1987). *Epistemic responsibility.* Hanover: Brown University Press.

Conceição, P., Gibson, D. V., Heitor, M. V., & Shariq, S. (Eds.). (2000). *Science, technology and innovation policy: Opportunities and challenges for the knowledge economy.* Westport: Quorum Books.

Corno, F., Reinmoeller, P., & Nonaka, I. (2000). Knowledge creation within industrial systems. *Journal of Management and Governance,* 3(4), 379–394.

Csikszentmihalyi, M. (1996). *Creativity: Flow and the psychology of discovery and invention.* New York: Harper Collins.

Csikszentmihalyi, M. (1999). The implications of a systems perspective for the study of creativity. In R. J. Sternberg (Ed.), *Handbook of creativity* (pp. 313–335). New York: Cambridge University Press.

Eflin, J. (2003). Epistemic presuppositions and their consequences. In M. Brady & D. Pritchard (Eds.), *Moral and epistemic virtues* (pp. 47–66). Oxford: Blackwell.

Ericsson, K. A., Krampe, R. T., & Clemens, T.-R. (1993). The role of deliberate practice in expert performance. *Psychological Review,* 103, 363–406.

Fairclough, N. (2003). *Analysing discourse: Textual analysis for social research.* London: Routledge.

Foucault, M. (1970). *The order of things: An archaeology of the human sciences.* London: Tavistock.

Foucault, M. (1972). *The archaeology of knowledge* (A. M. T. Sheridan Smith, Trans.). London: Tavistock.

Gosling, J., & Mintzberg, H. (2006). Management education as if both matter. *Management Learning,* 37(4), 419–428.

Graham, P., & Rooney, D. (2001). A sociolinguistic approach to applied epistemology: Examining technocratic values in global "knowledge" policy. *Social Epistemology,* 15(3), 155–169.

Hart, T. (2001). *From information to transformation: Education for the evolution of consciousness.* New York: Peter Lang.

Hearn, G., Rooney, D., & Mandeville, T. (2003). Phenomenological turbulence and innovation in knowledge systems. *Prometheus,* 21(2), 231–246.

Jacques, R. (1996). *Manufacturing the employee: Management knowledge from the 19th to 21st centuries.* London: Sage.

Kuhn, D. and Udell, W. (2001). The Path to Wisdom. Educational Psychology, 36(4), 261–264.

Luhmann, N. (1995). *Social systems* (J. John Bednarz & T. with Dirk Baeker, Trans.). Stanford, CA.: Stanford University Press.

Maxwell, N. (1984). *From knowledge to wisdom: A revolution in the aims and methods of science.* Oxford: Basil Blackwell.

McCarthy, E. D. (1996). *Knowledge as culture: The new sociology of knowledge.* London: Routledge.

Negus, K., & Pickering, M. (2004). *Creativity, communication and cultural value.* London: Sage.

Percy, W. (1958). Symbol, consciousness and intersubjectivity. *Journal of Philosophy,* 5, 631–641.

Phillips, N., Lawrence, T. B., & Hardy, C. (2004). Discourse and institutions. *Academy of Management Review,* 29(4), 635–652.

Ranciére, J. (2004). *The politics of aesthetics* (G. Rockhill, Trans.). London: Continuum.

Reich, R. (1991). *The work of nations.* New York: Knopf.

Rooney, D. (2005). Knowledge, economy, technology and society: The politics of discourse. *Telematics and Informatics,* 22(3), 405–422.

Rooney, D., & Mandeville, T. (1998). The knowing nation: A framework for public policy in a knowledge economy. *Prometheus,* 16(4), 453–467.

Rooney, D., & McKenna, B. (2005). Should the knowledge-based economy be a savant or a sage? wisdom and socially intelligent innovation. *Prometheus,* 23(3), 307–323.

Rooney, D., McKenna, B., & Liesch, P. (2010). *Wisdom and management in the knowledge economy.* London: Routledge.

Rooney, D., & Schneider, U. (2005). A model of the material, mental, historical and social character of knowledge. In D. Rooney, G. Hearn & A. Ninan (Eds.), *Handbook on the knowledge economy* (pp. 19–36). Cheltenham: Edward Elgar.

Simonton, D. K. (1990). Political pathology and societal creativity. *Creativity Research Journal,* 3(2), 85–99.

Simonton, D. K. (2003). Scientific creativity as constrained stochastic behavior: The integration of product, process, and person perspectives. *Psychological Bulletin,* 129, 475–494.

Somerville, M. (2006). *The ethical imagination: Journeys of the human spirit.* Melbourne: Melbourne University Press.

Stacey, R. D. (2001). *Complex responsive processes in organizations: Learning and knowledge creation.* London: Routledge.

Sternberg, R. J. (2001). Why should schools teach for wisdom: The balance theory of wisdom in educational settings. *Educational Psychologist,* 36(4), 227–245.

Sternberg, R. J. (2003). *Wisdom, intelligence and creativity synthesized.* Cambridge: Cambridge University Press.

Sternberg, R. J. (2004). Why smart people can be so foolish. *European Psychologist,* 9(3), 145–150.

Sternberg, R. J. (2008). How wise is it to teach for wisdom? A reply to five critiques. *Educational Psychology,* 36(4), 269–272.

Tsoukas, H. (2005). *Complex knowledge: Studies in organizational epistemology.* Oxford: Oxford University Press.

Vico, G. (1948). *The new science of Giambattista Vico: Revised translation of the third edition* (1744) (T. G. Bergin & M. H. Fisch, Trans.). Ithaca, NY: Cornell University Press.

Weick, K. E. (1998). Introductory essay: Improvisation as a mindset for organizational analysis. *Organization Science,* 9(5), 543–555.

Weisberg, R. W. (1999). *Creativity and knowledge: A challenge to theories.* In R. J. Sternberg (Ed.), *Handbook of creativity* (pp. 226–250). Cambridge: Cambridge University Press.

Part Two: Technology and Economy

10

Creativity, Openness, and User-Generated Cultures

Michael A. Peters

Creativity and the Global Knowledge Economy[1]

The global knowledge economy, comprised of increasingly integrated cross-border distributed knowledge and learning systems, represents a new stage of development that is characterized by a fundamental sociality—knowledge and the value of knowledge are rooted in social relations. More than any time in the past, the global economy and society are undergoing a massive transformation from an industrial age that was dominated by the logic of standardized mass production and epitomized by the assembly-line in the auto industry to a knowledge economy that is characterized by decentralized networked communications. These communication systems reflect "intellectual capital" in a range of information-service industries that are propelled by brainpower and the constant demand for innovation. These innovations do not mean the demise of the industrial economy but rather the development of a new relation between manufacturing and information services that permits the sharing of knowledge through open source models and the continuous redesign of flexible production regimes. It also means the rapid development of "mind-intensive" industries, especially in the software, media, healthcare, education, and other mind-intensive industries. Increasingly, the move to the knowledge economy redefines the value creation process, alters the organization and pattern of work, and creates new forms of borderless cooperation and intercultural exchange. This dynamic has led many national government and international organizations to plan for a restructuring of the economy that increasingly focuses on knowledge, education, and creativity. The New Club of Rome, for instance, calls this new era the paradigm of an "economy of the intangibles" and predicts "Third Phase Industries," "sustainable development" and the development of "intellectual capital":

- This trend means that the intellectual, social, and cultural issues require much higher attention. They are the determinants of Third Phase Industries based on creativity, software, media, finance, services, and, more generally, combined intelligence. These qualities are more representative of today's developed economies, and they produce more value than traditional manufacturing per se. They are of decisive importance to the development of all sectors, including traditional ones. Only through careful and sustainable utilization of the new, nonmaterial resources will we be in a position to better organize material and energy resources that are increasingly in short supply.

- More specifically the "Ever More" of the current economic model of the Western industrial society has outlived its legitimacy. What matters are not mere survival strategies or linear expansion but rather sustainable preservation so that we can retain our prosperity. In order to master the future, we need more intelligent modes of cultivation and exploitation and a new balance between material and nonmaterial resources.

- Intellectual capital (comprising assets such as human abilities, structural, relational, and innovation capital, as well as social capital) founded on clear, practiced values such as integrity, transparency, cooperation ability, and social responsibility, constitute the basic substance from which our future society will nurture itself.[2]

The postindustrial society, a term invented by Arthur Penty, a British Guild Socialist and follower of William Morris, at the turn from the nineteenth to the twentieth century, was based on the model of the craft workshop and decentralized units of government. The postindustrial society is marked by the change from a goods-producing to a service economy and the widespread diffusion of "intellectual technologies." For Daniel Bell (1973) the concept of post-industrialism dealt primarily with changes *in the social structure* including the shift from a goods-producing economy to a service economy, the centrality of theoretical knowledge for innovation, the change in the character of work, and the shift from a game against nature to a game among persons. His early account given in the

1970s—before the invention of the Internet and the spread of communications networks—did not foresee the phenomenon of virtualization or the emergence of personalization as a 24/7 totally person-centered, unique learning environment (Peters, 2009a).

Although there are different readings and accounts of the knowledge economy, it was only when the OECD (1996) used the label in the mid-1990s and it was adopted as a major policy description/prescription and strategy by the United Kingdom in 1999 that the term passed into the policy literature and became acceptable and increasingly widely used. The "creative economy" is an adjunct policy term based on many of the same economic arguments—and especially the centrality of theoretical knowledge and the significance of innovation. Most definitions highlight the growing relative significance of knowledge compared with traditional factors of production—natural resources, physical capital and low-skill labor—in wealth creation and the importance of knowledge creation as a source of competitive advantage to all sectors of the economy, with a special emphasis on R&D, higher education and knowledge-intensive industries such as the media and entertainment. At least two sets of principles distinguish knowledge goods, in terms of their behavior, from other goods, commodities, or services; the first set concerns knowledge as a global public good; the second concerns the digitalization of knowledge goods.

These features have led a number of economists to hypothesize the knowledge economy and to picture it as different from the traditional industrial economy, leading to a structural transformation. In *The Economics of Knowledge* (2004) Dominique Foray argues:

> Some, who had thought that the concepts of a new economy and a knowledge-based economy related to more or less the same phenomenon, logically concluded that the bursting of the speculative high-tech bubble sealed the fate of a short-lived knowledge-based economy. My conception is different. I think that the term 'knowledge-based economy' is still valid insofar as it characterizes *a possible scenario of structural transformations of our economies.* This is, moreover, the conception of major international organizations such as the World Bank and the Organisation for Economic Cooperation and Development (OECD). (p. ix, emphasis added).

In this scenario "the rapid creation of new knowledge and the improvement of access to the knowledge bases thus constituted, in every possible way (education, training, transfer of technological knowledge, diffusion of innovations), are factors increasing economic efficiency, innovation, the quality of goods and services, and equity between individuals, social categories, and generations." He goes on to argue that there is a collision between two phenomena—"a long-standing trend, reflected in the expansion of 'knowledge-related' investments" and "a unique technological revolution" (Foray, 2004, p. x).

Knowledge As a Global Public Good

The first set of principles concerning knowledge as an economic good indicates that knowledge defies traditional understandings of property and principles of exchange and closely conforms to the criteria for a public good:

1. knowledge is *non-rivalrous*: the stock of knowledge is not depleted by use, and in this sense knowledge is not consumable; sharing with others, use, reuse, and modification may indeed add rather than deplete value;

2. knowledge is barely *excludable*: it is difficult to exclude users and to force them to become buyers; it is difficult, if not impossible, to restrict distribution of goods that can be reproduced with no or little cost;

3. knowledge is not *transparent*: knowledge requires some experience of it before one discovers whether it is worthwhile, relevant, or suited to a particular purpose.

Thus, knowledge at the *ideation* or *immaterial* stage considered as pure ideas operates expansively to defy the law of scarcity. It does not conform to the traditional criteria for an economic good, and the economics of knowledge is, therefore, not based on an understanding of those features that characterize property or exchange and cannot be based on economics as the science of the allocation of scarce public goods. Of course, as soon as knowledge becomes codified or written down or physically embedded in a system or process, it can be made subject to copyright or patent and then may be treated and behave like other commodities (Stiglitz, 1999a).

Digital Information Goods Approximating Pure Thought

The second set of principles applies to digital information goods insofar as they approximate pure thought or the ideational stage of knowledge, insofar as data and information through experimentation and hypothesis testing (the traditional methods of sciences) can be turned into justified true belief. In other words, digital information goods also undermine traditional economic assumptions of rivalry, excludability, and transparency, as the knowledge economy is about creating intellectual capital rather than accumulating physical capital. Digital information goods differ from traditional goods in a number of ways:

1. Information goods, especially in digital forms, can be copied cheaply, so there is little or no cost in adding new users. Although production costs for information have been high, developments in desktop and just-in-time publishing, together with new forms of copying, archiving and content creation, have substantially lowered fixed costs.

2. Information and knowledge goods typically have an experiential and participatory element that increasingly requires the active co-production of the reader/writer, listener and viewer.

3. Digital information goods can be transported, broadcast, or shared at low cost, which may approach free transmission across bulk communication networks.

4. Since digital information can be copied exactly and easily shared, it is never consumed (see Varian, 1998; Morris-Suzuki, 1997; Davis & Stack, 1997; Kelly, 1998).

The implication of this brief analysis is that the laws of supply and demand that depend on the scarcity of products do not apply to digital information goods.

Creating the Creative Economy

Today there is a strong renewal of interest by politicians and policy-makers worldwide in the related notions of creativity and innovation, especially in relation to terms like "the creative economy," "knowledge economy," "enterprise society," "entrepreneurship," and "national systems of innovation" (Baumol, 2002; Cowen, 2002; Lash & Urry, 1994). In its most obvious form the notion of the creative economy emerges from a set of claims that suggests that the Industrial Economy is giving way to the Creative Economy based on the growing power of ideas and virtual value—the turn from steel and hamburgers to software and intellectual property (Florida, 2002; Howkins, 2001; Landry, 2000). In this context increasingly policy latches onto the issues of copyright as an aspect of IP, piracy, distribution systems, network literacy, public service content, the creative industries, new interoperability standards, the WIPO and the development agenda, WTO and trade, and means to bring creativity and commerce together (Cowen, 2002; Shapiro & Varian, 1998; Davenport & Beck, 2001; Hughes, 1988; Netanel, 1996, 1998;

Gordon, 1993; Lemley, 2005; Wagner, 2003). At the same time, this focus on creativity has exercised strong appeal to policy-makers who wish to link education more firmly to new forms of capitalism emphasizing how creativity must be taught, how educational theory and research can be used to improve student learning in mathematics, reading and science, and how different models of intelligence and creativity can inform educational practice (Blythe, 2000). Under the spell of the creative economy discourse, there has been a flourishing of new accelerated learning methodologies together with a focus on giftedness and the design of learning programs for exceptional children.[3] One strand of the emerging literature highlights the role of the creative and expressive arts, of performance, of aesthetics in general, and the significant role of design as an underlying infrastructure for the creative economy (Caves, 2000; Frey, 2000; Frey & Pommerehne, 1989; Ginsburgh & Menger, 1996; Heilbron & Gray, 2001; Hesmondhalgh, 2002). There is now widespread agreement among economists, sociologists, and policy analysts that creativity, design, and innovation are at the heart of the global knowledge economy: together creativity, design, and innovation define knowledge capitalism and its ability to continuously reinvent itself.[4] Together and in conjunction with new communications technologies, they give expression to the essence of digital capitalism—the "economy of ideas"—and to new architectures of mass collaboration that distinguish it as a new generic form of economy different in nature from industrial capitalism. The fact is that knowledge in its immaterial digitized informational form as sequences and value chains of 1s and 0s—ideas, concepts, functions, and abstractions—approaches the status of pure thought. Unlike other commodities, it operates expansively to defy the law of scarcity that is fundamental to classical and neoclassical economics and to the traditional understanding of markets. As mentioned above a generation of economists has expressed this truth by emphasizing that knowledge is (almost) a global public good: it is non-rivalrous and barely excludable (Stiglitz, 1999b; Verschraegen & Schiltz, 2007). It is non-rivalrous in the sense that there is little or only marginal cost to adding new users. In other words, knowledge and information, especially in digital form, cannot be consumed. The use of knowledge or information as digital goods can be distributed and shared at no extra cost, and the distribution and sharing is likely to add to its value rather than to deplete it or use it up. This is the essence of the economics of file-sharing education; it is also the essence of new forms of distributed creativity, intelligence and innovation in an age of mass participation and collaboration (Brown & Duguid, 2000; Tapscott & Williams, 2006; Surowiecki, 2004).

Openness and Creativity

There is a long established literature on openness and creativity in the field of personality psychology emphasizing the uniqueness of the individual. Prabhu et al. (2008, p. 53), for instance, report that four decades of work have generated more than 9,000 published studies. They also report that in the five-factor model of personality—based on openness to experience, conscientiousness, extraversion, agreeableness, and neuroticism—"openness to experience has the most empirical support as being closely related to creativity." In this context, openness is correlated with the appreciation for art, emotionality, sense of adventure, new ideas, imagination, curiosity, and variety of experience. On this psychological reading open people prefer novelty and change, and tend to be more aware of their feelings with a corresponding willingness to tolerate diversity and entertain new ideas. Those people with "closed" personality, by contrast, tend to exhibit more traditional and conventional interests and prefer familiarity over novelty and change. The five-factor personality psychology is purely descriptive rather than theory driven, and current research is testing the cross-cultural and social validity of the program. While it is still in progress, this research at least raises the strong possibility of the close correlation of openness with creativity at the level of individual personalities emphasizing the relation to concepts of measured intelligence, achievement, and political attitudes (Simonton, 2000; Aitken Harris, 2004; Dollinger, 2007).

Individualist approaches to the relation of openness to creativity can only take us so far. The National Academy of Sciences' (2003) report *Beyond Productivity: Information Technology, Innovation and Creativity*, began by recognizing the crucial role that creativity plays in culture and the way in which at the beginning of the twenty-first century, "information technology (IT) is forming a powerful alliance with creative practices in the arts and design to establish the exciting new domain of information technology and creative practices" (p. 1). Others such as Richard Florida (2004) have emphasized that the United States needs to invest more in the development of its creative sector as a basis to sustain its competitiveness from the rate of technological innovation and economic growth. Florida (2002, p. 21) argues "human creativity as the defining feature of economic life. . . . [New] technologies, new industries, new wealth and all other good economic things flow from it," and he goes on to write "[Human] creativity is multifaceted and multidimensional. It is not limited to technological innovation or new business models. It is not something that can be kept in a box and trotted out when one arrives at the office. Creativity involves distinct kinds of thinking and habits that must be cultivated both in the individual and in the surrounding society" (p. 22). Rutten and

Gelissen (2008) test Florida's creativity and diversity hypothesis for European regions, and their results indicate that regional differences in diversity are directly related to differences in wealth between regions.

The relation between openness and creativity is brought out even more forcefully through the concept and practice of open innovation. Peter Teirlinck and Andre Spithoven (2008) indicate that the increasing complexity of innovation has encouraged companies to use external knowledge sources to complement in-house activities, attempting to substitute a nonlinear feedback model for the old linear model, capturing the benefits of the learning process within and between firms and other organizations. As innovation networks grew even more complex, firms adopted the "new imperative" for creating and profiting from technology in the model of open innovation where innovation becomes increasingly distributed among various partners (Von Hippel, 1988). They write:

> The notion of open innovation is the result of the increasing complexity of innovation and how innovation management should cope with this complexity. It reflects an ever changing research environment (Chesbrough, 2001): the increasing mobility of knowledge workers; the applicability of research results of universities to enterprises; more widely distributed knowledge; erosion of oligopoly market positions; more deregulation and an increase in venture capital. This resulted in an open stage gate process with the following features: (1) the centralized inhouse R&D laboratory is no longer the main source of ideas or knowledge and is being complemented by other enterprises, new technology based start-ups, universities, and public research centres; (2) commercialization also occurs outside the traditional markets of the enterprise through licensing, spin-offs, and research joint ventures; (3) the role of the first mover advantage becomes more important than the development of a defensively orientated system of knowledge and technology protection. (p. 689)

This model of open innovation is made possible through "creativity support tools" that help to accelerate discovery and innovation. Ben Shneiderman (2007) notes that new "generations of programming, simulation, information visualization, and other tools are empowering engineers and scientists just as animation and music composition tools have invigorated filmmakers and musicians" (p. 20). He goes on to write:

> These and many other creativity support tools enable discovery and innovation on a broader scale than ever before; eager novices are performing like seasoned masters and the grandmasters are producing startling results. The accelerating pace of academic research, engineering innovation, and consumer product design is

> amply documented in journal publications, patents, and customer purchases. Creativity support tools extend users' capability to make discoveries or inventions from early stages of gathering information, hypothesis generation, and initial production, through the later stages of refinement, validation, and dissemination (p. 20).

The sustainability of "social creativity" depends upon a greater recognition of the importance of social and material surroundings. As Fischer and Giaccardi (2007) argue "Individual and social creativity can and must complement each other" (p. 28). They suggest:

> Environments supporting mass collaboration and social production such as annotated collections (GenBank), media sharing (Flickr, YouTube), wikis (Wikipedia), folksonomies (del.icio.us), and virtual worlds (Second Life) are other examples of social creativity. The diverse and collective stock of scientific content and artistic or stylistic ideas that individuals and communities share, reinterpret, and use as a basis for new ideas and visions constitutes the vital source of invention and creativity (p. 28).

They argue that creativity needs the "synergy of many" which can be facilitated by meta-design—"a sociotechnical approach that characterizes objectives, techniques, and processes that allow users to act as designers and be creative in personally meaningful activities" (p. 28), and they note a tension between creativity and organization. Organizational environments must be kept open to users' modifications and adaptations by technical and social means that empower participation to serve the double purpose: "to provide a potential source for new insights, new knowledge, and new understandings; and to provide a higher degree of synergy and self-organization" (p. 29).

The relationship between creativity and open systems especially in computing is growing in significance. Colin G. Johnson (2005) draws a strong set of connection between openness, creativity, and search processes. He begins by noting that "One characteristic of systems in which creativity can occur is that they are open. That is, the space being explored appears to be (theoretically or pragmatically) unbounded, and there is no easy way in which the structure of the space can be simply summarized" (p. 1). He suggests that evolutionary search processes (moving from one-to-point, using the information from previously visited sites) are seen as creative for one of three reasons:

> Firstly because the criteria for evaluation are not easy to capture in a rulebound fashion. An example of this is searching a space of melodies for 'interesting' or 'tuneful' melodies. Secondly because the search space is seen as having some complexity which belies 'easy' search. Examples of this [sic] ideas include the use of search to explore the space of designs for mechanical devices or electrical circuits. Even though an exhaustive search would turn up the same result as a 'creative' search, both the size of the search space and the complex structure thereof (e.g. it is not possible for a 'naïve' thinker to conceive of how to specify and order the 'all possible' designs). Thirdly, because the search space is seen as being extensible. Consider the idea of searching a space of melodies as discussed above. In order to search this space, we will need to give a description of what a 'melody' is, e.g. a sequence of notes in a particular key. However this definition has limitations: what about a melody that changes key half way through? So we expand the search space to include such melodies, then. . . . The search space can always be extended. It is these latter two characteristics which seem particularly to capture the idea of 'openness' in creativity (pp.1–2). (http://kar.kent.ac.uk/14358/1/VarietiesColin.pdf)

Open source in computing developed around Linux as an operating system where in such open systems intellectual property is seen as "open" and is made freely available, allowing people to use ideas and code without locking them up as private intellectual property. It is based on three essential features (Tippett, 2007, updated from Weber, 2004):

- source code is distributed with the software, or made available at no more cost than distribution (this means that users can see and change the actual mechanisms that makes the software work);
- anyone may distribute the software for free (there is no obligation for other users of the software to pay royalties or licensing fees to the originator);
- anyone may modify the software, or develop new software from the original product, and the modified software is then distributed under the same terms as the original software (e.g., it remains open) (p. 4).

As Weber comments, these concepts represent a fundamentally different concept of property, typically seen as:

> a regime built around a set of assumptions and goals that are different from those of mainstream intellectual property rights thinking. The principal goal of the open source intellectual property regime is to maximize the ongoing use, growth, development, and distribution of free software. To achieve that goal, this regime

> shifts the fundamental optic of intellectual property rights away from protecting the prerogatives of an author towards protecting the prerogatives of generations of users. (Weber, 2004, p. 84)

The idea of open source still retains concepts of copyright and the rights of the author or creator over their original work. As Tippett (2007) remarks: "It does thus not negate the concept of property within intellectual products, but rather shifts the view of the rights conferred by the property, so that the 'concept of property [is] configured around the right and responsibility to distribute, not to exclude' (Weber, 2004: 86)" (p. 4). Tippett also usefully documents the emerging field that applies open source to areas of scholarship and creative endeavor outside software:

> For example, open source has been explored as a valuable approach in scientific endeavour and making scientific information available (Jones, 2001; Mulgan, 2005; Schweik, C., Evans and Grove, 2005). Keats (2003) has explored open source in terms of developing teaching and learning resources for African universities'. In a series of articles looking at the 'Adaptive State', the potential value of open source ideas for public policy delivery are explored (Bentley and Wilsdon, 2003; Leadbeater, 2003; Mulgan, Salem and Steinberg, 2005). The ideas have been developed in product design, linked to ideas of open innovation, as companies engage with user communities (Goldman and Gabriel, 2005), one example being user-led innovation in sports gear (Fuller, Jawecki and Muhlbacher, 2007) (p. 4). http://www.aesop2007napoli.it/full_paper/track3/track3_298.pdf.

Digital technologies have become engines of cultural innovation, and user-centered content production has become a sign of the general transformation of organizational forms. However, the transformation of digital culture also transforms "what it means to be a creator within a vast and growing reservoir of media, data, computational power, and communicative possibilities." We are only now beginning to devise understandings of the power of databases, network representations, filtering techniques, digital rights management, and the other new architectures of agency and control and "how these new capacities transform our shared cultures, our understanding of them, and our capacities to act within them" (Karaganis, 2008, p. 1).

As Jean Burgess (2007) comments in *Vernacular Creativity and the New Media* "The manufacturers of content-creating tools, who relentlessly push us to unleash that creativity, using—of course—their ever cheaper, ever more powerful gadgets and gizmos. Instead of asking consumers to watch, to listen, to play, to passively consume, the race is on to get them to create, to produce, and to participate" (p. 7). She goes on to register the development of a new vocabulary that speaks of a participatory culture based on creation and user-generated content.

> In game environments particularly, terms like 'co-creators' (Banks, 2002) and 'productive players' (Humphreys, 2005) are increasingly gaining purchase as replacements for 'consumers', 'players', or even 'participants'. These reconfigurations force us to consider the 'texts' of new media to be emergent—always in the process of being 'made'; further, 'co-creation' is built around network sociality and the dynamics of community, prompting a reconsideration of the idea of the individual producer or consumer of culture—even as corporate content 'owners' continue, in varying degrees, to assert rights that have their basis in the romantic notion of the individual creative author (Herman et al., 2006). It is not only the 'who' of production that is transformed in contemporary digital culture, but the *how.* (pp. 7–8)

Furthermore, Burgess details three important structural transformations from the point of view of cultural participation implied by the Web 2.0 model. I summarize from Burgess as follows:

1. The shift from content "production," "distribution" and "consumption" to a convergence of all three, resulting in a hybrid mode of engagement called "produsage," defined as "the collaborative and continuous building and extending of existing content in pursuit of further improvement" (Bruns, 2005).

2. A shift from "user-generated content" to "user-led" content creation, editing, repurposing, and distribution; whereby the users of a given Web service increasingly take on leadership roles, and where designers and developers to some extent allow the emergence of communities of practice to shape the culture of the network—even to determine what the Web service or online community is "for." This dynamic represents a convergence of the "value chain" where users are simultaneously the producers, users, editors and consumers of the content, leading to "network effects."

3. The convergence of user-generated content and social software to produce hybrid spaces, examples of which are sometimes described as "social media" (Coates, 2006)—most clearly represented by MySpace, YouTube and Flickr (Burgess, 2007, pp. 10–11)

Burgess (2007) argues:

> It is this third feature of the new networks of cultural production that has the most profound implications for cultural participation, at least in potential, because this shift opens up new and diverse spaces for individuals to engage with a variety of aesthetic experiences at the same time as their participation contributes to the creation of communities. That is, the significance of "Web 2.0," from a cultural studies point of view, lies in its potential for a new configuration of the relations between the *aesthetic* and the *social* aspects of culture, developed at a grass-roots level. (p. 11)

As many scholars and commentators have suggested since the "change merchants" of the 1970s—Marshall McLuhan, Drucker, and Alvin Toffler—first raised the issue we are in the middle of a long-term cultural evolutionary shift based on the digitization and the logic of open systems that has the capacity to profoundly change all aspects of our daily lives—work, home, school—and existing systems of culture and economy. A wide range of scholars from different disciplines and new media organizations have speculated on the nature of the shift: Richard Stallman established the Free Software Movement and the GNU project[5]; Yochai Benkler (2006), the Yale law professor, has commented on the wealth of networks and the way that social production transforms freedom and markets; his colleague, Larry Lessig (2004, 2007), also a law professor, has written convincingly on code, copyright, and the creative commons[6] and launched the Free Culture Movement designed to promote the freedom to distribute and modify creative works through the new social media[7]; Students for Free Culture,[8] launched in 2004, "is a diverse, non-partisan group of students and young people who are working to get their peers involved in the free culture movement"; Michel Bauwens (2005) has written about the political economy of peer production and established the P-2-P Foundation[9]; Creative Commons[10] was founded in 2001 by experts in cyber law and intellectual property; Wikipedia[11] the world's largest and open-content encyclopedia was established in 2001 by Jimmy Wales, an American Internet entrepreneur, whose blog is subtitled Free Knowledge for Free Minds.[12]

One influential definition suggests

> Social and technological advances make it possible for a growing part of humanity to *access, create, modify, publish and distribute* various kinds of works—artworks, scientific and educational materials, software, articles—in short: *anything that can be represented in digital form.* Many communities have formed to exercise those new possibilities and create a wealth of collectively re-usable works. http://freedomdefined.org/Definition

By *freedom* they mean:

- the **freedom to use** the work and enjoy the benefits of using it
- the **freedom to study** the work and to apply knowledge acquired from it
- the **freedom to make and redistribute copies**, in whole or in part, of the information or expression
- the **freedom to make changes and improvements**, and to distribute derivative works[13]

This is how the Open Cultures Working Group—an open group of artists, researchers and cultural activists—describe the situation in their Vienna Document subtitled Xnational Net Culture and "The Need to Know" of Information Societies:

> Information technologies are setting the global stage for economic and cultural change. More than ever, involvement in shaping the future calls for a wide understanding and reflection on the ecology and politics of information cultures. So called globalization not only signifies a worldwide network of exchange but new forms of hierarchies and fragmentation, producing deep transformations in both physical spaces and immaterial information domains . . . global communication technologies still hold a significant potential for empowerment, cultural expression and transnational collaboration. To fully realize the potential of life in global information societies we need to acknowledge the plurality of agents in the information landscape and the heterogeneity of collaborative cultural practice. The exploration of alternative futures is linked to a living cultural commons and social practice based on networks of open exchange and communication.[14]

Every aspect of culture and economy is becoming transformed through the process of digitization that creates new systems of archives, representation, and reproduction technologies that portend Web 3.0 and Web 4.0 where all production, material and immaterial, is digitally designed and coordinated through distributed information systems. As Felix Stadler (2004) remarks "information can be infinitely copied, easily distributed, and endlessly transformed. Contrary to analog culture, other people's work is not just referenced, but directly incorporated through copying and pasting, remixing, and other standard digital procedures" (p. 7). Digitization transforms all aspects of cultural production and consumption favoring the networked peer community over the individual author and blurring the distinction between artists and their audiences. These new digital logics alter the logic of the organization of knowledge, education, and culture spawning new technologies as a condition of the openness of the system. Now the production of texts, sounds, and images is open to new rounds of experimentation and development providing what Stadler calls "a new grammar of digital culture" (p. 7). Furthermore, the processes of creativity are no longer controlled by traditional knowledge institutions and organizations but rather have emerged as platforms and infrastructures that encourage large-scale participation and challenge old hierarchies.

The shift to networked media cultures based on the ethics of participation, sharing, and collaboration, involving a volunteer, peer-to-peer gift economy has its early beginnings in the right to freedom of speech that depended upon the flow and exchange of ideas essential to political democracy, including the notion of a "free press," the market and the academy. Perhaps, even more fundamentally free speech is a significant personal, psychological, and educational good that promotes self-expression and creativity and also the autonomy and development of the self necessary for representation in a linguistic and political sense and the formation of identity (Peters, 2009).

References

Aitken Harris, J. (2004, March). Measure intelligence, achievement, openness to experience and creativity, *Personality & Individual Differences,* 36(4), 913

Banks, John (2002) 'Gamers as Co-Creators: Enlisting the Virtual Audience—A Report from the Net Face.' *Mobilising the Audience.* Mark Balnaves, Tom O'Regan & Jason Sternberg, Eds. Brisbane: University of Queensland Press.

Baumol, W. J. (2002). *The free-market innovation machine: Analyzing the growth miracle of capitalism.* Princeton, NJ: Princeton University Press.

Bauwens, M. (2005) The Political economy of Peer production', *C-Theory* at http://www.ctheory.net/articles.aspx?id=499

Bell, D. (1973). *The coming of post-industrial society: A venture in social forecasting.* New York: Basic Books.

Benkler, Y. (2006). *The wealth of networks.* New Haven, CT: Yale University Press.

Bentley, T., & Wilsdon, J. (Eds.). (2003). *The adaptive state—Strategies for personalising the public realm.* London, Demos. Retrieved from http://www.demos.co.uk/files/HPAPft.pdf.

Blythe, M. (2000). *Creative learning futures: Literature review of training & development needs in the creative industries.* Retrieved from http://www.cadise.ac.uk/projects/creative-learning/New_Lit.doc.

Bourdieu, P. (1986). The forms of capital. Richard Nice (Trans.). In J. F. Richardson (Ed.), *Handbook of theory of research for sociology of education* (pp. 241–258). Westport, CT: Greenwood Press.

Brown, J. S., & Duguid, P. (2000). *The social life of information.* Boston, MA: Harvard Business School Press.

Bruns, A. (2005) Gatewatching: Collaborative Online News Production. *Digital Formations.* Steve Jones, Ed. New York: Peter Lang.

Burgess, J. (2007). *Vernacular creativity and the new media* (Doctoral dissertation, Creative Industries Faculty, Queensland University of Technology). Retrieved from http://eprints.qut.edu.au/10076/1/Burgess_PhD_FINAL.pdf

Cardoso, G., & Castells, M. (Eds.). (2006). *The network society: From knowledge to policy.* Washington, DC: Brookings Institute Press.

Castells, M. (1996/2000). *The rise of the network society: The information age: Economy, society and culture.* Vol. I. Cambridge, MA: Blackwell.

Caves, R. E. (2000) *Creative industries: Contracts between art and commerce.* Cambridge, MA: Harvard University Press.

Coates, Tom (2006) 'What Do We Do With "Social Media"?' *Plastic Bag.* Retrieved 4 May 2006 from http://www.plasticbag.org/archives/2006/03/what_do_we_do_with_social_media.shtml.

Coleman, J. (1988). Social capital in the creation of human capital. *American Journal of Sociology, 94* Supplement, pp. S95-S-120.

Cowen, T. (2002). *Creative destruction: How globalization is changing the world's cultures.* Princeton, NJ: Princeton University Press.

Davenport, T., & Beck, J. (2001). *The attention economy: Understanding the new economy of business.* Cambridge, MA: Harvard Business School Press.

Davis, J. & Stack, M. (1997) Rethinking Globalization, *Race & Class, 40 (2/3).* Retrieved from http://www.gocatgo.com/texts/rc.global.v4.html .

DeLong, J. B., & Summers, L. H. (2002). 'The "new economy": Background, historical perspective, questions, and speculations.' Retrieved from http://www.kansascityfed.org/PUBLICAT/SYMPOS/2001/papers/S02delo.pdf

Dollinger, S. J. (2007, October). Creativity and conservatism. *Personality & Individual Differences,* 43(5), 1025–1035.

Drucker, P. (1969). *The age of discontinuity: Guidelines to our changing society.* New York: Harper & Row.

Drucker, P. F. (1993). *Post-capitalist society.* New York: Harper Business.

Fischer, G., & Giaccardi, E. (2007). Sustaining social creativity. *Communications of the ACM, 50*: 28–29.

Florida, R. (2004). America's looming creativity crisis, *Harvard Business Review, 82*(10), 122–136.

Florida, R. (2002). *The rise of the creative class.* New York: Basic Books.

Foray, D. (2004). The economics of knowledge. Cambridge, MA: MIT Press.

Frey, B. (2000). *Arts and economics: Analysis and cultural policy.* New York: Springer.

Frey, B. S., & Pommerehne, W. W. (1989). *Muses and markets: Explorations in the economics of the arts.* Cambridge, Mass: Blackwell.

Fuller, J., Jawecki, G., & Muhlbacher, H. (2007). Innovation creation by online basketball communities. *Journal of Business Research, 60*(1), 60–71.

Ginsburgh, Victor A., & Menger, P.-M. (Eds.). (1996). *Economics of the arts.* Amsterdam: North Holland.

Goldman, R., & Gabriel, R. P. (2005). *Innovation happens elsewhere: Open source as business strategy.* San Francisco: Morgan Kaufmann Publishers, Inc.

Gordon, W. J. (1993). A property right in self-expression: Equality and individualism in the natural law of intellectual property, 102 *Yale L.J.* 1533, 1568–1572.

Granovetter, M. (1973, May). The strength of weak ties. *American Journal of Sociology, 78*(6), 1360–1380.

Granovetter, M. (1983). The strength of weak ties: A network theory revisited. *Sociological Theory, 1,* 201–233.

Harvey, D. (1989). *The condition of postmodernity.* Oxford: Blackwell.

Hayek, F. (1937). Economics and knowledge. Presidential address delivered before the London Economic Club, November 10, 1936. Reprinted in *Economica IV* (new ser., 1937), 33–54.

Hayek, F. (1945, September). The use of knowledge in society. *The American Economic Review, 35*(4), 519–530.

Hearn, G., & Rooney, D. (Eds.). (2008). *Knowledge policy: Challenges for the twenty-first century.* Cheltenham: Edward Elgar.

Heilbrun, J., & Gray, C. M. (2001). *The economics of art and culture* (2nd ed.). New York: Cambridge University Press.

Herman, Andrew, Rosemary J. Coombe & Lewis Kaye (2006). Your *Second Life*? Goodwill and the performativity of intellectual property in online gaming. *Cultural Studies* 20(2–3): 184–210.

Hesmondhalgh, D. (2002). *The cultural industries.* Thousand Oaks, CA: Sage.

Howkins, J. (2001). *The creative economy: How people make money from ideas.* London: Allen Lane.

Hughes, J. (1988) The philosophy of intellectual property, 77 *Geo. L.J.* 287, 337–344.

Humphreys, S. (2005) Productive players: Online computer games challenge to conventional media forms. *Journal of Communication and Critical/Cultural Studies* 2(1): 36–50.

Johnson, C. (2005) Varieties of openness in evolutionary creativity. Retrieved from http://kar.kent.ac.uk/14358/1/VarietiesColin.pdf.

Jones, P. (2001). Open(source)ing the doors for contributor-run digital libraries. *Communications of the Association for Computing Machinery, 44*(5), 45–46. Retrieved from http://delivery.acm.org/10.1145/380000/374337/p45-jones.html?key1=374337&key2=9102569901&coll=ACM&dl=ACM&CFID=11722369&CFTOKEN=18823426.

Karaganis, J. (Ed.). (2008). *Structures of participation in digital culture.* New York: Columbia University Press.

Keats, D. (2003). Collaborative development of open content: A process model to unlock the potential for African universities. *First Monday,* **8** (3): Retrieved from http://firstmonday.org/issues/issue8_2/keats/index.html.

Kelly K. (1998). *New rules for the new economy*. London: Fourth Estate.

Landry, C. (2000). *The creative city: A toolkit for urban innovators*. London: Earthscan.

Lash, S., & Urry, J. (1994). *Economies of signs and space*. Thousand Oaks, CA: Sage.

Leadbeater, C. (2003.) Open innovation in public services. *The Adaptive State—Strategies for personalising the public realm*. T. Bentley & J. Wilsdon (Eds.), *Demos*: 37–49. Retrieved from http://www.demos.co.uk/files/HPAPft.pdf.

Lemley, M. A. (2005). Property, intellectual property, and free riding. 83 *Tex. L. Rev.* 1031, 1031.

Lessig, L. (2004) *Free Culture: How Big Media Uses Technology and the Law to Lock Down Culture and Control Creativity*. New York, The Penguin Press.

Lessig, L. (2006). *Code: Version 2.0*. New York: Basic Books.

Lessig, L. (2007)Foreword, in *Freedom of Expression: Resistance and Repression in the Age of Intellectual Property*, Kembrew McLeod, Minneapolis: University of Minnesota Press, 2007.

Lorenz, E., & Lundvall, B.-Å. (2006). (Eds.). *How Europe's economies learn*. Oxford: Oxford University Press.

Lundvall, B.-Å., & Archibugi, D. (2001). *The globalizing learning economy*. New York: Oxford University Press.

Lundvall, B.-Å., & Johnson, B. (1994). The learning economy. *Journal of Industry Studies*, *1*(2), 23–42.

Lyotard, J.F. (1984). *The postmodern condition: A report on knowledge*. G. Bennington & B. Massumi (Trans.). Manchester: Manchester University Press.

Morris-Suzuki, T. (1997). Capitalism in the computer age and afterward. In J. Davis, T. A. Hirschl, & M. Stack (Eds.), *Cutting edge: Technology, information capitalism and social revolution* (pp. 57–72). London: Verso.

Mulgan, G. (2005). A wiki way to work and spread scholarship. *The Times Higher Education Supplement*, 12, July 29.

Mulgan, G., Salem, O., & Steinberg, T. (2005). *Wide open: Open source methods and their future*. London, Demos. Retrieved from http://www.demos.co.uk/catalogue/wideopen/.

National Academy of Sciences. (2003). *Beyond productivity: Information technology, innovation and creativity*. Washington, DC: NAS Press.

Netanel, N. W. (1996). Copyright and a democratic civil society. 106 *Yale L.J.* 283, 347-–362.

Netanel, N. W. (1998). Asserting copyright's democratic principles in the global arena. *51 Vand. L. Rev.* 217, 272–276.

Nonaka, I., & Takeuchi, H. (1995). *The knowledge-creating company*. New York: Oxford University Press.

OECD. (1996). *The knowledge-based economy*. Paris: The Organization.

Pasquinelli, M.(2008) *'The Ideology of Free Culture and the Grammar of Sabotage'*. Retrieved from http://www.generation-online.org/c/fc_rent4.pdf. Chapter 14, this volume.

Peters, M. (2009) Degrees of openness and the virtues of openness: The Internet, the academy and digital rights, in J. Satterthwaite, H. Piper & P. Sikes, Eds. *Power in the academy.* London, Trenthan Books, 79–96.

Peters, M., & Besley, T. (A. C.) (2006). *Building knowledge cultures: Education and development in the age of knowledge capitalism.* Lanham, MD: Rowman & Littlefield.

Peters M., Murphy, P. & Marginson, S. (2009) *Creativity and the Global Knowledge Economy.* New York, Peter Lang.

Porat, M. (1977). *The information economy.* Washington, DC: US Department of Commerce.

Powell, W. W., & Snellman, K. (2004). The knowledge economy. *Annual Review of Sociology, 30,* 199–220.

Prabhu, V., Sutton, C. &, Sauser, W. (2008). Creativity and certain personality traits: Understanding the mediating effect of intrinsic motivation, *Creativity Research Journal 20*(1), 53–66.

Prusak, L. (1997). *Knowledge in organizations.* Boston, MA: Butterworth-Heinemann.

Putnam, R. (2000). *Bowling alone: The collapse and revival of American community.* New York: Simon and Schuster.

Quah, D. (2003a). Digital goods and the new economy. In Derek Jones (Ed.), *New Economy Handbook* (pp. 289–321). Amsterdam, the Netherlands: Elsevier.

Quah, D. (2003b). The weightless economy. Retrieved from http://econ.lse.ac.uk/staff/dquah/tweirl0.html (accessed 30 August 2008).

Renn, A. (1998) 'Free', 'Open Source', and Philosophies of Software Ownership'. Retrieved from http://www.urbanophile.com/arenn/hacking/fsvos.html.

Romer, P. M. (1990). Endogenous technological change. *Journal of Political Economy, 98,* 71–102.

Rooney, D., Hearn, G., Mandeville, T., & Joseph, R. (2003). *Public policy in knowledge-based economies: Foundations and frameworks.* Cheltenham: Edward Elgar.

Rutten, R., & Gelissen, J. (2008). Technology, talent, diversity and the wealth of European regions. *European Planning Studies, 16*(7): 253–270.

Schweik, C., Evans, T., & Grove, J. M. (2005). Open source and open content: A framework for global collaboration in social-ecological research. *Ecology and Society, 10*(1). Retrieved from http://www.ecologyandsociety.org/vol10/iss1/.

Shapiro, Carl and Hal R. Varian (1998) Information Rules: A Strategic Guide to the NetworK Economy. Boston, MA: Harvard Business Press.

Shneiderman, B. (2007). Creativity support tools: accelerating discovery and innovation. *Communications of the ACM, 50:* 20–32.

Simonton, D. K. (2000). Creativity: Cognitive, personal, developmental, and social aspects. *American Psychologist, 55*(1), 151–158.

Söderberg, J. (2002). Copyleft vs. copyright: A Marxist critique. *First Monday, 7*(3). Retrieved from http://www.firstmonday.org/issues/issue7_3/soderberg/ (accessed 30 August 2008).

Stadler, F. (2004) Open Cultures and the nature of networks. Retrieved from http://felix.openflows.com/pdf/Notebook_eng.pdf.

Stiglitz, J. (1999a). Knowledge as a global public good. Retrieved from http://www.worldbank.org/knowledge/chiefecon/articles/undpk2/ (accessed 30 August, 2008).

Stiglitz, J. E. (1999b, July). Knowledge as a global public good. *Global Public Goods, 19,* 308–326.

Surowiecki, J. (2004). *The wisdom of crowds: Why the many are smarter than the few and how collective wisdom shapes business, economies, societies and nations.* New York: Anchor.

Tapscott, D., & Williams, A. (2006). *Wikinomics: How mass collaboration changes everything.* Atlantic Books.

Tapscott, D., & Williams, A. D. (2007). *Wikinomics: How mass collaboration changes everything.* New York: Penguin.

Teirlinck, P., & Spithoven, A. (2008). The spatial organization of innovation: Open innovation, external knowledge relations and urban structure. *Regional Studies, 42*(5), 689–704.

Tippett, J. (2007). Creativity, networks and openness—The potential value of an open source approach to support practitioners in planning for sustainability. Retrieved from http://www.aesop2007napoli.it/full_paper/track3/track3_298.pdf.

Toffler, A. (1980). *The third wave.* New York: Bantam Books.

Touraine, A. (1971). *The post-industrial society: Tomorrow's social history: Classes conflicts & culture in the programmed society.* L. Mayhew (Trans.). New York: Random House.

Verschraegen, G., & Schiltz, M. (2007). Knowledge as a global public good: The role and importance of open access. *Societies Without Borders, 2*(2),157–174.

Von Hippel , E. (1998) 'Economics of Product Development by Users: The Impact of "Sticky" Local Information', *Management Science,* vol. 44, No. 5 (May) p. 629-644

Wagner, R. P. (2003). Information wants to be free: Intellectual property and the mythologies of control, 103 *Colum. L. Rev.* 995, 1001–1003.

Weber, S. (2004). *The success of open source.* Cambridge, MA: Harvard University Press.

Notes

1. This chapter draws on my Introduction to *Creativity in the Global Knowledge Economy* (Peters, Marginson & Murphy, 2009).
2. These statements are taken from the New Club of Rome's 2006 Manifesto at http://www.the-new-club-of-paris.org/mission.htm.
3. See The Center for Accelerated Learning at http://www.alcenter.com/; see e.g., The Framework for Gifted Education at http://education.qld.gov.au/publication/production/reports/pdfs/giftedandtalfwrk.pdf.
4. For innovation theory see the Swedish economist Bengt-Åke Lundvall's Web page at http://www.business.aau.dk/ike/members/bal.html and especially his concept of "the learning economy"; see also Globelics, The Global Network for the Economics of Learning, Innovation, and Competence Building Systems at http://www.globelics.org/.

5. See the GNU site http://www.gnu.org/gnu/initial-announcement.html, a 2006 lecture by Stallman entitled "The Free Software Movement and the Future of Freedom" and Aaron Renn's (1998) "Free," "Open Source," and Philosophies of Software Ownership at http://www.urbanophile.com/arenn/hacking/fsvos.html.
6. See his bestseller *Free Culture* http://www.free-culture.cc/freeculture.pdf.
7. See the videoblog "Free Culture, Free Software, Free Infrastructures! Openness and Freedom in Every Layer of the Network" at http://www.perspektive89.com/2006/10/18/free_culture_free_software_free_infrastructures_openness_and_freedom_in_every_layer_of_the_network_flo_fleissig_episo but see also Pasquinelli's (2008) "The Ideology of Free Culture and the Grammar of Sabotage" at http://www.rekombinant.org/docs-/Ideology-of-Free-Culture.pdf.
8. See the Web site http://freeculture.org/.
9. See the foundation at http://p2pfoundation.net/The_Foundation_for_P2P_Alternatives and the associated blog at http://blog.p2pfoundation.net/.
10. See http://creativecommons.org/.
11. See http://www.wikipedia.org/.
12. See http://blog.jimmywales.com/.
13. See http://freedomdefined.org/Definition.
14. See http://world-information.org/wio/readme/992003309/1134396702.

11

Symposium on *The Wealth of Networks*

Philippe Aigrain
Leslie Chan
Jean-Claude Guédon
John Willinsky
Yochai Benkler

PHILIPPE AIGRAIN: On the Economic Impact and Needs of The Wealth of Networks

When I drafted *Cause commune*[1] in 2003–04, my main motivation was to provide a comprehensive theoretical foundation for the growing information commons[2] movement in Europe. In the background was another motivation: helping continental European readers to better relate with American commons thinking. Several factors were limiting the ability of European readers to draw inspiration from or to build critical conversations with the generation of thinkers who have put information commons on the agenda in the USA. The absence in continental Europe of an equivalent to First Amendment thinking (despite commitments to freedom of expression); a stronger focus in Europe on the opposition between markets and State, with less consideration for the importance of societal action; the difficulty to translate terms such as commons in a modern sense, even though they are of Latin origin: all this contributed to possible misunderstandings.

Today, commons thinking has become global in scope and develops in original forms in non-English speaking areas. Translations of *commons* slowly percolate in various languages. In France, Members of Parliament speak about *libre* software in their debates and lawyers in university and official circles organize workshops on creative commons approaches. Brazil is at the forefront of commons-based creation, and despite some opposite trends, India hosts new approaches to commons-based innovation and some original commentators (sarai.net, Suman Sahai or Vandana Shiva). In North America, thinkers such as James Boyle, Lawrence Lessig, or Eben Moglen have gone to great efforts to create narratives with a more universal perspective. James Love and Knowledge Ecology International[3] bootstrapped worldwide non-governmental organization (NGO) and government coalitions that push a knowledge commons agenda forward in international organizations. The revision of the GNU GPL (General Public Licence) free software license has explicitly aimed at a more universal language and validity.

Yochai Benkler's *Wealth of Networks* takes a new step in this direction. Benkler's work is deeply rooted in American liberal philosophy. However, his interest in political philosophy globally has led him to express his views in a language that can be read from a European or global perspective. Central to this possibility is the bridge provided by Amartya Sen's theory of capabilities. Amartya Sen has emphasized[4] how freedom is necessary to development but also how health, education, or preventing excessive inequalities are necessary for the build-up of capabilities, without which freedom remains just an idea. By linking in an inseparable manner freedom, concrete capabilities of action, human development, and social justice, Amartya Sen has provided a theory of justice that can be heard across the Atlantic and elsewhere, in particular, of course, in India, his mother country, recognizing its calls for radical reform to our ways of thinking, for instance, about education, culture, or public health. Seeing education mostly as a transfer of existing knowledge (which is still a predominant view in many quarters of Europe) is clearly challenged by the evidence that the construction of critical individuals able to take initiatives in the world deserves more emphasis in our information era. The traditional European view of culture as the recognition of a distinguished elite by connoisseurs (also adopted in some circles in the USA) is challenged by the alternative model of a continuum of practice ranging from reception to professional creation of the highest quality. Meanwhile, approaches to health based on targeted technological medical acts and drugs and the associated economic and patent models exhibit poor performance in comparison to systems that put more emphasis on education, universal access to medical services, and drugs and the non-medical conditions of public health.

It is in the field of information and its technology that capability theory is most relevant.[5] When information and tools are available as commons for all, an unprecedented ability to express oneself, reach for others, criticize or praise, cooperate on all forms of achievements develops in individuals and groups that they form. *The*

Wealth of Networks is an enlightening tribute to this power. However, this recognition, and the ability of European and American analysts to share this common umbrella, is only the start of a new conversation. This conversation is likely to have many threads, but I will just initiate one by asking: How does the growth of information commons and related non-market activities interact with the monetary economy?

From Parallelism to Collision

Yochai Benkler, just like Lawrence Lessig,[6] proposes an optimistic view of the impact of commons-based societal production of information on the economy overall. There are strong arguments in favor of such an optimistic view as a long-term perspective. If a huge sphere of non- market activities develops, the provision of infrastructure and support services to these activities is itself a huge domain. In addition, more demanding consumers and more knowledge and innovation (for instance, on environmental issues) open new prospects for development in the material economy. However, what about the transition from our existing economy to this long- term perspective? Here is a small narrative to illustrate why this transition risks being a difficult and chaotic one, well beyond the choices whose need Yochai Benkler rightly stresses regarding our regulatory ecology.

Once upon a time, there came the information revolution. First, from 1945 to 1975, non-market societal production of freely exchangeable and usable information developed silently. It happened as a natural way of using information technology and information-based science in specialized circles (scientists, programmers), often without naming explicitly what was done. In this period scientists and engineers created the best part of software techniques and algorithmics, of network protocols, user interfaces and digital media concepts, of information science and how it can be applied in biology. This infrastructure set the basis for the next 30 years of human development, new markets and growth. However, very different ways of using information technology also matured during the early information age. Large organizations (both companies and public organizations), which were at the time the principal users of computers focused on process, profit and cost optimizing, with contrasted results on the latter. These early trends are now visible at full range, with the domination of finance on industry agendas and with a form of globalization where mechanisms[7] delay its positive effects on development. More recently, it was imitated by some emerging countries (mostly China) in which the lack of democracy makes it possible to exert lasting pressure against the build-up of local initiatives that serve human development. The control and surveillance aspects of information and communications technology (ICT)[8] have become more prominent: security and short-term profit optimizing have developed new synergies.

In a second phase, the two trends that earlier developed in parallel have started colliding. At first this collision was noticed only in specialized circles. Though the process preparing TRIPS (Agreement on Trade-related Aspects of Intellectual Property Rights administered by the World Trade Organization) can be traced back to the mid-1960s, who understood what was going on before its signature in 1994? Similarly, the expansion of information commons and societal production was visible only for its initiators until the mid-1990s. Today, it has become evident that there is a big battle to set the regulatory environment in favor of one or the other trend. However, this regulatory battle is only the perceptible part of deeper economic tensions. The two models are also fighting for money and for time. I mean capital and human time.

Capital and Profits

Let's start with capital. When one compares the ratio of stock capitalization to turnover, added value or profit[9] for major world companies in many sectors, one finds huge discrepancies. The ratio of capitalization to turnover varies from 0.11 (General Motors or Ford Motor) to 17 (Google) across companies, from 0.35 (automotive) to 3.55 (pharmaceuticals) across sectors. It ranges from 0.45 to 17 across the software sector where software service companies, proprietary software semi-monopolies and companies based on large network effects such as Google cohabit. Most companies that have high ratios are patent (mostly in pharmaceuticals and agro-food genetically modified organisms), copyright (in software and media) or trademarks (in food and luxury consumer goods) businesses.[10] These information capitalists[11] set the standards of desired return on capital. The effects of these standards are felt well beyond quoted (public) companies: research and development (R&D) funding or venture capital force many innovators to pose as future information capitalists rather than simply develop a sustainable and reasonably profitable activity that serves the information commons and non-market activities. In the present state of market organizations, the prospects of ROI (return on investment) for added-value intermediators in the non-market sphere are a huge question mark. In a given domain, one dominant collaborative medium can rest on advertising but serious economic works[12] have demonstrated that this model cannot scale up. Can the indirect funding by those who benefit from the non-market sphere (for instance, hardware manufacturers and telecommunications companies) and the recycling by some public interest-minded winners of financial games suffice to fuel the future wealth of networks? Can mutualizing between individuals provide enough of a complement? Is some form of government organization of mutual funds needed? All these questions deserve policy's and society's interest.

Human Time

The fight for access to capital is only a skirmish compared to the war that rages for human time. In a few decades, television has captured half of human free time in the developed world (31/2 hours per day in most countries). Time recently liberated from television in favor of ICT-mediated activities has unfortunately gone for a significant part to immersive activities, such as games and advertising-dominated digital media. What is less known is the degree to which the present economy has come to depend upon the capture of human attention. We are not speaking here of media alone. The demand for many consumer goods is sustained only through what one of the French television chief executive officers described as "available brain time" being provided by media to the providers of these goods. As brain time becomes less available, as more valuable endeavors compete for it, our present economy will go into a crisis. This crisis is also an opportunity, just as the environment and climate change challenges are an opportunity.

People becoming more independent, more critical, more able to choose how they use their time, becoming producers as well as consumers, members of the public in John Dewey's sense,[13] is an exciting perspective. It is infinitely valuable, whether or not it is good for the economy, but there is all reason to believe that it will lead to new forms of economic growth. However, we will go through chaotic paths before we are there. What will be apparently taken away from today's economy is actually fake; it is monetary face value that stands only on arbitrary monopolies, conventional beliefs or what will look retrospectively to be a strange consent to hypnotic consumption. We had better get our measures right as soon as possible, so they start showing what is being built during this process, with indicators that capture the many facets of the wealth of networks.

Notes

1. *Cause commune: l'information entre bien commun et propriété* [Common cause: information between commons and property]. Paris: Fayard, 2005.
2. In this article, information commons refers to productions that can be represented as information (creative works in all media, software, biological or other scientific information, information processing tools such as software, etc.) when they are given a commons status (are freely usable to relevant degrees). By extension, it also covers the collaborative activities that produce and value these productions.

3. Formerly the Consumer Project on Technology.
4. In particular in *Development as Freedom*. New York: Anchor, 2000.
5. See *The Wealth of Networks*, pp. 308–311, or my paper, "Capabilities in the Information Era," TransAtlantic Consumer Dialogue (TACD) Workshop on the *Politics and Rhetorics of Intellectual Property*, Brussels, March 2006. http://paigrain.debatpublic.net/docs/TACD-200306.pdf
6. This section is the product of a conversation with Lawrence Lessig initiated in Berlin after his *Read/Write Society* talk during the *Wizard of OS 4* Conference and pursued through later exchanges by email.
7. Globalization of uniform patent and copyright rules, division of production in small modules with no visibility of local workers—including technicians and engineers—on the overall strategies.
8. The aspects were always present, from World War II to the Cold War. The difference is the degree of confusion between domains: non-commercial exchange of information covered by copyright can be depicted as a form of terrorist cybercrime by copyright stock holders who successfully lobby for making it the object of criminal sanctions, while the creation of data retention for security purposes can be included in a European directive on privacy.
9. Cf. *Economic Impact of Open Source Software on Innovation and the Competitiveness of the ICT Sector in the EU,* study conducted by a consortium led by MERIT for the European Commission Enterprise and Industry General Directorate, pp.118–121. http://ec.europa.eu/enterprise/ict/policy/doc/2006–11–20-flossimpact.pdf (2004–05 data).
10. A small number are network effects companies, Google being the prototype of this category. They would deserve a specific treatment, as their activity is much more compatible with the development of commons-based non-market activities, though in a limited (in number of possible winners) and very unstable manner.
11. I use the word for all companies whose added value lies predominantly in the costless reproduction of an intangible entity, that can be information per se or some informational entity included or attached to their products.
12. See, for instance, Douglas A. Galbi, "Some Economics of Personal Activity and Implications for the Digital Economy," *First Monday*, 6(7). http://www.firstmonday.org/issues/issue6_7/galbi/index.html
13. Informed by their awareness of the interest of a greater community.

LESLIE CHAN: Human Development and Open Access 2.0 Overview

In the landmark book that is the subject of this symposium, Yochai Benkler lays out a grand vision of how non-market commons-based peer production of knowledge and culture is transforming the global economic and political systems, while simultaneously empowering individuals and citizens with new forms of personal and political autonomy. In a world where one billion people have the capacity to create, store, share and distribute information at minimal cost, the distinction and boundaries between the intellectual centre and the periphery are beginning to blur, as each node on the network has the potential to become the centre. This dramatic shift from the industrial model of production to distributed and decentralized knowledge production has the enormous potential to alter the course of human development and greatly reduce the huge North–South asymmetries in economic power and access to knowledge. Both are fundamental to the improvement of human well-being.

However, the connection between the seemingly abstract notion of commons-based knowledge production and human development is neither obvious nor straightforward. Some might ask, 'what has Wikipedia got to do with the 49 percent of the population of Congo that lacks sustainable access to improve water resources?' (Benkler, p. 321). Is a commons-based and non-market approach to development possible and how would it differ from the dominant development thinking based on the neo-liberal philosophy of market competition and global economic integration? Furthermore, what roles, if any, does open access to scientific literature play in international development, particularly with regard to poverty alleviation and the reduction of inequality in wealth and the general well-being of the citizens in less developed countries? Could open and collaborative research lead to substantive improvement in technical and research infrastructure, particularly in the areas of medicine, biotechnology and agriculture, in poorly resourced countries?

Benkler provides glimpses and examples for some of these questions in chapter 9 of his book, though he cautions against unbridled optimism, given that many of the commons-based initiatives are still in their infancy. Further, Benkler's interest is in demonstrating the normative aspects of commons-based productions and why they should work, given the appropriate conditions, and he is less concerned with the details of how the various models—such as open access publishing—could be carried out in practice and, more importantly, sustained. For those of us involved in fostering alternative models of scholarly publishing and, in particular, providing open access to research originating from developing countries, economic sustainability is one of the most pressing challenges. This is in part because funding bodies, universities, donors, and development agencies are not yet fully aware of the enormous benefits of open access to publicly funded research and, in particular, its effect on sustainable development.

In addition, universities and government funding bodies the world over are increasingly concerned with the economic return on research investment in terms of patents, spin-offs and intellectual property, while paying relatively little or diminished attention to the potential social and political return on research and public access.

Thus, another pressing need is the development of a framework for measuring social and intellectual capital and other benefits of open access that are non-market driven. Again, Benkler provides hints on how this can be accomplished, such as the development of new indicators, but further research and development in this area are much needed. There could also be an explicit link to Amartya Sen's vision of "Development as Freedom,"[1] which Benkler cites. In Sen's view, development should be seen not in terms of economic measures (e.g., gross domestic product [GDP] growth, average annual income), but in terms of the real freedoms that people can enjoy, such as educational facilities and social opportunities. Sen describes human freedom as both the primary end objective and the principal means of development, while economic measures are merely the means to this end. This view appears to be highly congruent with open access and this connection deserves further exploration.

In addition to economic sustainability and the need for alternative indicators, there are further challenges to the integration of open access and other forms of open and collaborative processes into current thinking and practices in the development arena. These include inclusive participation, *The Wealth of Networks* and multiple layers of interoperability. I highlight each of these and conclude with reasons why we should be optimistic about the future of human development in the networked economy.

Inclusive Participation

For researchers in developing countries, informed participation in global research agenda setting is often hampered by limited access to scientific information and essential data. Improved connectivity in many parts of the developing world is certainly improving access to the literature, but pricing and permission barriers are still significant impediments to the development of local research infrastructures. Programs such as the Health InterNetwork Access to Research Initiative (HINARI) supported by the World Health Organization and the sister programmes, AGORA (Access to Global Online Research in Agriculture, managed by the Food and Agriculture Organization) and OARE (Online Access to Research in the Environment, managed by the United Nations [UN] Environment Programme), are supposed to provide free access to researchers in qualified institutions in countries with gross national product (GNP) of $1000 or less per annum. These initiatives are being tied to the UN's Millennium Development Goals, and aim to "represent a truly global public private partnership for development, providing essential information for life to those who need it most."[2]

However, these programs are based on the implicit assumption that development is sufficient with the flow of knowledge or resources from the North to the South, as almost all the over three thousand journal titles are published in the North with only a small number of titles originating from the developing world. Are health and agricultural research conducted in America and Europe necessarily relevant to health workers, farmers and students in African countries, where disease profiles and food security are drastically different from the rich economies? Would work published in other developing regions of the world be more appropriate for researchers from those areas, particularly where development-related research is concerned?

Supporting scholarship in the global South must be a two-way street. In addition, the South–South exchange of scientific and traditional knowledge as well as common experiences may in fact be far more important for local development. Instead of just "donating" information to researchers in developing countries, international foundations and the public–private partnerships must provide researchers with a way to share knowledge with each other and participate in research opportunities with peers in the developed world. The integration of journals and research results from the South in the global knowledge base, made possible through the use of open access repositories, may be a simple route for achieving this goal.[3]

In his recent book *Convergence Culture: Where Old and New Media Collide,*[4] Henry Jenkins remarks,

> "Increasingly, the digital divide is giving way to concern about the participation gap. As long as the focus remains on access, reform remains focused on technologies; as soon as we start to talk about participation, the emphasis shifts to cultural protocols and practices" (Jenkins, 2006, p. 23). Until recently, development programs, particularly those initiated by the World Bank and the International Monetary Fund, have been top-down, bureaucratic, program and donor driven. However, we are now seeing more grass-roots driven initiatives based on participatory approaches so that decision making flows from the bottom up, rather than being driven from the top.

At the same time, we are still far from having a good understanding about what motivates participation in the new knowledge space. Numerous research universities in North America and Europe have set up institutional repositories, and an increasing number of repositories are also springing up in transitional countries.[5] Even so, most of these repositories remain largely empty despite convincing studies that show the higher number of citations and impact of materials deposited in these spaces.[6] We know even less about researchers' behavior, motivation, and institutional practices in the developing world. Again, much empirical research remains to be done.

Interoperability

The term interoperability is generally understood to be a technical practice of ensuring that different computing systems can communicate and that diverse digital objects could be easily exchanged and retrieved through a common protocol. In the open access environment, the Open Archive Initiative Protocol for Metadata Harvesting (OAI-PMH) has become the de facto standard for ensuring discovery and retrieval of OA objects. Interoperability is particularly important, as common-based production has been spreading rapidly across the various knowledge domains, from software to scholarly publications to educational materials. The Open Educational Resources (OER) movement is now a significant force in education and it also has great potential for transforming the nature of access to knowledge not only in the industrialized world but also in fostering endogenous development and South–South collaboration.

The OA movement and the OER movements have been developing somewhat independently, with different agendas, institutional affiliations, strategies, technical tools, and standards. Both movements, however, are also supported by the development of open source tools, reflecting the common philosophy of knowledge sharing and community building. These independent developments reflect to some extent the separation between teaching and research in most higher education institutions, where research tends to be more highly regarded and rewarded. It is time for a more coordinated effort between the two movements, and more emphasis on ensuring interoperability between open access scholarly repositories, learning management systems and learning objects repositories. The convergence of the two movements is natural given the deep interconnection between teaching, learning, and research, and it is surprising that little dialogue has taken place across the communities until recently.

While it is important to ensure technical interoperability, it is just as crucial to ensure social and institutional interoperability because "the institutional framework we use to manage the stock of existing information and knowledge around the world can have significant impact on human development" (Benkler, p. 310). In this regard, it is likely that developments such as the Creative Commons and Science Commons will play a key role in ensuring institutional interoperability.

In addition to technical and institutional interoperability, there is also social or organizational interoperability that needs attention. Even though the OA and OER communities are distinct, they are relatively permeable, as interest in the development of learning resources and the dissemination of knowledge is seldom driven by political or monetary concerns. This is not the case with many development-related initiatives, which are often dictated by the interest of public–private partnerships.

As an example, there are currently over 100 such partnerships operating in the "research for health" arena, including MMV (Medicines for Malaria Venture), the Stop TB Partnership, IAVI (International AIDS Vaccine Initiative), and many more. They have undoubtedly boosted research activity in providing medicines for neglected diseases in recent years and have created effective channels for joint international, philanthropic, and private funding efforts. There are, however, serious questions with regard to the governance of these multiple initiatives and the ways in which the partnerships are structured. There is often a duplication of efforts, as knowledge created is often kept in silos that are inaccessible. Indeed, the problem of reinventing the wheel is all too common in the development arena, and an open and transparent environment would ensure a more efficient funding and knowledge building environment.

As commons-based production becomes more widespread across various knowledge and cultural domains and as more organizations (both public and private) begin to ride the wave, the issues of technical, institutional and organizational interoperability and governance will become increasingly important. Will international governing bodies for OA and OER be necessary? Or should these initiatives be left on their own, to grow and to perish, depending on the demands and usage by the creators and the users? When would institutional policies be necessary and how would they affect participants' behavior? The network information economy is ripe with interesting research questions with practical and policy consequences. Benkler's book will keep students and scholars of the new economy busy for years to come.

Towards Global Partnership

In December 2007, hundreds of non-governmental organizations, major development organizations from around the world, UN bodies and officials, and many citizens, activists, and academics converged in Kuala Lumpur for the third Global Knowledge Partnership extravaganza.[7] In keeping with the transition from the industrial information economy to the networked economy, one of the stated goals of the Global Knowledge Partnership meeting was to examine the "need for a user driven approach to development and application of technologies." The meeting was also dedicated to the development of public–private partnerships in the use of information and communication technologies for development; to exploring emerging markets and business models in the increasingly open network environment, and to promoting social networking in a global development context. Open access to the scholarly literature was a topic of discussion at several major sessions at the conference, and the topic turned out to be a new one for many of the participants at this venue. However, the productive encounters with a multitude of grass-roots driven initiatives and networked based innovation have broadened the perspective and meaning of OA, transforming it from a debate among publishers, researchers, and librarians, to a topic that is increasingly seen as central to the future of knowledge driven human development. Though there is no easy answer to the question of what Wikipedia has to do with the 49% of citizens in the Congo who have no access to clean drinking water, Benkler has provided us a structured framework and a set of powerful ideas with which to debate and examine the role of open access and network in development.

Notes

1. Amartya Kumar Sen (2001) *Development as Freedom.* Oxford: Oxford University Press.
2. http://www.who.int/entity/hinari/Hinari-Oare-Agora%20Leaflet%204pp.pdf
3. See Leslie Chan, Barbara Kirsop & Subbiah Arunachalam (2005) Open Access Archiving: the fast track to building research capacity in developing countries, *SciDev.Net,* November. http://www.scidev.net/ms/openaccess/
4. Henry Jenkins (2006) *Convergence Culture: Where Old and New Media Collide.* Cambridge, MA: MIT Press.
5. See the Directory of Open Access Repositories. http://www.opendoar.org
6. See Steve Hitchcock's bibliography on open access citation advantage. http://opcit.eprints.org/oacitation-biblio.html
7. See http://www.gkpeventsonthefuture.org/

JEAN-CLAUDE GUÉDON: *Network Power and "Phonemic" Individualism*

Introduction

For the last 60 years, sometimes silently, sometimes noisily, computers have invaded ever larger segments of our lives. In the early phase of the digital age, the military came first; then management took advantage of the new technology. From defense (e.g., the Electronic Numerator Integrator and Computer—ENIAC—or, later, the Semi-Automatic Ground Environment or SAGE) to aviation (American Airlines SABRE), computers became the workhorse of vast managerial structures. Their presence was felt in ways that could not have been predicted: many people first met the digital world through the punched card that, for a time, adorned the billing processes of a number of utilities and other large companies. Destined to be treated by mechanical reading devices, their physical integrity became an issue: "do not fold, spindle or mutilate" was the stock warning. Ultimately, the injunction was ironically extended to the misuse of human beings in bureaucracies and in the Vietnam war.[1]

Meanwhile, the blinking lights of computer displays were hard to miss and predictably attracted Hollywood types. They often came to symbolize threatening and uncontrollable technologies (e.g., Stanley Kubrick's HAL in *2001: A Space Odyssey* in1968 and Joseph Sargent's *Colossus: The Forbin Project* in 1970). Although dealing with the future, these films reiterate a well-worn argument going back to at least Samuel Butler's utopia, *Erewhon* (1872), particularly its chapters on "Darwin Among the Machines" and to Karel Čapek's famous play *RUR* (1921) where the word *robot'* was first introduced[2]: human beings would have to compete with their own machines and would not necessarily win.

To be complete, the digital world needed to add another dimension—that of the network. In the early 1960s, J.C.R. Licklider came up with the startling notion that computers were communication tools and could communicate either among themselves or with humans.[3] Under the somewhat hyperbolic title of "Galactic network," Licklider envisioned a world where computers *and* information would be linked and accessible to anyone anywhere in the world. This vision was later implemented, first through ARPAnet after 1969 and a little later, by the Internet. At that point in history, most of the tools needed to connect computers together were in place. Connecting information through computer networks was the next step.

The dream of connecting documents together is actually an old one. Ramelli's famous book wheel is a clear example of the desire to compare and connect texts together.[4] Similar devices appear as late as the eighteenth century, thus demonstrating the staying power of a device that incorporated the collating, verifying, and stabilizing functions of the codex when it was redesigned into the Hexapla by Origen in the third century CE.[5] Diderot and d'Alembert's *Encyclopédie* incorporated a series of *renvois* allowing the reader to both navigate large amounts of information organized in alphabetical order and read this information as if it were a book but a book dispersed within a dictionary format.[6]

Closer to us, Vannevar Bush is often seen as a precursor of the contemporary concept of hypertext, and his celebrated article "As We May Think"[7] is often mentioned in this context, but it would be just as easy to see him as a distant disciple of Origen. However, a further and decisive step was made possible when Tim Berners-Lee, then of CERN near Geneva, developed the foundations for the World Wide Web, starting in late 1989. By providing a simple protocol (http or hypertext transfer protocol) and a simple "tagging" language (html or hypertext mark-up language), Tim Berners-Lee did manage to add the document-linking function to the machine-linking capacity of the TCP/IP protocols that defined the Internet. In 1993, when Mosaic, the prototype of our modern browsers, appeared, it can be said that the publishing tools of the World Wide Web had found their reading aid complement. It can also be argued that the "Network Age" had then reached its first complete implementation stage, however "incunabular"[8] it may already look to us because of its touching attempts to emulate the print world.

In the last dozen years, many scholarly and journalistic efforts have been expended with the objective of understanding or interpreting what, for lack of a better expression, could be termed the "Network Age." As a result, a considerable bibliography has developed around the nature of Internet communication, the creation of new "communities," the redefinitions of self, etc. Hypertexts, multimedia, interactivity, immersion, and virtuality are but a few of the terms that have focused the attention of countless scholars. The Internet has spawned intellectual cottage industries that have already produced several hundreds of titles in practically all the major languages of the planet. These studies vary from frantically enthusiastic to somber and dystopian. For example, the French, true to their skeptical form, have been very active in predicting all kind of disasters ranging from the loss of social cohesion through anarchic fragmentation (Dominique Wolton) to the birth of dark religious-like sentiments fed by some sinister cybernetic impulse (Philippe Breton), but they also harbor their enthusiasts, such as Pierre Lévy.[9]

One main feature characterizes most of these Internet studies: they tend to react to some aspect of the Network Age rather than seek its deeper essence, and they do so to such an extent that, in many ways, they tend rapidly to fall on the side of primary sources. It would not be an exaggeration to treat them as incunabular reactions to events accompanying the incunabular phase of the Network Age. They provide many important insights, much-needed documentation and, on occasion, they even manage to reach a useful level of generalization or synthesis. However, these partial, incomplete and not entirely satisfactory results *a contrario* demonstrate the absence of a theory of the Network Age while underscoring the need for it.

The importance of Yochai Benkler's book, *The Wealth of Networks*, lies precisely in the fact that it is the first book that attempts a coherent theoretical treatment of the Network Age. It provides a sound foundation for such a theory and the remainder of this small essay will try to demonstrate why this is the case and why it is important, including in some of its consequences.

What Sets The Wealth of Networks Apart?

Benkler's title is a sly wink at Adam Smith's classic, but it is also more than that. There is a foundational intent in the choice of this title, and it is important to try locating it as precisely as possible. In Adam Smith's study, "markets" and "division of labour" correspond to two of the most important concepts of the emerging science of economics. What are the equivalent concepts in Benkler's study?

To provide an answer to this question, we must turn to what I consider to be the crucial passage in the whole book: it starts with defining three kinds of storytelling societies, the Reds, the Greens, and the Blues.

> Each society follows a set of customs as to how they live and how they tell stories. Among the Reds and the Blues, everyone is busy all day, and no one tells stories except in the evening. In the evening, in both of these societies, everyone gathers in a big tent, and there is one designated storyteller who sits in front of the audience and tells stories. It is not that no one is allowed to tell stories elsewhere. However, in these societies, given the time constraints people face, if anyone were to sit down in the shade in the middle of the day and start to tell a story, no one else would stop to listen. Among the Reds, the storyteller is a hereditary position, and he or she alone decides which stories to tell. Among the Blues, the storyteller is elected every night by simple majority vote. Every member of the community is eligible to offer him- or herself as that night's storyteller, and every member is eligible to vote. Among the Greens, people tell stories all day, and everywhere. Everyone tells stories. People stop and listen if they wish, sometimes in small groups of two or three, sometimes in very large groups. Stories in each of these societies play a very important role in understanding and evaluating the world. They are the way people describe the world as they know it. They serve as testing grounds to imagine how the world might be, and as a way to work out what is good and desirable and what is bad and undesirable. The societies are isolated from each other and from any other source of information.[10]

Benkler starts from an interesting anthropological basis: societies exist through what he calls "story-telling" and the ways in which the social fabric is woven and maintained is related to the mode of storytelling adopted by a particular society. This raises a number of subsidiary issues, such as: how many modes of storytelling exist? How are modes of storytelling selected by a particular society? How long do the storytelling modes last? What makes them last? What makes them disappear? How does a society move from one mode to another? What are the implications of each mode of story-telling for the corresponding social system? Leaving aside these questions for the moment, the storytelling hypothesis offered in *The Wealth of Networks*—itself presented as a story—provides a conceptual framework that shifts the analysis away from the reactive stance that has characterized most of the Internet studies alluded to earlier: to the market concept that Adam Smith constructed in the *Wealth of Nations*, Benkler adds a market of ideas that can emerge only if a number of conditions are satisfied. Otherwise, the communication modes that prevail are based on power, and, in turn, serve to maintain that power.

As any good fable, Benkler's is a *roman à clef*, and it is not difficult to decipher: absolute monarchies and most forms of religion would fall in the red category, as would (appropriately red) Marxist countries and all kinds of fascistic governments where freedom of expression has been or is severely controlled. From Benkler's perspective, the famous, if somewhat disingenuous, distinction between "authoritarian" and "totalitarian" regimes that Jeane Kirkpatrick advanced when she was U.S. ambassador to the United Nations in the Reagan years, appears completely irrelevant: both types are red, solidly so. Even formally democratic regimes could be easily classified as red states if their control over the press, radio and television is a little too tight—a point which Ithiel de Sola Pool explored some years ago.[11]

The blue version reserves a few more surprises. The vote described in Benkler's text refers to the way in which the storyteller is chosen, and it corresponds to the set of thousands and even millions of decisions made to tune radio or television, or it resides in the choices of printed materials. Bob, member of a blue society, sees his autonomy constrained not so much by the storyteller as by the choices of his contemporaries. As Benkler continues, "If the majority selects only a small group of entertaining, popular, pleasing, or powerful (in some other dimension, like wealth or political power) storytellers, then Bob's range of options will appear only slightly wider than Ron's, if at all."[12] The majority rules, even though, in principle, anyone may tell a story. The problem is that only a few stories will be available on the main and most accessible channels. As for the other stories, only a few, in the best of hypotheses, will make the effort to seek them out. And the perspective of a very small audience contributes to limiting offerings that do not conform to the majority's wish. The constraints in this case will emerge out of practical rather than doctrinal considerations, and for this reason, the blue world is a lot more stable than the red world because its constraining machinery is built on acquiescence rather than forcible or violent repression. In fact, the blue world is better at controlling the scope of dominant themes than the red world. In so describing the blue world, Benkler brings back to the surface some of the critiques of media as found in the Frankfurt school, and particularly in Herbert Marcuse's work. Intimations of Michel Foucault's theory of power can also be discerned in this context.

Gertrude's world, the "Green" world, may well deserve its ecological connotation by the fact that, in it, everyone can be a storyteller and everyone can tell a story wherever and whenever he/she wants. Benkler's main point, however, is that the disappearance of the "big tent" and of the "evening session" provides a far more open context for the circulation of a wide range of stories. Unexpected and unusual stories may emerge from anywhere and reach unintended audiences. In short, all of society would begin to work as if all the stories obeyed diffusion and transmission mechanisms similar to those of jokes and rumors.

How does a society choose between a particular mode of storytelling and another? Benkler's answer, for the present world, is largely an economic one: in what he calls the "industrial information economy," two characteristics contribute to the tight control over who can tell stories: high entry barriers and large economies of scale. In other words, to become a storyteller, one needs a lot of capital; once that hurdle is past, the economies of scale contribute to maintaining the achieved system in place. Newspapers and other print products, radio and television stations, as well as movie studios all obey these two fundamental rules. They fundamentally explain why Orson Welles's famous *Citizen Kane* was more of a citizen—"more equal" would quip Orwell—than his contemporaries.

To pursue the exploration of subsidiary questions already listed earlier, the three modes of storytelling refer back to three basic social systems—centralized and hierarchical, apparently decentralized but recentred around a majority, and distributed. The centralized and hierarchical model maps easily onto the holistic vision of the universe that prevailed through various neo-platonic doctrines that dominated the Mediterranean world until at least the Renaissance.[13] The decentralized, majority rule maps onto the modern political theories that emerge around Locke and other thinkers such as Montesquieu that still make up the philosophical foundation of modern liberal democracies. In it, the individual, either as citizen (or actor in a market), is a fully informed being *à la* Descartes that always makes rational choices. In the end, this individual behaves very much like the social and moral equivalent of a physical atom: in this view, the social world, like the physical world, is built up from "simples." This way of thinking reflects the world of both the American and French Revolutions.

As for Benkler's green world, it is a social form that presently exists only in niche areas of our societies. The world of "free software" provides him with both empirical ammunition and theoretical fodder. However, this is not the only available example. Scientists conform to this distributed form of social structure. Robert K. Merton, the well-known sociologist of science, described the scientific ethos as significantly different from society at large and resting on a number of values that, taken together, are specific to scientists—organized skepticism, disinterestedness, universalism, and communalism. The scientific ethos thus distinguished the collective behavior of scientists and functionally contributed to its social success.

Earlier, we raised the issue of how does a society move from one form of storytelling to another? Yochai Benkler's answer is interesting because if there is indeed a link between the mode of storytelling and social order, then a new mode of storytelling should correspond to a new form of social structure with associated forms of power and economic systems. What sets Yochai Benkler's book apart from other studies of the network age is not that he sees it as revolutionary (although he does)—many others have made similar claims; it is not that he distinguishes it from other periods of history—again, several authors have used phrases such as "network age" or *The Wealth of Networks'* "Internet epoch." It is really based on the second element also taken from the foundational contributions of the *Wealth of Nations*: the division of labor. From there, Benkler invites us to revisit the whole idea of individual.

Towards "Phonemic" Individualism

Whereas division of labor is seen by Smith as the result of a top-down, managerial intent, as a production master plan that sets everyone in a well-defined role, Benkler, when he deals with the "green" world, sees the division of labor as an emergent phenomenon stemming from interactions between individuals: out of the constant dialogues, discussions, and debates fluid roles arise. Like eddies in a stream, these roles enjoy relative, but only relative, stability. Individuality, in this perspective, sums up the possible role shifts one person may live through.

In the "Green" world, individuals are found positioning themselves temporarily in one role or another according to the relations they develop with other individuals. In other words, in the green world, individuality is no longer built like an atom, in full self-sufficiency. It is no longer an individual simply endowed with "properties"—the whole polysemic wealth of the term is needed here—but rather an individual whose very essence, paradoxically, depends on his/her relations with other individuals. More precisely, existence depends on distinguishing oneself from others.[14] A form of individuality that necessarily rests on the individuality of others calls for a general interpretative scheme that goes beyond what earlier theories of society have contributed. It goes beyond an "emanation" or holistic theory of individuals, based on divinities and their human proxies, leading to a feudal vision of society.[15] It cannot limit itself to the self-sufficient atom-like individual that stands as the foundation of the liberal age (where "liberal" here means adherence to the tenets of classical economics). We must, therefore, reach beyond emanation and atom-like individuals to reach for a third kind of individuality. Let us call this third way the "phonemic" approach. Although as powerful in its reach as the holistic or atomistic approaches, it has not been used nearly as much until now.

What is a "phonemic" approach? It is based on the concept of phoneme, of course.[16] Here, it is adduced as, in a sense, a synthesis of the holistic and atomistic explanatory modes: imagine a universe where every existing entity would have the appearance of an atom, but, simultaneously, would appear to emanate from a number of these other apparent atoms. Let us add that the emanation is not a transitive, transparent process: the link between two phonemic entities is not guided by some form of analogy, but, on the contrary, by some distinctive characteristic. The total result could be described as a "peer-to-peer emanation system." Phonemes, in the field of phonology, behave precisely in this manner. They exist only by being distinct from other phonemes. The existence of one entity depends on the existence of all, and it also depends on maintaining a distinctive uniqueness with respect to all of the other entities. Their existence marks the fact that their difference makes a difference—precisely the definition of information according to Gregory Bateson.[17] They offer, therefore, a powerful metaphor to think beyond atomistic or emanation-based individualism.

What Yochai Benkler is founding with his important book is not only a revision of the market concept or of the division of labor that accompanies it. What Yochai Benkler is really inviting us to do is to revisit our understanding of markets and division of labor in terms of a new form of individuality that cannot be thought within the atom category or denied on account of a divine hierarchy out of which everything emanates (and to which it must return).

What remains difficult to apprehend with social phenomena such as the free software movement, Wikipedia, and other peer-to-peer processes that seem to fly in the face of long-accepted notions of "human nature" becomes far more comprehensible if we begin to look at human beings behaving like phonemes. If we remember that phonemes relate to language and that human beings do speak, the metaphor appears far less contrived. On the other hand, the reasons why human beings should be apprehended as emanation of some wholeness can only be based on faith. Moreover, if human beings chose to apprehend themselves as the similes of atoms, it may simply have been a reaction to that faith. Neither emanation nor atoms need language incidentally, but human beings distinguish themselves through language. Indeed, the full deployment of language requires the existence of phonemic individuals. The wealth of networks, therefore, lies in phonemic individuality. Any other approach to human beings will simply be suboptimal and that is the fundamental thesis of Yochai Benkler's crucial work.

Notes

1. Steven Lubar (1991) "Do not Fold, Spindle or Mutilate": A Cultural History of the Punch Card. http://ccat.sas.upenn.edu/slubar/fsm.html
2. In Czech, *robota* means boring and unpleasant labor. The word was re-used by Doug Chiang and Orson Scott Card for the title of an illustrated sci-fi book of this title (2003).
3. J. C.R. Licklider & W. Clark (1962) On-Line Man–Computer Communication, *Proceedings of the AFIPS SJCC,* 21, 113–128.
4. Agostino Ramelli (1588) *Le diverse et artificiose machine del Capitano Agostino Ramelli* [The various and ingenious machines of Captain Agostino Ramelli]. Paris: in casa del'autore.http://www.sil.si.edu/Exhibitions/Science-andthe ArtistsBook/76–14435.jpg
5. The excellent anthology, *Books and the Sciences in History,* ed. Marina Frasca-Spada & Nick Jardine (Cambridge: Cambridge University Press, 2000) contains an illustration, p. [169] taken from the *Recueil d'ouvrages curieux de mathématiques et de mécanique* . . . by Gaspard Grollier de Servière (Lyon: David Forey, 1719). The same illustration can be found online at http://cnum.cnam.fr/CGI/fpage.cgi?4PO3/204/110/223/31/21 3. On the Hexapla and its potential history for the history of the codex and of reading, see Anthony Grafton & Megan Williams (2006) *Christianity and the Transformation of the Book.* Cambridge, MA: Belknap Press of Harvard University Press.
6. This point is well made by Richard Yeo in his "Encyclopedic Knowledge," in *Books and the Sciences in History,* pp. 207–224. My own PhD thesis tried to demonstrate this point in the case of chemistry, albeit without any reference to the possibility of a history of reading. See Jean-Claude Guédon (1974) The Still-Life of a Transition: Chemistry in the *Encyclopédie,* PhD thesis, University of Wisconsin-Madison.
7. *Atlantic Monthly* (July 1945). The article can be accessed at http://www.theatlantic.com/doc/194507/bush
8. The term incunabula applied to digital documents was introduced by Gregory Crane. See Gregory Crane, David Bamman, Lisa Cerrato, et al., Beyond Digital *Incunabula:* Modeling the Next Generation of Digital Libraries? http://www.cs.umass.edu/~mimno/papers/ecdl2006.pre.pdf
9. Dominique Wolton (2000) *Internet et après ? Une théorie critique des nouveaux média.* Paris: Flammarion; Philippe Breton (2000) *Le culte de l'Internet. Une menace pour le lien social?* Paris: *La Découverte;* Pierre Lévy (2002) *Cyberdémocratie.* Paris: Odile Jacob.
10. *Wealth of Networks,* p. 162.
11. Ithiel de Sola Pool (1983) *Technologies of Freedom.* Cambridge, MA: Belknap Press.

12. *Wealth of Networks,* p. 163. In Benkler's fable, Ron is Red, Bob is Blue and Gertrude is Green.
13. Arthur O. Lovejoy, (1976) *The Great Chain of Being: A Study of the History of an Idea.* Cambridge, MA: Harvard University Press.
14. Although Pierre Bourdieu does not use the "phonemic" terminology, many of these arguments are present in his aptly titled study: *La distinction. Critique sociale du jugement* (Paris: Minuit, 1980).
15. Neo-platonic philosophy has provided a great deal to this general vision. In Europe, it controlled the understanding of the natural and social order until the end of the sixteenth century. See A.O. Lovejoy (1936) *The Great Chain of Being.* Cambridge, MA: Harvard University Press.
16. On the phoneme, see http://en.wikipedia.org/wiki/Phoneme
17. Gregory Bateson (2000) *Steps to an Ecology of Mind,* 457–459. Chicago: University of Chicago Press.

References

2001: A Space Odyssey. (1968). Dir. Stanley Kubrick. (Metro-Goldwyn-Mayer).
Colossus: The Forbin Project. (1970). Dir. Joseph Sargent. (Universal Pictures).

JOHN WILLINSKY: The Wealth of Networks

The Educational Implications of Networks

With *The Wealth of Networks,* Benkler takes on Adam Smith's epoch-defining work, first published in 1776, at the very point in history when the economic system that Smith so carefully describes in *An Inquiry into the Nature and Causes of the Wealth of Nations* appears to have finally realized its global destiny, with market economies having now taken root around the world. It may seem an odd moment, then, for Benkler to turn the tables on Smith's vision; that is, to displace nations with networks and transform markets through social production (into *nonmarkets,* as it turns out). Although Benkler does not anywhere else in his book make such direct use of Smith's influential book, *The Wealth of Networks* establishes the economic viability of what is, at many points, much the opposite of what Smith was describing then as a new economic regime and what has subsequently taken on the qualities of natural law.

In the process, Benkler takes hold of capitalism's two dearest concepts, wealth and freedom, and gives them both a second economic life. He identifies project after project which is driven not by national and personal self-interest—which figured so prominently in Smith's work, as well as the continuing stream of economic theory following that tradition—but operates instead cooperatively through global, collaborative networks. These networks represent for Benkler a revolution in individual autonomy and democratic action, given how they freely distribute the means of participation to others, and those two concepts have a certain resonance with other events from 1776.

However, if Benkler's book plays off the *Wealth of Nations*, concept by concept, it still resembles Smith's book in form. Both books describe new developments by identifying the logic and economic benefits in each case. Both give name and shape to what are already growing segments of the economy; both deploy prime instances, like the pin factory and open source software, leading to improvements in quality and increases in productivity and creative application. By rendering these developments sensibly and visibly part of a larger development, Smith and Benkler accelerate their take-up by others over the longer term, if Smith's success is anything to go by.

To begin at the beginning: when Smith introduces on the first page of the *Wealth of Nations* the "division of labour" as the new best hope of "the productive powers of labour," Benkler opens with "the networked information environment," which represents the evolution of "liberal markets and liberal democracies" that have prevailed since Smith's day (p. 1). To stay with Benkler's key term, the *networked information environment* brings to the fore what is most valuable and what might otherwise be overlooked in "the Internet Revolution" (p. 1). At a time when, as he rightly points out, academics are dismissing such revolutionary talk as "positively naïve," Benkler compresses into a triple-decker phrase like *networked information environment* the pervasive and encompassing flow of information through our lives and work, whether in call centres or college campuses. However, if Benkler had left it at that, we would have little that was not already well known and often stated. Instead, he follows this initial portmanteau of a phrase with, in quick succession, the new terms of this revolution, marked by "cooperative nonmarket production" (p. 2), "decentralized individual action" (p. 3), "nonproprietary strategies" (p. 4), "large-scale cooperative efforts" (p. 5), and so on. Recombinant possibilities soon emerge, with the likes of "networked information economy" (p. 3), "radically distributed nonmarket mechanisms" (p. 3), and "nonmarket, nonpropriety production" (p. 106). Each of Benkler's phrases has its own way of rewriting one or more of Smith's basic economic principles, whether one thinks of Smith's sense of *market, exchange value, self-interest, nation,* or the *division of labour.*[1]

Benkler's forceful linguistic turn makes him a strong candidate for what the late philosopher Richard Rorty identified as the transformative poet. Benkler makes no pretence to being a poet, but he is certainly a writer capable of generating "increasingly useful metaphors," in Rorty's term, who thus changes how the world is viewed and read (1989, p. 9). Benkler does appear to have an inexhaustible ability, again in Rorty's seeming simplification of things, "to redescribe lots and lots of things in new ways," leading to "a pattern of linguistic behavior which will tempt the rising generation to adopt it" (pp. 7, 9). For Rorty, there is no greater intellectual or poetic power than this particular knack; the "talent for speaking differently, rather than arguing well, is the chief instrument of cultural change" (p. 7).

Now in addition to speaking differently, Benkler also argues these cultural changes, and exceptionally well, to my way of thinking. He makes fine distinctions, sets up sensible categories, and marshals myriad on-the-ground instances to substantiate them, from Free High School Science Texts in South Africa (p. 101) to NASA's use of the public to mark crater maps and undertake other scientific work (p. 69). However, this particular talent for naming what these various projects have in common contributes, in its own way, to "an increasingly robust ethic of open sharing," as Benkler names what many of us hope will indeed carry the spirit of the age (p. 7).

By naming this economic model, if only in the negative terms, as both *nonmarket* and *nonproprietary*, Benkler makes it clear that the creation and distribution, for example, of free software code is not simply a circumvention or aberration in what is software's rightful market.[2] Rather, open source software represents a highly productive way for people to work together toward a public good. In addition, while Benkler allows that people work on developing open source software because it provides people with access to what has become one of our basic communication systems, he also holds that it is about more than an ethics of openness. It is also about efficiency and productivity, those two critical wealth factors. Benkler very clearly sets out how cooperative approaches are contributing to "the greatest improvement in the productive powers of labor" since the division of labour, to borrow from the opening from Adam Smith's first chapter (2006, p. 5). By demonstrating the effectiveness of cooperative ventures, such as open source software and *Wikipedia*, Benkler undermines what might otherwise have seemed, at the close of the twentieth century, to be the ubiquitous triumph of the market.

In the context of *Policy Futures for Education*, it makes sense to ask what the new terms of this alternative economy mean for the schools. When Benkler writes of the Internet's democratic spirit—in terms of how the "network allows all citizens to change their relationship to the public sphere" as "creators and primary subjects"—he could as easily be addressing what the public schools have long promised, if not always delivered (p. 272). This overlap is nowhere more clearly at issue than with the educational challenge posed by *Wikipedia*.

Benkler regards this multilingual free encyclopedia, not surprisingly, as a leading instance of an "open, peer-produced model," and "one of the most successful collaborative enterprises that has developed in the first five years of the twenty-first century" (pp. 70, 71). Even so, *Wikipedia* is not like anything taking place in the schools today. It is the exact opposite. Ask yourself, as I have done more than once in the face of *Wikipedia*'s heart-felt learning, what in today's schools can be said to really prepare students to collaborate anonymously, without credit or deadlines, on a drop-in basis, at the risk of being overwritten and vigorously attacked by equally anonymous strangers, as they press together collectively in the name of a "neutral point of view" (as *Wikipedia* puts it), while being governed by a loosely organized (and enforced) series of principles having to do with verification and structure? *Wikipedia* demonstrates what a life of learning outside of school, for the sake of learning, can mean. It is a demonstration for the schools that continues to grow daily on a global scale and in a remarkably organic way. For all of its shortcomings, *Wikipedia* serves for most people as the primary educational gateway into this networked information environment. That this open and vibrant model of learning is so removed from the everyday world of schooling surely has implications for the policy futures for education.

I am not suggesting, however, that Benkler has fallen short in addressing the educational implications in *The Wealth of Networks.* He is above reproach on this count. He has done more than enough by pausing for a moment and offering a brilliantly sweeping educational vision based on "the possibility that teachers and educators can collaborate, both locally and globally, on a platform model like *Wikipedia*, to coauthor learning objects, teaching modules, and more ambitiously, textbooks that could then be widely accessed by local teachers" (p. 315). It seems only fair to say that the onus for pursuing this book's educational implications falls on those who profess education for a living, at least in so far as they are persuaded by this book.

Indeed, that would be me, as I am an obvious enthusiast for Benkler's approach, and I have already been involved in opening access to knowledge online (through work over the last 10 years on the Public Knowledge Project). This symposium may not be the appropriate place to pursue all of the educational implications of this work, although a few initial observations do seem in order, especially as it seems to me that Benkler's particular rhetorical casting of the new non-Smithean economics cannot be directly applied to the public schools. The schools may already be an information commons of sorts, operating outside of the commercialized world of markets. Nevertheless, schools that are going to have their students actively contributing to the intellectual commons within their communities will have to teach these students many orders of propriety and property. In other words, the need is not, then, for a *nonproprietary* program in the schools or a program that engages in the nonmarket production of knowledge, per se. This is because, unlike the open revolt against the restrictive marketing of intellectual property represented by open-source software, there has never been a market for the intellectual work coming out of the schools. In fact, some thought needs to be given to cultivating such a market, to finding ways for students to direct their learning toward work that serves others.

The educator, entering the school with *The Wealth of Networks* in hand, has now to assemble a curriculum that provides opportunities for learning about proprieties and properties, including the different forms of producing and utilizing intellectual property. It is not simply that one must learn the rules in order to break them. It is to understand that Benkler's nonproprietary economics is *nonproprietary* in very particular ways. It says *no* to only certain limited aspects of this broad concept. For instance, with open access research, another of Benkler's leading instances of markets transformed, we are seeing a number of scholars and librarians challenging an extremely damaging proprietary element of scholarly publishing, namely, the exorbitant pricing of scholarly journals that results in reduced access (which is further compounded by the impossibility of being able to subscribe to all journals even if they were reasonably priced). However, the open access movement in scholarly publishing leaves untouched the proprieties of intellectual ownership that demand that authors duly credit those whose work they draw on, just as open access is not about the proprieties of grammar, genre, bibliographic formatting, graphic representation, and on and on.

Then, though, when it comes to Benkler's particular focus on *nonproprietary* forms of cooperation and production, what seems clear is that schoolwork is already all too nonproprietary, in the sense that students' work lacks any value as intellectual property. At a time when schools seem increasingly like training grounds for large-scale test-score production, there are few opportunities for students to engage in working on something that has value in its own right. The preparation for, and writing of, such tests has taken on such importance that it can end up teaching the students that their learning has nothing to do with creating a property. In this sense, the test-driven school is entirely a nonmarket and nonproprietary entity, and discouragingly so, given Benkler's sense that such entities otherwise are leading to increases in individual creativity and autonomy, as well as democratic responsiveness.

However, there has always been a river running through the schools that is given to the cooperative, collaborative production—as students gather with paints, paper, glue, scissors, and computer—and it is now time to think about the market for what these students could produce, as they are gathered at perhaps the sole centre in their community engaged in non-commercialized intellectual production. The schools need to begin to think of the work that students do, as a result of their learning, as having value and interest for others, as itself one of the *properties* of intellectual work. Students can indeed, as Benkler suggests, help others in their learning, by developing resources for teachers and students; they need to contribute to *Wikipedia*, creating intellectual properties that begin with the local.

This suggests that students will have first to learn about their own capacity to produce intellectual properties of value to others, as well as learn, in the process, about the qualities (and proprieties) that such properties entail. Once students are thinking about producing intellectual properties of potential interest (and thus of value) to others, they could then take their first lesson in nonproprietary production by selecting one of the various Creative Commons licences for their work. In helping people select a licence, the Creative Commons provides a clear and readily comprehensive introduction to such issues as attribution, derivatives, share-alike, non-commercial use, and so on. In thinking about their own work, students would be in a good position to learn about how properties of this sort—whether for photographs, maps, music, and so on—are marketed within and outside of traditional corporate economies. If terms such as nonproprietary are indeed metaphors, in the spirit of Rorty, then a basic concept like *property* can be further stretched and turned, rather than simply negated, as if it referred but to one thing, even within the economic realm.

By the same token, Benkler's use of the term *nonmarket* for this new economy is directed at negating but one aspect of *market*, by which goods are distributed on a commercial basis, with the goal of maximizing profits and, in the case of public corporations, increasing shareholder value. The nonmarket of (nonproprietary) open source software exists within the well-defined market of operating systems, which is dominated by Microsoft, while the growing success of Linux, Apache, and other open source software is measured in their "market share." That is, the nonmarket is itself a portion of the market that has grown out of the refusal of the current commercial model. It operates within an existing market of users.

In addition, in the world of scholarly publishing, open access could be said to be creating a new manner of marketing research among authors and readers, one that ensures that the ability to find and read the relevant research on a topic is no longer unduly influenced by price structures and profit margins. That said, though, scholarly publishing, as infused as it is today with various open access models, is no less a marketplace of ideas governed by long-standing proprieties. So, before the schools foster students of the nonmarket and nonproprietary aspects of this new economy, they would do well to consider using the very idea of *market* as a way of thinking about how students could direct some part of their learning toward the interests of those within their community, who could benefit from the sort of intellectual work that students are capable of producing, whether one thinks of local history, language services, performing arts.

Such work would still entail the proprieties of both student accountability and audience expectations, in terms of how this work is marketed within school districts and communities. On the question of accountability, for example, some educational jurisdictions have been experimenting for some time with students assembling portfolios for evaluation purposes, which represent a range of projects to which they have contributed (Tierney et al., 1998). These portfolios can capture the nature of the students' contributions, as well as—following the model of open source software development—provide evidence of the students' growing reputations for a certain quality of work (which will have been enhanced and developed in the hands of inspired teachers). In the process, the students learn the value of responding to the expectations of the market for their work.[3] By virtue of their work in the community, the students would have a stake in the game and be able to see that their work within a public institution, like the schools, is already part of this knowledge economy.[4]

Progressive forces within the schools have long sought to embrace the commons and take full advantage of the John Dewey movement that Benkler champions: "There is emerging a broad practice of *learning by doing* that makes the entire society more effective readers and writers of their own culture" (p. 299, emphasis added). Today, and in light of Benkler's book, what needs to be *learnt by doing* is how to direct one's learning toward sharing with others, even as learning how to establish a market for one's writing is exactly what being an *effective* writer is all about. Inspired by Benkler, educators have their own part to play in learning to do, by going back to public education's basic democratic promise, in examining how the schools can do more to ensure that, in fact, "a networked information economy overcomes some of the structural components of continued poverty" (p. 307). To return, finally, to the title of Benkler's book, the schools should be able to use this model of social production, which they are so well suited for, to transform the current market for achievement-test scores into a new account of what students are learning and achieving in school, and they could do so in ways that would very much support the "thickening of preexisting relationships with friends, families and neighbors" that Benkler notes as another effect of this new economy (p. 357). Such could be the wealth of networks when it comes to policy futures for education.

Notes

1. Part of the power of a compound concept like *networked information environment* is how each term shares equally in the idea and any one of the three terms can come to the fore, while the other two proximate terms can be hyphenated (i.e., networked-information environment; networked information-environment).
2. The refusal to hyphenate *non* in *nonmarket* and tying it to *nonproprietary* suggests that this negation is already commonplace.
3. There is a parallel here with what is known in educational circles as "service learning" (e.g., Wolfson & Willinsky, 1998).
4. Another point of connection among progressive educators is with the "see for yourself" political culture, which Benkler notes is superseding the sole reliance on mainstream media (p. 218), that is found in those social studies classrooms that have set aside the textbooks (read mainstream media) and taken up the study of the primary sources around historical events (Wineburg, 2007).

References

Rorty, R. (1989). *Contingency, irony, and solidarity.* Cambridge: Cambridge University Press.

Smith, Adam (1776/1910) *An inquiry into the nature and causes of the wealth of nations.* London: J. M. Dent.

Tierney, R. J., Clark, C., Fenner, L., Herter, R. J., Staunton Simpson, C., & Wiser, B. (1998). Theory and research into practice: Portfolios: Assumptions, tensions, and possibilities. *Reading Research Quarterly, 33*(4), 474–486.

Wineburg, S. (2007). Opening up the textbook: And offering students a "second voice." *Education Week*, 26(39), 28–29.

Wolfson, L., & Willinsky, J. (1998). What service learning can learn from situated learning. *Michigan Journal of Community Service Learning,* 5, 22–31.

YOCHAI BENKLER: Response

Educating for Participation in the Networked Environment

There is something humbling about being in conversation with four insightful discussants simultaneously. I will try to do justice to these four very generous and entirely distinct interventions by asking a basic question: What question does each of these four ask about the future of education? Jean-Claude Guédon asks: What kind of person is it who comes into the conversation that is the educational relationship, and what kind of person ought we imagine coming out of that conversation as a lifelong participant in learning conversations? John Willinsky asks how we (educators) should engage individuals such as these, to create in them a sense of meaning and efficacy in their educational process. Philippe Aigrain and Leslie Chan both ask us what are the constraints on our ability to pursue an educational agenda focused on lifelong engagement in peer production and open interactions, reminding us of the difficulties imposed by the limitations of resources. Aigrain highlights the fierce competition that the practices of peer production and commons-based production face for both capital and human attention. Chan complements his concern from competition in wealthy economies with her emphasis on the practical constraints faced by attempts to implement the promise of the networked environment in poor societies, for example, in providing access in the global South to research done in places that possibly have the most useful insight—elsewhere in the South.

Let me begin by telling four very short stories of a particular type of educational intervention in diverse contexts. The first I already mentioned in *Wealth of Networks* (p. 353): an initiative by William Scott, a chemistry professor at the University of Indiana and Purdue University, Indianapolis, who proposed teaching basic chemistry to undergraduates by having them synthesize molecules identified through computational biology as potential targets for developing world disease treatment. The idea consisted of developing low-cost experimental kits that could be used in classrooms across multiple institutions to teach chemistry and deploying them in a network of institutions, so that the multiple redundant classrooms and institutions could provide quality control for each other. The second story concerns my twelve-year-old nephew, who was matched up by his piano teacher as a tutor for another of her students, an eight-year-old. The change was immediate—here was a near-teenager converting from a student who has to be persuaded to count or do his scales, to becoming the teacher, patiently explaining and practicing the benefits of counting, and going slowly, and doing one's scales. The third is the case of law school clinical education programs, through which students take on real clients, selected by full time clinical faculty to present cases of manageable length and effort, through which students learn how to research, think, present, and innovate legal arguments in the context of providing legal services to people who simply are too poor to afford legal services, in contexts to which the resource-poor government-sponsored legal aid bureaus simply do not extend services—be it in employment and 169 immigration law help to immigrants, inmates' rights in prisons, or tenants facing eviction. The fourth and last story concerns a seminar I ran for students who were engaged in building a student network aimed at persuading universities to leverage their patent portfolios to influence pharmaceutical industries to either produce or allow generic drugs manufacturers to produce otherwise-patented drugs for distribution in poor countries at affordable prices. They spent the semester reviewing the literature on university innovation and patenting, interviewing academics and administrators, developing substantial insights into the economics and politics of university patentable science and patents management and the relationships to the pharmaceuticals industry, and concluded with a program of action for their organization. The seminar occurred opportunistically—the students asked my substantive advice on their organization, and as we were talking about their interests we organized an *ad hoc* seminar on the subject, for which they worked much harder than in a usual seminar, but out of which they learned and achieved an efficacy in the world much greater than in the normal seminar.

How do these stories respond to the questions of Who? What? and Under What Constraints? that are presented by the four essays in this symposium?

Who?

Jean-Claude Guédon has put his finger on a core problem that I found myself presented with in writing the book. The argument is situated in the American liberal tradition, driven by a search for political efficacy within my own society. However, I also tried to provide a theoretical framework that would be available throughout parts of the world that have largely congruent but nonetheless distinct ideas about freedom, justice, and society. This created the basic problem of characterizing what kind of individual human being would be both structured and constrained by practical systems of affordance and constraint and capable of at least a practical autonomy worthy of respect as a creative, expressive individual and a participant in democratic discourse and cultural creation. Guédon's construct, the "phonemic" individual, captures this continuous need for duality well. It offers us a way of thinking about individuals as both necessarily individual and distinct, as well as always in relation to others, and as capable of changing meaning and effect through continuous recombination with others. It also evokes our ambiguous relationship to structure—both enabled by it, and able to be creative by breaking it; as slang, or poetic license.

Easiest to map on to this shift is my story about my nephew. A rearrangement of the conversational role in turns rearranges the meaning of practice. From skill earned by rote achieved through discipline, the piano becomes a platform for conversation as mentor; a platform through which a pre-teen begins to differentiate himself from childhood through constructive enactment of the adult role of mentor and guide to a child. At the same time, the pre-teen comes to be respected by stranger adults (the mentee's parents) who make room for the tutor as a surrogate adult better able than they to fulfill this role; and the younger child comes to see maturity as within grasp, rather than as all powerful and authoritative. Practice as discipline, experienced by both children as impinging on their autonomy, shifts to becoming a form of serious play and transition to maturity and the acquisition through interaction with another. The clinical training has a similar structure, where students begin to practice in collaboration with peers, in conditions where they can use their emerging skills to provide empathetic help to others, and experience themselves through these relationships of caring as effective individuals and participants in a community. Their commitment as students in this context shifts from performing to test, and toward being successful in the real world, for the benefit of a real human being in real need. The meaning of what they learn shifts. Consistently, students who sign up for clinical programs rank them as among their most significant experiences in law school.

To emphasize, then, the critique of liberal individualism, on the one hand, or of the elimination of the individual in favor of some structure, deity, or entity like the nation or class, on the other, is not new. Finding some place outside of this unproductive binary has been a project of the intellectual center-left for a very long time. The rise of the networked information environment, and in particular of commons-based and peer production, however, creates new opportunities for large scale cooperative behaviors among remote strangers that simply will not be explained by either of the two ends of the binary. They create new practical urgency to develop in *The Wealth of Networks* such an understanding of the self in context and new domains of observation and reflection from which to develop it. The educational practices that are capable of utilizing persons of such a character, and building them up as both individual and connected in conversation are also not new, stretching back to Dewey at least. The networked information economy does, however, provide new avenues for communication and action by pupils and students, so that it offers not only a new urgency to educate people who are inquiring, cooperative, and creatively engaged but also new means of doing so within one institution and across institutions.

What?

I take John Willinsky's core claim to be that we need to assure, as we embrace the role of nonmarket, distributed, nonproprietary production, that children and young adults whom we educate retain a sense of efficacy in their own work. He does so by, on the one hand, pushing back on the "nonmarket" language I use and, on the other hand, re-appropriating the term "market" to mean not only commercial, but more generally effective and valuable. I agree that efficacy and value are important to maintain and to render visible to children and students. I would offer a caution as to whether locating our concern with individual and collective efficacy in the metaphor of the market, by trying to extend that concept, or that of properties, is in fact the best strategy. My concern is that the market has long been the domain of not only commercial practice but also of the instrumental view of human beings and their interactions. My preference is therefore to emphasize and legitimate the language of society, humanity, and efficacy and to characterize these as practically usable, behaviorally and psychologically realistic, and analytically tractable constructs, rather than to try to expand the meaning of the market while naturalizing and legitimating the expansion of its domain beyond the priced and instrumental. While I think the language is important, I want to emphasize that I do fundamentally agree with Willinsky on the importance of effective action, or the fact and sense of efficacy that students can experience, in the educational program. "Learning by doing" isn't learning by *doing* if the outputs of the action are meaningless. The question becomes how one speaks of the value of the doing in terms that do not depend on the market and on property as the core metaphors. Again, I think the stories I tell offer an intuitive way of talking about meaning and efficacy that are not dependent on marketplace metaphors or on market value as the touchstone of effective doing as a modality of learning.

All four stories involve an instance of effective action. Be it synthesizing molecules whose quality and consistency actually matter, because they can be used to test compounds to battle neglected diseases, seeing a younger child improve his piano playing, seeing one's clients' interests protected, or seeing your hard work as a seminar student translate into political action about which you care. Any of these examples represents a direct means of doing meaningful and effective work as the platform for learning. The different stories evoke very different levels of action, from the minutely local and personal, to the grand global scale of global health. From the educational perspective, it is the efficacy that matters, and in particular efficacy in a domain that carries meaning for the students.

How one translates the need for, and commitment to, effective learning by doing suggests several paths. At a minimum, allowing and enabling students and even younger children to pursue self-selected goals seems likely to support the sense of efficacy in a domain meaningful to them. Doing so risks, of course, many projects related to sports or celebrities. However, even here, depending on context, this context may be sufficiently useful to permit effective learning. Learning skills, like statistics, or the formulation of hypotheses and tests, may be as easily transmitted through putting together a presentation on sports statistics as it can for disease burden. Beyond pure selection, the approach in law school clinical training suggests structured choice—cases and clients can be selected for a combination of social impact and educational value, and then students can (at least in some instances), choose from among several projects and clients with whom they will work. Moving yet one level further in emphasizing efficacy and value, as understood socially, is to begin to connect schools with local public goods providers—beginning with local government, and moving to local social organizations and networks. Here, enlisting high school students at least, and perhaps even younger kids, in exploring solutions to local problems; learning how to diagnose problems, formulate solutions, identify resources and combinations of work, and proposing them for implementation by the local government, organization, or school itself provides perhaps the highest form of effective engagement and the production of "goods" that adults in their community value. Interestingly, this may be part of a broader move to harness peer production and open, collaborative models of social production to solving public goods problems well beyond the reach of the digitally networked environment. In this case, specifically online action offers a relatively simple avenue for developing visible and valued projects. The most obvious is developing enough research to correct or extend Wikipedia articles and engage the other participants in conversation about why one's own summary of knowledge in this particular field is the one that ought to be respected, at least for now. However, collecting information and producing well organized arguments about matters students care about, and then going about in social networks and other online mechanisms and trying to make that intervention visible and discursively significant for the relevant target audience or interlocutors is a more general approach. It is important to emphasize, however, that I see the networked environment as a domain in which technological-economic conditions have provided greater efficacy to actions that follow generally prevalent aspects of human motivation and social behavior. The critical educational intervention then is selection of domains of practically feasible effective action in the world, whether the feasibility is born of networked connections or of localized concern and focus.

Under What Constraints?

Philippe Aigrain and Leslie Chan both represent concerns about the limits of peer production and collaboration as solution spaces, each from a very different perspective. Aigrain speaks from the hard experience of free and open source software. Here, the historical arc from open to proprietary to embattled open again suggests to him two distinct domains of competition between the open models of innovation and knowledge production and the proprietary, market-based models. These are, first, competition over capital, driven by the high expectations of returns on investment (ROI) generated by proprietary models, which suck capital out of peer production, and, second, competition over human attention, whose capture and manipulation is the core focus of many Internet businesses and which is a genuinely scarce resource on the Net. Chan focuses less on competition, and more on poverty. Much of my argument about how peer production and sharing are enabled in wealthier economies emphasizes the role of peer production in harnessing excess capacity—of computation cycles, storage, or bandwidth as well even more importantly human creativity, wisdom, attention, and insight. To what extent is this abundance that I rely on in wealthier contexts still applicable in poorer economies?

I am more confident about my answer for wealthier economies than for poorer ones. First, the problem of capital accumulation is precisely the problem that the networked environment largely solved, because of the economics of personal computers. To me, the critical shift represented by the networked information economy is the fact that for the first time the most important material inputs, into the core economic activities, of the most advanced economies, are widely distributed in the population. Competition over capital occurs against the background of the widespread distribution of capitalization and the emerging social practice of pooling human and material capital in large scale collaboration. These background facts shape the competitive environment for commercial organizations no less than the latter shape the environment for peer production. When venture capitalists have to decide where to put their money, they need to think not only of facilities for exclusion—say, through patents or copyrights—but also, perhaps mostly—on opportunities for growth, innovation, and scale. These have for several years come from platforms for peer production, not from platforms that rely on exclusionary practices. TripAdvisor is the leading site for tourist information, not Priceline. The former is a platform for users to comment on hotels and restaurants; Priceline famously received one of the first Internet business process patents for its reverse auction model. From Red Hat to Facebook, platforms for non-controlled, non-commodified interactions have fared better than many of the proprietary models, and there is neither need to think nor obvious evidence to suggest that open productive practices are losing out to proprietary models in the competition to attract capital. The same can be said for human attention—users appear to gravitate more toward sites that allow them either to engage in their own expression, alone or with others, or to use facilities to passively consume the outputs of the peer production of others. While I do not think that we have reason to believe that peer production will dominate the networked economy, we certainly have reason to be reasonably secure in its sustainability.

The answer for poorer countries is more complex. Here, much depends on access to minimal physical capabilities—from computers to mobile phones and network connections as well as software and most importantly skills and training of much larger proportions of the population than currently prevalent. Taking the "simple" problem Chan raises—South-South open access publication—we can more or less hold general education to one side, because we are talking about communication among educated professionals who publish and read professional and academic journals. Constraints here are time and effort necessary to prepare and upload manuscripts to open access databases; maintenance and storage costs; developing search and archiving software; sufficiently fast connections to upload and download; sufficiently open platforms to allow use of the materials; and sufficiently open licensing models to permit these actions where technically feasible. The last question is perhaps most difficult, as journals are resistant to open publication, and open access journals are rare, and often rely on "author pays" models with only limited ability to fund unfunded authors. Of the other questions, the most significant is raised by the fact that the most likely information platform in poorer countries is the mobile phone, not the computer. This means that the primary communications and information platform is the descendant of an appliance, not a general purpose machine, and runs on networks optimized for billing and control, not for innovation at the edges. The question, which for now remains open, is whether competition from ever smaller laptops in wealthier countries will drive mobile phone providers to develop more open systems that enable their users to be as flexible as computers now can be. Given open, widely distributed physical capabilities, and a legal regime that permits it, there is no obvious reason to think that the motivational profiles of scientists in poorer countries are any different from those of scientists elsewhere. The opportunity to publish in widely available open access resources, as long as it is easy to do and effectively at no additional cost beyond the sunk costs of computers and network connections, should suffice to achieve South-South open access publication.

To conclude, the basic questions we face are how we understand the human being who is revealed by the new practices of large scale, distributed cooperation in the networked environment and how we educate such human beings as they are and become. My answer is that we must see, with increasing clarity, that human beings are basically diverse in their motivational profiles, proclivities to sociality, backgrounds, insights, and creativity, and that networks allow us to pool these individual capabilities in an ever-wider range of combinations and institutional frameworks, well beyond those that were available in the past, to a new and ever-growing set of effective social tasks. As educators we need to emphasize the creative and social capabilities for children and students in later years, as individual explorers and inquirers who can take risks, fail, learn, and teach others about their successes and failures; we need to harness the practical efficacy made possible by the network to place students in consistent relations of cooperation and communication among themselves, but also with the world outside them, as effective agents and participants in social problem solving exercises. Through this process, we will make them better suited to a production system that increasingly depends on innovation, exploration, and learning by individuals who are only loosely constrained and afforded by the systems they inhabit, and who are expected to take on an ever greater role in defining their own task environment and the human and material resource base they must pool to solve always new problems. More importantly, perhaps, we will make them better able to be engaged citizens in a networked public sphere which, while far from perfect, still allows for much greater visibility and organizational capacity in the hands of loosely connected individuals and cooperative efforts than was possible in the past, when the public sphere was dominated by the mass media. Early efficacy in action and communication, in an educational environment that is itself networked to, and has permeable boundaries with, the "real world," online and off, can be, and ought to become, the training ground for such cooperative and effective political and social action, no less than economic, later in life.

12

Catalyst

John Howkins

Talk about old laws and people hear the words boring, dusty, and out-of-date. The oldest English law is the Magna Carta, 1215, although it is a mark of an English gentleman not to know what the Magna Carta actually says. However, one English law, now close to 300 years old, still wins respect and even affection. Indeed, lawyers still read it.

England's 1710 Statute on Copyright was the world's first copyright law. It reveals its extraordinary quality in its opening line: "An Act for the Encouragement of Learning." Till then, negotiations between writers and printers, which is what the statute dealt with, were guided only by the King's monopoly powers to license printers, and the printers' often unscrupulous dealings with writers. However, in the spring of 1709, the writers decided to fight back. After some vigorous debates, Queen Anne's government agreed to regulate writing and publishing according to the public interest. Seventy years later, the American constitution extended the principle to inventions by giving Congress powers "To promote the Progress of Science and Useful Arts, by securing for limited Times to Authors and Inventors the exclusive Right to their respective Writings and Discoveries," known as the "progress" clause. The essence of both laws is that intellectual property rights should benefit society.

Three hundred years later, some governments still struggle with this concept. Many assume that the sole purpose of copyright and patents is to make owners rich.

It is not that governments are opposed to learning or to progress. Rather it is that they do not understand the importance of ideas in the modern economy; they do not understand how creativity and innovation operate, and they do not understand the nature of intellectual property. It is, therefore, not surprising that they sometimes formulate laws that hinder, not help, learning and progress.

In the rich industrialized economies it is accepted that a country's competitive edge comes from creativity and innovation, but few policy-makers know the principles or the workings of IP. In the developing world, people are familiar with commodities, trade balances, direct investment and import substitution and know little of creativity and innovation (and their knowledge is often contentious). Their knowledge of how IP could help development is minimal.

Many would gain by reading the 1710 Statute. Most of the terms of today's discussions—the discussions that most governments shirk—were first raised then, including ethics, human rights, free speech, ownership, fair competition, money, and the public domain.

Gillian Davies, chairman of a European Patent Office appeal court and author of "Copyright and the Public Interest," has described the Statute as "the foundation upon which the modern concept of copyright in the Western world was built." *Halsbury's Laws of England,* edited by Lord Hailsham, said, "In changing the conceptual nature of copyright, it became the most important single event in copyright history." Barbara Ringer, U.S. Registrar of Copyrights, said, "It is the mother of us all, and a very possessive mother at that."

Why does it have such a grip on us today? First, remarkably, the English Parliament realized what was at stake. Members did not treat the arguments between writers and printers as a small local dispute. They recognized that the ways in which writers were published and paid were not only of importance to them but affected the health of society.

Second, the Parliament focused on the key issue, then as now. This is the balance between an individual's right to own his or her words (in order to make money) and the public's right to have access to those words (in order to learn and express themselves better).

Parliament was reflecting the mood of the times. Isaac Newton who had been a Member of Parliament a few years earlier (it is sadly impossible to imagine such a world-ranking scientist being a member of any parliament today) said that he had managed to develop the concept of gravity "by standing on the shoulders of giants," a phrase he himself had borrowed from earlier writers. The purpose of the 1710 Statute was to enable the giants to stay standing while making it possible for others to look even further ahead.

The third reason is that this tension between ownership and access, between private restrictions and public freedom, now affects vast swathes of modern society. Intellectual property rights are no longer only the concern of writers but regulate most sectors of the modern economy. They are the lynch-pin of branding consumer goods, media, entertainment, technology, drugs, computer software, R&D, and education. Copyright covers every novelist and anyone who writes a line of a genome sequence (though not a mathematical formula which is specially barred). Trademarks cover food and commodities as seen in the Ethiopian coffee initiative. These laws have moved centre stage of the global economy.

In 2006, the U.S. Federal Reserve Bank data suggested that over 45% of U.S. corporate assets are vested in intellectual property rights. According to the McKinsey consultancy firm, in 2007 over 46% of jobs in Britain and over 40% of American jobs required the employee to exercise his or her creativity in ways that qualify for intellectual property. Even more striking is McKinsey's estimate that 70% of *new* American jobs require such creative judgment. The figure refers not to 70% of jobs in the arts and sciences but to all jobs. This is the reality of what I call the "creative ecology."

The great Indonesian thinker, Soedjatmoko, had a wonderful phrase, "the learning capacity of nations." He knew many factors were important in determining a country's ability to develop, including natural resources, free speech, education, and immigration. However, he believed the most fundamental factor was the people's ability to learn. By this he meant their ability to discover other people's thoughts, to question, to explore alternatives, to experiment, and to adapt their own behavior.

In my work on the creative ecology it has become clear that learning is the most common and widely shared characteristic of creative people, from the genius to the journeyman. Creative people may differ in everything else, but they are all persistent, endless learners.

The laws that regulate how we get access to other people's ideas and inventions, whether we can use and share them, and how we can make money out of them, are obviously crucial. When I wrote *The Creative Economy* in 2000, I wanted to identify the principles that guided governments in setting their intellectual property policies. I imagined that national governments would have formulated principles for intellectual property as they routinely do so for foreign affairs, education, health, housing, and all other areas of public policy. Otherwise, how could they decide whether to allow genetic forms to be patented or whether to allow music to be downloaded from the Internet?

I was mistaken. I discovered that no one had bothered to do it: neither any national government, nor the World Intellectual Property Organisation (WIPO), nor any think-tank. There were some useful statements on traditional knowledge and academic research but, as far as we were aware, no general statement. National patent, trademark, and copyright offices generally had some "mission" statement about rewarding invention and encouraging economic development, and some vague declaration about social benefits. A few national organizations said they had principles. However, in fact, their principles simply affirmed the rights of rights-holders and did not include users. In a reversal of the emphasis on learning and understanding, economic growth has always been the priority and social development always an afterthought.

All the national offices (generally known as Patent Offices, because registering patents is their main job, although they typically cover all varieties of intellectual property) were proceeding step by step, responding to each new technology, each new situation, in what might seem the best way at the time but usually turned out to be less than optimal. I asked the UK Minister for intellectual property, "What is the government's policy? What are your objectives?" She gave a valiant answer and then laughed: "Good question. I need to do a bit of work." Her answer was one of the better ones.

What was needed was a statement of general principles that covered all possible eventualities. I kept in mind two examples of contrasting strategies. In 1980 Bill Gates agreed with IBM to design a new operating system for its new computer. He bought a basic package for $75,000 and developed it to become MS-DOS and what is now known as Windows. Microsoft vigorously uses copyrights and trademarks to protect its assets. Today, Microsoft software provides global standards that, in the world of computing, are rare. It is immensely successful.

Twenty years later, Tim Berners-Lee wrote an 11-page document specifying the basic protocols of what became known as the World Wide Web. Reportedly he considered calling it the "Mine of Information" (moi), possibly an ironic reference to Miss Piggy in *The Muppets*, or "the information mine" (tim) which shows he has a sense of humor, a trait Bill Gates appears to lack, but not much skill as a writer. He did not claim copyright. Today, the Web offers a global standard in telecommunications and information sharing that is virtually unique. It is immensely successful.

The choice of which of these two diametrically opposed strategies to follow, blending morality and money, depended on the personal inclinations of the two men involved, and also, in Gates's case, his colleagues' inclinations. However, do we have any idea as to which is more appropriate for social and economic development? Is it sensible to leave such decisions about the ownership of ideas to personal whim?

Surely, a set of principles for intellectual property should guide us in deciding whether these two completely different strategies are a good thing or a bad thing. The truth is, all current laws favor Bill Gates over Tim Berners-Lee.

When I used this example in a series of meetings in China in May and suggested that a university researcher in 2008 would be expected to assign all copyright to the university for commercial exploitation, everyone nodded vigorously. It would be the same in Britain, where universities are desperate to own every scintilla of intellectual property to show to government they are commercially savvy. It would be the same in America and in most other countries. Thus, if Tim Berners-Lee, who was employed by CERN, had invented the Web ten years later, we would be paying to use it just as we pay for TV and telephones.

Is This a Good Idea or a Bad Idea? Does It Matter?

My second example comes from scientific R&D. In 2001 two organizations published a map of the human genome, a U.S.-UK university team led by John Sulston and a private U.S. company led by Craig Venter. John Sulston, who mapped every sequence and won the Nobel Prize, adamantly insisted that his team's results should be published on the Web free of copyright and free of the European Union's notorious database law. His team used Free and Open Source (FOSS) principles to encourage sharing and collaboration. At the same time, American Craig Venter applied for patents for his gene sequences that he believed would be valuable and did not bother with the remainder.

Which of these two approaches is more valuable? Are they equally valuable? How should governments design not only the laws but encourage the social attitudes that would enable us to make sense of all this? Nobody seemed to have worked it out.

Faced with this deafening silence, I approached the Royal Society of Arts in London, and together we set up The Adelphi Charter on Creativity, Innovation and Intellectual Property (www.adelphicharter.org). We appointed a 20-strong international commission of artists, scientists, librarians, lawyers, Internet experts, consumer representatives, and business people. They included as well as Sir John Sulston, the famed musician Gilberto Gil, who was and is Brazil's Minister of Culture; Louise Sylvan, Chairman of Australia's Consumer Association; Indian environmentalist activist Vandana Shiva; Lawrence Lessig, constitutional lawyer and Chair of Creative Commons; Lynn Brindley, Chief Executive of the British Library; and James Love of Knowledge Ecology International.

The title was carefully chosen. We wanted to put creativity before intellectual property. We also wanted to brand the Charter with a snappy name. The full name is a bit of a mouthful. Fortuitously, the RSA is based in Adelphi House, named because it was built by the Adam brothers who used the Greek word to describe their own working partnership. "Adelphi" is usually translated as "brothers," but it is not gender-specific and actually means cousins in the wider sense than siblings. It means people whose collaboration is based on shared assumptions and emotions.

As well as a snappy title, I wanted to keep it short. My target audience was the politician who had a faint feeling that intellectual property was important but found it too complicated and legalistic. Following Churchill's maxim, I wanted to limit ourselves to one side of a sheet of paper. That way, nobody could have an excuse not to read it. The final text was 460 words long.

It still took two years of thinking and travelling and researching and talking and writing. For obvious reasons, the Internet was a headline issue. The Internet is one of the most remarkable tools the world has ever known for sharing information and knowledge and for allowing us to make contact with other people and to discover what they are saying. It continually offers new possibilities, new ideas, new friendships, new networks, and new businesses.

It works because it is a massive copying machine. It allows us to upload and download, copy and share, on a massive scale. Moreover, if we apply the laws that regulate the copying printed books to copying Web pages, then we will strangle the Web.

It is vital to protect the Internet's essential freedoms. However, we must also enable people to be rewarded for their work and investment. What is the right balance between freedom and enforcement? How do we answer that question?

We kept two things in kind. The first is that the Internet is largely a copyright-free zone. I do not mean everything is pirated but that the people who developed its basic software decided, like Tim Berners-Lee, not to claim protection. The Internet and the Web are therefore almost infinitely flexible and can be adapted very quickly to new circumstances. Second, the Internet is an open network. Most users are not professionals; even fewer are lawyers. Thus, it needs a regulatory system that ordinary people can understand and use. This concept rules out most copyright laws. Fortunately a bunch of lawyers came up with the idea of Creative Commons, which is basically copyright for ordinary people on the assumption that most authors (and filmmakers and musicians) want their works to be shared rather than restricted. For both these reasons, the Internet's assumption of collaboration, openness and flexibility is a model for the future for how we learn and share information.

In addition to deciding what should be copyrightable or patentable, whether a computer code, a drug, or the genome, we have to decide how long this protection should last. A trademark lasts as long as the holder goes on trading, which seems reasonable. Other kinds have widely varying terms. Patents, if renewed, continue for a maximum of 20 years. Copyright in a literary work, which encompasses everything written from poetry to computer code, lasts for the author's lifetime plus 70 years in Europe. Is this sensible? The American entertainment industries frequently try to extend copyright terms, and they are usually successful. The Supreme Court ruled in 2003 that the Constitution's "progress" clause that said Congress can give rights for "limited times" does not preclude frequent extensions even if, in practice, these extensions mean that works never fall out of copyright. While Congress was deliberating the matter, eleven Nobel Laureates in economics signed a letter saying American copyright was now effectively infinite. The Supreme Court ignored the evidence. In 2008, the European Commission was similarly reluctant to listen to the evidence in its deliberations on music rights.

The progress clause has become known as the Mickey Mouse clause because, it has been claimed, there seems to be an uncanny coincidence between the imminent ending of copyright in Mickey Mouse and the government's decision to extend copyright. In America, the "progress" clause is now interpreted as a rights-holders' clause.

No one would argue that Mickey Mouse should never be in copyright (otherwise Walt Disney could not have paid his rent). However, few would argue that the person who invented Mickey should have total, permanent control over the use of his name (otherwise I could not quote him in this article). Where, though, should we draw the line?

One issue that the 1710 Statute did not cover was a country's differing needs at different stages of its development. It dealt with Britain's domestic concerns. However, we now know that countries at different stages of development require different kinds of laws. The rich countries who own the majority of intellectual property rights are lobbying for stronger rights more rigorously enforced. In contrast, a large number of developing countries, led by Argentina, Brazil, and Chile, with the support of several American and global NGOs, says that WIPO, reflecting the UN Millennia Goals, should take more account of the needs of developing countries. Last October, WIPO agreed to do so.

The question we have to ask is this: What is the right way to regulate the ownership of ideas in the twenty-first century? There is the belief that we have a basic right to our ideas and that we have a right to charge others compensation if they want to use our ideas. In this world, incentives and rewards must always take priority, must always trump access.

This argument has a sound economic base. As I have mentioned, an increasing percentage of global business depends on these powers. The evidence is compelling not only in companies' profit-and-loss accounts but in their balance sheets. It is understandable that governments, who are keen to make their economies more competitive and protect jobs, want these intellectual assets to be protected as much as possible and at all costs. This attitude can be summed up in the phrase, "the more the better" (that is, the stronger the rights, the stronger the economy).

Intellectual property laws provide a means to establish and protect one's exclusive rights. We need them to assert our reputation as the author and, in the moral rights attached to copyright, to protect the integrity of the work. They provide incentives and rewards which, as everyone knows, are an essential part of the economic value chain. We need them to ensure our business contracts are solid and robust. When someone licenses a film on DVD, or a right to manufacture a patented drug, both the owner and the licensee need to have a common understanding that underpins what is being licensed and how the license will be enforced.

There is a second purpose which is built in to every law but which some people find counter-intuitive. This issue involves the fact that the law also enables people to have access to what has been created. For example, all patent laws require the details of the invention to be published so that others can see how it works. All copyrights come with what are called "Limitations" and "Exceptions" that restrict what the rights holder can do.

This approach puts access above incentives and rewards. It is based on three principles. First, we need existing data, ideas and, knowledge in order to have *new* ideas. Second, Europe, the United States, and Japan industrialiszd successfully in the nineteenth and twentieth centuries when their copyright and patent laws were weak, and many developing countries claim that they would benefit from similarly weak laws. Third, many major initiatives benefit from weak laws or from a refusal to claim any protection, as we have seen. This argument suggests that while intellectual property offers incentives and rewards, it does so at the cost of slowing down and inhibiting other work.

Which is the best way forward? I want to suggest a new answer based on what we know about the creative economy. What has emerged in a few countries, and what is sought by many others, though not all, is a new freedom for the individual to have, share, and enjoy his or her ideas, i.e., a freedom to make their ideas central to their lives; to use their ideas to build up their own personality, identity and status; to build up their earning power; and to turn these assets into their own creative capital.

It is risky to generalize about creative people, but it is truthful to say they are independent thinkers who are willing to challenge conventional wisdom, let alone conventional stupidities. They are curious about novelty for its own sake. They are determined to learn; at least, they are if they are successful.

They are sometimes criticized in the same manner as the Confucians described the Taoists for being "irresponsible hermits" (a description that was not intended to be a compliment). Are they irresponsible? I am reminded of William Butler Yeats's remark, "In dreams begins responsibility." He means, I believe, that only when we explore dreams and fantasies at a deep, private, personal, level, and when we know what is possible, can we really assume responsibility for our choices. Creative people need to fantasize, need to be aware of all possibilities. In addition, yes, creative people do like to break the rules. They have to break the rules. Without rule-breaking, nothing new happens. Hermits? Sometimes. However, they can be very sociable and gregarious when they want to be.

These Things Have Always Been True. So What Has Changed?

Today, creativity is no longer the preoccupation of a few people living privileged lives in special circumstances. It is now the default activity of millions of people and can be found everywhere: at home, at work, in schools, in small groups, on the street and, of course, in cyberspace. The numbers of people thinking about and using other people's ideas and creating their own ideas—ideas that may be copyrightable or patentable—can no longer be counted in thousands but in many millions. Creativity is now part of daily life for millions of people.

The scope and scale of today's creative economy is significantly greater than in previous eras. It encompasses a wider rage of activities and involves substantially more people.

There are three concentric spheres of creativity. First, there is the business of producing and distributing commercial work, which requires large financial investments. The creative work of the writer has not changed much in 300 years, but the potential for investment and commercial exploitation has exploded. The printer of Queen Anne's time now works for a publisher which is likely to be part of an international conglomerate.

However, these full-time, commercially minded creative people are not alone. They are now a small (but highly significant) part of a wider ecology, which includes a sphere of people, often working collaboratively, who are willing for others to use their work for non-commercial purposes; and an even larger group of countless people who are exploring ideas, sounds, and images, and are creating work with little thought of its commercial value or, to be more precise, of claiming any exclusive rights over it.

These three spheres, taken together, must be the basis for a sensible policy in the twenty-first century. We need to recognize each sphere's characteristics and the differences between them. Each must accept the other. Professionals must accept users not merely as consumers but as people with basic rights and inclinations to create. We need a system that maximizes access, which is everyone's interest, and which also enables rightsholders to have a reasonable reward from their work.

Laws on intellectual property should not be seen as ends in themselves but as means of achieving social, cultural and economic goals.

How can we produce such an enlightened situation? For years, IP policy has been concocted according to the business interests of Europe, the USA and, recently, Japan. That might well have suited the situation at the time.

Today, we face a different situation. We recognize that creativity is not the reserve of a few exceptional people but the result of millions of people, often working collaboratively in groups. The pursuit of learning, as indeed the pursuit of happiness, often depends upon one's freedom to pursue one's own ideas.

This dynamic operates differently in various cultures. I currently spend half my time in China where I am aware of the differences between the Western traditions of the eighteenth-century Enlightenment that are dominant in Europe and America and the Asian traditions of social harmony that are prevalent in China. Each, in its own way, exemplifies a different way of using ideas and knowledge. How we regulate ideas thus affects how we use our creative imagination.

When will the politicians realize this? When will they realize that their country's "capacity to learn" and "encouragement of learning" is just as important, if not more so, as its capacity to trade in commodities and manufactures and even financial services?

13

Reconceptualising Copyright Law for the Creative Economy through the Lens of Evolutionary Economics

Brian Fitzgerald and Sampsung Xiaoxiang Shi

I. Introduction: Can We Do Better?

Modern innovation theory posits the notion of "information flow" as a key ingredient of innovation. In turn copyright law is a key determinant of "information flow." While copyright law is meant to incentivize creativity in order to promote the dissemination or flow of information, in recent years copyright law has acted to inhibit information flow. In this chapter we argue that copyright law needs to be recast in a way that promotes information flow.

The birth of modern copyright law is normally traced back to the *Statute of Anne* 1709 (UK), a law informed by notions of crown patronage, vested interest, censorship, and monopoly business practices. Control of knowledge flow is a central theme in its creation and implementation. It is little wonder that such a law and its progeny are challenged by the social and economic context of the twenty-first century wherein participative Web applications,[1] read/write culture,[2] and social network markets,[3] and the democratization of creativity and user-led innovation[4] rise to the fore. If we are to harness to the affordances of our age, copyright law must do better at understanding and facilitating the flow of information.

As a starting point for our discussion we introduce the notion of knowledge growth and economic evolution articulated by evolutionary economists. Building on and linked to the work of Joseph Schumpeter, regarded by many as the author of modern innovation theory, evolutionary economists highlight the critical link between wealth creation and knowledge growth.[5] For evolutionary economists, the nature and causes of the wealth of nations "lie not in social governance, nor in national or even private resources, but in the human mind's ability to originate, adopt and retain generic rules."[6]

II. Evolutionary Economics
A. Overview

Evolutionary economics is in sharp contrast to classical or neoclassical economic reasoning, which begins with the presumption of scarcity of economic resources and rationality of economic agents.[7] Evolutionary economics, inspired by evolutionary biology,[8] focuses on the processes that transform the economy from within and their implications for many issues such as firms and institutions,[9] production,[10] competition,[11] science and technical advance,[12] and human nature.[13]

B. The Principle of Knowledge Growth

From an evolutionary perspective, knowledge growth is a meso trajectory and a population process by which new ideas are originated in the minds of individuals and then actualized into a carrier population. In other words, the growth of knowledge originates from the invention of novel ideas generated in individual human minds; however, it eventually results from the use and reuse of the novelties by a population of people.

In the micro-meso-macro analytical framework, evolutionary economics "endeavours to explain how macroeconomic systems evolve through generic change, that is, via the re-coordination of meso rules in consequence of micro generic choice."[14] Knowledge defined by evolutionary economists is composed of generic rules and ideas that organize actions or resources into operations.[15] Accordingly, the growth of knowledge (and thus the evolution of economy) is a recurrent "three-phase meso trajectory" that is "the process of a novel rule becoming actualized into a carrier population."[16] It is also a process of "the origination of a rule as a discovery (invention), its adoption into a population of carriers as evolutionary dynamic, and its retention by that population as an (evolved) institution."[17]

The meso trajectory of knowledge growth is "composed of a series of 'micro trajectories' that represent the process by which the novel rule is originated, adopted and retained into each individual carrier composing the population."[18] The micro trajectories involve tremendous amounts of individual inventors and carriers,[19] who compose the general reading public. The reading public today, of course, are not only reading, but listening and watching through multimedia empowered by the ICTs. In evolutionary economics, the generic micro analysis is the study of how generic rules are carried by economic agents (individuals) and agencies (organizations and firms), and the process by which a novel rule is originated, adopted, and retained by such carriers.[20] It is "a process of imagination, planning and experimental endeavour, for example, or of learning, habituation and other individual behaviours, including of course rationality."[21]

In both the micro and meso sense, the invention of novel ideas and knowledge is merely a segment of the entire dynamic process of growth. In the whole trajectory of knowledge growth and economic evolution, the value of invention can only be achieved after the accomplishment of the process of adoption and retention. In other words, the significance of knowledge is not only about the invention itself but more importantly about the use and reuse of the invented knowledge. Accordingly, the growth of knowledge can only come from the adoption and retention of a novel idea into a population of individual carriers instead of being from the invention of the idea alone.

To make the adoption and retention possible, the invented knowledge must be communicated among economic agents (carriers of rules and knowledge), and for the purpose of communication, it must first of all be encoded as information by an agent and then decoded into knowledge by other agents. In most cases, it is a process of *production*, *distribution*, and then *interpretation* of expressive works.

Evolutionary economics highlights the importance of sharing ideas to knowledge growth and the opening up of new endeavors and markets. The free flow of information and knowledge is essential because communication of novelties among individual agents is a prerequisite to the trajectory of knowledge growth. To make the dynamic process of knowledge growth possible, the invented novelty must be communicated among the economic agents and agencies. The creative works are the most dominant medium into which the novelty is fixed and by which the knowledge can be disseminated and communicated. These works play a critical part in the trajectory of the growth of knowledge in that they make it possible for novel ideas to be diluted and actualized into a population of agents and agencies. If we could posit a copyright law that promoted and facilitated access to and sharing of knowledge as well as incentivised creativity then we would be heading in the right direction. The traditional paradigm of controlling access to and reuse of creativity stands in the way. What we aim to do is highlight some key examples of how copyright control is thwarting the opportunity to share ideas and grow knowledge in the digital landscape of today and then propose some solutions for the future.

III. Recent Controversies

There are many instances over the last ten years where established industries have attempted to use copyright law in order to limit the impact of new digital technologies on existing business models. These campaigns are of concern because in most cases they are aimed at reducing the capacity of the (Internet based) "network," which has been built over the last ten years.[22] This network represents a truly unique and monumental change in social interaction. Never before have we seen communication on this scale and with such informality; millions of people forming an instantaneous and worldwide network for sharing knowledge. This is the very engine of creativity that an innovation system would crave, yet established industries are quick to try and limit its significance. To use copyright law and copyright ownership as a means for convincing courts to judicially modify the architecture of the network is dangerous. However, this kind of "copyright overreach" remains a real possibility the longer we perpetuate the myth that a copyright law based on controlled distribution can work in harmony with a network model that is driven by access and use.

The actions of established industries in trying to neutralize new and disruptive technologies is to some degree expected. As Joseph Schumpeter suggests this is the natural course toward the evolution of new opportunities and markets; it is the very substance of a capitalist economy. He explains: "But in capitalist reality . . .it is not the kind of competition which counts but the competition from the new commodity, the new technology, the new source of supply, the new type of organisation . . . competition which commands a decisive cost or quality advantage and which strikes not at the margins of the profits and the outputs of the existing firms but at their foundations and their very lives."[23]

This is also very much the history of copyright over the last 100 years. The move to each new publishing or communication technology or format has met with resistance by the established industry. The battles between publishers and recording companies, recording companies and broadcasters and broadcasters and cable companies make up the story of twentieth-century copyright law. Now at the dawn of the twenty-first century we see this battle again. Established industries firstly software, but now film and recording industries seek to control the emergence of Internet based technologies and services. Key distributors have long seen the value in tying copyright to their distribution models. The publishing, recording, and film industries have traditionally controlled reproduction and communication (= distribution) through their copyright in the underlying product.[24]

New Internet based technologies and the vast network of the Internet have challenged established industries and most particularly their approach to distribution and reuse of their material. The response of established industries has been to engage in copyright litigation. However the tide of social practice suggests that litigation (whether successful or not) will have little real impact on user behavior.

Starting with *A & M Records Inc v Napster Inc* 239 F. 3d 1004 (9th Cir. 2000) (Napster) through to *Universal Music Australia Pty Ltd v Sharman License Holdings Ltd* (2005) 65 IPR 289; [2005] FCA 1242 (Kazaa) and *MGM Studios Inc v Grokster Ltd* 545 US 913 (2005) (Grokster), we have seen the recording industry successfully pursue intermediaries that developed and/or distributed P2P file sharing technology or software. The defendants in these cases for the large part were not knowingly or intentionally reproducing or communicating unauthorized copies of songs but rather provided the facilities and services for others to do so. However, under copyright law one can infringe on someone's copyright not only by actually "doing" the infringing act (primary liability) but also by authorizing, inducing, or assisting another person to do an infringing act (authorization or secondary liability). In the United States secondary liability is spoken of in terms of contributory, inducement or vicarious liability. In a user generated distributed Web 2.0 world, the notion of intermediaries assisting end user infringement seems a little far-fetched.[25] The whole idea of this network model is to allow the end user to drive the system. The value of this network is in its "flow"; its decentralized and distributed nature allows all kinds of technologies to be connected.

The frightening aspect of the P2P litigation has been the ease with which the recording industry has got its way and the inability of the judges (with the exception of Justice Stephen Breyer in the US Supreme Court in *Grokster* 545 US 913 at 949–966 (2005)) to see the big picture. If the law is to sponsor creativity in the vast networks of the Internet, we need to see a much more sensible approach. There is no better example of this than in the multibillion dollar lawsuit Viacom (representing the interests of Hollywood) has taken against YouTube (owned by Google) for alleged copyright infringement.[26] Google Inc., who has suggested that this litigation will determine the future of the Internet,[27] is the leader of a new breed of what we might term "access corporations" that profit from greater access to knowledge—the more access there is, the more money they make. YouTube is a classic example of this concept in that it is built around freely accessible short user generated videos that are situated in a giant advertising scheme that earns Google enormous amounts of revenue. Should Google through YouTube be able to provide these services regardless of the fact that the user generators are appropriating material from Hollywood? This is a difficult dilemma for the law to resolve. However, if as practice shows that we are moving from a control mode of distribution to an access model, how much value can the law add by constantly denying this shift in the way we live and act.

How might we solve this dilemma? On the one hand, we have a tremendous new network or technology driven by people all over the world that can provide wide ranging and economically efficient distribution. On the other hand, we have an established industry saying we do not want to play in this new space unless we have control. We will sue whomever we have to sue in order to keep the status quo, regardless of the damage that may cause to the purity and operation of the network. One answer appears to be that we should decouple production (and copyright ownership) from the right of people to distribute copyright products. In other words, we should let copyright products (on publication) flow unhindered in the network in a way that promotes information flow but also provides revenue streams for creators and those that invest in the production of copyright material. We should not only increase creativity, but we should also encourage competition in distribution in order to open up new opportunities that networked technology can provide. Is this idea possible? Is it too radical?

IV. Reforming Copyright Law for the Network

If we rose above the current landscape of vested interest and imagined a digital utopia, then the notion of decoupling production and distribution would seem sensible. It would allow the creator to reach the broadest possible online audience through the fact that anyone can distribute anything. However, such an approach would require a revenue model for both the creator and the entity that invests in the creative product that is more appealing than what they have now. In a digital utopia, we could imagine technologies that would produce a way of identifying revenue streams and returning money to the relevant parties. Privacy might become an issue, but let us assume we can deal with that appropriately. This all sounds doable but a little unrealistic at least in this point in time.

What then if we revert to the here and now and the subject matter of the online distribution of music. Today it is suggested that 80% to 90% of the peer-to-peer file sharing market is beyond the reach of the recording industry.[28] If someone is making money directly from the "darknet," then it certainly is not the recording industry, although they may be gaining rewards from associated services or products. Apple's iTunes is held out as the leader of the authorized mp3 market, yet it is supposedly tapping into only 2% of the market. Hence the area of online music creates a space where people should be encouraged to explore new distribution models.

What if we suggested that in relation to the online distribution of music, copyright law should decouple production and distribution rights. To some extent in the past it has attempted to do this through compulsory licensing. For nearly 100 years copyright law has permitted the recording of a musical work (the "mechanical right") and for over 40 years the broadcasting of sound recordings under compulsory licenses subject to the payment of a prescribed fee. They are supplemented by a range of voluntary "blanket" licenses negotiated and managed by collecting societies. Even so, all of these licenses have not solved the issues raised by the Internet environment, nor are they necessarily the perfect instrument of the future. If we simply said that anyone could distribute recorded music (e.g., mp 3 files) online, that would legitimize the P2P market and allow information to flow for a range of lawful purposes. However, it would not necessarily create revenue streams for the creator and investor in the creative product. Copyright law would need to delineate on what basis revenue should flow to the relevant parties (more like a compulsory license)[29] or somehow convince parties to engage in a market based and collaborative benefit sharing of profits.

One direction to consider would be to set in place a structure where access intermediaries are allowed to access and distribute music online on the basis of revenue sharing. These access intermediaries would provide music online in a way that suited their business model, but one would suspect that their base business model would be free distribution with revenue generated from associated services. The trick here would be to encourage the parties to join together in a "communication strategy" for the digital age. The collaborative effort should reap rewards for each party and provide the consumer and society with an information flow that matches the capacity of the technology. Here the law and technology would flow in the same direction.

Bringing about the "communication strategy" might seem difficult, but we have some tools that might be employed. We could set up a compulsory license as the default mechanism, yet this mandate might create a disincentive for the access intermediary to seek a market based (rather than statutorily imposed) solution. Therefore, the statutory license would need to have limitations embedded in it. To get the copyright owner on board, we might remove or water down the right to control distribution online. The lure of monetizing the "darknet" should also act as a strong incentive to do a deal.

The only other way we could conceive of moving the recording industry into the digital age would be to impose obligations for the negative externalities it produces in trying to "hinder" the network. A clear analogy exists in real property law. One hundred years ago real property or land owners had the right to use their property as they wished. The rise of environmental law over the last 60 years has seen this sovereign right of the landowner subjected to a series of obligations to ensure land use does not pollute the existing environment to the detriment of the general public. Large entertainment companies holding intellectual property (particularly copyright) have steadfastly refused to promote new modes of exchange. They have asserted their sovereign right to exercise their property rights in any way they wish regardless of negative externalities. However the information environment like the natural environment is an ecosystem. As the argument would go, by trying to stifle the emergence of new communication structures, established industries have polluted the stream of the information ecosystem.[30] Although we need to capture the sentiment of this argument, its immediate acceptance (in its crudest form) is unlikely to happen in the near future.

While we might like to see the implementation of our vision for a digital utopia in which all content can be distributed online by anyone so long as it is adequately monetized, we are realistic in suggesting that copyright law could at least facilitate a new approach to distribution in the case of online music. It might also be possible to include films, yet film inclusion raises the issue with any or all of the content in question whether a period of exclusive distribution should be negotiated. Books are another possibility and now that Google has digitized over 10 million books in its Google Book project, perhaps we should be turning to them to support this new model of non-exclusive distribution in the area of "digitized in copyright but out of print" books.

To reiterate, an access intermediary would implement a revenue sharing model with the content owner supported by a copyright law that provides freedom to access and distribute the creative product. However, the access intermediary would need to meet certain requirements to be eligible to be an "authorized access provider" including some security for revenue streams. The model would need to be careful in not excluding small, leading edge and diverse actors. This exclusion would be difficult but not impossible to address. We imagine content producers/aggregators would arise such as the existing recording and publishing industries and new content producers/aggregators such as Google Books.

As we wrote this book chapter our attention was drawn to a new approach to online music that was emerging in China known as Google Music China. This project achieves through a privately negotiated agreement the reform we anticipate here. Our suggestion would be that this example will provide a firm basis on which to argue for copyright reform in this area. It shows us a working model in one of the most dynamic and populous markets in the world. The Google Music China project is based on a system of access intermediaries and has strengthened our belief that such a model could work on a global scale. As outlined below it shows what we have known all along. There are real benefits for content owners joining with the specialists in online distribution to provide new products and meet the growing demands of consumers.

V. Google Music China

In March 2009, Google along with its partner, the Whale Music Network (www.top100.cn), launched an online music search service at www.music.google.cn, providing free access to licensed music for users in China.[31] It is an advertising supported service allowing users to search, stream, and download high-quality songs free of charge. The service offers a catalogue of 1.1 million tracks from more than 140 labels, including the world's four biggest: Warner Music Group Corp., Vivendi SA's Universal Music, EMI Group Ltd., and Sony Corp.'s Sony Music Entertainment.[32] The service earns revenue from advertising on pages that let users stream or download songs. The income generated from the Web page advertisements is split (50:50) between the record labels and Top100, while Google presumably benefits from the increase of traffic on its site.[33]

The available music files are embedded with a digital watermark that enables the record companies to track how often and which of their songs are downloaded; however, for users, the files are totally free of Digital Right Management (DRM) or Technological Protection Measures (TPMs). It means that the songs can be played, copied, and shared through any computer or MP3 digital device. During the negotiation for this collaboration with Top100.cn and the record labels, the issue of DRM was a key issue. Google believed that DRM which limits copying and sharing of music files would be inconvenient for users and damage user experience. Google therefore required Top100 to obtain authorization for the music to be distributed without DRM.[34] Consequently, the only technological restriction of the service is the use of IP geo-location, making the service only accessible from inside China. Nevertheless, the IP geo-location can be circumvented and thus users outside China may also access the service.

It is a "content, access and enhancement" business model and a paradigm for the distribution of digital content on the Web. The Whale Music Network (www.top100.cn) is an online digital store maintaining an extensive music catalogue, which covers licensed songs from a large number of record labels. In this case, Whale provides a searchable catalogue of content. Google music search merely takes the role of an access enhancement platform; in this way, the service enriches the way people can find desirable music with Google's advanced search engine and technology.

What Google Music Search offers is the enhanced access to and facilitated distribution of music files. Apart from Google's, brand which is an advantage to attracting more users, the Google Music Search Engine also provides a significant enhancement in music access and user experiences. First of all, the Google service provides a specified and enhanced music searching capacity and a variety of ways in which users can search, choose, listen to or download the song. The search page displays a blank search box and lists of top songs and their artists, along with links enabling users to stream or download the tracks. Users can find a song through the search box and also through the names of artists, title of the song, album, sentence of a lyric, and so on.

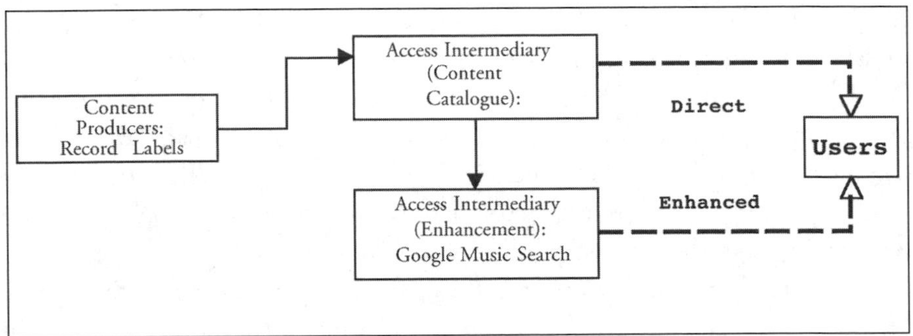

Figure 1: Google Music China and Enhancement of Access to Music

In addition, the Google Music Player and Playlist not only allow users to stream and listen to the tracks online, but they also enable users to choose, archive, and save their favorite songs online. If integrated into users' Google accounts and the next generation of service, Google Wave,[35] it will generate tremendous potential for users to enjoy and share music through the Web.

Even more importantly, the system can also recommend similar songs according to the nature and the difference of the tempo, tone, timber, genre, language, of specific songs and a singer's gender. It also has a "song screener," which is an automated system and suggests new music based on a listener's preferences for tempo or sound saturation.[36] This is important as Lachie Rutherford, president of Warner Music Asia Pacific, says: "You have to realize that not all consumers are musically knowledgeable. A lot of people need help to find out what they want."[37]

This project evidences a win-win reality, which makes it possible (and most importantly, legal) for copyright owners to benefit from their copyright while allowing other and complementary businesses to manage and enhance information flow through an access based rather than control methodology.

VI. Conclusion—The Future of Copyright in the Creative Economy

As we have shown, the key question that evolutionary economics asks of copyright law—especially in light of the behavior of copyright owners over the last ten years in trying to deny the value of information flow through the network—is to what extent copyright law should allow copyright owners the right to control reproduction and communication to the public?[38]

We argue that copyright law must ultimately not only facilitate the opportunity to create but that it must also provide the opportunity to distribute and communicate creative material to the broadest possible audience. For the last ten years, we have been held ransom to the legacy business models of established industries, and this situation has chilled new opportunities and markets.

We consider that at least in the area of online music, copyright law could provide much more incentive for information flow while still retaining a sensible and workable model for remunerating creators and investment in creativity.

Notes

1. See generally, OECD, OECD Information Technology Outlook 2006 (Paris: OECD, 2006).; OECD, Participatory Web and User-Created Content Paris: OECD, 2007).
2. H. Jenkins, et al. *Confronting the Challenges of Participatory Culture: Media Education for the 21st Century (John D. and Catherine T. MacArthur Foundation Reports on Digital Media and Learning)* (Boston: MIT Press, 2009); L. Lessig, *Remix: Making Art and Commerce Thrive in the Hybrid Economy* (New York: Penguin, 2009).
3. See generally, J. Hartley, "From the Consciousness Industry to Creative Industries: Consumer-Created Content, Social Network Markets and the Growth of Knowledge." In *The Media Industry Studies Book,* ed. J. H. Alisa Perren. New York: Routledge, 2008; J. Potts, S. Cunningham, J. Hartley, & P. Ormerod, "Social Network Markets: A New Definition of the Creative Industries." *Journal of Cultural Economics* 32, no. 3 (2008): 167–185.
4. E. V. Hippel, *Democratizing Innovation* (Boston: MIT Press, 2005).
5. See generally, A. Marshall, *Principles of Economics* (8th ed.) (London: Macmillan and Co., Ltd., 1920); J. Schumpeter, *Capitalism Socialism and Democracy* (London: Routledge, 1943); J. Schumpeter, *The Theory of Economic Development* (Cambridge MA: Harvard University Press, 1934); F. A. Hayek, *Individualism and Economic Order* (Chicago: University of Chicago Press, 1948); F. Jameson, *Postmodernism, or, the Cultural Logic of Late Capitalism* (Durham, NC: Duke University Press, 1991).
6. K. Dopfer, & J. Potts, *The General Theory of Economic Evolution* (New York: Routledge, 2008).

7. See generally, H. A. Simon, *Models of Bounded Rationality: Empirically Grounded Economic Reason* (Boston: MIT Press, 1997).
8. See generally, H. A. Simon, "Darwinism, Altruism and Economics." In *The Evolutionary Foundations of Economics,* ed. K. Dopfer, pp. 89–104 (Cambridge: Cambridge University Press); G. M. Hodgson, "Decomposition and Growth: Biological Metaphors in Economics from the 1880s to the 1980s." In *The Evolutionary Foundations of Economics,* ed. K. Dopfer, 105–150 (Cambridge: Cambridge University Press).
9. See generally, U. Witt, "The Evolutionary Perspective on Organizational Change and the Theory of the Firm," In *The Evolutionary Foundations of Economics,* ed. K. Dopfer, 339–366 (Cambridge: Cambridge University Press, 2005); R. R. Nelson, "Why Do Firms Differ, and How Does It Matter?" In *The Source of Economic Growth,* ed. R. R. Nelson, 100–119 (Cambridge, MA: Harvard University Press, 2000).
10. S. G. . Winter, "Towards an Evolutionary Theory of Production." In *The Evolutionary Foundations of Economics,* ed. K. Dopfer, pp. 223–254 (Cambridge: Cambridge University Press, 2005).
11. See generally, R. R. Nelson, & S. G. Winter, *An Evolutionary Theory of Economic Change* (Cambridge, MA: Belknap Press, 1982).
12. See generally, R. R. Nelson, ed. *The Source of Economic Growth* (Part III Science and Technical Advance)(Cambridge, MA: Harvard University Press); E. S. Andersen, "Knowledges, Specialization and Economic Evolution: Modelling the Evolving Division of Human Time." In *Evolution and Economic Complexity,* ed. J. S. Metcalfe pp. 108–150 (Cheltenham, UK: Edward Elgar, 2004).
13. See generally, J. Laurent, *Evolutionary Economics and Human Nature* (Cheltenham, UK: Edward Elgar, 2003).
14. K. Dopfer & J. Potts *The General Theory of Economic Evolution* (p. 25) (London: Routledge, 2005). For further discussion on the micro-meso-macro framework, see J. Foster & J. Potts, "A Micro-Meso-Macro Perspective on the Methodology of Evolutionary Economics: Integrating History, Simulation and Econometrics." In *Schumpeterian Perspectives on Innovation, Competition and Growth,* ed. U. C. et al. (pp. 53–68) (Berlin, Germany: Springer).
15. K. Dopfer & J. Potts, *The General Theory of Economic Evolution* (p. 6) (London: Routledge, 2008).
16. Ibid., 12.
17. Ibid., 46–50.
18. Ibid., 12.
19. Ibid.
20. Ibid., 27–44.
21. Ibid., 27.
22. L. Lessig, *The Future of Ideas* (New York: Random House, 2002).
23. J. Schumpeter, *Capitalism Socialism and Democracy* (p. 84) (London: Routledge, 1943).

24. See generally, T. Wu, "Copyright's Communications Policy." *Michigan Law Review, 103*, 278–366.
25. *CBS Songs Ltd v Amstrad Consumer Electronics PLC* [1988] 1 AC 1013.
26. *Viacom International Inc., v YouTube, Inc.,* 2007, S.D. NY., filed 13/3/2007. Viacom's complaint is available at www.paidcontent.org/audio/viacomtubesuit.pdf, and YouTube and Google's response is available at http://news.com.com//pdf/ne/2007/070430_Google_Viacom.pdf
27. N. Weinstein,"Google Says Copyright Suit Threatens the Internet [Electronic Version]. *ZDNet Australia.* 28 May 2008, Retrieved 14 October 2009, from http://www.zdnet.com.au/news/software/soa/Google-says-copyright-suit-threatens-the-Internet/0,130061733,339289364,00.htm?omnRef=1337
28. See further IFPI. Digital Music Report 2009 [Electronic Version]. 2009; Retrieved 14 October, from http://www.ifpi.org/content/library/DMR2009.pdf
29. For example, see V. Grassmuck, The World Is Going Flat(-Rate) [Electronic Version]. 2009. Retrieved 10 October 2009, from http://www.ip-watch.org/weblog/2009/05/11/the-world-is-going-flat-rate/; On a proposal for a voluntary license see: EFF. A Better Way Forward: Voluntary Collective Licensing of Music File Sharing [Electronic Version]. Retrieved 10 October 2009, from http://www.eff.org/wp/better-way-forward-voluntary-collective-licensing-music-file-sharing
30. Compare J. Boyle, "A Politics of Intellectual Property: Environmentalism for the Net?" *Duke Law Journal,* 47 no. 1(1997): 87–116.
31. J. McDonald,. Google, Music Labels Launch China Download Service [Electronic Version]. *ABC News.* 30 March 2009. Retrieved 10 October 2009, from http://abcnews.go.com/Business/wireStory?id=7205789
32. A. Back and L. Chao, "Google Begins China Music Service" [Electronic Version]. *The Wall Street Journal.* 31 March 2009. Retrieved 10 October 2009, from http://online.wsj.com/article/SB123841495337969485.html
33. Ibid.
34. M. L. Zhang, "When Google Gets Musical," *Global Entrepreneur Magazine.* Beijing, China. The article was originally published in Chinese and its English translation is available at *Google Blogoscoped,* 18 September 2009, Retrieved 10 October 2009, from http://blogoscoped.com/archive/2009–09–18-n38.html
35. Google Wave is an online tool for real-time communication and collaboration. A wave can be both a conversation and a document where people can discuss and work together using richly formatted text, photos, videos, maps, and more. See the potential Google Wave at www.wave.google.com.
36. L. Chao, "Music-Industry Execs Weigh in on Google's China Service," [Electronic Version]. *The WSJ Blogs,* 31 March 2009.Retrieved 10 October 2009, from http://blogs.wsj.com/digits/2009/03/31/music-industry-execs-weigh-in-on-googles-china-service/

37. Ibid.
38. See further: R. E. Spoo, "Ezra Pound's Copyright Statute: Perpetual Rights and the Problem of Heirs." *UCLA Law Review,* 56(2009): 1775v1834; M. Lemley and P. Weiser, "Should Property or Liability Rules Govern Information?" *Texas L Review,* 85 no. 4 (2007): 783–841.

14

The Ideology of Free Culture and the Grammar of Sabotage

Matteo Pasquinelli

> The parasite invents something new. He obtains energy and pays for it in information. He obtains the roast and pays for it with stories. Two days of writing the new contract. He establishes an unjust pact; relative to the old type of balance, he builds a new one. He speaks in a logic considered irrational up to now, a new epistemology and a new theory of equilibrium. He makes the order of things as well as the states of things—solid and gas—into diagonals. He evaluates information. Even better: he discovers information in his voice and good words; he discovers the Spirit in the wind and the breath of air. He invents cybernetics.
>
> —Michel Serres, *The Parasite*

The Living Energy of Machines: Michel Serres and the Cybernetic Parasite

Below technology, there is energy—living energy. In *The Accursed Share* Bataille described society as the management of energy surplus that constantly reincarnates itself in new forms of state and economy.[1] Being consequent with his intuition, even the contemporary mediascape can be framed as an ecosystem driven by the growth of natural energies. Media are indeed feral habitats, whose underground belly is crossed daily by large torrents of pornography and whose surface provides the battlefield for geopolitical warfare. Media are fed by the same excess of energy that shapes economy and social conflicts. However, has the energy surplus of media ever been described in an effective way? If not, which understanding of energy is unconsciously employed by the schools of media criticism? What is the role of technology in the production, consumption, and sacrifice of surplus? Furthermore, exactly what kinds of surplus are involved: energy, libido, value, money, information? Looking at today's media discourse, Bataille is enrolled only to justify a sort of *digital potlatch*—a furious but sterile reproduction of digital copies. On the contrary, under his "general economy," energy seems to float around and inside the machines, crossing and feeding a multitude of devices. To overcome an endogamic destiny, media culture should be redesigned around a radical understanding of surplus. Bataille himself considered technology as an extension of life to accumulate energy and provide better conditions for reproduction. Like "tree branches and bird wings in nature" technology opens new spaces to be populated.[2] However, something new happened when information networks entered the biosphere. What kind of energy do digital machines incarnate? Are they a further extension of biochemical energy like the classical technologies that Bataille had in mind? Digital machines are a clear bifurcation of the *machinic phylum*: semiotic and biologic domains represent two different strata. The energy of semiotic flows is not the energy of material and economical flows. They interact but not in a symmetrical and specular way as propagated by the widespread digital ideology (to be introduced as *digitalism* below).

Energy always flows one way. Acquainted with the scenario of the network society and the celebration of its *space of flows*,[3] a safari with Bataille along the ecosystems of excess is useful to remind the dystopian nature of capitalism. In Bataille economic surplus is strictly related to libidinal excess, enjoyment, and sacrifice. Even so, between endless fluxes and their "glorious expenditure,"[4] a specific model that explains how surplus is accumulated and exchanged is missing. In his inspiring and seminal book *The Parasite* Michel Serres catches the asymmetry of universal life in the conceptual figure of the *parasite*: there is never an equal exchange of energy but always a parasite stealing energy and feeding on another organism. At the beginning of the computer age (the book was published in 1980), the parasite inaugurates a materialistic critique of all the forms of thought based on a binary model of energy: Serres's semiconductors steal energy instead of computing. "Man is a louse for other men. Thus man is a host for other men. The flow goes one way, never the other. I call this semiconduction, this valve, this single arrow, this relation without a reversal of direction, 'parasitic.'"[5]

If Bataille calls attention to the expenditure of energy after its production, Serres shows how "abuse" is at work since accumulation: "abuse appears before use." Serres introduces an *abuse-value* preceding both use-value and exchange-value: "quite simply, it is the arrow with only one direction." The parasite is the asymmetrical arrow absorbing and condensing energy in a natural continuum from small organisms to human beings: "the parasite parasites the parasites." The parasite is not binary but ternary. The concept of parasite appears like a dystopian version of Deleuze and Guattari's desiring machines, as it is focused more on surplus exploitation than on endless flows. Serres shares the same vitalism of Bataille, but Serres provides in addition a punctual model to understand the relation between material and immaterial, biologic and semiotic, economy and media. In this sense, the *organic* model of the parasite should be embraced as the core concept of a new understanding of media ecosystems.[6] Indeed, Serres prophetically introduced cybernetics as the latest manifestation of the parasitic food chain (as the opening quote of this text reminds).

Moreover, Serres uses the same parasitic model for intellectual labor and the network itself (as Technology is an extension of the deceptive nature of Logos): "This cybernetics gets more and more complicated, makes a chain, then a network. Yet it is founded on the theft of information, quite a simple thing." Serres's opportunistic relation between intellectual and material production may sound traditionalist, but even when Lazzarato and Negri started to write in 1991 about the "hegemony of intellectual labour,"[7] the exploitive dimension of capital over mass intellectuality was clear. Today the immaterial parasite has become molecular and endemic—everybody is carrying an intellectual and cybernetic parasite. In this scenario what happens to the notion of multitude when intellectual labor enters the political arena in the form of a parasite? What happens to network subcultures when the network is outlined as a massive cybernetic parasite? It is time to re-introduce a sharp asymmetry between the semiotic, technological, and biological levels, between material and immaterial.

By the conceptual figure of the *immaterial parasite* I name precisely the exploitation of the biological production through the semiotic and technological domain: material energy and economic surplus are not absorbed and consumed by digital machines but simply allocated. The immaterial flow extracts surplus from the material flow and through continuous exchanges (energy-commodity-technology-knowledge-money). The immaterial parasite functions first as a spectacular device: simulating a fictional world, building a collaborative environment or simply providing communication channels, it accumulates energy through and in favour of its physical substratum. The immaterial parasite belongs to a diverse family, where rents seem to be the dominant form of metabolism. It survives in different kinds of habitat. Its tentacles innervate the metropolis (real estate speculation through the Creative Industries hype), the media (rent over material infrastructures and monopoly of online spaces), software (exploitation of Free Software to sell proprietary hardware), knowledge (revenues on intellectual property), financial markets (stock exchange speculation over collective hysteria) to mention a few examples.

Digitalism: the Impasse of Media Culture

Digitalism is a sort of modern, egalitarian, and cheap gnosis, where knowledge fetishism has been replaced by the cult of a digital network.[8] Like a religious sect, it has its peculiar theology. *Ontologically* the dominant techno-paradigm believes that the semiotic and biologic domains are perfectly parallel and specular to each other (like in the Google utopia of universal digitization). A material event can be easily translated on the immaterial plane, and conversely the immaterial can be embodied into the material. This second passage is the passage of a millenary misunderstanding, and anthropology has a lot to say about the relation between magic and logocentrism. *Economically* digitalism believes that an almost energy-free digital reproduction of data can emulate the energy-expensive material production. For sure the digital can dematerialize any kind of communication, but it cannot affect biomass production. *Politically* digitalism believes in a mutual gift economy. The Internet is supposed to be virtually free of any exploitation and tends naturally toward a social equilibrium. Here digitalism works as a disembodied form of politics with no acknowledgment of the offline labor that is sustaining the online world (a class divide that precedes any digital divide). *Ecologically* digitalism promotes itself as an environmentally friendly and zero emission machinery against the pollution of the old Fordism. However, it seems that an avatar on Second Life consumes more electricity than the average Brazilian.[9]

As Marx spotlighted commodity fetishism right at the beginning of *Capital*, a fetishism of code should be put at the basis of the network economy. "God is the machine" was the title of Kevin Kelly's digitalist manifesto whose points proclaimed distinctly: computation can describe all things; all things can compute; all computation is one.[10] Digitalism is one of those political models inspired by technology and not by social conflicts. As McLuhan once said, "We shape our tools, and afterwards our tools shape us."[11] The Internet in particular was fuelled by the political dreams of the American counter-culture of the 1960s. Today, according to the Autonomist Marxist tradition,[12] the network is at the same time the structure of the Empire and the tool for the self-organization of the multitudes. However, only the Anglo-American culture conceived the faith in the primacy of technology over politics. If today activists apply the Free Software model to traditional artefacts and talk of a "GPL society"[13] and "P2P production,"[14] they do so precisely because they believe in a pure symmetry of the technological realm over the social one. In this sense, the definition of Free Culture gathers all those subcultures that shaped a quasi-political agenda around the free reproduction of digital file. The kick-off was the slogan "Information wants to be free"[15] launched by Stewart Brand at the first Hackers' Conference in 1984. Later the hacker underground boosted the Free Software movement and then a chain of new keywords was generated: Open Source, Open Content, Gift Economy, Digital Commons, Free Cooperation, Knowledge Sharing, and other do-it-yourself variants like Open Source Architecture, Open Source Art, and so on. "Free Culture" is also the title of the book of Lawrence Lessig, founder of Creative Commons. Without mentioning the social improvements and crucial battles of the Free Software movement within the digital sphere, what is questioned here is the off-line application of these paradigms.

An old saying still resounds: *the word is made flesh.* A religious unconscious element is at work behind technology. Florian Cramer in his book *Words Made Flesh*[16] provides a genealogy of code culture rooted in the ancient brainframes of the Western world belonging to Judaism, Christianity, Pythagoreans, and Hermeticism. However, as Serres may suggest, the primordial saying must be reversed: *the flesh is made code.* The spirit itself is a parasitic strategy of the flesh. The flesh is first, before the Logos. There is nothing digital in any digital dream. Merged with a global economy, each bit of "free" information carries its own micro slave like a forgotten twin.

The Ideology of Free Culture

Literature on *free culturalism* is vast but can be partially unpacked through focusing the lens of surplus. Reading authors like Stallman and Lessig, a question rises: where does profit end up in the so-called Free Society? Free Culture seems to focus only on the issue of immaterial property rather than production. Although given a closer look, the ghost of the surplus reappears. In his book *Free Culture,* Lawrence Lessig connects the Creative Commons initiative to the Anglo-American libertarian tradition where *free speech* always rhymes with *free market.*[17] Lessig takes inspiration from the copyleft and hacker culture quoting Richard Stallman,[18] but where the latter refers only to software, Lessig applies that paradigm to the whole spectrum of cultural artefacts. Software is taken as a universal political model. The book is a useful critique of the copyright regime and at the same time an apology of a generic digital freedom, at least until Lessig pronounces the evil word: taxation. Facing the crisis of the music industries, Lessig has to provide his "alternative compensation system"[19] to reward creators for their works. Lessig modifies a proposal coming from Harvard law professor William Fisher:

> Under his plan, all content capable of digital transmission would (1) be marked with a digital watermark [. . .]. Once the content is marked, then entrepreneurs would develop (2) systems to monitor how many items of each content were distributed. On the basis of those numbers, then (3) artists would be compensated. The compensation would be paid for by (4) an appropriate tax.

In the "tradition of free culture" the solution is paradoxically a new *tax*. Tracking Internet downloads and taxation implies a public and centralized intervention quite unusual for the United States and imaginable only in a Scandinavian social democracy. The question remains unclear. More explicitly, another passage suggests the sacrifice of intellectual property to gain a larger Internet. Here Lessig's intuition is right (for capitalism). Lessig is aware that the market needs a dynamic and self-generating space to expand and establish new monopolies and rents. A dynamic space is more important than a lazy copyright regime.

> Is it better (a) to have a technology that is 95 percent secure and produces a market of size x, or (b) to have a technology that is 50 percent secure but produces a market of five times x? Less secure might produce more unauthorized sharing, but it is likely to also produce a much bigger market in authorized sharing. The most important thing is to assure artists' compensation without breaking the Internet.

In this sense Creative Commons licences help to expand and lubricate the space of market. As John Perry Barlow puts it: "For ideas, fame is fortune. And nothing makes you famous faster than an audience willing to distribute your work for free."[20] Despite its political dreams, the friction-free space of digitalism seems to accelerate toward an even more competitive scenario. In this sense Benkler in his *The Wealth of Networks* is absolutely wrong when he writes that "information is nonrival." The nonrivalry of information is another important postulate of free culturalism: Lessig and Benkler take for granted that the free digital reproduction does not cause more competition but only more cooperation. Of course, rivalry is not produced by digital copies but by their friction on real space and other limited resources. Benkler celebrates "peer production," but actually he is merely covering immaterial reproduction. Free Software and Wikipedia are extensively over-quoted as the main examples of "social production," but these examples actually only point to online social production.

Against the Creative Anti-Commons

After an initial honeymoon, the Creative Commons (CC) initiative is facing a growing criticism that comes especially from the European media culture. Scouting articles from 2004 to 2006, two fronts of critique can be distinguished: those who claim the institution of a real commonality against Creative Commons restrictions (non-commercial, share-alike, and so on) and those who point out Creative Commons complicity with global capitalism. An example of the first front, Florian Cramer provides a precise and drastic analysis:

> To say that something is available under a CC license is meaningless in practice. [. . .] Creative Commons licenses are fragmented, do not define a common minimum standard of freedoms and rights granted to users or even fail to meet the criteria of free licenses altogether, and that unlike the Free Software and Open Source movements, they follow a philosophy of reserving rights of copyright owners rather than granting them to audiences.[21]

Berlin-based Neoist Anna Nimus agrees with Cramer that CC licences protect only the producers while consumer rights are left unmentioned: "Creative Commons legitimates, rather than denies, producer-control and enforces, rather than abolishes, the distinction between producer and consumer. It expands the legal framework for producers to deny consumers the possibility to create use-value or exchange-value out of the common stock."[22] Nimus claims the total freedom for consumers to produce use-value out of the common stock (as in Free Software) but more important to produce even exchange-value (that means commercial use). For Nimus a commons is defined by its productive consumers and not merely by its producers or passive consumers. She claims that CC licences close the commons with many restrictions rather than opening it to real productivity. In a new nickname, they are "Creative Anti-Commons."

Both Nimus and Cramer's critiques remain closer to the libertarian tradition with few accounts of the surplus-value extraction and large economy behind IP (in any form: copyright, copy left or CC). On the contrary, among post-Autonomist Marxists a stronger criticism is moved against the ideology implicitly pushed by CC and other forms of *digital-only commonism.* For instance activist Martin Hardie thinks that "The logic of FLOSS seems only to promise a new space for entrepreneurial freedom where we are never exploited or subject to others' command. The sole focus upon 'copyright freedom' sweeps away consideration of the processes of valorisation active within the global factory without walls."[23] Hardie criticizes FLOSS precisely because it never questions the way it is captured by capital and its relations with the productive forces.

In conclusion, a tactical notion of autonomous commons can be imagined to include new projects and tendencies against the hyper-celebrated Creative Commons. In a schematic way, autonomous commons, first, allow not only passive and personal consumption but even a productive use of the common stock—implying commercial use by single workers; second, they question the role and complicity of the commons within the global economy and place the common stock out of the exploitation of large companies; third, they are aware of the asymmetry between immaterial and material commons and the impact of immaterial accumulation over material production (e.g., IBM using Linux); fourth, they consider the commons as a hybrid and dynamic space that dynamically must be built and defended.

Toward an Autonomous Commons

Among all the appeals for "real" commons only Dmytri Kleiner's idea of "Copyfarleft" condenses the nodal point of the conflict in a pragmatic proposal that breaks the flat paradigm of Free Culture. In his article "Copyfarleft and Copyjustright,"[24] Kleiner notices a *property divide* that is more crucial than any digital divide: 10% of the world population owns 85% of the global assets against a multitude of people owning nearly nothing. This material dominion of the owning class is consequently extended owing to the copyright over immaterial assets, so that these can be owned, controlled, and traded. In the case of music, for example, intellectual property is more crucial to the owning class than musicians, as they are forced to resign author rights over their own works. On the other side, the digital commons do not provide a better habitat: authors are sceptical that copyleft can earn them a living. In the end the authors' wage conditions within cognitive capitalism seem to follow the same old laws of Fordism. Moving from Ricardo's definition of rent and the so-called "Iron Law of Wages,"[25] Kleiner develops the "iron law of copyright earnings."

> The system of private control of the means of publication, distribution, promotion and media production ensures that artists and all other creative workers can earn no more than their subsistence. Whether you are biochemist, a musician, a software engineer or a film-maker, you have signed over all your copyrights to property owners before these rights have any real financial value for no more than the reproduction costs of your work. This is what I call the Iron Law of Copyright Earnings.

Kleiner recognizes that both copyright and copyleft regimes keep workers' earnings constantly below average needs. In particular copyleft helps neither software developers nor artists as it reallocates profit only in favor of the owners of material assets. The solution advanced by Kleiner is *copyfarleft*, a license with a hybrid status that recognizes *class divide* and allows workers to claim back the "means of production." Copyfarleft products are free and can be used to make money only by those who do not exploit wage labor (like other workers or co-ops).

> For copyleft to have any revolutionary potential it must be Copyfarleft. It must insist upon workers [sic] ownership of the means of production. In order to do this a license cannot have a single set of terms for all users, but rather must have different rules for different classes. Specifically one set of rules for those who are working within the context of workers ownership and commons based production, and another for those who employ private property and wage labour in production.

For example, "under a copyfarleft license a worker-owned printing cooperative could be free to reproduce, distribute, and modify the common stock as they like, but a privately owned publishing company would be prevented from having free access." Copyfarleft is quite different from the "non-commercial" use supported by some CC licences because they do not distinguish between endogenous (within the commons) commercial use and exogenous (outside the commons) commercial use. Kleiner suggests introducing an asymmetry: endogenous commercial use should be allowed while keeping exogenous commercial use forbidden. Interestingly, this is the correct application of the original institution of the commons, which were strictly related to material production: the commons were land used by a specific community to harvest or breed their animals. If someone cannot pasture cows and produce milk on it, the farm will not be considered a real common. Kleiner says that if money cannot be made out of it, a work does not belong to the commons: it is merely private property.

Rent Is the Other Side of the Commons

How does cognitive capitalism make money? Where does a digital economy extract surplus? While digerati and activists are stuck to the glorification of peer production, good managers—but also good Marxists—are aware of the profits made on the shoulders of the collective intelligence. For instance, the school of post-Operaismo has always had a dystopian vision of the *general intellect* produced by workers and digital multitudes: it is potentially liberating but constantly absorbed before turning into a true social autonomy. The cooperation celebrated by *freeculturalists* is only the last stage of a long process of socialization of knowledge that is not improving the life conditions of the last digital generations: in the end online "free labour"[26] appears to be more dominant than the "wealth of networks." The theory of rent recently advanced by the post-Operaist school can uncover the digital economy more clearly.

Autonomist Marxism has become renowned for shaping a new toolbox of political concepts for the late capitalism (such as multitude, immaterial labour, biopolitical production, and cognitive capitalism to name a few). In an article[27] published in 2007 in *Posse,* Negri and Vercellone make a further step: they establish rent as the nodal mechanism of contemporary economy, thus opening a new field of antagonism. Until then Autonomist Marxism had been used to focus more on the transformations of the labor conditions than on the new parasitic modes of surplus extraction. In classical theory, rent is distinguished from profit. Rent is the *parasitic* income an owner can earn just by owning an asset, and it traditionally refers to land property. Profit on the contrary is meant to be *productive,* and it refers to the power of capital to generate and extract surplus (from commodity value and workforce).[28] Vercellone criticizes the idea of a "good productive capitalism," pointing to the potential rent of profit as the driving force of the current economy: below the hype of technological innovation and creative economy, the whole of capitalism is breeding a subterranean parasitic nature. So Vercellone's motto becomes "rent is the new profit" in cognitive capitalism. Rent is parasitic because it is orthogonal to the line of the classic profit. Parasite means etymologically "eating at another's table," sucking surplus not directly but in a furtive way. If we produce freely in front of our computers, certainly somebody has his hands in our wallet. Rent is the other side of the commons—once it was over the common land, today over the network commons.

Becoming rent of profit means a transformation of management and the cognitive workforce too. The autonomization of capital has grown in parallel with the autonomization of cooperation. Today managers are dealing more and more often with financial and speculative tasks, while workers are in charge of a distributed management. In this evolution, the *cognitariat* is split into two tendencies. On one side the high-skilled cognitive workers become "functionaries of the capital rent"[29] and are co-opted within the rent system through stock options. On the other side, the majority of workers face a declassing (*déclassement*) of life conditions despite skills getting richer and richer in knowledge. It is not a mystery that the New Economy has generated more McJobs. This model can be easily applied to the Internet economy and its workforce, where users are in charge of content production and Web management but do not share any profit. Big corporations like

Google, for instance, make money over the attention economy of the user-generated content with its services Adsense and Adwords. Google provides just a light infrastructure for Web advertisement that infiltrates Web sites as a subtle and mono-dimensional parasite and extracts profit without producing any content. Part of the value is shared with users of course, and the Google coders are paid in stock options to develop more sophisticated algorithms.

The Four Dimensions of Cognitive Capitalism

The digital revolution made the reproduction of immaterial objects easier, faster, ubiquitous and almost free. However, as the Italian economist Enzo Rullani points out, within cognitive capitalism, "proprietary logic does not disappear but has to subordinate itself to the law of diffusion."[30] Intellectual property (and so rent) is no longer based on space and objects but on time and speed. Apart from copyright, there are many other modes of extracting rent. In his book *Economia della conoscenza,* Rullani writes that cognitive products that are easy to reproduce have to start a process of diffusion as soon as possible in order to maintain control over it. As an entropic tendency affects any cognitive product, it is not recommended to invest on a static proprietary rent. More specifically, there is a rent produced on the multiplication of the uses and a rent produced on the monopoly of a secret. Two opposite strategies: the former is recommended for cultural products like music, the latter for patents. Rullani is inclined to suggest that free multiplication is a vital strategy within cognitive capitalism, as the value of knowledge is fragile and tends to decline. Immaterial commodities (that populate any spectacular, symbolic, affective, cognitive space) seem to suffer from a strong entropic decay of meaning. At the end of the curve of diffusion, a banal destiny is waiting for any meme, especially in today's emotional market that constantly tries to sell unique and exclusive experiences.

For Rullani the value of knowledge (extensively of any cognitive product, artwork, brand, information) is given by the composition of three drivers: the value of its performance and application (v); the number of its multiplications and replica (n); the sharing rate of the value among the people involved in the process (p). Knowledge is successful when it becomes self-propulsive and pushes all the three drivers: first, maximizing the value; second, multiplying effectively; third, sharing the value that is produced. Of course, in a dynamic scenario a compromise between the three forces is necessary, as they are alternative and competitive to each other. If one driver improves, the others get worse. Rullani's model is fascinating precisely because intellectual property has no central role in extracting surplus. In other words, the rent is applied strategically and dynamically along the three drivers, along different regimes of intellectual property. Knowledge is therefore projected into a less fictional cyberspace, a sort of invisible landscape where cognitive competition should be described along new space-time coordinates.[31] Rullani describes his model as 3D but actually it is 4-dimensional as it runs especially along time.

The dynamic model provided by Rullani is more interesting than, for instance, Benkler's plain notion of "social production" but it is not yet employed by radical criticism and activism. What is clear and important in his perspective is also that the material cannot be replaced by the immaterial despite the contemporary hypertrophy of signs and digital enthusiasm. There is a general misunderstanding about the cognitive economy as an autonomous and virtuous space. On the contrary, Rullani points out that knowledge exists only through material vectors. The nodal point is the friction between the free reproducibility of knowledge and the non-reproducibility of the material. The immaterial generates value only if it grants meaning to a material process. A music CD, for example, has to be physically produced and physically consumed. We need our body and especially our time to produce and consume music. And when the CD vector is dematerialized because of the evolution of digital media into P2P networks, the body of the artist has to be engaged in a stronger competition. Have digital media galvanized more competition or more cooperation? This question resonates with aptness for today's Internet criticism.

A Taxonomy of the Immaterial Parasites

A taxonomy of rent and its parasites is needed to describe cognitive capitalism in detail. Taxonomy is not merely a metaphor as cognitive systems tend to behave like living systems.[32] According to Vercellone, a specific form of rent introduced by cognitive capitalism is the cognitive rent that is captured over intellectual property such as patents, copyrights and trademarks. More precisely Rullani contextualizes the new forms of rent within a speed-based competitive scenario. He shows how rent can be extracted dynamically along mobile and temporary micro-monopolies, skipping the limits of intellectual property.

The possibility of the cognitive rent has been strictly determined by the technological substratum. Digital technologies have opened new spaces of communication, socialization and cooperation that are only virtually "free." The surplus extraction is channelled generously along the material infrastructure needed to sustain the immaterial "second life." Technological rent[33] is the rent applied on the ICT infrastructures when they established a monopoly on media, bandwidth, protocols, standards, software or virtual spaces (including the recent social networks: MySpace, Facebook, etc.). It is composed by different layers: from the materiality of hardware and electricity to the immateriality of the software running a server, a blog, a community. The technological rent is fed by general consumption and social communication, by P2P networks and the activism of Free Culture. The technological rent is different from the cognitive one as it is based on the exploitation of (material and immaterial) spaces and not only knowledge. Similarly also attention economy[34] can be described as an attention rent applied on the limited resource of the consumer time-space. In the society of the spectacle and pervasive media the attention economy is responsible for commodity valorization to a great extent. The attention time of consumers is like a scarce piece of land that is constantly disputed. In the end, the technological rent is a large part of the metabolism sustaining the techno-parasite.

It is well known how the new economy hype was a driving force of the speculation over stock markets. The dot-com bubble exploited a spiral of virtual valorization channelled across the Internet and new spaces of communication. More generally, the whole finance world is based on rent. *Financialisation* is precisely the name of rent that parasites domestic savings.[35] Today even wages are directly enslaved by the same mechanism: workers are paid in stock options and so fatally co-opted in the destiny of the owning capital. Finally, even the primordial concept of land rent has been updated by cognitive capitalism. As the relation between artistic underground and gentrification shows, real estate speculation is strictly related to the "collective symbolic capital" of a physical place (as defined by David Harvey in his essay "The Art of Rent"[36]). Today both historical symbolic capital (like in Berlin or Barcelona) and artificial symbolic capital (like in Richard Florida's marketing campaigns[37]) are exploited by real estate speculation on a massive scale.

All these types of rent are immaterial parasites. The parasite is immaterial as rent is produced dynamically along the virtual extensions of space, time, communication, imagination, desire. The parasite is indeed material as value is transmitted through physical vectors such as commodities in the case of cognitive rent and attention rent, media infrastructure in the case of technological rent, real estate in the case of the speculation over symbolic capital, and so forth (only financial speculation is a completely dematerialized machine of value). The awareness of the parasitic dimension of technology should inaugurate the decline of the old digitalist *media culture* in favour of a new *dystopian cult* of the techno-parasite.

The Bicephalous Multitude and the Grammar of Sabotage

Many of the subcultures and political schools emerged around knowledge and network paradigms (from Free Culture to the "creative class," and even many radical readings of these positions) do not acknowledge cognitive capitalism as a conflictive and competitive scenario. Paolo Virno is one of the few authors to underline the "amphibious" nature of the multitude, that is cooperative as well as aggressive if not struggling "within itself."[38] The *Bildung* of an autonomous network is not immediate and easy. As Geert Lovink and Ned Rossiter put it: "Networks thrive on diversity and conflict (the notworking), not on unity, and this is what community theorists are unable to reflect upon."[39] Lovink and Rossiter notice that cooperation and collective intelligence have their own grey sides. Online life especially is dominated by passivity. Digitalism itself can be described as a sublimation of the collective desire for a pure space and at the same time as the grey accomplice of a parasitic mega-machine. A new theory of the negative must be established around the missing political link of digital culture: its disengagement from materiality and its uncooperative nature. Networks and cooperation do not always fit each other. Geert Lovink and Christopher Spehr ask precisely this: when do networks start to not work? How do people start to un-cooperate? Freedom of refusal and not-working are put by Lovink and Spehr at the very foundation of any collaboration (an echo of the Autonomist *refusal to work*).[40]

"Free uncooperation" is the negative ontology of cooperation and may provide the missing link that unveils the relation with the consensual parasite. Furthermore, a new right and freedom to sabotage must be included within the notion of *uncooperation* to make finally clear also the individualistic and private gesture of "illegal" file-sharing. Obfuscated by the ideology of the Free, a new practice is needed to see clearly beyond the screen. If the positive gesture of cooperation has been saturated and digitalized in a neutral space, only a sharpened tool can reveal the movements of the parasite. As profit has taken the impersonal form of rent, its by-effect is the anonymity of sabotage. As rent changed its coordinates of exploitation, a new theory of rent demands a new theory of sabotage before aiming to any new form of organization. Which kind of sabotage is affecting the *social factory*? In cognitive capitalism competition is said to be stronger, but for the same reasons sabotage is easier, as the relation between the immaterial (value) and the material (goods) is even more fragile.

The grey multitude of online users are learning a simple grammar of sabotage against capital and its concrete revenues along the immaterial/material conflict. To label as Free Culture the desolate gesture of downloading the last Hollywood movie sounds rather like armchair activism. If radical culture is established along real conflicts, a more frank question is necessary: does "good" digital piracy produce conflict, or does it simply sell more hardware and bandwidth? Is "good" piracy an effective hazard against real accumulation, or does it help other kinds of rent accumulation? Alongside and owing to any digital *commonism*, accumulation still runs. Nevertheless in the contemporary hype, there is no room for a critical approach or a negative tendency. A pervasive density of digital networks and computer-based immaterial labor is not supposed to bring about any counter-effect. Maybe as Marx pointed out in his "Fragments on Machines," a larger dominion of the (digital) machinery may bring simply an entropy and slowing down of the capitalistic accumulation, i.e., a more clouded and dense parasitic economy. A therapeutic doubt remains open to a dystopian destiny: is cognitive capitalism simply tending to slow down capitalism instead of fulfilling the self-organization of the general intellect?

A breaking point of capitalist accumulation is not found only in the cognitive rent of the music and movie corporations. The previous taxonomy of cognitive parasites has shown how the symbolic and immaterial rent affects daily life on different levels. The displaced multitudes of the global cities are starting right now to understand gentrification and how to deal with the new symbolic capital. In his novel *Millennium People* Ballard prophetically describes the riots originating within the middle class (not the working class!) and targeting cultural institutions like the National Film Theatre in London. Less fictionally and less violently, new tensions are rising today in East London against the urban renovation in preparation of the 2012 Olympics. In recent years in Barcelona a big mobilization has been fighting against the gentrification of the former industrial district Poble Nou following the 22@ plan for a "knowledge-based society."[41] Similarly in East Berlin, the Media Spree[42] project is trying to attract big media companies in an area widely renowned for its cultural underground. The Kafkaesque saga of Andrej Holm—an academic researcher at Humboldt University who was arrested in July 2007 and accused of terrorism because of his research around gentrification and radical activism in Germany—is not a coincidence.[43] As real estate speculation is one of the leading forces of parasitic capitalism, these types of struggles and their connections with cultural production are far more interesting than any Free Culture agenda. The link between symbolic capital and material valorisation is symptomatic of a phenomenon digitalists are not able to track and describe. The constitution of autonomous and productive commons does not pass through the traditional forms of activism and for sure not through a digital-only resistance and knowledge-sharing. The commons should be acknowledged as a dynamic and hybrid space that is constantly configured along the friction between material and immaterial. If the commons becomes a dynamic space, it must be defended in a dynamic way. Because of the immateriality and anonymity of rent, the grammar of sabotage has become the *modus operandi* of the multitudes trapped into the network society and cognitive capitalism. The sabotage is the only possible gesture specular to the rent—the only possible gesture to defend the commons.

References

Bataille, Georges. *La part maudite*, Paris: Minuit, 1949. Trans.: *The Accursed Share.* Vol. I. New York: Zone, 1988.

Benkler, Yochai. *The Wealth of Networks: How Social Production Transforms Markets and Freedom.* New Haven, CT: Yale University Press, 2006.

Carr, Nicholas. *Does IT matter? Information Technology and the Corrosion of Competitive Advantage.* Cambridge, MA: Harvard Business School Press, 2004.

Cramer, Florian. *Words Made Flesh: Code, Culture, Imagination.* Rotterdam: Piet Zwart Institute, 2005.

Davis, Erik. *Techgnosis: Myth, Magic, Mysticism in the Age of Information.* London: Serpent's Tail, 1999.

Hardie, Martin. "Change of the Century: Free Software and the Positive Possibility." Mute, 9 Jan. 2006. Retrieved from http://*www.metamute.org/en/Change-of-the-Century-Free-Software-and-the-Positive-Possibility*

Hardt, Michael and Negri, Antonio. *Multitude: War and Democracy in the Age of Empire.* New York: Penguin, 2004.

Harvey, David. "The Art of Rent: Globabalization and the Commodification of Culture." In *Spaces of Capital.* New York: Routledge, 2001.

Kelly, Kevin. "God Is the Machine." *Wired*, Dec. 2002. Retrieved from *www.wired.com/wire d/archive/10.12*

Kleiner, Dmytri. "Copyfarleft and Copyjustright." *Mute*, 18 July 2007. Retrieved from http://*www.metamute.org/en/Copyfarleft-and-Copyjustright*

Kleiner, Dmytri and Richardson, Joanne (alias Anna Nimus). "Copyright, Copyleft & the Creative Anti-Commons," Dec. 2006. Retrieved from http://*subsol.c3.hu/subsol_2/contributors0/nimustext.html*

Lazzarato, Maurizio and Negri, Antonio. "Travail immatériel et subjectivité." in *Futur Antérieur* n. 6, Summer 1991, Paris.

Lessig, Lawrence. *Free Culture: How Big Media Uses Technology and the Law to Lock Down Culture and Control Creativity.* New York: Penguin, 2004.

Lovink, Geert. "The Principles of Notworking." Amsterdam: Hogeschool van Amsterdam, 2005. Retrieved from http://*www.hva.nl/lectoraten/documenten/ol09–050224-lovink.pdf*

Lovink, Geert and Rossiter, Ned. "Dawn of the Organised Networks." *Fiberculture* 5, 2005. Retrieved from http://*journal.fibreculture.org/issue5/lovink_rossiter.html*

Lovink Geert and Spehr, Christoph. "Out-Cooperating the Empire?" In Geert Loving and Ned Rossiter (Eds), *My Creativity Reader: A Critique of Creative Industries.* Amsterdam: Institute of Network Cultures, 2007.

Marazzi, Christian. *Capitale e linguaggio: Dalla New Economy all'economia di Guerra.* Roma: Derive Approdi, 2002.

Negri, Antonio and Vercellone, Carlo. "Il rapporto capitale/lavoro nel capitalismo cognitivo." In *Posse*, "La classe a venire," Nov. 2007.

Parikka, Jussi. "Contagion and Repetition: On the Viral Logic of Network Culture." *Ephemera: Theory and Politics in Organisation,* 7, no. 2 (2007).

Rossiter, Ned. *Organized Networks: Media Theory, Creative Labour, New Institutions.* Rotterdam: NAi Publisher, Institute of Network Cultures, 2006.

Rullani, Enzo. *Economia della conoscenza: Creatività e valore nel capitalismo delle reti.* Milano: Carocci, 2004.

Serres, Michel. *Le parasite,* Paris: Grasset, 1980. Trans.: *The Parasite.* Baltimore: Johns Hopkins University Press, 1982.

Terranova, Tiziana. *Network Culture: Politics for the Information Age.* London: Pluto Press, 2004.

Vercellone, Carlo. "La nuova articolazione salario, rendita, profitto nel capitalismo cognitivo." In *Posse,* "Potere Precario," 2006.

Notes

1. Georges Batailles, *The Accursed Share,* vol. I (New York: Zone, 1988).
2. Georges Batailles, *The Accursed Share,* 36. "The space that labor and technical know-how open to the increased reproduction of men is not, in the proper sense, one that life has not yet populated. But human activity transforming the world augments the mass of living matter with supplementary apparatuses, composed of an immense quantity of inert matter, which considerably increases the resource of available energy."
3. "Space of flows" is a concept introduced by Manuel Castells in *The Informational City* (1989).
4. See Georges Batailles, "The Notion of Expenditure," in *Vision of Excess* (University of Minnesota Press, 1985); Batailles, *The Accursed Share.*
5. Michel Serres, *The Parasite* (Baltimore: Johns Hopkins University Press, 1982), p. 5.
6. Parikka offers an example of "parasitic media analysis" but focusing only on "(nonorganic) ways of network life." See Jussi Parikka, "Contagion and Repetition: On the Viral Logic of Network Culture," *Ephemera: Theory and Politics in Organisation* 7 (2).
7. Maurizio Lazzarato and Antonio Negri, "Travail immatériel et subjectivité," *Futur Antérieur,* no. 6 (1991, summer).
8. As similarly Erik Davis shows in his book *Techgnosis: Myth, Magic, Mysticism in the Age of Information* (London: Serpent's Tail, 1999).
9. Nicholas Carr, "Avatars consume as much electricity as Brazilians," draft (5 Dec. 2006), www.roughtype.com/archives/2006/12/avatars_consume.php
10. Kevin Kelly, "God Is the Machine," *Wired,* Dec. 2002. www.wired.com/wired/archive/10.12/holytech.html
11. Marshall McLuhan, *Understanding Media* (New York: McGraw-Hill, 1964).
12. See Antonio Negri and Michael Hardt, *Multitude: War and Democracy in the Age of Empire* (New York: Penguin, 2004); see also Ned Rossiter, *Organized Networks: Media*

Theory, Creative Labour, New Institutions (Rotterdam: NAi Publisher, Institute of Network Cultures, 2006).

13. "GPL society means a formation of society, which is based on the principles of the development of Free Software," Project Oekonux definition, www.oekonux.org.
14. See Michel Bauwens, "The Political Economy of Peer Production," *Ctheory*, June 12, 2005, http://www.ctheory.net/articles.aspx?id=499
15. See Roger Clarke, "Information Wants to be Free," 2000; www.anu.edu.au/people/Roger.Clarke/II/IWtbF.html
16. Florian Cramer, *Words Made Flesh: Code, Culture, Imagination* (Rotterdam: Piet Zwart Institute, 2005).
17. Lawrence Lessig, *Free Culture* (New York: Penguin, 2004): "We come from a tradition of 'free culture'—not 'free' as in 'free beer' (to borrow a phrase from the founder of the free-software movement), but 'free' as in 'free speech,' 'free markets,' 'free trade,' 'free enterprise,' 'free will,' and 'free elections.'"
18. Richard Stallman, *Free Software, Free Society* (GNU Press, 2002); www.gnu.org/doc/book13.html
19. See www.crosscommons.org/acs.html,en.wikipedia.org/wiki/Alternative_Compensation_System
20. John Perry Barlow, "The Next Economy of Ideas," *Wired*, Oct. 2000; www.wired.com/wired/archive/8.10
21. Florian Cramer, "The Creative Common Misunderstanding," 2006; Web: www.nettime.org/Lists-Archives/nettime-l-0610/msg00025.html
22. Anna Nimus (alias Dmytri Kleiner & Joanne Richardson), "Copyright, Copyleft & the Creative AntiCommons," Dec. 2006. Web: subsol.c3.hu/subsol_2/contributors0/nimustext.html
23. Martin Hardie, "Change of the Century: Free Software and the Positive Possibility," Mute, 9 Jan. 2006. www.metamute.org/en/Change-of-the-Century-Free-Software-and-the-Positive-Possibility
24. Dmytri Kleiner, "Copyfarleft and Copyjustright," Mute, 18 Jul. 2007. Web: www.metamute.org/en/Copyfarleft-and-Copyjustright
25. See en.wikipedia.org/wiki/Iron_law_of_wages
26. As Tiziana Terranova states: "It is important to remember that the gift economy, as part of a larger digital economy, is itself an important force within the reproduction of the labor force in late capitalism as a whole. The provision of 'free labor' [. . .] is a fundamental moment in the creation of value in the digital economies." Tiziana Terranova, "Free Labor: Producing Culture for the Digital Economy," in *Network Culture* (London: Pluto Press, 2004).
27. Antonio Negri and Carlo Vercellone, "Il rapporto capitale/lavoro nel capitalismo cognitivo," in *Posse,* "La classe a venire," Nov. 2007. Web: www.posseweb.net/spip.php?article17

28. Carlo Vercellone, "La nuova articolazione salario, rendita, profitto nel capitalismo cognitivo," in *Posse,* "Potere Precario," 2006, translated by A. Bove: "The new articulation of wages, rent and profit in cognitive capitalism"; Web: www.generation-online.org/c/fc_rent2.htm
29. See Antonio Negri, Carlo Vercellone.
30. A. Corsani and E. Rullani, "Production de connaissance et valeur dans le postfordisme," *Multitudes,* no. 2, May 2000, Paris [translation mine]. Web: multitudes.samizdat.net/Production-de-connaissance-et.html. Original version in Italian in: Y. Moulier Boutang (ed.), *L'età del capitalismo cognitivo* (Verona: Ombre Corte, 2002).
31. See also the notion of time-space compression as noted by David Harvey, *The Condition of Postmodernity* (Oxford: Basil Blackwell, 1989).
32. On living systems and cognitive systems see: Rullani, *Economica della conoscenza,* p. 363.
33. For a definition of infrastructural technologies see Nicholas Carr, *Does IT Matter? Information Technology and the Corrosion of Competitive Advantage,* Harvard Business School, 2004. Web: www.nicholasgcarr.com/doesitmatter.html: "A distinction needs to be made between proprietary technologies and what might be called infrastructural technologies. Proprietary technologies can be owned, actually or effectively, by a single company. A pharmaceutical firm, for example, may hold a patent on a particular compound that serves as the basis for a family of drugs. [. . .] As long as they remain protected, proprietary technologies can be the foundations for long-term strategic advantages, enabling companies to reap higher profits than their rivals. Infrastructural technologies, in contrast, offer far more value when shared than when used in isolation. [. . .] The characteristics and economics of infrastructural technologies, whether railroads or telegraph lines or power generators, make it inevitable that they will be broadly shared—that they will become part of the general business infrastructure. [. . .] In the earliest phases of its buildout, however, an infrastructural technology can take the form of a proprietary technology. As long as access to the technology is restricted—through physical limitations, intellectual property rights, high costs, or a lack of standards—a company can use it to gain advantages over rivals."
34. See Herbert Simon, "Designing Organizations for an Information-Rich World," in M. Greenberger, ed., *Computers, Communication, and the Public Interest* (Baltimore: Johns Hopkins Press, 1971). See also T. Davenport and J. Beck, *The Attention Economy: Understanding the New Currency of Business* (Boston: Harvard Business School Press, 2001).
35. See Christian Marazzi, *Capitale e linguaggio* (Roma: Derive Approdi, 2002) see also Randy Martin, *Financialization of Daily Life* (Philadelphia: Temple University Press, 2002).
36. David Harvey, "The Art of Rent: Globalization and the Commodification of Culture," in *Spaces of Capital* (New York: Routledge, 2001).

37. See Matteo Pasquinelli, "Immaterial Civil War," in Geert Lovink and Ned Rossiter, *My Creativity Reader.*
38. Among his recent texts is Paolo Virno, "Anthropology and Theory of Institutions," in *Trasversal* "Progressive Institutions," May 2007, Wien: Eipcp. Web: transform.eipcp.net /transversal/0407/virno/en ; See also Paolo Virno, "La multitud es ambivalente: es solidaria y es agresiva," interview, *Pagina* 12 (25 Sept. 2006), Buenos Aires, www.pagina12.com.ar/diario/dialogos/21-73518-2006-09-25.html
39. Geert Lovink and Ned Rossiter, "Dawn of the Organised Networks," *Fiberculture 5-Precarious Labour,* 2005. Web: journal.fibreculture.org/issue5/lovink_rossiter.html
40. Geert Lovink, "The Principles of Notworking," Inaugural speech, Hogeschool van Amsterdam, 2005. See also: Geert Lovink, "Out-Cooperating the Empire? Exchange with Christoph Spehr," in Geert Lovink and Ned Rossiter, *My Creativity Reader.*
41. See, www.22barcelona.com: "22@ Barcelona project transforms two hundred hectares of industrial land of Poblenou into an innovative district offering modern spaces for the strategic concentration of intensive knowledge-based activities. This initiative is also a project of urban refurbishment and a new model of city providing a response to the challenges posed by the knowledge-based society."
42. See www.mediaspree.de
43. See en.wikipedia.org/wiki/Andrej_Holm

15 Toward a P2P Economy

Michel Bauwens

The recent financial crisis has been described as the most serious global financial crisis since the Great Depression. Many causes have been proposed for this crisis, but what are the long-term solutions? This chapter consists of three parts. The first part is a general presentation of the nature of the present crisis and how we can possibly/realistically expect a renewed period of growth. The second part explains the role of peer to peer (P2P) dynamics in this re-orientation of our political economy, while the third part explains its political implications, and the possibilities for a phase transition toward a post-capitalist society, centered around peer production.

Part One: Understanding the Present Crisis

Our Key Hypothesis

One cannot understand peer-to-peer dynamics without the context of the general evolution of our political economy and the structural crisis inherent in an infinite growth economy. The paradox of peer production and P2P infrastructures and practices in general is that they are both immanent to the system, in fact helping it and assisting in its survival, while also transcending it and pointing to a phase transition beyond the current models. Before the P2P models can become more dominant, their emergence will become a structural part of the post-meltdown capitalism. Green capitalism cannot emerge without substantially adopting and adapting to open and participatory social practices that are enabled by P2P infrastructures. Just like emergent capitalist practices saved and strengthened the crisis-ridden feudal system after the sixteenth century, today, P2P practices save and strengthen capitalism. However, just as the new capitalist practices eventually emerged as the core of a new system, because of their hyper-productive nature, so does the hyper-productive nature of peer production point to its own eventual dominance in a new sustainable political economy and civilization.

In this transition period, peer-to-peer learning will become more and more important, supplementing and strengthening formal institutional forms of education and learning. If our hypothesis is correct, they may one day also become the core method of learning, supplemented and aided by formal institutional learning capabilities. Thus, in the context of the future of education, the practices that are currently at the periphery of learning will become core practices, and the practices that are now at the core of education will move to the periphery.

Before explaining why peer-to-peer is such an important new paradigm of social practice, let's examine the nature of the current political and economic conjuncture.

The Nature of the Present Crisis

My understanding of the present crisis is inspired by the works on long waves by Kondratieff, and how it has been updated in particular by Carlota Perez, in her work: *Technological Revolutions and Financial Capital.*[1] This work has recently been updated and re-interpreted by Badalian & Krivorotov (2008).[2]

The essential understanding of these approaches that economic history can be understood as a series of long waves of technological development, embedded in a particular supportive institutional framework.[3] These long waves inevitably end up in crisis, in a Sudden System Shock, a sign that the old framework is no longer operative.

Why Is That So?

According to Perez, these waves have a certain internal logic. They start with a period of gestation, in which the new technology is established, creating enthusiasm and bubbles, but they cannot really emerge because the institutional framework still reflects older realities. A deep economic and institutional meltdown is required as shock therapy. This period is followed by a period of maturation, marked by institutional adaptation, massive investment by the state, and productive investment by business, leading to a growth cycle. Finally, a period of decline and saturation, in which the state retreats, business investments become parasitic, leading to a contraction cycle with speculative financial bubbles.

With Perez, and our current experience with the last meltdown, we can date Sudden Systemic Shocks: 1797, 1847, 1893, 1929, and 2008.

To understand the current period in this framework, some dates are important:

- 1929 as the Sudden Systemic Shock ending the previous long wave
- 1929–1945: gestation period of the new system
- 1945–1973: maturation period, the high days of the Fordist system based on cheap domestic oil in the United States
- 1973: inflationary oil shock, leading to outward globalization but also speculative investment and the downward phase ending in the Sudden Systemic Shock of 2008

Interestingly, every long wave of approximately 50 to 60 years has been based on a combination of different structural developments in production and distribution. While modern economics is totally focusing on the monetary aspect, the crisis is only explainable if there is also a focus on the physical side.

The Cycles

Each long wave cycle was an interplay of, first, a new form of energy (e.g., the UK domination was based on coal; the US domination was based on oil); in the beginning of a new wave, the newly dominant power has particular privileged access to a cheap domestic supply, which funds its dominance; when that cheap supply dries up, an inflationary crisis ensues, forcing that power outwards to look for new supplies in the rest of the world. This development results in both dynamic globalization and in the awakening of a new periphery. Because the last phase is linked to globalization and the control of external energy supplies, it is also strongly correlated to military overstretch, which is a crucial factor in weakening the dominance of the main player.

Second, some radical technological innovations are combined into a new system; for example, the three last ones were—1830: Steam and railways; 1870: Heavy engineering; 1920: Automotive and mass production.

Third, is a new "hyper-productive" way to "exploit the territory." This is where land use comes in. For the last period, though the overall benefits are contested industrial agriculture and the "Green Revolution" did lead to a jump in agricultural production capacity. The last "parasitic" phase of a long wave cycle is then also marked by hyper-exploitation of existing land base. The example of the Dust

Bowl in the American mid-West is an example. This accumulation of problems in turn leads to the search for new methods of land-use that can be used to develop new types of land for the next up cycle.

Fourth, an appropriate financial system, i.e., the new type of public companies, and New Deal type investments (such as the Marshall Plan) in the growth cycle phase, morphing into the parasitic investments of casino capitalism in the second phase. Importantly, Badalian and Krivorotov note that each new financial system was more socialized than the previous one as, for example, the joint stock company allowing a multitude of shareholders to invest.

In the growth phase, the newly expanded financial means fund the large infrastructural investments needed to create the new integrated accumulation engine; in the declining phase, the financial system overshoots the capabilities of the productive economy, becomes separated from it, and starts investing in parasitic investments.

Fifth, a particular social contract. Here also, we can see waves of more intensive "socialization." For example, the Fordist social contract created the mass consumer in the first phase, based on social peace with labor, while in the second parasitic phase, the part going to workers was drastically reduced but replaced by a systemic indebtedness of consumers, leading to the current Sudden System Shock.

Sixth, a particular way of conceiving of the organization of human institutions, in particular the conception of the types of businesses and the management-workers relations, but also internally, the types of collaboration among employees and between employees and management.

Seventh, as we mentioned above, each wave has been dominated by a particular great political power as well, and in the second phase of expansion, a new periphery is awakened, creating the seeds for a future wave of dominance by new players. For example, the U.S. was peripheral for the long wave occupied by the British Empire but became dominant in the next phase.

Roots of the Current Crisis

It is important to remember the essential characteristics of the contraction cycle: what enables growth in a first phase becomes an unproductive burden in the second, declining phase of the wave.

If we review the different factors mentioned above, it is easy to see where the problems are:

1. The era of abundant fossil fuels is likely coming to an end; after Peak Oil, oil is bound to become more and more expensive, making oil-based production uneconomical. Nuclear power is no real replacement for oil, as its own raw material is equally subject to depletion, and it poses many long-term problems through its waste products.

2. The era of mass production, based on the fossil-fuelled car, requires a too heavy environmental burden to be sustainable, and is/was heavily dependent on cheap energy for transportation.

3. Industrial agriculture destroys the very soils that it uses and is mainly based on depletable petroleum-derivates.

4. The financial system is broken, and the massive bailout sums drain productive investments toward unproductive parasitic investments.

5. The Fordist social contract, broken in the 1980s, has led to the increased weakening of the Western middle class and a generalized precariousness, which no longer functions after Sudden System Shock.

6. The old dominant power, the United States, can no longer afford its dominance and has awakened the periphery, most likely East Asia. The powers that see the opportunity to compete are looking for new societal structures that help them emerge. They cannot rely on the strategies of the dying long wave to achicvc these goals but must invent new ones.

Seeds of the New

What innovations can we expect if a new wave is to occur?

1. The technology for renewable energy has been developed but needs at least $150b annual investments in the U.S. alone,[4] in order to become economical. A Green New Deal would jumpstart the new energy era. The wasteful heavy energy usage of the fossil fuel era will need to be replaced by smart precision-based energy usage. Solar energy will probably be the backbone of renewable forms of energy but can be supplemented by other forms.

2. The era of mass production is ready to be replaced by more local production in small series, based on developments such as flexible and rapid prototyping, based manufacturing, mass customization, personal fabrication and additive

fabrication, and multi-purpose machinery.[5] This flexible system of manufacturing is faster, cheaper, more adaptive, more compatible with solar and renewable energy, and can only thrive by deepening participative engagement, thus requiring the re-awakening of production intelligence and personal initiative, concepts that were discouraged by the various forms of the industrial system, including the systems based on central planning.

3. Post-industrial organic agriculture has already proven more productive[6] than destructive industrial agriculture, but it needs to be generalized; land use needs to be re-expanded within cities where vertical agriculture can be developed more intensively. This form of agriculture uses diversity as its backbone and works with the most sophisticated feedback cycles of nature. It saves also human labor time.

4. The seeds of the new financial system, based on increased socialization toward civil society, have been developed in the last few decades: first, sovereign wealth funds re-insert the public good in investment decisions; second, Islamic banking and similar mechanisms avoid the hyper-leveraging that destroyed the Wall Street system; third, microfinance broadens entrepreneurship and financing to the "base of the pyramid"; fourth, crowd funding mechanisms,[7] social lending, and various credit commons approaches expand the availability of credit; fifth, flow money approaches through a circulation charge to discourage parasitic investments.

5. The periphery of newly emergent countries has been awakened and will in all likelihood lead to the increased role of the East-Asian region. However, opportunities for other emergent players are still open, provided they find the appropriate local integration of the productive resources of the new long wave. In this context, we can see the emerging success of Brazil, while Russia has its enormous landmass as immense and under-exploited productive resource.

6. Social media and the Internet, now used primarily by civil society and networked individuals, will profoundly change the nature of businesses and other human organizations. Business and work organization needs to go through a profound redesign process to incorporate the hyperproductive benefits of social media.[8]

Peer-to-Peer and the New Social Contract

A new long phase has been historically associated with an upsurge of the role of the state and the public sector, which alone can undertake the necessary investments that private investment cannot take up in the early phases.

However, we need to be aware of one of the fundamental characteristics of the new period, which is a revival of the role of civil society. The Internet is enabling the self-aggregation of civil society forces in the creation of common value, i.e., through peer production. Global communities have shown themselves capable of being hyper-productive in the creation of complex knowledge products, free and open source software, and increasingly, open design associated with distributed manufacturing.

In other words, a hybrid form of production has emerged that combines the existence of global self-managed open design communities, for-benefit associations in the form of foundations that manage the infrastructure of cooperation, and an ecology of associated businesses that benefit from and contribute to this commons-based peer production.

These companies that enable and empower the social production of value have become the seeds for the dominant companies of the future (Google, eBay, and so on . . .). Companies will need to open up to co-design and co-creation, while the distribution (miniaturization) of the means of physical production liberates the possibilities for smaller, more localized production units to play more essential roles. The role of solely profit driven multinational companies, without any roots in local communities, is reaching its historical end, and it will be replaced increasingly by new models of entities combining profit with the realization of social and public goods. Socially conscious investment, sovereign wealth funds, micro-finance, social entrepreneurship, fair trade, and the emergence of for-benefit entities point to this new institutional future of entrepreneurship. For the state form, this means morphing from the welfare or neoliberal state models, to that of the Partner State to facilitate and and to empower social production.

The new social contract therefore will mean:

1. Expanding entrepreneurship to civil society and the base of the pyramid

2. New institutions that do well by doing good (outcome based enterprise)

3. Social financing mechanisms based on peer-to-peer aggregation

4. Mechanisms that sustain social innovation (co-design, co-creation) and peer production by civil society

5. Participatory businesses and other human organizations

6. Focus on more localized precision-based physical production in small series but linked to global open design communities

The new long wave that we are hypothesizing is of course speculative and needs some caveats.

First of all, the new wave cannot occur without a long period of disruption and adaptation, also needed for the deleveraging of debt of the previous period.

Second, though long waves have structurally occurred in the last two centuries, the severe crises related to the depletion of fossil fuels, but also the impact of climate change, could possibly derail such a scenario. It may also be that, as the current infinite growth system is incompatible with the survival of the biosphere, that these cyclic tendencies may be overturned and interrupted by a more fundamental crisis, involving the very survival of capitalism.

Nevertheless, there is a real possibility of a next long wave, based on a new social contract, where netarchical capitalists and peer producing communities will play a larger role. This long wave may likely be interrupted half-way, however, given that the severity of the climate and ecological crisis may make the dominance of market mechanisms unacceptable, though they are likely to persist in a subordinate role.

What we deem likely is the following: to begin, a period of deleveraging and restructuration followed by a new upturn cycle of the new wave.

Therefore, when the upturn hits the first halfway crisis of a Kondratieff Wave, in the context of deepening resource and climate change related crises and challenges, that the crisis of the present system will become systemic, and open up the possibility of a further phase transition, to a form of post-capitalism, which is compatible with the survival of the biosphere.

The new modality that has been emerging since before the crisis as a new social, political, and economic practice is the peer-to-peer dynamic.

There are two possible scenarios. In the first scenario, enlightened global leadership integrates the new peer-to-peer demands and practices as part of its social contract, and the upturn is relatively smooth. In the second scenario, the shortsightedness of the global system leadership fails to do this, and the resulting disruption of the global system, forces peer-to-peer practices into the resilient mode at the more local level.

However, in any case, its uptake will speed up during the deleveraging and adaptation crises in order to become a new part of the new social contract during the new upturn of the Kondratieff cycle. At the end of this half-cycle, when peer-to-peer may achieve some form of parity, the systemic crises may then lead to the new system becoming the dominant meta-system, while the market system may be the new subsystem integrated in the new system.

With this context set, we can now explain the importance of the peer-to-peer dynamic itself.

Part Two: The Economics of P2P
General Introduction

Peer-to-peer social processes are bottom-up processes whereby agents in a distributed network can freely engage in common pursuits without external coercion, i.e., undertake actions and relations without permission or force. This dynamic requires not just "decentralized" systems but "distributed" systems[9] that enable individuals to cooperate. Distributed networks do have constraints, forms of internal coercion, that are the conditions for the group to operate, and they may be embedded in the technical infrastructure, the social norms, or legal rules. Despite these caveats, we have here a remarkable social dynamic, which is based both on voluntary participation in the creation of common goods, which are made universally available to all.

Peer-to-peer processes are emerging in literally every cranny of social life and have been extensively documented in 10,000-plus pages at the Foundation for Peer to Peer Alternatives and many other places on the Web.

P2P social processes more precisely engender:

1. **Peer production:** wherever a group of peers decides to engage in the production of a common resource;

2. **Peer governance:** the means the peers choose to govern themselves while they engage in such pursuit

3. **Peer property:** the institutional and legal framework they choose to guard against the private appropriation of this common work; this usually takes the form of non-exclusionary forms of universal common property, as defined through the *General Public License*, some forms of the *Creative Commons licenses*, or similar derivatives.

Peer governance combines the free self-aggregation between individual skills and universally broadcast tasks, processes for communal validation of excellence within the broader pool of input, and defense mechanisms against private appropriation and sabotage. Peer governance differs from hierarchical allocation of resources, from allocation through the market, and even from democracy, as these are all mechanisms for dealing with scarce resources. Peer governance essentially aims, and often succeeds, at making sure that no formal "representative group" can make decisions separate from the community of peer producers.

These new property forms have at least three characteristics: The first one is to prevent the private appropriation of the commonly created value. The second feature is aimed at creating the widest possible usage; in other words, there are universal common property regimes. The third characteristic aims to keep the sovereignty with the individual. More specifically, this last aspect shows why peer property fundamentally differs both from private property and collective property.

Private property is individual but is exclusionary in that it implies that what belongs to one person does not belong to another. However, collective property is also exclusionary in the sense that an item belongs to everyone, but the individual has no sovereignty over it. It is from us, regulated by a bureaucracy or representative democracy but without private property rights. The collective has taken over from the individual, and more often than not, coercion is involved.

The general public license, though, or the creative commons licenses are different. Common property is not collective property. Using these properties, the individual gets full attribution, i.e., the recognition of his personal property. Individuals are freely sharing their sovereignty with others. This type of relationship is especially clear in the creative commons licensing schemes, where the individual gets a whole gamut of options for sharing. Each person remains fully in control, i.e., "sovereign," and there is no coercion involved.

It is important to note that peer production is a form of "generalized," or nonreciprocal, exchange. It is not a gift economy, based on direct exchange or obligation. So peer production is not to be equated by cooperative production for the market: participation has to be voluntary; there is no direct reward—though there are many indirect rewards—in the form of monetary compensation. The process itself is participative, and the outcome is similarly free in the sense that anyone can access and use the common resource. In reality, most peer production projects are intertwined with a smaller core of people who may get paid and use finances to create an infrastructure so that the peer production may occur.

If we look at peer production as a mode of production, as a process involving a input, "processing," and output phase, then we can say that it requires the following:

- Open and free raw material that can be used without permission. Thus, peer production either requires the creation of such open and free raw material by the producers themselves, or materials that are in the public domain or in a commons format already.

- The process is participatory with a design that is geared toward inclusion and *a posteriori* validation, not exclusion through *a priori* filtering of the participants.

- The output is universally available and, therefore, uses peer property formats or in other words: a Commons.

As the Commons-oriented output creates a new layer of open and free input for further transformation and processing, we have here the requirements for social reproduction of the system, called the Circulation of the Common by Nick Dyer-Witheford (2006).[10]

Looking at these three inter-related paradigms of open and free participation and the Commons, we can then easily understand why movements striving for these conditions and social practices are arising in almost every single field of human activity. The conditions for peer production to emerge are essentially abundance and distribution. Abundance refers to the abundance of intellect or surplus creativity, to the capacity to own means of production with similar excess capacity. Distribution is the accessibility of such abundant resources in fine-grained implements, what *Yochai Benkler has called modularity or granularity.* Again we could talk about the distribution of intellect, of the production infrastructure, of financial capital.

It is important to distinguish two spheres. In one sphere, our digitally enabled cooperation, reproduction of non-rival knowledge goods, such as software, content, open designs, takes place at marginal costs, and there is no loss by sharing, but actually a gain, through network effects. Such free cooperation can only be hindered "artificially," through either legal means (intellectual property regimes) or through technical restrictions such as Digital Rights Management, which essentially hinder the possible social innovations. In this sphere, a non-reciprocal mode of production becomes dominant since individuals are not competing for resources, are not rivals, and they are not losing but gaining through giving. In the sphere of material production, where the costs of production are higher, and we have rival goods, we still require regimes of exchange, or regimes of reciprocity. Thus, in a sphere of virtual abundance, where copying is trivial, there is no tension between supply and demand and hence no need for a market.

Postcapitalist Aspects of Peer-to-Peer

Peer production, though embedded in the current political economy and essential for the survival of the cognitive forms of capitalism, is therefore itself essentially postcapitalist because it is outside wage dependency, outside the control of a corporate hierarchy and does not allocate resources according to any pricing or market mechanism.

Similarly, peer governance could be said to be postdemocratic because it is a form of governance that does not rely on representation, but where participants directly co-decide. Morever, it is not limited to the political field but can be used in any social field. Peer governance is non-representational, and this is essentially so because the networked communication facilitates the global coordination of small groups, and, therefore, the peer-to-peer logic of small groups can operate on a global scope.

Hierarchies, the market, and even representative democracy, are all only means to allocate scarce resources and do not apply in the context where abundant resources are allocated directly through the social process of cooperation. However, since the pure peer-to-peer logic only fully functions in the sphere of abundance, it will always have to insert itself in the forms that are responsible for the allocation of resources in the sphere of material scarcity. Peer governance based leadership seems a combination of invitational leadership, i.e., the capacity to inspire voluntary cooperation, and *a posteriori* arbitrage based on the reputational capital thus obtained. However, the process of production itself is an emergent property of the cooperating networks. Finally, peer property is a post-capitalist form of property because it is non-exclusionary, and it creates a commons with marginal reproduction costs. There are two main forms of peer property. One is based on the individual sharing of creative expression and is dominated by the Creative Commons option that allows an individual to determine the level of sharing. The other is applied to commons-based peer production and takes the form of the general public license or its derivatives or alternatives and requires that any change to the common also belongs to the common.

The Hyper-Productive Nature of Peer-to-Peer

Pre-capitalist class societies are based on coercive extraction of surplus value and hierarchical allocation of resources. Capitalism is based on the part-real and part-fictional process of equal exchange of value. In other words, we can say that coercive societies are based on the extrinsic motivation of fear, while capitalism is based on the extrinsic motivation of self-interest.

Peer production structurally eliminates extrinsic motivation and replaces it with intrinsic motivation or, in one word, passion. It is psychologically the most potent and productive form of human motivation. In addition, the market only allows, at best, for win-win scenarios of mutual interest, but it is structurally designed to ignore externalities. Corporate firms can only strive for relative quality in a competitive environment, but peer producing communities strive structurally for absolute quality. As an object-oriented sociality based on the construction of universally available common value, peer production inherently strives for positive externalities and lacks much of the motivation to create negative externalities for the sake of profit. The combination of all these characteristics creates a hyper-productive mode of production and an asymmetrical competition with pure for-profit firms relying on wage labor and closed intellectual property.

This explanation facilitates the formulation of the bold hypothesis of the law of asymmetrical competition:

- Any for-profit company based on closed IP that is faced with the competition of a peer producing community, a for-benefit association managing the infrastructure of cooperation, and an ecology of businesses based on a commons, will lose that competitive race.

Indeed, this hypothesis explains the gains of Linux over Microsoft, the rise of Wikipedia as compared to *Britannica* as models for many other examples of asymmetrical completion.

An entity based on innovation-impeding intellectual property, appropriation of common social value that discourages free contributions, and striving for relative quality (hence consciously substandard products), cannot in the long run survive the challenge of an open competition based on peer production.

However there is an important corollary to this first law that explains the necessity of hybrid forms and why peer production can be embedded within an overall capitalist context.

The corollary law is this:

- Any peer production community that creates a sustainable management for its infrastructure of cooperation and a ecology of businesses that can fund it will be more competitive than a community that fails to do so.

Pure non-reciprocal production can only occur within a sphere of relative abundance, characterized by the free aggregation of human brains, ownership or easy access to computers, and socialized access to the networks, such as the Internet. However, if peer production is collectively sustainable as long as it can maintain a similar level of volunteerism (offsetting departures with newcomers), it is not so for the individuals concerned. In addition, it also requires an additional infrastructure of cooperation that may have to operate in addition to the Internet. For example: it may need costly servers in case of success. Peer production cannot, therefore, fully escape the monetary sphere or its requirements; in other words, it demands hybrid formats.

In brief, successful peer projects combine:

1. The freely self-aggregating community

2. A for-benefit association, usually in the form of a nonprofit foundation that funds and manages the infrastructure of cooperation

3. A ecology of businesses that practice benefit-sharing, returning part of the profit obtained from selling back to the commons on which their value-creation is based. Such businesses therefore fund the infrastructure of cooperation, hire many of the participants, and thereby maintain the viability and sustainability of their respective Commons.

Adaptation of Cognitive Capitalism to Peer-to-Peer

So far, empirical evidence suggests three emerging forms of adaption between the sphere of peer-to-peer cooperation, and the institutional and market fields:

- The sphere of individual sharing, e.g., YouTube, where sharers have relatively weak links to each other, creates the Web 2.0 business model. In this model, an ethical economy of sharing co-exists with proprietary platforms that enable and empower such sharing in exchange for the selling of the aggregated attention.

- The sphere of commons-oriented peer production, based on stronger links between cooperators, e.g., Linux or Wikipedia, usually combines a self-governing community, with for-benefit institutions (Apache Foundation, Wikimedia Foundation, and so on . . .) that manage the infrastructure of collaboration, and a ecology of businesses that create scarcities around the commons, and in return support the commons that provide their value.

- Finally, crowd sourcing occurs when it is the institutions themselves that attempt to create a framework, where participation can be integrated in their value chain, and this can take a wide variety of forms. This is generally the field of co-creation.

There is a mutual dependence of peer production and the market. Peer production is based on the achievements and surplus of the existing market-dominated society, and on the income that can be generated through participation in the market; on the other hand, market players are increasingly dependent and profiting from social innovation.

Because of the law of asymmetrical competition, i.e., the hyper-productive nature of peer production, corporations are driven to adapt substantially to the new practices, and new players emerge who are based on an alliance with peer production. The companies that do so are more competitive than those who do not, creating a new sector of "netarchical capitalism" that facilitates social innovation and peer production.

Because of their contradictory nature, corporations have a dual role in this dynamic. They have to sustain cooperation and sharing, i.e., the openness that creates value but also have to enclose part of the value, as they are competing with others in a scarcity-based marketplace.

Monetary value that is being realized by the capital players, is—in many if not most of the cases—not of the same order as the value created by the social innovation processes. The user-producers-participants are creating direct use value, videos in YouTube, knowledge and software in the case of commons-oriented projects. This use value is put in a common pool, freely usable, and, therefore, does not consist of scarce products for which pricing can be demanded. The sharing platforms live from selling the derivative attention created, not the use value itself. In the commons model, the abundant commons can also not be directly marketed, without the creation of additional "scarcities."

What does all of this mean for the market sphere?

It is now possible to create all kinds of use value without, with only a minimal, or with only an a posteriori intervention of capital. We are dealing with post-monetary, post-capitalist modes of value creation and exchange, that are immanent, i.e., embedded in the market but also transcendent to it, i.e., operating outside its boundaries. Capital is increasingly dependent and profiting in all kinds of ways, from the positive externalities of such social innovation.

Thus, the challenge can be described as follows: first, we have a process of social innovation that creates mostly non-monetary value for the participants; second, we may have an increasingly huge gap between the possibility of creating post-monetary value, and the derivative exchange values that are realized by enterprise; third, the participants engaged in such passionate production and innovation frequently cannot find in such processes an answer to their own sustainability.

Hence, it is impossible to realize more than just a small partial monetary value, from the point of view of most commercial players. This development means an increasing precariousness for the participants of social innovation. In other words, the current market model does not have a reverse process of redistribution for the value that is being created. This crisis may of course be a temporary crisis, but it is doubtful that it is. The reason is that the market can only indirectly and partially provide monetary compensation for processes that are not motivated by such compensation. More general redistributive processes are needed to allow society and the market to give back part of the value that is being created.

One possibility is the further development of transitional labor market measures (protect the worker, not the job) that recognize the flexibility and mobility of contemporary careers. However, this concept needs an important add-on development: the realization that contemporary workers are moving not just from job to job, but also from jobs to non-jobs, and that in fact, what is most useful and meaningful for them (and the market, and society) are not the paid jobs for the market, but the episodes of passionate production. It seems, therefore, that a more general measure, not linked to the job, but conceived as a repayment for, and enabler of, social innovation, is needed. The name of that general measure is most probably some form of basic income.

Likely Expansion of Peer Production Principles to Material Production

Peer production naturally occurs in the sphere of immaterial production. In this sphere, the access to distributed resources is relatively easy. Large sections of the population in the Western countries are educated and the many people who live there can have a computer at their disposal—and the costs of reproduction are marginal.The expansion of peer production is dependent on cultural/legal conditions. It requires open and free raw cultural material to use; participative structures to process it; and commons-based property forms to protect the results from private appropriation. Hence, a circulation of the common is obtained, through which peer production virally expands.

However, peer production is not limited to the sphere of immaterial production. First of all, any physical production process needs to be immaterially designed, and open design is not fundamentally different, though it is more complex, than collaborative knowledge or free software production. So, peer production can work for the design phase of physical production, provided a good infrastructure is available for such co-design.

Physical resources can be shared, if they are available in a distributed format, e.g., computers and their files and processing power. Cars can be pooled. Money can be pooled as in the P2P financial exchanges such as *Zopa* or through mutual credit systems. Wealth acknowledgment procedures can be the basis of the creation of complementary currencies. Rapid tooling and prototyping, desktop manufacturing, personal fabricators and 3D printers, multi-purpose machinery, and other similar developments may and will lower the threshold of participation, creating more modularity and granularity in new fields. In fact, we may observe that the same tendency to miniaturization, which led to the networked computer, is taking place in the domain of physical machinery. Given the decrease in the cost of physical capital, it becomes easy to imagine the combination of open design communities with cooperative forms of relocalized physical production.

Such expansion is not just a natural extension of technical evolution, but it has structural and therefore political impediments. The centralized capital formats of contemporary neoliberal anti-markets obviously impede such expansion. However, even with such constraints, the scope for the expansion of peer production is significant. Again, we will make the following caveat. In the immaterial sphere, non-reciprocal peer production is likely to become dominant. In the field of scarcity, there will be a rise of peer-informed modes of production. This dynamic means that market forms are starting to change from a logic of pure capitalism (making commodities for exchange, so as to increase capital) to logics where the logic of exchange is subsumed to the logic of partnership. Think about fair trade (a market subjected to peer arbitrage), social entrepreneurship (profit used to sustain social goals), base of the pyramid inclusive capitalism, and the many political-social movements that aim to divorce market forms from the infinite growth logic of capitalism, such as the natural capitalism movement in the United States.

Since about 2006, there has been a renewed emergence and rapid growth of craft communities, a maker movement, distributed desktop manufacturing through commercial platforms, and a free and open hardware movement. Open hardware is growing very fast, with companies such as Arduino and Buglabs providing living exemplars and role models. These companies are inventing their own platforms and infrastructures such as the Open Source Hardware Bank. The latter is particularly significant as it shows that open hardware producing communities, such as the ones around the Arduino electronic circuit boards, are creating their own business ecologies. They are combining the existing triarchical commons model (community, foundation, business) with a solution to the cost recovery problem typical for physical production. Thus, they are emerging as viable alternatives to the traditional corporate models, and owing to the inherent hyper-productivity discussed above, they are slated to play an increasingly dominant role.

To prosper and expand beyond its current confines in the sphere of immaterial production, more distributed infrastructures will be necessary, complementing the already existing communication infrastructures:

- Distributed energy: this requires a move away from centralized energy production based on depletable fossil fuels and a move toward a home and neighborhood based infrastructure producing renewable energy.

- Distributed and multiple currency systems: meta-currency platforms will allow local and virtual (affinity-based) communities to produce exchange mechanisms that are not based on compound interest and fractional reserve banking and can both promote specialized in-community exchange, protect from globalized dislocation, and create an alternative infrastructure of inter-community and inter-individual exchange.

- Open and distributed manufacturing: distributed capital goods with radically lower thresholds, such as the ones being developed today, need to be reconfigured and integrated in a vision of relocalized production in the context of a global cooperation with open design communities

Part Three: The Politics of P2P
P2P Theory as the Emancipatory Possibility of the Age

Our current political economy is based on a fundamental mistake. It is based on the assumption that natural resources are unlimited and that there exists an endless drain. Furthermore, it creates artificial scarcity for potentially abundant cultural resources. This combination of quasi-abundance and quasi-scarcity destroys the biosphere and hampers the expansion of social innovation and a free culture.

In a P2P-based society, this situation is reversed: the limits of natural resources are recognized, and the abundance of immaterial resources becomes the core operating principle.

The vision of P2P theory is the following:

1. the core intellectual, cultural, and spiritual value will be produced through nonreciprocal peer production;

2. it is surrounded by a reformed, peer-inspired, sphere of material exchange;

3. it is globally managed by a peer-inspired and reformed state and governance system, a "partner state which enables and empowers the social production of value."

Because of these characteristics, peer-to-peer can be said to be the core logic of the successor civilization and is an answer and a solution to the structural crisis of contemporary capitalism.

Indeed, because an infinite growth system is a logical and physical impossibility with a limited natural environment, the current world system is facing a structural crisis for its extensive growth. Currently consuming resources at the rate of 'two planets," it would need four planets if countries like China and India obtained equity with the current Western levels of consumption. Because of the ecological and resource crisis that this possible development might cause, the system is ultimately limited in its extensive expansion.

However, its dream for intensive development in the immaterial sphere is equally blocked since the sphere of abundance and direct social production of value through peer production creates a kind of "exponential" growth in use value but only a kind of "linear growth" of the market opportunities in its margins.

The current world system is facing a similar crisis to that of the slave-based Roman Empire, which could no longer grow extensively (at some point the cost of expansion is greater than the benefits of added productivity) but could not grow intensively either since that would have demanded autonomy for the slaves. Hence, the feudal system emerged, which refocused on the local level, where it could become much more productive and grow "intensively." Serfs, who were tied to the land but now had families, a fixed part of what they produced and a much lighter taxation load, were substantially more productive than slaves. The domain-based lords took a substantially smaller part of the surplus. Today, extensive growth is ultimately blocked, but intensive growth in the immaterial sphere requires a substantial reconfiguration that largely transcends the current system's logic.

Similarly, the current structural crisis causes a reconfiguration of the two main classes (just as the slave owners had to become feudal lords and the slaves had to become serfs). At present, we see the emergence of a netarchical class of capital owners, who are renouncing their dependence on the present regime of immaterial accumulation through intellectual property, in favor of a role as facilitators of social participation through proprietary platforms, which cleverly combine open and closed elements so as to ensure a measure of control and profit, while knowledge workers are reconfiguring from a class that was dissociated from the means of production, to one that is no longer dissociated from its means of production, as their brains and the networks are now their socialized means of production. (However, they are still largely dissociated from autonomous means of monetization.) It would be fair to say that currently peer production communities are collectively sustainable, but not individually, leading to a crisis of value and widespread precariousness among knowledge workers.

The solution would in my opinion point in the following direction:

1. The private sector recognizes its increasing dependence on the positive externalizations of social cooperation, and together with the public authorities, agrees to a new historical compromise in the form of a basic income; this allows the sphere of cooperation to thrive even more, creating market benefits.

2. The sphere of the market is dissociated from infinite-growth capitalism (how this can be done would require a separate article, but the key would be a macro-monetary reform such as those proposed by Bernard Lietaer, associated with a new regime that extends the production of money from private banks to the social field, through open money systems).

3. The sphere of peer production creates appropriate "wealth acknowledgement systems" to recognize those that sustain its existence, and systems exist that can translate that reputational wealth in income.

Peer Governance and Democracy

As peer-to-peer technical and social infrastructures such as sociable media and self-directed teams are emerging to become an important if not dominant format for the changes induced by *cognitive capitalism*, the peer-to-peer relational dynamic will increasingly have political effects.

As a reminder, the P2P relational dynamic arises wherever there are distributed networks, i.e., networks where agents are free to undertake actions and relationships, and where there is an absence of overt coercion so that governance modes are emerging from the bottom up. It creates processes such as peer production, the common production of value; peer governance, i.e., the self-governance of such projects; and peer property, the auto-immune system that prevents the private appropriation of the common.

It is important to distinguish the peer governance of a multitude of small but coordinated global groups that choose nonrepresentational processes in which participants co-decide on the projects, from representative democracy. The latter is a decentralized form of power-sharing based on elections and representatives. Since society is not a peer group with an *a priori* consensus but rather a decentralized structure of competing groups, representative democracy cannot be replaced by peer governance.

However, both modes will influence and accommodate each other. Peer projects that evolve beyond a certain scale and start facing issues of decisions about scarce resources will probably adapt some representational mechanisms. In fact, there are a few observations we can already make about the emerging templates of peer governance. In the sharing mode, proprietary third-party platforms are responsible for the setting of design rules that facilitate sharing and demand some form of openness that creates the value. However, they are balanced by their need to capture that value, with the existing possibilities and mobilization power of the sharing communities acting as a counterweight. In the commons-oriented form of peer production, as seen in free software, for example, a triarchical model emerges that comprises a self-aggregating "permission-less" and self-governed community; with a for-benefit association (usually a NGO in the form of foundations) that manages the infrastructure of cooperation, and subjected to formal democratic rules; and an ecology of businesses creating market value on top of the commons, while returning some of its profit in the form of benefit sharing towards the Foundation or community, thereby insuring the continuation of the Commons on which they depend. These form templates will be increasingly used in the expanding field of social production but are not as such applicable to the polis as a totality.

Representative and bureaucratic decision-making can and will in some places be replaced by global governance networks which may be self-governed to a large extent, but in any case, it will and should incorporate more and more multi-stakeholder models, which it strives to include as participants in decision-making, all groups that could be affected by such actions. This group-based partnership model is different, but related in spirit, to the individual-based peer governance, because they share an ethos of participation.

Toward a Partner State Approach

Partner state policy is an approach in which the state enables and empowers user communities to create value themselves and which also focuses on the elimination of obstacles. The fundamental change in approach is the following. In the modern view, individuals were seen as atomized. They were believed to be in need of a social contract that delegated authority to a sovereign in order to create society and in need of socialization by institutions that addressed them as an undifferentiated mass. In the new view, however, individuals are always already connected with their peers, and looking at institutions in such a peer-informed way. Institutions therefore, will have to evolve to become support ecologies, devising ways to create infrastructures of support.

The politicians become interpreters and experts, which can guide the issues emerging out of civil society based networks into the institutional realm. The state becomes an at least neutral (or better yet, commons-favorable) arbiter, i.e., the meta-regulator of the the realms, and retreats from the binary state/privatisation dilemma to the triarchical choice for an optimal mix between government regulation, private market freedom, and autonomous civil society projects. A partner state recognizes that the law of asymmetric competition dictates that it has to support social innovation to its utmost ability.

An example I recently encountered was the work of the municipality of Brest, in French Brittany. There, the "Local Democracy" section of the city, under the leadership of Michel Briand, makes available online infrastructures, training modules, and physical infrastructure for sharing (cameras, sound equipment, and so on . . .), so that local individuals and groups can create cultural and social projects on their own. For example, the Territoires Sonores project allows for the creation by the public of audio and video files to enrich custom trails, which are therefore neither produced by a private company, nor by the city itself. In other words, the public authority in this case enables and empowers the direct social production of value.

The peer-to-peer dynamic, and the thinking and experimentation it inspires, does not just present a third form for the production of social value, it also produces also new forms of institutionalization and regulation, which could be fruitfully explored and/or applied. Indeed, from civil society emerges a new institutionalization, the commons, which is a distinct new form of regulation and property. Unlike private property, which is exclusionary, and unlike state property, in which the collective "expropriates" the individual; by contrast in the form of the commons, the individual retains his sovereignty but has voluntarily shared it. Only the commons-based property approach recognizes knowledge's propensity to flow everywhere, while the proprietary property regime requires a radical fight against that natural propensity. This makes it likely that the commons-format will be adopted as the more competitive solution.

In terms of the institutionalization of these new forms of common property, Peter Barnes, in his important book *Capitalism 3.0*,[11] explains how national parks and environmental commons (such as a *proposed Skytrust*[12]), could be run by trusts, who have the obligation to retain all (natural) capital intact, and through a one man/one vote/ they would be in charge of preserving common natural resources. This could become an accepted alternative to both nationalization and deregulation/privatization.

I would surmise that in the hypothetical successor civilization, when the peer-to-peer logic is the core logic of value creation, the commons is the central institution that drives the meta-system, and the market is a peer-informed sub-system that deals with the production of rival physical products, in the context of a pluralist economy that is augmented with a variety of reciprocity-based schemes.

A Renewed Progressive Policy Centered Around the Sustenance of the Commons

What does it mean for the emancipatory traditions that emerged from the industrial era?

I believe it could have 2 positive effects:

1. a dissociation of the automatic link with bureaucratic government modalities (which does not mean that it is not appropriate in certain circumstances); proposals can be formulated that directly support the development of the commons;

2. a dissociation from its alternative: deregulation/privatization; support for the Commons and peer production means that there is an alternative from both neoliberal privatization, and the Blairite introduction of private logics in the public sphere.

The progressive movements can thereby become informational rather than a modality of industrial society. Instead of defending the industrial status quo, it becomes again an offensive force (say, striving for an equity-based information society), more closely allied with the open/free, participatory, commons-oriented forces and movements. These three social movements have arisen because of the need for an efficient social reproduction of peer production and the common.

Open and free movements want to insure that there is raw material for free cultural production and appropriation and fight against the monopoly rents accorded to capital, as it now restricts innovation. They work on the input side of the equation. Participatory movements want to ensure that anybody can use his specific combination of skills to contribute to common projects and work on lowering the technical, social and political thresholds; finally, the Commons movement works on preserving the common from private appropriation, so that its social reproduction is insured, and the circulation of the common can go on unimpeded, as it is the Commons which in turn creates new layers of open and free raw material.

These various movements come in the usual three flavors:

1. transgressive movements, such as young and old filesharers, which show that the legal regime has to be changed

2. constructive movements, which create a framework for new types of social relationships, such as the Creative Commons movement, the free software movement, etc. . . .

3. reformist or radical attempts to change the institutional regime and adapt it to the new realities

I personally believe that these movements will not create new political parties, but that these networks of networks will indeed look for political liaison. While peer-to-peer is a regime that combines equality and liberty and therefore potentially combines elements from various sides of the political spectrum, I believe the left is particularly apt to forge an alliance with the new desires and demands of these movements. It remains to be seen whether new political and cultural expression of the emerging free culture, such as the Swedish Pirate Party, will change that expectation by creating a new kind of political force, more directly in tune with peer production communities.

There is also a connection with the environmental movement. On one side, the culturally oriented movements fight against the artificial scarcities induced by the restrictive regimes of copyright law and patent law; on the other side, the environmental movement fights against the artificial abundance created by unrestricted market logics. The removal of pseudo-abundance and pseudo-scarcity are exactly what needs to happen to make our human civilization sustainable at this stage. As has been stressed by Richard Stallman and others, the copyright and patent regimes are explicitly intended to inhibit the free cooperation and cultural flow between creative humans and are just as pernicious to the further development of humanity as the biospheric destruction.

Finally, restoring the balance between a scarcity-recognizing material regime, and a abundance-recognizing immaterial regime, cannot be seen as separate from the efforts of social forces to obtain more social justice, thereby linking the new open/free, participatory and commons-oriented forces with emancipatory social movements. There is, therefore, a huge potential for such a renewed movement for human emancipation to become aligned with the values of a new generation of youth, and achieve the long-term advantage that the Republicans had achieved since the 1980s.

References

Badalian L., & Krivorotov, V. (2008). Technological shift and the rise of a new finance system: the market-pendulum model. *European Journal of Economic and Social Systems, 21*(2), 231–264.

Barnes, P. (2006). *Capitalism 3.0: Enriching ourselves by enhancing our commons.* San Francisco: Berrett-Koehler.

Bauwens, M. (2006). The political economy of peer production. *CTheory,* October 2, 2006. Retrieved from http://www.ctheory.net/articles.aspx?id=499

Bauwens, M. (2008). The social web and its social contracts: *Re-public.* Retrieved from http://www.re-public.gr/en/?p=261

Bauwens, M., with Arvidsson, A., & Peitersen, N. 2008. The crisis of value and the ethical economy. *Journal of Futures Studies, 12*(4), 9–20. Retrieved from http://www.jfs.tku.edu.tw/12–4/A02.pdf

Dyer-Witheford , N. (2006). The circulation of the common. *Autonomedia.* Retrieved from http://slash.autonomedia.org/node/5259

Klepper, R. et al. (1975). A comparison of the production, economics, returns, and energy-intensiveness of corn belt farms that do and do not use inorganic fertilizers and pesticides, CBNS Report AE 4, St. Louis.

Perez, C. (2002). *Technological revolutions and financial capital: The dynamics of bubbles and golden ages.* Cheltenham, UK: Edward Elgar.

Pollin, R. (2009). Green reconstruction vs. speculative capital: How a green economy is an antidote to casino capitalism. SolidarityEconomy.net. Retrieved from. http://www.solidarityeconomy.net/2009/05/21/green-reconstruction-vs-speculative-capital/

Posner, J. L., Baldock, J. O., & Hedtcke, J. L. (2008). Organic and conventional production systems in the Wisconsin integrated cropping systems trials: I. Productivity 1990–2002. *Agronomy Journal 100,* 253–260.

UNEP-UNCTAD (2008). Capacity Building Task Force on Trade, Environment, and Development, Organic Agriculture and Food Security in Africa. New York and Geneva, 2008.

Notes

1. Perez, C. (2002). *Technological Revolutions and Financial Capital. The Dynamics of Bubbles and Golden Ages.* Edward Elgar, Cheltenham, UK, 2002
2. Badalian L., Krivorotov V. (2008), "Technological Shift and the Rise of a New Finance System: the Market-pendulum Model", *European Journal of Economic and Social Systems,* Vol. 21, No. 2, 2008, p. 231–264
3. Kondratieff begins and ends with systemic shocks, with an high growth up-wave, followed by a low growth down-wave, with usually also a major crisis in the middle. So, the last wave starts with the crisis of 1929, culminates its high growth phase in the oil crisis of 1973–74, and enters the low growth period of neoliberalism. But Carlota Perez starts her own wave patterns, which she calls 'great surges of development,' in the middle of the Kondratieff waves, where a new set of technologies emerges at the very moment when the Kondratieff cycle enters into the mid-term crisis which announces the low-growth period. She explains this by noting that financial capital, faced with low profitability of the mainstream system, seeks alternative outlets and starts funding new possibilities. The making of the Intel chip in 1971 is such a beginning. But it requires the Sudden System Shock ending Kondratieff to create the new institutions that can favour the growth of the new paradigm.
4. Budget calculation cited in: Robert Pollin, "Green Reconstruction vs. Speculative Capital. How a Green Economy Is an Antidote to Casino Capitalism." SolidarityEcono

my.net. Accesssed on November 25, 2009. At http://www.solidarityeconomy.net/2009/05/21/green-reconstruction-vs-speculative-capital/

5. We are monitoring these developments here at http://p2pfoundation.net/Category:Manufacturing
6. Evidence demonstrating organic agriculture's productive potential has been steadily accumulating. The roots of this research can be traced at least back to the seminal 1970s study conducted by Washington University's Center for the Biology of Natural Systems under the direction of Barry Commoner. See Robert Klepper et al., "A Comparison of the Production, Economic Returns, and Energy-intensiveness of Corn Belt Farms That Do and Do Not Use Inorganic Fertilizers and Pesticides," CBNS Report AE 4 (St. Louis, 1975). More recent studies include Catherine Badgley et al. "Organic Agriculture and the Global Food Supply," *Renewable Agriculture and Food Systems* 22 (2007): 86-108; Joshua L. Posner, John O. Baldock, and Janet L. Hedtcke. "Organic and Conventional Production Systems in the Wisconsin Integrated Cropping Systems Trials: I. Productivity 1990–2002," *Agronomy Journal* 100 (2008): 253–60; 3. UNEP-UNCTAD Capacity-Building Task Force on Trade, Environment, and Development, Organic Agriculture and Food Security in Africa (New York and Geneva, 2008).
7. For details, see http://p2pfoundation.net/Crowdfunding
8. This is the core of a new proposed practice of Social Business Design, see http://p2pfoundation.net/Social_Business_Design
9. See the description at http://p2pfoundation.net/Distributed_Systems
10. The 'Circulation of the Common' is an analytical concept proposed by Nick Dyer-Witheford in a landmark essay of the same title. It refers to the social reproduction mechanism of peer production, in a process analogous with the circulation of capital described by Marx. See: http://slash.autonomedia.org/node/5259
11. Peter Barnes. *Capitalism 3.0: Enriching Ourselves by Enhancing Our Commons,* Berrett-Koehler, 2006. It is also available at the excellent On the Commons blog, where it is downloadable for free in pdf format. URL = http://capitalism3.com/
12. See at http://www.skyhook.org/skytrust.html

16

Creative Economies and Research Universities

Peter Murphy

After the Culture Wars, Now Come the Economy Wars

When the world recession in 2008 began, the economy wars, which had been dormant for two decades, flared again. After thirty years of the culture wars, this came as a bit of a relief. In one corner, we had the followers of John Maynard Keynes (1883–1946), who were filled with a kind of self-belief that we had not seen since the 1960s. They had a few scores to settle. In another corner were the market-friendly followers of Friedrich Hayek (1899–1992) and Milton Friedman (1912–2006). They were looking a bit bloodied after having dominated public policy for two decades. Looking on skeptically from outside the ring was another cohort, the admirers of Joseph Schumpeter (1883–1950). These were, as usual, less combative than the other fighters, and had a quizzical eye trained on all of the pugilists.

Part of the skepticism of the Schumpeter camp was a wariness of public policy *tout court.* It did not matter whether this was a policy bent on big government or one in love with small government. Schumpeter had been a student of the great Austro-Hungarian Empire Finance Minister, Eugen von Böhm-Bawerk. Schumpeter himself was the first Minister of Finance of the modern Republic of Austria. He seemed to take away from that unusually intimate experience of public policy a strong sense of the need for economists to look beyond the policy cycle and explore the deeper structures and long-run temporalities of economies. Schumpeter was a great economist who at the same time understood the power of history and society in shaping economies. He also appreciated the power of the imagination. He observed that modern capitalist economies were driven as much by creative impulse and imaginative insight, as they were by the more commonplace behaviors that arose out of greed, interest, need or calculation. It was not that societies could not—or should not—control such behaviors or encourage them, depending on prevailing economic philosophy. It was just that some of the most decisive economic outcomes could not be determined by such policy tools. Somewhere beyond them, in a larger social-historical zone, lay the human drive to innovate and create.

This view is at odds with both Keynesianism and the contending philosophies of Friedrich Hayek and Milton Friedman. It sits at a tangent to both "liberal" and "neo-liberal" views of the world. Whether it is the social liberalism of the Keynesian or the classic market liberalism of the anti-Keynesian, each exemplifies the manner in which economists became enthralled by the temporal horizons of public policy and indifferent to the deeper cultural and historical causes of economic and social prosperity. Economic crashes, such as the one that occurred in 2008, trigger a stock set of responses. Keynesians suppose that capitalist economies tend to stagnation and that the motive force of these economies is immoral. Economies accordingly must be stimulated by government spending in order to return an economy to prosperity, and then must be regulated with a sure hand. Thus, contracts for public works are used to sustain businesses. Government bail-outs rescue firms from insolvency. In a recession, with declining revenues, a government can still spend more if the state increases tax levels, borrows from banks, or prints money. All economic policy tools, however, have limited and negative effects. Higher taxation means less consumer spending and less investment. Government borrowing competes with private borrowers, restricting business access to credit and pushing up the price of money. The repayment of high levels of public debt is a long-term drain on the economy. Printing money on the other hand causes rampant inflation and government spending is often wasteful. Neo-liberals are a much more optimistic breed than Keynesians. They assume that capitalism tends to prosperity, politics is a primary cause of recession, competition is effective, and self-interest is not immoral. Market failures are caused by too much regulation, too much taxation, and too much government borrowing. Yet market liberals on the whole show only a muted interest in the roles of management, technology, and industrialization in securing the success of markets. The firm is peripheral to their explanation of economic dynamism.

Schumpeter's understanding of capitalism differed in significantly interesting ways from both Keynes and Hayek. He thought that the capitalism that he observed was dynamic not stagnant, but that its dynamism came not from markets in general but from the power of innovation that had been unleashed by modern industrial capitalism. Schumpeter took a long-term view of economies. From this historical viewpoint, economic crashes are a normal part of the dynamics of modern capitalist economies. Periods of genuine prosperity and long-term increases in wealth and general standards of living are followed, as night follows day, by a sequence of speculative boom, slide, panic, crash, and recovery. Boom-time actors never predict, and cannot predict, the time of the crash. They always think the good times will last forever. In fact, though, business cycles trend in waves, up and down. These waves cycle over the short, medium, and long term. Schumpeter was most interested in the long-term dynamics of capitalist economies because these, he observed, had the most important effects of all. Public policy, in contrast, is concerned principally with short-term effects. Public policy instruments have moderately foreseeable impacts that run over periods of eighteen months to three years. Very few tax or spending policies have observably sustainable effects beyond that. However, as Schumpeter outlined in his classic work *The Theory of Economic Development* in 1911, the most powerful drivers of modern capitalism work over periods of twenty, thirty, sixty years and more. These are the forces of innovation that create new industrial sectors.

The first chapter of *The Theory of Economic Development* set out a model of a static capitalist economy. Its paradigm reflects the tradition of economics from Adam Smith to John Stuart Mill to John Maynard Keynes and what they considered the components of a capitalist economy. Like all economies hitherto, it had no real endogenous driver of growth. Schumpeter noted that a handful of economies, beginning with Britain in the 1820s and Germany in the 1840s, behaved differently. They had a built-in source of expansion. Schumpeter set out to explain its parameters in the brilliant second chapter of *The Theory of Economic Development.* "Development" referred to those changes in economic life that are not forced from without but that arise from within triggered by their own initiative (Schumpeter, 2008, p.63). In this economy, change does not occur continuously but in fits and starts. This type of economy tends not toward a homeostatic equilibrium but rather toward a dynamic equilibrium.[1] This form of equilibrium is mildly enigmatic and suggests a kind of balance that is slightly off-balance all the time. Schumpeter explained the discontinuous change, the periodic ruptures, and the disturbances in the economic equilibrium of modern capitalist economies with one word: innovation. Periodically, the most advanced industrial economies go through a phase of intensive innovation. At the heart of these innovations are new combinations of economic materials and forces. What follows from these new combinations are new goods, new methods of production, new markets, new sources of supply, and new kinds of organization. They in turn create new leading industrial sectors. The Manchester cotton industry in 1780s, the railroads in the 1830s, Pittsburgh steel in the 1870s, the Detroit car industry in the 1910s, and the Silicon Valley information industries in the 1980s exemplify this phenomenon.

The ICT industries reached maturity around 2000. The pricking of the dot.com stock market boom symbolized this situation. Thirty years hence, the ICT companies will probably resemble the car companies of the 1970s—far removed from their glory days. At the point of a serious market recession, a compelling question surfaces: what new leading industrial sector will emerge? Unfortunately, it is difficult to predict who and what will be the shakers and makers of the next economic boom. Certainly, though, it is not the "known quantities" that will constitute a new sector. If they did, how easy it would be to foretell the future. In reality, it is "factor X"—the factor that is not known—that is most important. From the standpoint of the unknowable future, capitalism's "new wave," whatever it proves to be, will not be "green technology," the pop economics obsession of 2008.[2] Versions of that neologism have been commonplace since 1973 when the economist E. F. Schumacher (1911–1977) published his influential volume of essays *Small Is Beautiful*.[3] Schumacher, a young protégé of Keynes, was deeply influenced by Catholic mysticism. While the intuition of the mystic is arguably a better cognitive model than rationalist prediction when dealing with the tricky matter of social creation, Schumacher's insight was original in the 1970s but not today. Whatever will form the leading economic sector in 2038 is unknown, and it is only now being conceived in obscurity. Indeed, it is the uncanny conjunctions of the imagination that create the figments of a new economy. Such conjunctions are like the punch lines of truly funny jokes. They are not predictable. Less so are they clichéd conjunctions like "the green automobile."[4] When personal computers first appeared, the typical reaction was that "they won't catch on." Most observers did not say: "oh let's trade in the mainframe computer for the PC." IBM certainly did not say that, and it nearly destroyed the company. Similarly when technology becomes a favorite of public policy ("a computer on every school child's desk"), it is already closer to being a sunset than a sunrise industry. Just think of fashion: by the time it comes to Walmart, it is yesterday's trend.

New industry sectors provide the basis of sustained periods of economic and social prosperity. Orthodox policy instruments such as state taxes or budgets play only a minor role in economic innovation. Cities and regions are much more central to such innovation, a point made very clearly by the urban economist Jane Jacobs (1969, 1986) and, later on, by the urban sociologist, Richard Florida (2002). The most robust economy in the world after the 2008 global slump was Australia's. In 2009, it held the rate of unemployment to 5.8%, and the economy grew in the first two quarters of that year (Uren & Hohenboken, 2009). In comparative terms, this growth was remarkable. It was the result of three factors: a flexible *national* labor market, *global* export growth, and powerful *local* urban economies. In other words, Australia's economy displayed strong "glonacal" characteristics.[5] It fused global-export, national-flexible, and local-urban features in efficacious and uncanny ways. Firms reduced labor hours (thereby reducing the unemployment rate); low interest rates compensated for the income loss represented by flexible lower working hours; the international demand for natural resources and Australian undergraduate higher education places (Australia's number three export industry) remained high, and Australia's urban economies continued to be a powerful source of demand.[6]

Public policy is a contributor, but only one contributor, to long-term economic well-being. Australian national economic policy through the years of the Hawke-Keating (1983–1996) and Howard-Costello (1996–2007) administrations explains in part, but only in part, the capacity of the country to withstand the worst of the economic downturn in 2008. Policy-makers de-regulated the labor market and re-regulated the universities. Such policies, though, were meaningless without the support of flexible firms, first-class urbanism, and internationally focused universities. This combination laid the foundation for the emergence of Australia as the "Switzerland of the Asia-Pacific Region." Even so, like everywhere else, these developments still begged the question of "what next." We will know the answer to that question in thirty years from now. Much about creation can only be understood in retrospect. We understand the future by its past. All industry sectors, we know, eventually mature. That will apply to Australia's higher education export industry. Higher education for export became Australia's prime "new industry sector" as the country emerged from the 1980s. It was not clear at the time that higher education as an export good was Australia's answer to Silicon Valley, and the extent of the growth of this new sector only became widely understood as late as the 2000s. By 2009, Australia, with a population approaching twenty-two million people, had a half million foreign students in residence.[7] Conversely the sector was showing distinct signs of maturation and the inevitable strains that accompany incipient sector maturity.[8]

When the Australian education export sector reaches maturity, what will then serve as the next new powerful industry sector? What will supplement, and in part succeed, the quaternary information, education, research, and development (IERD) sector? The most that we can reliably predict, based on past experience, is that cities and city-regions will continue to be the crucible of new sector creation.[9] They are the point of intersection of art, science, production, and distribution. Perhaps, given the speed of state-directed urban creation that we see in China and elsewhere, the template-like "manufacture of cities" might even emerge as the quinary sector of the future. However, in spite of the fact that we can imagine this possible development, the dynamics of large-scale urban economies remain far from being fully understood because economic factors are invariably overlain with unpredictable aesthetic factors. In urban economies, aesthetic, design, and taste cultures intersect powerfully with housing and infrastructure demand. The discipline of cultural economics that might explain this phenomenon remains undeveloped. In addition, the mutual suspicion of art and economics does not help this state of affairs. Even an economist with bohemian connections like Keynes held the two at arms' length. Keynes's view of economies echoed that of Edwardian elites—namely that capitalism was a failure that proved itself only insofar as it generated wealth for Bloomsbury-style art. That art was intrinsic to modern capitalism was an idea that was inconceivable for elites raised in pre-capitalist cultures, as it is equally for elites steeped in post-1960s anti-capitalist cultures of complaint.[10]

A cultural economics would explain the relationship between the arts and sciences, on the one hand, and economies, on the other hand. The city, historically, has played the key mediating role in this relationship. Cities do several things. First, they are the place where the arts and sciences flourish. Second, they create aesthetically mediated demand. Third, they introduce science into everyday economic and social life through technology. Modern economies grow through aesthetically mediated and technologically mediated demand. Art and science do not create this demand directly. Rather their works are conveyed in a series of steps from artistic and scientific discovery through various institutional media, notably universities, galleries, and laboratories, and then via firms and organizations, into the familiar products, processes, forms, and artifacts of daily social and economic life. The chain of discovery-innovation-firm-organization-product-process-artifact is a long one. It is also one that is not continuous. Entropy commonly happens at all points along this chain. Correspondingly, established markets and firms play little role in the creation of new industry sectors. Schumpeter observed that it is new firms at the leading edge of new industries that are the core of capitalist innovation. Alternatively, as he quipped, "*add as many mail coaches as you please, you will never get a railroad thereby.*"[11] These new firms are created by entrepreneurs, a class of business leaders who notably are distinct from both owners and managers of business. The business class of entrepreneurs is perhaps best understood in terms of what the philosopher Hannah Arendt called "action" (1958). Action is the human capacity to initiate and lead—to bring things into the world. The business class of entrepreneurs creates new firms that create new types of goods, technologies, markets, supply chains, and forms of organization that provide the basis for new industry sectors.

Innovation and Invention

In the wake of *The Theory of Economic Development*, much of the most interesting work of twentieth-century economists was devoted to rethinking the neoclassical formula that land, labor, and capital are the key factors of production. In the nineteenth century, Alfred Marshall already had added "organization" to the neoclassical list. Information, knowledge, technology, cities, arts, and sciences followed Schumpeter's theory of the role of the entrepreneur. Fritz Machlup (1902–1983) and Robert M. Solow (1921–) observed, respectively, that information and technology were as important factors of production as the trinity of land, labor, and capital.[12] Machlup was a friend of Hayek's from their days at the University of Vienna; Solow was briefly a student of Schumpeter at Harvard University and later a close associate of the great American Keynesian economist Paul Samuelson, another one of Schumpeter's students. Machlup coined the term "the information society," and by the end of the twentieth century, Machlup's and Solow's ideas had spawned the popular notion of the knowledge economy, which crystallized for understandable reasons in the wake of the rise of the information technology industries. As California's Silicon Valley grew into an economic powerhouse, the literature on knowledge economies ballooned. One of the central institutions of the knowledge economy was the university. Both Machlup and Solow were cited by Daniel Bell in 1973 when Bell prophesized "the coming of the post-industrial society." One of Bell's many canny observations concerned the central role of the research university in postindustrial societies. The research university played an economic and ideological role similar to that of the church in medieval society. The sociologist's prognosis would eventually be echoed by professional economists. Indeed, such was the popularity of this idea that the American liberal political economist Jeffrey Sachs in 2005 even included the funding of universities, laboratories, and research as a key developmental step for nations seeking a way out of poverty (Sachs, 2005, pp. 58, 259).

Schumpeter was more cautious. When he wrote his classic work in 1911, he was well aware of the role that the arts and sciences played in modern economies. In fact, the theory of the arts and sciences as an economic driver goes back to eighteenth-century philosophers and political economists like Nicolas de Condorcet (1743–1794).[13] They observed the centrality of inventive knowledge ("the advancement of the arts and sciences") to modern capitalism—in the same way that Adam Smith (1776/1970, pp. 483, 502, 506–520) had noted the key part that "foreign commerce" cities play in dynamic economies. However, Schumpeter also drew a distinction between innovation and invention. Innovation was the function of the entrepreneurial class. Invention was the responsibility of the creative class. There was a division of labor between the two. Schumpeter noted (2008, p. 88) that it was not part of the role of entrepreneurs to find or create new possibilities. "These are always present, abundantly accumulated by all sorts of people. Often they are also generally known and being discussed by scientific or literary writers." The function of the entrepreneur was not to find or create "the new thing" but rather to lead others to accept or adopt it. However, as Schumpeter also accepted, this was not a strict 'division of labor' between business, on the one hand, and the arts and sciences, on the other. Schumpeter was aware that leadership was just as important in the arts and sciences as it was in business and that the acceptance of significant new ideas is just as difficult in a university as it is in a company, possibly more so. He observed that the history of science is one great confirmation of the fact that individuals find it exceedingly difficult to adopt a new scientific view or method (2008, p. 86). Thus, by Schumpeter's own hands, his carefully crafted distinction between invention and innovation begins to break down. As in all of the great works of creation, there is instability at the heart of things. Identities generate distinctions, and distinctions generate identities. That is the very nature of the process of creation that Schumpeter was trying to understand.

Interestingly, Schumpeter thought that innovation was more difficult to achieve than invention—because innovation is the enemy of habit. Habits, including the habits of thinking, are very efficient. Rather than having to consciously think through every task that we do, we form habits and act subconsciously on them in a time-efficient manner. One cost of this practice, though, is that when someone wants to implement change, the forces of habit rise up—Schumpeter noted—to bear witness against the embryonic project. An entrepreneur is a person with the will and the drive to wear down the forces of habit and side-line the naysayers who cry out that "this is the way it has always been done." Consequently, an entrepreneur must possess a series of distinctive traits: a desire to struggle against well-worn ways, to enjoy getting things done, to seek out difficulties, and engage in ventures (2008, pp. 93–94). In fact, Schumpeter was saying, in effect, that if Andrew Carnegie (1835–1919), who invented the idea of the vertical integration of a company, had not had the ability to impress that idea on his associates and wear down their opposition to it, his idea would have meant little. He would never have reaped a massive fortune from the steel business. While Carnegie's story is exemplary, it is just as true that the inventor also must struggle mentally against well-worn ways, enjoy getting things done, seek out difficulties, and engage in new ventures. Thus, in the end, Schumpeter's distinction does break down. Invention and innovation share common characteristics.

Appositional Thinking

Given the number of times words such as "new," "change," and "innovation" occur in his work, it may be a surprise to note that Schumpeter described himself as a conservative. It is certainly surprising insofar as the role of the entrepreneur is to struggle against ingrained habit. One of the definitions of being conservative is to stand for habit against change. However, just as most of the revolutionaries of the modern age created systems of sclerotic reaction, perhaps it is less surprising that Schumpeter, the self-declared conservative, also became the prophet of innovation. The situation resembles the Big Bang, the moment of the creation of the universe when nothing switched into something. If habit is the first economy of the human species, a recipe for the efficient use of energies, then habit turned against inefficiency is a powerful force for change. If that is a paradox, then so is the act of creation that allows economies to defy stasis and grow.

Everything is its opposite. In that idea lies the core conception of creation. Schumpeter once said that he had long planned to write a book on conservatism. If he had written it, it might have begun with a meditation on the idea of value-free science. The phrase "value-free" tends to be met with bemusement by social scientists today. However, Schumpeter thought of value-freedom in an interesting way. A value-free science was a science that embodied all of the contradictory values of a society—by being one step removed from them. That was conservative in the sense that the conservative is, in a subtle manner, a sharp critic of all forms of ideology. Schumpeter belongs to a class of twentieth-century intellectuals and writers who include G. K. Chesterton, Evelyn Waugh, Marshall McLuhan, Kenneth Burke, Saul Bellow, Daniel Bell, Hannah Arendt, Agnes Heller, Christopher Lasch, Cornelius Castoriadis, Roger Scruton, Christopher Hitchens, John Carroll, and Peter Berger. Each one of this group defies simple ideological classification. Some began, but none ended their intellectual careers as socialists or liberals ordinarily understood. Some were not camp followers even to begin with. Often they are best identified not by any kind of "ism" at all but rather by a tone that either is wry, ironic, comic, or skeptical. Tone replaces ideology. It is notable that many among this group either wrote humorous works or else wrote books or essays about comedy.[14] Arthur Koestler, in his illuminating treatise on the creative act, *The Act of Creation* (1964), observed at great length the structural parallel between comedy and creativity.

A person can be a conservative of the left as well as of the right. That is not incongruous, for the very nature of the conservative is to deal in incongruities. Wry tone rises above the bellows of modern politics. As Chesterton put it so well: "The whole modern world has divided itself into Conservatives and Progressives. The business of Progressives is to go on making mistakes. The business of the Conservatives is to prevent the mistakes from being corrected."[15] The aspiration to be free of the hum-bug of ideology, including the hum-bug of conservatism, might be another way of understanding Schumpeter's sense of himself as a conservative. He promised for a long time to write a book on conservatism, but did not, which might be the best kind of book on the topic. The attitude of the conservative is one of dry humor. It is marked by a gleeful insistence in deviating from any right direction in thinking. It is executed in witty observations that deliver unexpected twists and turns or in the screwed-up face that signifies impatience, disgust, or discomfort with human folly. The conservative and the humorist deal in ways of marrying incongruities. This might appear to be a useless talent excepting that the most successful societies in human history have been riddled with the most amazing contradictions and yet managed them with grace. Here we see explained the conservative prophet of innovation. What Schumpeter shared with other conservatives was an unusual sensitivity to appositions. Appositions are what drive dynamic economies.

Schumpeter's sense of his own self as a conservative was intimately bound up with his view of modern capitalist economies. He observed that what kept those economies growing were periodic bursts of innovation. Fundamental to these spectacular cloud bursts of ineffable creativity was the ability of entrepreneurs to think in new ways about products, markets, and organizations. These new ways were always new combinations, unprecedented conjunctions of concepts that people conventionally thought of as different and unrelated. To achieve this goal, the mind could not be too partisan or too fixated on one side, one thing, or one approach. Ideology means the fixation on one value or set of values in a world that is subject to multiple and irreducible value currents. Schumpeter wrote generously about Marx and Keynes and Marshall, and many other economists of many different outlooks because he understood that great ideas come out of an uncanny confluence of often very contradictory precepts.

The conservative stance is to take a skeptical view of all of these in order to see what can be done with each of them. The underlying impulse is to conserve them all in order to overcome them by marrying them together. Overcoming is not an act of abolition but an act of conjuration that takes opposing qualities and, through uncanny tactics, forges new ideas from old precepts. Andrew Carnegie took the lateral-horizontal-procedural (what today is often called the "network") idea of a market economy and fused it with the vertical-hierarchical-personalized forms of the medieval and pre-capitalist imagination that the Social Darwinists, whom Carnegie admired, loathed. This scenario may have been very contrary—but it was also, so far as the act of creation was concerned, entirely consistent. Carnegie laid the template, or part thereof, of modern organizations. In the same spirit, it may have been paradoxical that the conservative Schumpeter was the great modern prophet of innovation, but this was for a very good reason. The kind of skeptical conservatism typified by Schumpeter illuminates the dynamism of modern capitalism because it grasps the kind of appositions that make it possible. It is difficult to over-estimate how peculiar these appositions are.

Appositional thinking is helpful to explain the dynamic mutating forms of successful modern societies and economies, without falling into the trap of idolizing pyrrhic fashions. The cult of the new is conspicuously mindless. Ironically, it requires a conservative instinct to explain innovation. What matters in acts of creation is not so much what is new, which often is uninteresting, but rather the surprising takes on what is old.[16] That in a nutshell is the problem of the creative economies. They exist, but what drives them is difficult to identify, let alone to subject to public policy prescriptions. The simplistic equation of "the new" and "the creative" can be very misleading. Schumpeter was the first to distinguish between creative industries and mature industries. Creative industries appear dramatically as if out of nowhere. They capture appositions, unlikely combinations of ideas that are seized upon by mercurial entrepreneurs. Eventually with the passage of time, creative industries slow down, as invention idles and innovation turns into convention, and the profits of innovation decline. However, at their peak, these industries race ahead on the back of startling ideas. They prove themselves to be much more dynamic than other industry sectors. There is always an element of "the new" in this. However, one should also be wary of overstating the significance of the new.

As Schumpeter often observed, creation comes through the unlikely combinations of what exists. The word "unlikely" is important. The unlikely character of protean combinations requires exceptional insight. The act of conjuration underlying them is very unusual. Terms like progress, contemporary, modern, up-to-date, and so on are not always very helpful in understanding these conjurations.

Words like these point to the temporal dynamic of creation, but what they screen out are the appositional structures of innovation and invention. It is not time that explains creation but rather the finding of similarities in what is dissimilar. Creation connects the unconnected. This process is much closer in nature to poetic analogy than it is to social progress. The assembly line radically changed the methods of industrial production. Henry Ford's car assembly technique had a significant impact on the organization of labor in the twentieth century. Someone sometime along the way looked at the dis-assembly techniques used in the Chicago slaughter houses and meat packing plants of the late nineteenth century. Not every person's way of looking at things is the same. Someone looked at the dis-assembly line and imagined it in reverse where the parts of the animal were not pulled apart but were put together, this time as an automobile. Later in the 1960s, Andy Warhol, who grew up in then industrial Pittsburgh, reworked this idea into "the Factory," a multi-medium, output-driven art loft studio in New York City. This in turn was echoed in the early twenty-first century business model of the "art firm." From the slaughter house to the aesthetic company, we see the analogical power of the mind at work. The analogy drawn is not a literary one per se, but it is no less powerful for that.

Creative Achievement in Real Terms

One of the great laboratories for understanding the "breath of capitalism"—the diaphragm-like growth-and-recession pattern of modern capitalist economics—was the 1980s. That period illustrates a number of very Schumpeter-type issues—the role of ideas-production in economic life, and the very interesting matter of where those ideas come from. The 1980s saw the start of what became known as "post-industrialism." Post-industrialism is an imperfect term. It implies that the driving forces, the catalysts, of this era were fundamentally different from the industrial age, whereas in fact it is the symptoms of what those catalytic forces produced that was different. New information and communication technologies saw the rise of new industrial sectors. That was spectacular in its way, but it was not different in ultimate type from what had created a previous series of leading industrial sectors and that had driven capitalist economies since the latter part of the eighteenth century. In every case, the driver was the application of ideas to production, or perhaps more precisely the new ways of conceiving goods, markets, and organizations. New in this case always meant contradictory or uncanny ideas—like the idea of a soft industry or an item of software as opposed to the older notion of hard ware, or the imagining of a computer as something personal rather than institutional.

This is at the point where we see Schumpeter exceed all of his students. The best of them grasped that knowledge, information, technology—all of those iconic words that defined the tail end of the twentieth century—were metaphors for the act of creation. Schumpeter, however, saw that creation was an act of metaphor. He saw that words like soft or hard, industrial or service were not just metaphors for economies, but that the engines of economies were metaphors. He was not suggesting that economics was a kind of literature, but rather that the serious entrepreneur and the serious artist, both rare birds, were comparable in nature. Science, technology, the social sciences, and so on, are important to economies not just because they invent useful, expedient, and efficient ways of doing things, but because they are capable of harnessing the act of thinking which, at its core, where it is most powerful, is metaphoric. A metaphor is a combination, and as Schumpeter repeatedly observed, innovations come out of combinations. When innovations are in the phase of discovery, they emerge out of metaphors. Even the most utilitarian innovation is poetic in its origins.

The year 1980 was very depressing in the United States. Inflation was running at 13%, and the unemployment rate was 7.8%. The economy was in deep recession. The old powerhouse industries of the American Mid-West had become rust-belt industries. Once the epitome of industrial power, dynamism, and innovation, they were now mature or over-mature industries struggling to avoid bankruptcy. America elected Ronald Reagan as president (1981–1989). The 1980s saw America return to economic prosperity. In 1989, inflation was 4.0% and unemployment 5.4%. The official policy prescriptions of the Reagan era were neo-liberal, small-government policy inspired by the theories of Friedrich Hayek, Milton Friedman, and Arthur Lafter. In fact, though, government spending per capita continued to rise throughout the Reagan years as did government deficits and government debt. Spending to win the Cold War drove this, as did the fact that the conservative Reagan had a large streak of liberalism in his soul. He was a man of interesting contradictions. He had begun political life as a Democrat before switching to the Republican Party. Personal income tax fell dramatically in the Reagan years, but social security taxes rose. Reagan was a man with a grasp of the economics of laughter. He promised that as taxes went down, tax revenues would rise. Liberal economists guffawed. However, in truth, economic policies often have quantum effects of this kind. As Austrian finance minister, Schumpeter had experienced that reality at close quarters, and the experience of it had made him skeptical of the efficacy of public policy.

What really made the Reagan years an economic success story was the beginning of the rise of the new information and communication (ICT) industries that would transform the face of the American economy. The genius of the Reagan Administration was to do nothing to throttle this new industry sector in its crucial early phase of growth. The ICT industry followed a classic Schumpeter script. It emerged from the heat of recession. It was pioneered by entrepreneurial figures (Bill Gates, Michael Dell, Steve Jobs, and others). It generated super-profits. It developed separately from existing industries and firms. Nevertheless, its technologies and methods of organization spread to existing industries and firms, transforming them. Then gradually it ran out of creative energy. Its pioneering figures lost interest in innovation. They took their profits and turned to social activism and philanthropy. Gates became the Carnegie of his time. As Schumpeter might have observed, it is a pattern; it has been done before. As the ICT industry took off, sociologists began to talk about "post-industrialization." In fact, looking backward, the emergence of the digital communications sector was part of the normal process of industrial capitalism at work. What happened in the 1980s was one of the periodic re-energizing phases of modern capitalism as a new and unpredicted industry sector took off. American GNP per capita, in Year 2000 dollars, rose from $22,346 in 1982 to $27,514 in 1988.[17] In 1974, 1975, 1980, and 1982 U.S. real GDP per capita had actually fallen. It rose steadily thereafter through to 2008 with the exception of 2001. In 2008 it stood at $38,262 in Year 2000 dollars.

Universities played a part in the ICT-fueled resurgence in the 1980s. However, as in the case of all invention, the number of university actors was very small. Discovery in a measurable sense is overwhelmingly the preserve of a small number of research universities, and a small number of professors and graduate students from those institutions. The decisive fact about research, as about culture creation generally, is that it concentrates. The rise of the ICT industry illustrates this dynamic perfectly. The principal technology building blocks of ICT were devised by a very small cohort of professors and PhDs from the universities of California, MIT, Harvard, Brown, Stanford, Illinois, Duke, Washington, and Oxford, along with contributions from the IBM, RAND, and BBN corporations, the Swiss CERN lab, and the US Defense Department's Advanced Research Projects Agency.[18] This high level of concentration is characteristic of invention generally across the arts and sciences. As Daniel Bell noted in 1973, 100 of the 2,500 accredited colleges and universities in the United States—or 4% of the total—carried out more than 93% of higher education sector research.[19] Moreover, of that tiny group, 1% of them—21 universities—carried out 54% of the total of the sector's research output, and 10 universities were responsible for 38% of the total research output. Today there are 2,618 accredited four-year colleges and universities in the United States.[20] In 2009, The Carnegie Foundation for the Advancement of Teaching classified 96 universities as "research universities with very high research activity," essentially the same as Bell's 1973 figure.[21] If we look at the top twenty research universities in the world today, defined by output and citation, we find that not only are they all American universities, but that they are concentrated in specific geographical locations, principally on the Eastern and Western seaboards of the United States and around the Great Lakes, and in the orbit of major nodal city-regions, some border-hopping.[22] New York City and Boston together with the strips and arcs connecting Los Angeles-San Diego, San Francisco-San Jose, Madison-Chicago-Detroit-Toronto, Portland-Seattle-Vancouver, and Baltimore-Washington, DC,-Durham-Atlanta are especially prominent. The Houston-Austin-Dallas-Tampa-Miami arc might one day be competitive with the others.

Research and culture creation generally not only concentrate in space but also in time. The rise of the ICT industries was a notable phenomenon in the second half of the twentieth century. However, it was neither the most measurably creative period in American history nor was it time-unlimited. Per capita rates of copyright and patent registrations are a good indicator of national innovation. In the case of the United States, the peak year for patents registered per capita in the United States was 1916.[23] The rate trended downward till 1985 where it stood at 50% of the 1916 peak. It rose again, as would be expected, in step with the information technology boom from 1985 to the present day. However, even at its renewed highest level in 2005, it was still only 95% of the 1916 per capita figure. Nationally American registrations of copyrights per capita slightly increased between 1900 and today but only because the number of categories of copyrightable objects increased markedly in the same period—meaning that copyright registration per capita in real terms actually fell. The period 1890-1910 appears to be the peak time for copyright creation in the United States once we take into account the increase in copyrightable objects during the twentieth century.[24] In 1871, 12,688 copyrights were registered in the United States, which then had a population of 50 million.[25] That is the equivalent of 0.03 registrations per 100 Americans. In 1900 that figure had risen to 0.13. In 1925, it was 0.15, 1950, 0.14, 1978, 0.15. After this plateau, it rises in 1988 to 0.23, and then falls away again to 0.20 in 1994, then 0.18 in 2000 and 2007. Not only had the nominal figure per capita risen only marginally in a hundred years, but in the period since 1909 many new categories had been added to the schedule of protected works.[26] In spite of all the additional copyrightable works that this represents, copyright productivity per capita expanded negligibly in a century. In real terms, in effect copyright activity shrank. As with patents, the peak of copyright registrations in nominal terms (i.e., not accounting for additional copyrightable objects) occurred at the turn of the century, around 1907, with a nominal rate of 0.14 registrations per one hundred Americans.[27]

"Creativity" became a buzz-word in the later part of the twentieth century. The rise of the ICT industries encouraged this development. Policy makers rushed to embrace labels like the knowledge economy, the information society, and the clever country. Universities hopped on the bandwagon. Even so, there is little evidence that the late twentieth century was especially creative. In retrospect, the rise of a new industry sector is not something extraordinary. It is rather the norm of modern capitalism. That is how industrial capitalism functions, as Schumpeter reiterated ad nauseam. Without such invention, we are all dead. Why should we regard it as special? The evidence from copyright and patent registrations is that there was no explosive moment of innovation in the late twentieth century, even if ICT did manage to recover a badly faltering technology momentum that had reached a bleak bottom during the 1970s.

Achievements in fundamental discovery are even less impressive when we step back and look at them in historical perspective (Murray, pp. 309–330). Per capita measures of fundamental discovery in Europe and North America strongly suggest that the golden age of the visual arts was between the mid-1400s and mid-1500s with a second peak in the mid-1600s. Music creation peaks in the early 1700s and sustains a moderate high through to the middle of the 1800s. Western literature peaks in the early 1600s and again in the middle of the 1800s. Scientific creativity peaks in the later 1600s and then again for a remarkable period from the mid- 1700s to the late 1800s. Huebner calculated that high-level technology discovery peaked in 1873.[28] Similarly, after 1870, the rate of major achievement—that is, the number of outstanding figures, works and events per capita, in the United States and Europe—in mathematics, visual arts, and literature also declines (Murray, 2003, pp. 312–320). There were some countervailing trends: an upswing in the number of significant figures (though not works and events) in literature, science, and visual art from 1900 to 1920 and an upswing in technology advances in the period from 1920 to 1950. The film arts flourished in the 1940s and 1950s, as did recorded music from the mid-1960s to the mid-1970s. However, overall since 1870 there has been a long-term downturn in creativity.

The Economics of Laughter

The dynamic of creativity in the last 140 years has trended down with punctuated up-swings. In the United States, the turn-of-the-century, the late 1920s and the late 1980s were relative high spots. The presidential eras of Theodore Roosevelt (1901–1909), Calvin Coolidge (1923–1929), and Ronald Reagan (1981–1989) were the most creative in the American twentieth century.[29] This pattern of punctuation, though, poses an interesting conundrum. In the last 50 years the overwhelming majority of academics in American research universities have identified with the liberal wing of the Democratic Party.[30] However, the peak of American creation in the last 100 years occurred during Republican presidencies.[31] Few American researchers or cultural figures today would identify with Teddy Roosevelt, Calvin Coolidge, or Ronald Reagan. Most would blanch at the very thought of that. However, such a thought may help us better understand one of the primary social conditions for creativity. Dean Keith Simonton posed the interesting question: what social factor most strongly correlates with periods of peak creation in societies generally? The answer that he drew from extensive historico-metrical data was, in a nutshell, political decentralization—the division of an overarching political world into autonomous states (1984, pp. 143–146). Correlated with this phenomenon is what Philip Tetlock and his colleagues dubbed "integrative complexity"—the ability to tolerate ideological polarities and synthesize them.[32] High-functioning enigmatic political regimes—ones that internalize high levels of opposing views and yet at the same time exhibit high levels of integration of those competing perspectives—are crucibles of peak creation (Murphy, 2010). A society that can cope with opposition at the same time as it can function in an integrated manner is a society that is able to meld incongruous values into a rich and uncanny culture. On paper such a culture might be expected not to work. In practice, such cultures can and do work—wonderfully.

The ancient and Renaissance city states are classic examples.[33] The federal-state forms and distinctive city-regions of the United States resemble them in a structural sense.[34] However structural patterns, no matter how powerful, do not in themselves explain the conundrum of why it is that creative peaks in the United States correlate with Republican presidencies. This historical pattern contradicts the common assumption that liberal culture best supports research. Tetlock's conclusion that moderate liberalism best aligns with cognitive complexity is widely cited, though the underlying studies do have their critics.[35] Sometimes in these kinds of matters, especially where the interpretation of data is contested, it is worth going back to basics. About one matter at least there seems to be consensus. A defining characteristic of the imagination is that it comprehends concepts simultaneously in multiple dimensions. The imagination is ambidextrous—and integrative complexity, like value freedom, is an expression of that. However, the very condition of multi-dimensionality begs serious questions about the equation of liberalism and complexity. The psychologist Jonathan Haidt conducted a number of survey studies. From these he concluded that liberals are politically responsive on the dimensions of protection/care and fairness/reciprocity—a commonsensical conclusion.[36]

He observed that the same principle applies to conservatives but that conservatives are also responsive to three further dimensions: in-group/loyalty, authority/respect, and purity/sanctity. If the integration of dimensions is a key indicator of imaginative thinking, which very likely it is, then the conservative curiously has an edge over the liberal. It might be countered that the values of order or authority (for example) are not valid values but that then defines complexity out of the equation of integrative complexity. The imagination *stretches* to integrate contrary dimensions. Can a high-functioning contemporary society be "Millian" without being "Durkheimian" at the same time? Can such a society function without an ironic, even comic, relation to what to the great American sociologist Talcott Parsons (1970) called the AGIL dimensionality of modern society—the adaptive (economic), goal-orientated (political), integrative (normative), and latent pattern maintenance (cultural) aspects of these societies?

Ambidextrousness and paradox are characteristics of strong cultures, and strong cultures in their turn are the principal drivers of knowledge (Murphy, 2010). Comedy and tragedy are iconic forms of strong culture. They meld the antithetical and incongruous. Shakespeare imagined history in this way (Murphy, 2009). Shakespeare could be cutting toward rebels yet damning of tyrants in the same breath. The vocation of science that Max Weber appealed to is similar in nature. Its key tenet, value-freedom, is double-edged in the same way that history and tragedy and comedy are. The double-edge of creation exhibits itself in paradoxes—in which nothing is something, division and integration are identical, reduced taxes mean greater tax revenues, cats are simultaneously alive and dead in the thought experiments of science, and warfare economics coexists with welfare economics. Without Eisenhower's Advanced Research Projects Agency and the Cold War, the Internet—the research medium par excellence—would not exist. The military-industrial economy stands to the welfare economy as Spencer Dryden's bed-rock martial drum-beat does to Grace Slick's possessed singing on Jefferson Airplane's 1967 classic hippie-psychedelic masterpiece *White Rabbit*.[37] As California governor, Ronald Reagan had many testy battles with the 1968 generation of students and faculty at the University of California. Confronted on one occasion by protestors carrying banners saying "Make love, not war," he quipped that they probably didn't know how to do either. However, for all of Reagan's impatience with the baby-boom generation, it was his successor—the ascetic Democrat Jerry Brown—who slashed university budgets and made professors teach longer hours, while Reagan's America saw a jump in R&D spending as a share of GDP from 2.1 % in 1979 to 2.7 % in 1984. It has remained around that level ever since (Carlsson, Acs, Audretsch & Braunerhjelm, 2007).

The lesson learned is that, sometimes, one's worst political enemy is in fact one's best friend. Lessons in irony, in principle, should find a ready audience among researchers. After all, in matters of the mind nothing is more profound than the economics of laughter. What most becomes the human imagination is wit, and the brevity of wit is the mind at its sharpest. However, while much is said in theory in the defence of irony, wit, and paradox, in practice, earnestness and complaint are often allowed to brush them aside. The dangers of acting that way are not political. The ideology of researchers has a miniscule impact on politics. Researcher bias is like media bias. It has inconsequential effects on the political system. Journalists might be a very liberal cohort, but elections are not decided by their political preferences. As Paul Lazarsfeld (1901–1976) concluded in the 1950s, the media have a weak influence and minimal effects on the political system.[38] Universities have even less influence on the political system. However, arguably, the political system or more precisely political symbolism has a significant influence on the universities. This influence may not always be positive. Max Weber (1946) observed the stifling effect that politics can have on research. This phenomenon does not occur because politics is capable of controlling the life of the mind. The ancient Greek Stoics already knew that was nonsensical. One can imprison a person's body but not a person's mind. Much more important are the subtle and indirect effects of political atmospherics. Certain common styles of politics have a sullen effect on the mind. These styles are ideological, moralistic, and non-ironic. They exhibit few signs of integrative complexity. They inspire priggishness and pomposity. They lack value-freedom and the kind of wit that accompanies it. The joke, like the metaphor, transports us from one idea or one value to another. Wit and analogy are conducted by the twists, turns, leaps, and jumps of the imagination (Davis, 2007; Murphy, 2009). The act of imagination—the act of creation—causes us mentally to "switch" sides. This ability is indispensable to the scientist who is able thereby to imagine light as a wave and a particle simultaneously. It is not amenable, though, to the political ingénue who feels a deep urge to "take sides" without any sense of irony. One wonders whether the triumph of the ingénue is reflected in falling rates of discovery and innovation measurable in copyright and patent registrations per capita and in the long-term decline in the production of great works per capita over the past 140 years in most areas of the arts and sciences. If so, the absence of laughter might turn out to be no laughing matter after all.

References

Alvarez, G. C. (2009). *Study of the effects on employment of public aid to renewable energy sources.* Instituto Juan de Mariana, Universidad Rey Juan Carlos. Retrieved from http://www.juandemariana.org/pdf/090327-employment-public-aid-renewable.pdf. Accessed 6 October 2009.

Arendt, H. (1958). *The human condition.* Chicago: University of Chicago.

Bell, D. (1976/1996). *The cultural contradictions of capitalism.* New York: Basic Books.

Bell, D. (1973/1999). *The Coming of Post-Industrial Society.* New York: Basic Books.

Bellow, S. (2001). *The adventures of Augie March.* Introduction by Christopher Hitchens. London: Penguin.

Berger, P. (1997). *Redeeming laughter: The comic dimension of human experience.* Berlin: Walter de Gruyter.

Boldrin, M., & Levine, D. K. (2008). *Against intellectual monopoly.* Cambridge: Cambridge University Press.

Burke, K. (1989). On symbols and society. J. R. Gusfield (Ed.). Chicago: University of Chicago Press.

Carlsson, B., Acs, Z. J., Audretsch, D. B., & Braunerhjelm, P. (2007). The knowledge filter, entrepreneurship, and economic growth. *CESIS Electronic Working Paper Series,* Paper 104. Retrieved from http://www.infra.kth.se/cesis/documents/WP104.pdf. Accessed May 15, 2009.

Carnegie Foundation for the Advancement of Teaching. (1989). *The condition of the professoriate: Attitudes and trends.* Princeton, NJ: The Carnegie Foundation for the Advancement of Teaching.

Chen, S. (2009). Dirty reality behind solar power. *South China Morning Post,* September 10.

Chesterton, G. K. (1924, April 19). Column. *Illustrated London News.*

Colebatch, T., & Lahey, K. (2009, September 23). Melbourne's population hits 4 million. *The Age.*

Davis, P. (2007). *Shakespeare thinking.* London: Continuum.

Florida, R. (2002). *The rise of the creative class.* New York: Basic Books.

Graham, J., Haidt, J., & Nosek, B. A. (2009). Liberals and conservatives rely on different sets of moral foundations. *Journal of Personality and Social Psychology, 96,* 1029–1046.

Gruenfeld, D. H. (1995). Status, ideology and integrative complexity on the U.S. Supreme Court: Rethinking the politics of political decision making. *Journal of Personality and Social Psychology, 68,* 5–20.

Haidt, J. (2008, September 9). What makes people vote republican? *The Third Culture.* Retrieved from http://www.edge.org/3rd_culture/haidt08/haidt08_index.html. Accessed August 10, 2009.

Haidt, J. (2009a, April 27). Conservatives live in a different moral universe—and here's why it matters. *Mother Jones.* Retrieved from http://www.motherjones.com/politics/2009/04/conservatives-live-different-moral-universe8212and-heres-why-it-matters. Accessed August 10, 2009.

Haidt, J., & Graham, J. (2009b). The planet of the Durkheimians, where community, authority and sacredness are foundations of morality. In J. T. Jost, A. C. Kay, & H. Thorisdottir (Eds.), *Social and psychological bases of ideology and system justification* (pp. 371–401). New York: Oxford University Press.

Hall, P. (1998). *Cities in civilization: Culture, innovation and urban order.* London: Weidenfeld & Nicolson.

Hambridge, J. (1967) *The elements of dynamic symmetry.* New York: Dover Publications.

Hamilton, R. F., & Hargens, L. L. (1993). The politics of the professors: Self-identifications, 1969–1984. *Social Forces, 71,* p. 3.

Heller, A. (1982). *A theory of history.* London: Routledge and Kegan Paul.

Heller, A. (2005). *Immortal comedy: The comic phenomenon in art, literature and life.* Lanham, MD: Rowman and Littlefield.

Hitchens, C. (2004). *Scoop* and *The Adventures of Augie March.* In *Love, Poverty, and War: Journeys and Essays.* New York: Nation Books.

Huebner, J. (2005). A possible declining trend for worldwide innovation. *Technological Forecasting & Social Change, 72,* 980–986.

Hughes, R. (1994). *Culture of complaint: The fraying of America.* New York: Warner Books.

Jacobs, J. (1969). *The economy of cities.* New York: Random House.

Jacobs, J. (1984/1986). *Cities and the wealth of nations.* Harmondsworth: Penguin.

Katz, E., & Lazarsfeld, P. F. (1955). *Personal influence: The part played by people in the flow of mass communication.* Glencoe, IL: Free Press.

Kenner, H. (1947). *Paradox in Chesterton.* New York: Sheed & Ward.

Klein, D. B., & Stern, C. (2005). Professors and their politics: The policy views of social scientists. *Critical Review, 17*(3–4), 257–303.

Koestler, A. (1964). *The act of creation.* New York: Dell.

Ladd, E. C., & Lipset, S. M. (1975). *The divided academy: Professors and politics.* New York: McGraw-Hill.

Lazarsfeld, P., Berelson, F., & McPhee, W. (1954). *Voting.* Chicago: University of Chicago Press.

Machlup, F. (1962/1973. *The production and distribution of knowledge in the United States.* Princeton: Princeton University Press.

Marginson, S. (2007, October 16). The global positioning of Australian higher education: Where to from here? The University of Melbourne Faculty of Education Dean's Lecture series. Retrieved from http://www.cshe.unimelb.edu.au/people/staff_pages/Marginson/MarginsonDeansLecture161007.pdf. Accessed August 10, 2009.

Marginson, S. (2009, August 3). Providers rule the roost. *The Australian Financial Review.*

Marginson, S., & Rhoades, G. (2002). Beyond national states, markets, and systems of higher education. *Higher Education,* 43, 281–309.

McLuhan, M. (1936). G. K. Chesterton: A practical mystic. *Dalhousie Review, 15*(4).

Murphy, P. (2001). *Civic justice.* Amherst, NY: Humanity Books.

Murphy, P. (2006). American civilization. *Thesis Eleven: Critical Theory and Historical Sociology, 81,* 64–92.

Murphy, P. (2010). The limits of soft power. In D. Black, S. Epstein, & A. Tokita (Eds.), *Complicated currents: Media production, the Korean wave, and soft power in East Asia.* Melbourne: Monash E-Press.

Murphy, P. (2009). The power and the imagination: The enigmatic state in Shakespeare's English history plays. *Revue Internationale de Philosophie, 63*(1), 41–64.

Murray, C. (2003) *Human accomplishment.* New York: HarperCollins.

Parsons, T. (1970). *The social system.* London: Routledge & Kegan Paul.

Rothman, S., Lichter, S. R., & Nevitte, N. (2005). Politics and professional advancement among college faculty. *The Forum: A Journal of Applied Research in Contemporary Politics, 3*(1), Article 2. Retrieved from http://www.bepress.com/forum/vol3/iss1/art2/. Accessed August 10, 2009.

Sachs, J. (2005). *The end of poverty: Economic possibilities for our time.* New York: Penguin.

Schumacher, E. F. (1973). *Small is beautiful: A study of economics as if people mattered.* London: Blond and Briggs.

Schumpeter, J. (2008). *The theory of economic development.* New Brunswick, NJ: Transaction.

Scruton, R. (2007). *Modern culture.* London: Continuum.

Shepherd, G., & Shepherd, G. (1994). War and dissent: The political values of the American professoriate. *The Journal of Higher Education, 65*(5), 587–614.

Silverberg, G., & Verspagen, B. (2003). Breaking the waves: A Poisson regression approach to Schumpeterian clustering of basic innovations. *Cambridge Journal of Economics, 27*(5), 671–693.

Simonton, D. K. (1984). *Genius, creativity and leadership.* Cambridge, MA: Harvard University Press.

Smart, J. (2005). Measuring innovation in an accelerating world: Review of "A possible declining trend for worldwide innovation." Acceleration Studies Foundation. Retrieved from http://www.accelerating.org/articles/huebnerinnovation.html. Accessed August 10, 2009.

Smith, A. (1776/1970). *The wealth of nations.* A. Skinner (Ed.). Harmondsworth: Penguin.

Solow, R. M. (1956, February). A contribution to the theory of economic growth. *Quarterly Journal of Economics, 70,* 65–94.

Suedfeld, S., & Tetlock, P. (1977). Integrative complexity of communications in international crises. *Journal of Conflict Resolution, 21,* 169–184.

Suedfeld, P., Tetlock, P., & Streufert, S. (1992). Conceptual/integrative complexity. In C. P. Smith (Ed.), *Motivation and personality: Handbook of thematic content analysis.* Cambridge: Cambridge University Press.

Tetlock, P. (1998). Close-call counterfactuals and belief-system defenses: I was not almost wrong but I was almost right. *Journal of Personality & Social Psychology, 75,* 639–652.

Tetlock, P. E. (1984). Cognitive style and political belief systems in the British House of Commons. *Journal of Personality and Social Psychology: Personality Processes and Individual Differences, 46,* 365–375.

Tetlock, P. E. (1986). Integrative complexity of policy reasoning. In S. Kraus & R. Perloff (Eds.), *Mass media and political thought.* Beverly Hills: Sage.

Tetlock, P. E., Bernzweig, J., & Gallant, J. L. (1985). Supreme Court decision making: Cognitive style as a predictor of ideological consistency of voting. *Journal of Personality and Social Psychology: Personality Processes and Individual Differences, 48,* 1227–1239.

Theall, D. F. (2006). *The virtual Marshall McLuhan.* Montreal: McGill-Queen's University Press.

Tierney, J. (2004, November 18). Republicans outnumbered in academia, studies find. *New York Times,* 23.

Uren, D., & Hohenboken, A. (2009, August 7). Business is booming among cafe society. *The Australian.*

Waugh, E. (2000). *Scoop: A novel about journalists.* Introduction by Christopher Hitchens. London: Penguin.

Weber, M. (1946). Science as a vocation. In H. H. Gerth & C. W. Mills (Eds.), *From Max Weber: Essays in sociology* (129–156). New York: Oxford University Press.

Notes

1. Most of the language of economic equilibrium derives from aesthetics. The case of dynamic equilibrium is no exception. The term was coined by the American artist and mathematician, Jay Hambridge (1867–1924). See Hambridge's *The Elements of Dynamic Symmetry* (1967). The paintings by Paul Klee or the architecture of Ludwig Wittgenstein and Frank Lloyd Wright are aesthetic examples of dynamic equilibrium.
2. From 1997 onwards, Spain invested heavily in "green jobs." Each job cost the equivalent of $US 800,000, representing the effective loss of 2.2 jobs in other areas of the economy (Alvarez, 2009).
3. Schumacher (1973).
4. The following, reported by the *South China Morning Post,* is a classic example of the triumph of green rhetoric over intellectual substance. "A beaming Tony Blair posed for television cameras holding a sleek, shiny solar panel as smiling officials and film star Jet Li looked on. They announced an ambitious plan to bring modern, clean power to the world's poor. In the next five years, the programme would bring solar-powered street lamps to 1,000 villages in China, India and Africa, where people are so poor they still do not generate any of the greenhouse gases blamed for global warming. The plan was announced at a factory in Guizhou in southwestern China—one of its poorest provinces. However, would Blair, the former British prime minister, and Li have been smiling if they had known a factory must burn more than 40kg of coal to produce the panel—one metre by 1.5 metres—they were holding? Forty kilograms might not

sound like much. Nevertheless, even the country's least efficient coal-fired power plant would generate 130 kilowatt-hours of electricity burning that amount—enough power to keep a 22 watt LED light bulb beaming 12 hours a day for 30 years. A solar panel is designed to last just 20 years. Jian Shuisheng, a professor of optical technology at Beijing Jiaotong University, estimates it takes 10kg of polysilicon to produce a solar panel with a capacity of one kilowatt—just enough to generate the energy to keep a fridge cool for a day. To make that much polysilicon on the mainland would require the burning of more than two tonnes of coal. That amount of coal could generate enough electricity to keep the fridge running for two decades." Stephen Chen (2009).

5. On "glonacal" structures, see Marginson and Rhoades (2002).
6. Marginson (2007) notes that "In 2005 the industry generated $11.3 billion in fees and other spending by students, with more than $4 billion in fees, two thirds in higher education. Australia commanded 6 per cent of the world market in foreign students. Education is our third or fourth largest export sector, after iron ore and coal and on par with tourism."
7. The number of resident foreign students in Australia doubled in the five years from 2004 to 2008 (Colebatch & Lahey, 2009).
8. Like all industries, the education export industry had its problems, notable in 2009 with various scandals connected to sub-standard supplies, quality control, security of consumers, and so on. See Marginson (2009).
9. On the role of cities in the history of creation, see Peter Hall (1998).
10. On the adversary culture of mid and late twentieth-century cultural elites, see the critical assessment of Daniel Bell (1976/1996) and Robert Hughes (1994).
11. Schumpeter (2008), 64.
12. See Machlup (1962/1973) and Solow (1956).
13. On the role of knowledge as a factor in modern developmental philosophies of history, see Heller (1982), chapter 15.
14. Waugh, *Scoop: A novel about journalists* (2000); Bellow, *The adventures of Augie March* (2001); Burke, *On symbols and society* (1989), 261–267; Heller, *Immortal Comedy* (2005); Hitchens, "*Scoop* and *The adventures of Augie March.*" In *Love, poverty, and war* (2004); Berger, *Redeeming laughter* (1997). G. K. Chesterton's body of work in both fiction and non-fiction is peppered with the comic structure of paradox. Marshall McLuhan was inspired to write by his early encounter with Chesterton. McLuhan published an article on him ("G. K. Chesterton: A Practical Mystic") in the *Dalhousie Review,* 15(4), 1936. McLuhan's student, Hugh Kenner, contributed an excellent introduction to Chesterton, *Paradox in Chesterton* (1947). McLuhan built his understanding of communication on brilliant paradoxes like "the medium is the message," "the typographical essay," "knowing is making," "the mechanical bride" and the "global village." He also observed that the good humour needed to enter into fun and games is the mark of sanity and reason. McLuhan was the classic joker intellectual. A conservative Catholic, he was sceptical of moralists and moralism. He combined a love of

satire with a joker's intellectual tool kit. He explored paradox by rummaging through mysticism, Pythagoreanism, hermeticism, modernism, cynicism, stoicism, new criticism, and the heterodox orthodoxy of Gilbert Chesterton's Catholicism (Theall, 2006). McLuhan had a deeply satirical and paradoxical cast of mind. In his view, good communication was a kind of appositional poetics. This was a view shared by many of McLuhan's peers, ranging from the New Critics William Wimsatt and Cleanth Brooks through to Kenneth Burke. This was a tradition of thought enchanted by what McLuhan's Cambridge teacher William Empson once delightfully described as "knotted duality." It is a state, Empson explained, "where those who have been wedded in the argument are bedded together in the phrase." This is a state that, long ago, was recognized by the ancient Stoics. It is the state of antilogy, and its model is the *dissoi logi* or the double argument of the speaker who combines two opposing arguments into a single argument. McLuhan reduced such arguments to brilliant catchphrases. In a larger sense, McLuhan and his kindred spirits exemplify the flourishing of a strand of culture in North America that has its roots in the Renaissance and the Elizabethan world picture. Kenneth Burke (1989) called this cultural current the comic corrective. It is fascinated by phrases or scenes that have an agonistic logic. These phrases and scenes anchor sense in the non-sense of self-contradictory mottos.

15. "The whole modern world has divided itself into Conservatives and Progressives. The business of Progressives is to go on making mistakes. The business of Conservatives is to prevent the mistakes from being corrected. Even when the revolutionist might himself repent of his revolution, the traditionalist is already defending it as part of his tradition. Thus we have two great types—the advanced person who rushes us into ruin, and the retrospective person who admires the ruins. He admires them especially by moonlight, not to say moonshine. Each new blunder of the progressive or prig becomes instantly a legend of immemorial antiquity for the snob. This is called the balance, or mutual check, in our Constitution." G. K. Chesterton, "Column," *Illustrated London News,* 19 April 1924.
16. As the English philosopher Roger Scruton notes, so many of the great modernist artists of the twentieth century (Stravinsky, Moore, Matisse, and so on) were traditionalists. What makes something original, he suggests, is not defiance of the past or a rude assault on settled expectations, but the element of surprise that a given work invests the forms and repertoire of tradition. Scruton (2007), 45, 82–83.
17. Retrieved from http://www.measuringworth.org/usgdp/index.php
18. The key figures were Paul Baran (UCLA master's in engineering graduate and RAND corporation employee), J. C. R. Licklider (MIT Professor and Head of the Information Processing Techniques Office at ARPA), Douglas Engelbart (University of California PhD graduate and Stanford University), Theodore (Ted) Nelson (Master's in Sociology, Harvard University), Wes Clark and Larry Roberts (MIT PhD in Electrical Engineering [Clark & Roberts] and chief scientist in the ARPA Information Processing Techniques Office [Clark]), Ray Tomlinson (MIT Master's of Science graduate and employee of

the technology company of Bolt Beranek and Newman), Vinton Cerf and Robert Kahn (Stanford University mathematics graduate [Cerf] and ARPANET administrator and MIT graduate [Kahn], Bill Gates and Paul Allen (Harvard University dropouts), Randy Suess and Ward Christensen (Chicago computer hobbyists; IBM employee [Christensen]), Tom Truscott and Jim Ellis (Duke University graduate students), Tim Berners-Lee (Oxford University graduate and CERN, Switzerland, employee), Marc Andreessen and Eric Bina (University of Illinois undergraduate [Andreessen] and programmer [Bina]), Brian Pinkerton (University of Washington graduate student, later PhD in Computer Science), Larry Page and Sergey Brin (Stanford University Master of Science [Page] and PhD [Brin]).

19. Bell (1973/1999), p. 245.
20. The Association of American Colleges and Universities.
21. Retrieved from http://www.carnegiefoundation.org/ The difference between an American research university with very high research activity (A) and a regular doctoral-granting university that carries out research (B) is indicated by the following 2009 Carnegie figures based on 2002–2004 data. The mean number of humanities doctorates for A is 45, the mean number of social science doctorates is 38. In comparison, the mean number of humanities doctorates for B is 9 and the mean number of social science doctorates is 10.
22. MIT, California, Stanford, Harvard, California Institute of Technology, Chicago, Washington, Yale, Johns Hopkins, Columbia, Duke, Michigan, North Carolina, Northwestern, New York University, Boston, University of Pennsylvania, Washington University St. Louis, Emory, Vanderbilt. This list is based on data from the 2008 Leiden University index of world research universities.
23. The figures cited are drawn from Jonathan Huebner (2005, pp. 984–985) and John Smart (2005).
24. Copyright registrations today cover a remarkable spectrum of creative works including non-dramatic literary works, works of the performing arts, musical works, dramatic works, choreography and pantomimes, motion pictures and filmstrips, works of the visual arts, including two-dimensional works of fine and graphic art, sculptural works, technical drawings and models, photographs, cartographic works, commercial prints and labels, works of applied arts, and sound recordings.
25. The figures cited are from the U.S. Copyright Office and the U.S. Census Bureau, *Current Population Reports.* See also M. Boldrin & D. K. Levine (2008), 100.
26. Categories added to the U.S. schedule of copyright protected works: Motion pictures (1912), Recording and performance of non-dramatic literary works (1953), Computer programs (1980), Semi-conductor chips (1984), Architectural works (1990), Vessel hulls (1998). Unpublished works were covered in 1909 and sound recording protection was expanded in 1972.
27. Boldrin & Levine (2008), chapter 5.

28. Huebner (2005). Silverberg & Verspagen (2003) offer a somewhat different medium-term picture but the same long-term conclusion. Their quadratic analysis shows a higher level of innovation during 1850 and 1900 that levels off around 1930, or in the case of patents, 1920. Silverberg and Verspagen's general assessment is that the rate of basic innovation slowed down in the twentieth century after a period of relatively rapid increase in the later half of the nineteenth century. The authors' caution about this analysis stemmed from the fact that the data they analysed extended only to the end of the 1970s.
29. The history of U.S. copyright registration is graphically represented in the statistical chart in Boldrin and Levine (2008, p. 100). For U.S. patent registration, see Figure 3 in Jonathan Huebner (2005, p. 895). For graphic depictions of comparable trends across both the United States and Europe, as they affect major works and inventions, see the charts in Charles Murray (2003, pp. 428, 437, 441).
30. Ladd and Lipset (1975); Carnegie Foundation for the Advancement of Teaching (1989); Shepherd and Shepherd (1994); Hamilton and Hargens (1993); Tierney (2004); Rothman, Lichter, & Nevitte (2005); Klein & Stern (2005).
31. What about the case of the liberal *bête noire,* the George W. Bush administration? Patent registrations grew dramatically during the Bush years. Compared to the rate of 0.07 registrations per capita in 1994 and 0.09 in 1998, the rate in 2002 was 0.12 and in 2007 had risen to 0.15. The record in copyright registrations was less impressive, with a steady 0.18 registrations per 100 in the same years, compared with 0.20 and 0.21 in 1994 and 1998. Notably the latter figures lag well behind the rates of copyright registration achieved during the Reagan and George H.W. Bush years.
32. Suedfeld and Tetlock (1977, pp. 169-184); Suedfeld, Tetlock, & Streufert (1992); Tetlock (1998, pp. 639-652). The toleration supposed by integrative complexity is more closely aligned with creativity than measures of social toleration that Richard Florida uses in constructing his indexes of creative cities. Interesting as the latter are, they are not epiphenomena of the act of the creation in the same way that the capacity to integrate conflicting cognitive perspectives is.
33. Simonton (1984, p. 144); Murphy (2001), chapters 1–3, 5–7.
34. Murphy (2001), chapters 10 and 11; Murphy (2006, pp. 64–92).
35. Tetlock (1984, pp. 365–375); Tetlock (1986), Tetlock, Bernzweig, & Gallant (1985, pp. 1227–1239). Critics include Gruenfeld (1995, pp. 5–20). Gruenfeld's key point is that high-low integrative complexity on the US Supreme Court aligns not with the liberal-conservative divide but with whether the opinion writer is writing for a Court majority (higher complexity) or a Court minority (lower complexity).
36. Graham, Haidt, & Nosek (2009, pp. 1029–1046). In the article "What makes people vote republican?" (2008), Haidt reports that: "In several large internet surveys, my collaborators Jesse Graham, Brian Nosek and I have found that people who call themselves strongly liberal endorse statements related to the harm/care and fairness/reciprocity foundations, and they largely reject statements related to ingroup/loyalty,

authority/respect, and purity/sanctity. People who call themselves strongly conservative, in contrast, endorse statements related to all five foundations more or less equally." See also Haidt (2009a) and Haidt & Graham (2009b).

37. "White Rabbit" (Grace Slick), single by Jefferson Airplane from the album *Surrealistic Pillow* (RCA Victor, 1967).
38. Katz & Lazarsfeld (1955); Lazarsfeld, Berelson, & McPhee (1954). It was the work of Edward Shils, Talcott Parson's collaborator at the University of Chicago, that suggested this line of inquiry to Lazarsfeld.

17

The Creative Ecology of the Creative City: A Summary

Charles Landry

Preface

"The Creative Ecology of the Creative City" is a summary of ideas from several of my writings including "The Creative City: A Toolkit for Urban Innovators" and "The Art of City Making" and to some extent "The Intercultural City." Initially this was an extended think piece written for the British Council to help them define their broader global strategies. It attempts to assess what really matters for our cities to survive well and how creative thinking and new forms of learning might play a role.

Setting the Scene

There is an increasing frenzy for places to become and be creative cities, and for many persons, this idea means taking artists far more seriously as contributors to great place-making. Some even talk of a Creative City Movement. Everyone is responding to the fact that the world has changed dramatically so that it feels like a paradigm shift, and whilst many things seem the same, they are different in terms of their underlying operating dynamics. This is sometimes summarized as the move to the new knowledge intensive economy.

The Creative City notion for some seems like the answer to coping with this transition. Indeed the creativity focus is like a rash; it is everywhere. At times people try to get creativity to solve more problems than it can cope with. The creative city debate is sometimes focused on the arts and culture, but urban creativity is much more and affects the way cities are organized and managed. To be creative means changing the organizational culture of a city. It means creating the conditions within which people can think, plan, and act with imagination. It means there need to be creative individuals, organizations, and communities, as well as creative education and training. This environment can then establish a creative milieu and develop a creative ecology. As one moves from the creative individual to the organization and then the creative city, the levels of complexity expand exponentially as one is then dealing with an amalgam of organizations with differing cultures that in some sense need to be aligned. In the city of greater imagination, there are many new forms of organizations that bring together creative ideas, people, and projects—these might be called connectors. Open source innovation is an example of how this dynamic might happen. Seen in this way, there are more resources for a city to play with and many potential niches allow any city, in principle, to get on the global radar screen.

The city of the future depends much more on the creative economy. This statement is the platform for both developing the economy and even the city. Three main domains revolve around its core: the arts and cultural heritage, the media and entertainment industries, and perhaps most importantly creative business-to-business services and the way they can add value to every product or service. In the latter sphere, especially in design, advertising, and entertainment, the creative industries often act as drivers of innovation in the broader economy and shape the so-called "experience economy." This concept elevates the city, and artists are increasingly used to provide the imagination.

For some time now the discussion has been going on about what makes cities move forward, and the idea of the creative class has emerged. The economy, so the argument goes, is changing, thus highlighting how companies and especially innovative companies offering high-value employment are attracted to regions where the creatives like to live. As a result of this insight, the question emerges regarding who these people are and the key attributes of a region or place that are what attracts the Creative Class and how can one encourage the "creatives" to cluster. Artists are one group at the core of this class as are creative economy people, scientists, and knowledge nomads. Cities then need a people climate as well as a business climate.

In this new world of cities, the artistic imagination is far more important and in principle artists can contribute to making great places. However, it is also artistic thinking that might be the most important element, and developers, planners, and many others should enhance their imaginative frame of mind.

At this juncture in world history it is advisable for creative cities not to compete against one another to be the most creative city ***in the world*** or region or state. They should strive to be the best and most imaginative cities ***for the world.*** This one change of preposition—from "in" to "for"—has dramatic implications for a city's operating dynamics. It gives city-making an *ethical* foundation. It helps the aim of cities to become *places of solidarity* where the relations between the individual, the group, outsiders to the city, and the planet are in better alignment. These can be cities of passion and compassion.

In order to move onto this new terrain, we need to shift away from the old intellectual architecture, which has sedimented itself into our minds like a geological formation, especially since change has been happening so fast. In other words, it is important to address issues such as the difference between the complicated and the complex or understanding the dynamics of global change at a deeper level. For instance, the paradox of a risk culture and creativity agenda is emerging simultaneously. The former often reduces the impact of the latter. A conceptual framework is offered to unscramble this urban complexity by assessing the strategic fault lines, battlegrounds, and drivers for change.

I want to focus on being as lofty as possible. I will loop around these topics, and the way the argument unfolds will, I hope, guide readers to come up with ideas for their city or ideas for a program they would like to develop. Throughout the text, questions will be asked as a way of encouraging the readers to assess whether they are operating comprehensively as creative places or whether the blockages revealed are in fact part of the agenda of what has to be done.

Changing Circumstances

The graph on p. 358 encapsulates visually the rationale, reason, and drive of the creativity movement.

Note the intense speed within which the last four phases have occurred. We were an agrarian society for millennia, an industrial one for 200 years, a society whose wealth creation was primarily driven by information for 30 years, and ever more speedily we have moved through phases where the dominance of a particular driving asset has changed.

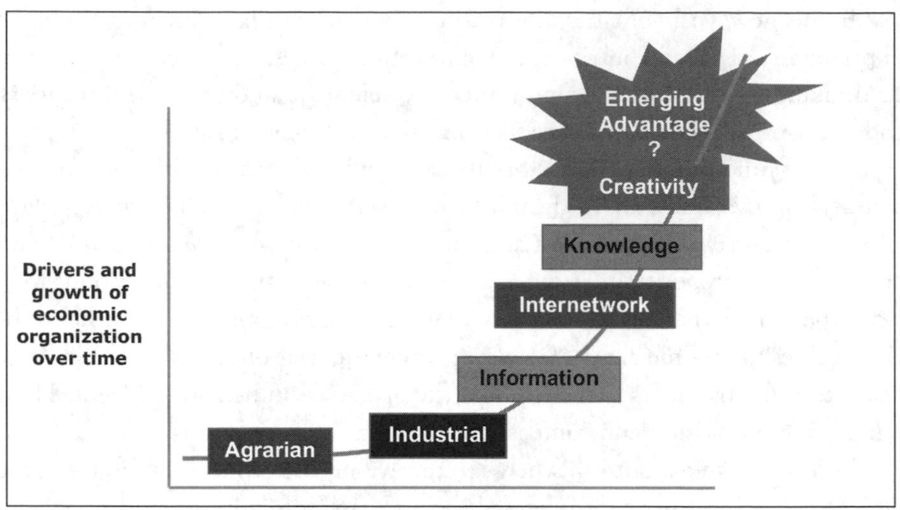

Shifts in the Evolution of Economic Organization

Each metaphor such as the "Knowledge Economy" provides a helpful lens from which to understand and gauge the shift in the primary means of wealth creation and the basis of competition. Now we have reached a stage where creativity and the capacity to imagine are key. Every shift in the means of economic wealth creation creates a new social order, new ways of learning and new subject matters to grasp, and new settings in which learning takes place. It requires different cultural capabilities.

A sense of certainty and predictability clung about the foreseeable results of the former phases. Predicting exactly the "'emerging advantage" from creativity will be less easy. However, it is possible to build capability and encourage the mindset of communities to be ready to identify the "advantage" when it starts to emerge, and thus to have the creative capacity to respond. Therefore, educational systems must be aware of these new demands and be willing to adapt to these new challenges.

Inevitably, the following question emerges: "What are the conditions within which it is possible to think, plan, and act with imagination?" Is it how people think? Is it how organizations behave? Are there classes of people who display creativity, and, if so what qualities do they display? what jobs do they do? What do they like to do? With whom do they want to live? How do they want to spend their time and where? Alternatively, is it the way the city is organized and manages itself? Are there cultures or cultural settings that make acting with imagination more likely or is creativity context and time specific? What may be seen as creative in one time and place may seem very ordinary somewhere else. Indeed, is creativity itself a term that only has validity in certain cultural environments? Some societies decry it by definition as a superior power that has already created everything that is worthy of creation. Being creative implies movement, a dynamic, and a constant re-assessment of the now. Change seems to be ever present. What, in this context, are the anchors that hold a place together and that provide predictability, certainty, and stability? Alternatively, is it possible that concepts like stability are old-fashioned? Why should people, organizations, and cities be creative when creativity destabilizes yet can stimulate and when many still think "if it ain't broke don't fix it"?

Finally, what is the role of artistic thinking, the arts, or artists? Are they in some sense a unique category, or is this unique category just a desire for special status? Are they at the core of the creative environment required? What of the creative economy, built largely on the skills and attributes we find in arts focused industries, such as film, music, and design? What of the cultural setting itself? How does it foster the movements of the imagination in people and organizations and turning parts into tangible coherent objects?

Approximately twelve key texts discuss the landscape of these ideas and movements of change (see the references to this chapter). They detail many issues and concerns, and thus it is important to understand the different approaches of each author because what each of them writes affects the aims and intents of policies, as well as the dynamics of how these policies are carried out.

The Same and the Different

The world is changing dramatically, in some sense, to the extent of a paradigm shift. Appearances have remained superficially the same in that there are roads, cars, houses, offices, and shops that still look roughly the same as 100 years ago, and people continue to meet, interact, trade, and share emotions. Material goods from food to cars are still produced, and thus the old wealth creation systems of agriculture and industry have survived. However, their operating systems are shaped, conditioned, and determined by the dynamics of the new. Behind the veneer of what seems familiar and recognizable are subtle differences that with ever greater force and impact will etch themselves into reality sharply. The driving factors are known only too well, primarily the interconnected, largely integrated IT platform, which was the contribution of the information and Internet revolutions. In the process "value" and "worth" have changed. For instance, global finance has remained one force driving the modern economy in spite of the credit crunch caused by turbo-charged casino capitalism and its consequences. However, that form of capitalism has been exhausted and can be superseded by a form where values are more than greed, and financial systems are supported by better regulatory structures.

Book titles like *The Weightless World* or *Living on Thin Air* aptly describe the new dynamic where ideas, knowledge, and imagination are just as pervasive and powerful as financial capitalism. What value is money when there is no idea to match it? In addition, as Tom Peters noted: "Screwing up is more important than ever in strange times. The screw-up rate is the best indicator of sufficiently rapid adaptation. The leader must 'manage' the screw-up process—literally." The world is less like a linear track where one step leads to another in a seamless logic.

Tom Peters's insight highlights the shifts of the relative importance of the tangible to the intangible—from the hardware to the software; from a dominance of finance to that of ideas. Two things are happening at once. First, if we can impregnate products and services with an idea, a set of associations, emotions, flair, a personality, and the senses—and make it an experience in the jargon—then ordinary products have a different lease on life. The ordinary feels creative, and the economy then becomes more like a performance with the skills needed like those of an artist. The dilemma for them is whether this process is one of selling out.

Second, the main protagonists focus on creativity as the dynamo of development. We have evolved from a world where prosperity depended on *natural advantage* (arising from access to more plentiful and cheap natural resources and labor than other countries) to a world where prosperity depends on *creative advantage*, arising from being able to use creativity to innovate in areas of specialized capability more effectively than other places.

The world looks different if you think differently

Various people have highlighted creativity in different ways. I proposed the idea of the Creative City in a short book with Franco Bianchini in 1995 and more extensively on my lengthier book in 2000, where the idea of cultural resources and the creative milieu was central. This publication was followed by the *Art of City-Making* in 2006. John Howkins wrote *The Creative Economy* in 2001, Charles Leadbeater *Living on Thin Air* in 1999, Richard Florida *The Rise of the Creative Class* in 2002. Collectively, these authors set and reflected an analysis and mood of thinking about cities, culture, creativity, and, to some extent, art .

The period they describe is different. This statement implies that if we think differently we do things differently, and sometimes we do different things. The old verities are no longer true. Minds, though, are fearful and different systems live side by side, but the cusp is bending—inexorably.

The Creative City

The Creative City argues that "Cities have one crucial resource—their people. Human cleverness, desires, motivations, imagination and creativity are replacing location, natural resources and market access as urban resources. The creativity of those who live in and run cities will determine future success. Of course this has always been critical to cities' ability to survive and adapt. As cities became large and complex enough to present problems of urban management so they became laboratories that developed the solutions—technological, conceptual and social—to the problems of growth." What are the unique forms of creativities in East Asian cities? What examples are there? Are they any different from those in Britain, and what reciprocal learning could take place?

The idea of the Creative City emerged from the late 1980s onwards along a number of trajectories, which both enrich what the creative city means today, yet they also cause confusion because of the city's diversity. More specifically, it was a response to the fact that, globally, cities had been flailing as they searched for new answers to create a purpose for themselves, an economic base, and jobs for their citizenry. This activity meant re-assessing and realizing that the old ways and the basis on which the old economy had worked had passed. Hierarchical management systems were insufficiently effective, education systems did not seem to prepare students for the demands of the "new" world. Schools remained factories to drill in knowledge rather than communities of enquiry; they taught specific facts rather than how to acquire higher order skills such as learning how to learn, to create, to discover, innovate, problem solve, and to self-assess. These traits are all features of artists. More generally, inquiry is more likely to trigger and activate wider ranges of forms of intelligence in that it fosters the adaptability to allow the transfer of knowledge between different disciplines and how to understand the essence of arguments rather than just recall facts in or out of context. In other words, talent was not sufficiently unleashed, explored, and harnessed.

The question is whether this is just a Western development. How does this global dynamic affect Asia, the Middle East, or Africa, and do any of these trends look intrinsically different through the eyes of non-Westerners? What spectrum of industries will the emerging powerhouses of the East have in 2050? Will they be any different from those in Europe? What is your view? When the creative city notion was introduced in the early 1990s, it was seen as aspirational, as a clarion call to encourage open-mindedness and imagination. Its intent was to have a dramatic impact on organizational culture. The philosophy was that there was always more potential in any place than any of us might think at first glance, even though very few cities, perhaps London, Tokyo, New York, or Amsterdam are comprehensively creative. Conditions need to be created for people to think, plan, and act with imagination in harnessing opportunities or addressing seemingly intractable urban problems. These issues might range from addressing homelessness to creating wealth or enhancing the visual environment, developing a new design template for a city, or getting artists to unsettle conventional attitudes. It is a positive concept; its assumption is that ordinary people can make the extraordinary happen if given the chance and that if everyone were just 5% more imaginative and open about what they do, the impact would be dramatic.

Creativity in this context is applied imagination, using qualities like intelligence, inventiveness, and learning along the way. In the "Creative City," it is not only artists and those involved in the creative economy who are creative, although they play an increasingly important and specific role. Creativity can come from any source including anyone who addresses issues in an inventive way, be it a social worker, a business person, an engineer, a scientist, or a public servant. Even so, creativity is legitimized in the arts, and the organization of artistic creativity has specific qualities that chime well with the needs of the idea-driven economy. Interestingly, in the urban context teams comprised of individuals with different insights, backgrounds, and personalities generate the most compelling ideas and projects. Do you agree with this statement or is your experience different?

The creative city advocates the need for a culture of creativity to be embedded in how the urban stakeholders operate. By encouraging creativity and legitimizing the use of imagination within the public, private, and community spheres, the "ideas bank" of possibilities and potential solutions to any urban problem will be broadened. Divergent ways of thinking generate multiple possibilities; these then need to be analyzed and narrowed down via reality check so that subsequently urban innovations can emerge. The creative city is a place that is comprehensively imaginative. It has a creative bureaucracy, creative individuals, organizations, schools, universities, and so on.

In the physical realm, the creative city requires infrastructures beyond the hardware, such as buildings, roads, or sanitation. Creative infrastructure is a combination of the hard and the soft, including the mental infrastructure, the way a city approaches opportunities and problems; the environmental conditions it creates to generate a pleasant atmosphere, and the opportunities it offers incentives and regulatory structures.

To be a creative city the soft infrastructure needs to include a highly skilled and flexible labor force; dynamic thinkers, creators, and implementers as creativity is not only about having ideas; a large formal and informal intellectual infrastructure considering that the old-fashioned empire building tendencies of universities are more like production factories; being able to give maverick personalities space; strong communication linkages internally and externally and an overall culture of entrepreneurship regardless of whether it is applied to social or economic ends. This attitude establishes a creative rub as the imaginative city stands at the cusp of a dynamic and tense equilibrium.

This creative city of imagination must identify, nurture, attract, and sustain talent so it is able to mobilize ideas, talents, and creative organizations in order to keep their young and gifted. Being creative as an individual or organization is relatively easy, yet to be creative as a city is a different proposition given the amalgam of cultures and interests involved. The characteristics tend to include taking measured risks, widespread leadership, a sense of going somewhere, being determined but not deterministic, having the strength to go beyond the political cycle and being strategically principled and tactically flexible. Maximizing this potential requires a change in mindset, perception, ambition, and will. Moreover, it requires an understanding of the new competitive urban tools such as a city's networking capacity, its cultural depth and richness, the quality of its governance, design awareness, and understanding of how to use the symbolic and perceptual understanding and eco-awareness This transformation has a strong impact on organizational culture and can only be achieved by going beyond the business as usual approach.

Instituting the creative city requires thousands of changes in mindset, creating the conditions for people to become agents of change rather than its victims, seeing transformation as a lived experience and not as an event beyond one's control. It demands invigorated leadership. What is your view of cities you know in the context of the comments above? Do they display this comprehensive, cross-cutting creativity, or are they more locked in silos?

The built environment—the stage, the setting, the container—is crucial for establishing a milieu. It provides the physical preconditions or platform upon which the activity base or atmosphere of a city can develop. A creative milieu is a place that contains the necessary requirements in terms of hard and soft infrastructure to generate a flow of ideas and inventions. A milieu can be a building, a street, an area, or a city, such as the Truman's Brewery in Brick Lane, London; Rundle Street East in Adelaide, or Queen Street in Toronto; Soho in New York, though, stands out as a milieu.

Essentially the city is seen as a complex adaptive system where a holistic, "whole systems" approach creates "systemic creativity," and where creativity is leveraged into the entire community. This milieu creates the mood of the city—how would you describe the mood of the cities you are involved in, are they open in the way implied above?

Creativity and Education

Mary Kalantzis and Bill Cope in their book *New Learning: Elements of a Science of Education* describe the changing characteristics of educational contexts through time and the necessary qualities required from each. First there is what they call "the recent past: mass-institutional education," then "more recent times: progressive educational modernization," and finally "new learning innovations anticipating the near future." This outline highlights how educational structures reflect the cultural, economic, and social frame they operate in. Tom Bentley and James Wilsdon expand the context in *The Adaptive State* by asserting that "There is a need to revisit the purpose and shape of government itself, and to explore models of organizational change for which the state is not currently equipped. Radically different models of service, organisation and value are required. But these must also be compatible with the effort of sustaining and managing existing commitments in the here and now." In other words, to take advantage of the creative cities notion, municipalities and governments will need to play a central role as catalysts and motivators for innovation within government, industry, education, and the community by encouraging decentralization. They need to be "creative bureaucracies." How well is your city doing on this front?

Local governments need to play a central role in changing the school curriculum to reflect new forms of learning that develop creativity, which means changing the predefined curriculum from learning topics at the expense of depth of understanding and breadth of application. As Seltzer and Bentley note in *The Creative Age* creative learners need four key qualities. They need to be able to

- identify new problems rather than depending on others to define them;
- to transfer knowledge from one context to another;
- to treat learning as an incremental process, in which repeated attempts will eventually lead to success;
- and to pursue a goal.

The range of skills required include: self-organization; being interdisciplinary; having the capacity to gather and manage information; having personal and interpersonal abilities; being able to reflect, and evaluate and manage risk, and to handle stress. Do our learning institutions reflect these views, and is their curriculum geared to us developing these newer skills?

Creativity and Resources

In this city that focuses on the imagination, there are far more tangible and intangible resources that range from the obvious to the nonobvious because such a place prioritizes authenticity, originality, uniqueness, and distinctiveness. These traits of each individual city distinguish it from the others and one place from the next. Cultural resources are the raw materials of the city and its value base; its assets replace coal, steel, or gold. Creativity is the method of exploiting these resources and helping them grow. Taking a broad sweep of a city's economy, social potential, and political traditions one can see how cultural assets can be turned to economic and other advantages. An old skill in carpentry or metal working could be linked with new technology to satisfy a new market for household goods or a tradition of learning and debate could be used to market a city as a conference venue. By considering the "senses" of the city including color, sound, smell, and visual appearance, or taking a sweep through mutual aid traditions, associative networks, and social rituals, further features emerge that can make a city special as well as more broadly competitive. As the world of cultural resources opens up, it is clear that every place can have unique niches, and ugly cities, cold or hot cities, or marginal places can get on the radar screen. Every city can be a world center for something if it is persistent. In identifying urban resources we see they are embodied in people's skills and talents. They are not only the buildings, but also symbols, activities, and the repertoire of local products, but also ideas, desires, and the imagination.

Creativity Is Like a Rash

Everyone is now in the creativity game, and this development poses its own challenges. Creativity has become a mantra of our age endowed almost exclusively with positive virtues without sufficient acknowledgment of its flipsides or its negative manifestations, such as its tendency to create new divisions between the supposedly "creatives" and the "non-creatives," the lack of stability, continuity, and predictability.

Twenty British cities alone at the last count call themselves creative: from Creative Manchester to Bristol to Plymouth to Norwich and, of course, Creative London. In addition, I have counted about 60 cities worldwide, e.g., in Canada Toronto, with its Culture Plan for the Creative City; Vancouver and the Creative City Task Force; London Ontario's similar task force, and Ottawa's plan to be a creative city. In the United States there is Creative Cincinnati, Creative Tampa Bay, and the welter of creative regions such as Creative New England. In Australia we find the Brisbane Creative City strategy, and then there is Creative Auckland. Partners for Livable Communities launched a Creative Cities Initiative in 2001 and Osaka set up a Graduate School for Creative Cities in 2003 and launched a Japanese Creative Cities Network in 2005. Even the somewhat lumbering UNESCO through its Global Alliance for Cultural Diversity launched its Creative Cities Network in 2004, anointing Edinburgh as the first for its literary creativity and others such as Santa Fe the creative city for folk arts and Shenzhen, next door to Hong Kong that mushroomed out of nowhere into a city of 13 million, as a "city of design."

On closer examination, most of the strategies and plans are in fact concerned with strengthening the arts and cultural fabric, such as support for the arts and artists, and the institutional infrastructure to match. In addition, they focus on fostering the creative industries. These initiatives are fine as far as they go.

If I were to advise you now, I would suggest you do not brand yourself a "creative city," although I would push you to be far more creative and imaginative than you ever thought you could be. I would encourage you to push at the boundaries of the tried and tested in disparate fields. However, I would let others who reflect on what you do call you a creative city.

Creative for the World

My main recommendation would be that you should think of your creative city as not seeking to be the most creative city *in the world* or region or state. It should strive to be the best and most imaginative *city for the world.* This one change of word—from "in" to "for"—has dramatic implications for a city's operating dynamics. It gives city-making an *ethical* foundation.[1] This approach supports the goal of cities to become *places of solidarity*, where the relations between the individual, the group, outsiders to the city, and the planet are in better alignment. These can be cities of passion and compassion.

This concept reflects a re-focus. It takes us away from continually thinking of creativity as predominantly focused on new media or the glamor of the advertising industry, film-making, the music business, or financial innovations that seem to create money out of nothing. The interesting question is how these groups would respond to this challenge. Are they up for it?

This approach would go with the grain of local cultures and their distinctiveness yet be open to outside influences. It would balance the local and global. It would learn from what others have done well but not copy them thoughtlessly. Cities focused mainly on best practices are followers not leaders and do not take the required risks to move themselves forward. It would encourage projects that add value economically and reinforce ethical values simultaneously. To accomplish this goal, it is necessary to revisit the balance between individual wants and collective and planetary needs relevant in the twenty-first century. Too often value is defined narrowly in terms of financial calculus. This understanding is limiting and naïve. The new economy requires an ethical value base to guide action. It will imply behavior change to meet value-based goals such as putting a halt to the exploitation of the environment. Indeed, protection of the environment is in itself the major cultural project of our times.

This philosophy fosters *civic creativity* as the ethos. Civic creativity is imaginative problem-solving applied to public sound objectives. It involves the public being more entrepreneurial within accountability principles and the private sector being more aware of its responsibilities to the collective whole.

What Is a Creative Idea?

What, in this context, is a good, creative, catalytic idea that can drive a process and that can become a trajectory on which to move forward? A great idea needs to be simple but complex in its potential. A good idea is instantly understandable, it resonates, and it communicates iconically—you grasp it in one. A good idea needs to have layers, depth, and be able to be interpreted and expressed creatively in many ways. As such it can involve many people who feel it can be theirs, and they have something to offer. A good idea connects and suggests linkages. It is dynamic. It breathes and implies multiple parents and possibilities. With a good idea, creativity and practicality come together. A good idea is usually grounded in the cultural realm, thus making an impact on the culture. It solves economic as well as social problems, for example. It has to embody issues beyond the economic. If it is just economic it can become mechanistic, too self-interested, and self-focused, and insufficiently inspiring. Ideally it should touch the identity of a place and so feel culturally relevant. Indeed it should support, build on, and contribute to the identity of a city, and, therefore, it may well be artistically expressed. In this way it should speak to deeper values and ambitions. It is significantly powerful and can be implemented in many ways.

Let's look at some ideas by focusing on education and talent issues. Many cities around the world say they are going to become the "Education City." This idea is narrow; it implies and feels as if it is only the education sector that is involved. It excludes everyone else. "A talent strategy for. . . ." idea would be better in that it is easy to understand. Clearly, many people would need to be engaged, and they can see their involvement from the arts to education to business providing professional development. It can be layered to focus on identifying, harnessing, attracting, sustaining, or exploiting talent. Alternatively, it can focus on the stages of talent from getting people to be curious, enterprising, entrepreneurial, or innovative. Its weakness is that it could apply anywhere. It needs a creative twist to make it unique to a city. To say as Memphis is beginning to say that it is "The City of Second Chances" is quite strong. It projects a positive ethos; openness, the willingness to listen, and tolerance. It recognizes the city has many disadvantaged people without going on about it. It acknowledges business start-up records are not too good. It opens out to the future, and, ideally, in a decade the slogan will be less relevant, because enough second chancers will have succeeded.

From the Creative City to the Learning City

Let's switch—I think beyond the idea of the creative city the notion of the "learning city" is even more important. A learning city is more than a city of education and well-educated people. A learning city is a clever city that reflects upon itself, learns from failure, and is strategic. In addition, it is open to all kinds of resources such as its culture. It is a place where individuals and organizations are encouraged to learn about the dynamics of where they live and how their home is changing; the city is then a learning field. The dumb city, on the other hand, does not reflect on itself and has a tendency to imitate and repeat past mistakes. Does that notion resonate? Probably not for many because it sounds too vague, worthy, and all embracing, whatever its merits. Thus, is it possible to embed this core idea, but to find a completely different expression, ideally with a sense of the imaginative at its core?

What would it mean to be a city of "creative inspiration" or "energy and imagination" or "artistic creativity"? As they stand the terms still sound weak. How could they be rephrased to have drive, purpose, encourage co-operation, and passion and to help the long-term future of any of your cities? Clearly there are many ideas to be had. The key is to project them, so they are relevant to large swathes of the population such as the poor as well as the well-to-do. In addition, can they be projected in a paced and purposeful way with an action plan to match? Many ideas that have a real, deep drive have a bigger purpose. In fact, they are likely to come from ecology because they link idealism with economics, behavior change, and sustainability as well as culture. If a smaller city really wanted to become the world's first "zero emissions city" or "solar city," it would be a powerful statement and provide a mass of business opportunities. This idea has an implied economic agenda and speaks powerfully to green issues. In addition, it would give creatives from eco-activists to artists an incredible canvas on which to play. Freiburg, in Germany, is one city that has moved forcefully in this direction.

Capturing Territory on the Global Imagination

Most importantly, any city can capture territory in the global imagination. Just as imperialists captured territories to secure trade routes or raw materials, and corporations think of capturing markets by selling products, we can think of capturing ideas and networks by making a city the center. An audit of creative potential, in any place would quickly reveal where it is uniquely positioned. What are the top 10 niches where the words "your city . . . . . . and . . . . . ." are typed into a search engine and where your city comes up. This method of connecting concepts could result in projects that link science and art or an obscure research discipline. By building on these ideas, economic and cultural linkages can be made, and by assessing in which networks or on which topics your city can take a prime position, it can reflect back to the world a sense of centrality.

One idea is never enough to put a city on the global radar screen and to create the necessary associational richness. Many ideas need to be drawn from the very large pool that represents driving themes and that coalesces many activities to the very small. The big themes chosen need to imply an unfolding, emergent story where everyone can see their role and be part of transforming your old city into shaping and creating your new city. Essentially we are here in the realm of urban storytelling, but these stories are more difficult to write than a book.

Individual, Organizational and Urban Creativity

Let us consider individual, organizational, and urban creativity. We can grasp quite easily the essence of creative individuals; for instance, their capacity to make interesting connections, to think out of the box, and to have sparks of insight. They have energy and some sense of where they are going, although it is often unclear how. The same principle applies to creative organizations. However, already the priorities are different, and this difference adds a layer of complexity, instilling a different dynamic. A creative organization probably has mavericks and creative individuals, but for the organization to work it needs other types too: Consolidators, sceptics, solidifiers, balancers, and people with people skills. Some people consider them as less interesting, but doing so is misguided because for the creative organization to work it needs mixed teams. How teams work together becomes significant. One needs to achieve a series of balances, such as a balance between being collaborative internally and perhaps using external competition to push the team forward. In addition, the organization needs a story and a proposed trajectory to give itself purpose in an attempt to become more internally cohesive. The task is to align inside to face an outside world. Indeed, it may be the case that a creative organization has quite ordinary people in it, but because its spirit or ethos is open, exploratory, and supportive, this sense of unity maximizes the organization's potential, thus possibly leading to greater sustained organizational achievements. This dynamic is typical of sport teams where a team with no supremely gifted individuals wins because the team as a cohesive unit knows how to make the most of its parts. The key is its open, collaborative ethos. The value of ethos is incalculable.

With the move to the next layer of complexity—the creative city—issues become very difficult as complexity rises exponentially as one involves a mass of individuals and an amalgam of organizations with different cultures, aims, and attitudes. These different elements of the whole can push in opposing directions. For example, it may be that some are pushing for urban expansion and extension, whereas others are focusing on the sustainability agenda and want to contract rather than expand. In another scenario, one organization may display great cultural understanding, whereas another may basically not get it. The challenge then is to discover where the lines of strong agreement can flow and to build on these so that similarities become more important than differences.

The Creative Connector

The creative city notion does not engage in the fantasy of a cosy consensus. Instead, it stresses how rules of engagement between differences can be negotiated to move forward akin to a mediation process. The overarching skill needed for a creative city, therefore, is that of the connectors, enablers, and facilitators. These can be individuals or organizations who can stand above the nitty gritty of the day-to-day, important as it is, and look at what really matters instead of getting stuck in detail. Many of the cultural diplomacy organizations like the British Council, the Alliance Française, or intermediary organizations in cities such as economic development agencies are examples of connector organizations.

The connector can try to stand above the fray and focus on the big picture by bringing people, organizations, and ideas together. It keeps on eagle eye on things and roves over concerns. It looks for the common agenda. For instance, it can see issues that many organizations view as quite important, but not as of prime importance—often because it is not their main *raison d'être*. As most organizations work in silos, many issues therefore slip through under-acknowledged, yet those under-acknowledged matters may be the most important task for a city. The global competitive position of a city is one example, or the definition of the 20-year-vision can be another, and simply the need to be more creative can be a third. Equally at times the connector can look at an issue such as the attraction and retention of talent and see that it is viewed too narrowly.

Most cities have to attract highly skilled, interesting, innovative, even edgy people—the mobile crowd of the talented. In addition, they need to create conditions for more home grown talent to emerge. This is a task well beyond the educational sector, although they play an important part. Education cannot solve the problems of education on its own. After all, for instance, school or university occupies only five to seven hours a day, yet we behave as if it were 24 hours. People could still be learning in the other 19 hours. Some of the most effective learning outcomes happen outside. When we look at the talent journey this dynamic becomes even more evident. Education, business, the arts, volunteering networks, and many other elements have a role in the learning outside the immediate learning environment.

Open Source Innovation: An Example of a Creative Milieu

By looking at how some knowledge industries are innovating, we can understand better how a creative milieu in a city as a whole could work. We have lived with the idea that inventions and innovations should be protected by patents as they guard the income stream that repays the effort, research, the resources expended, and the risk. Even so, patent protection, it now appears, has a flipside. It can reduce creative capacity and innovation potential because it locks in ideas within the domain of rights holders, monopolizes it, and blocks others from developing an idea, product, or process beyond its original concept. It seeks to optimize income by controlling the development process. However, as Kevin Kelly noted in his *New Rules for the New Economy*: "Wealth in this new regime flows directly from innovation, not optimization. That is, wealth is not gained by perfecting the known, but by imperfectly seizing the unknown."

At its best the Open Source Innovation process is participatory in that it involves, empowers, and seeks to encourage widespread creativity within groups or communities, which are usually communities of interest, such as the disparate group that developed the Linux platform. It provides a framework that allows a wide cross-section of institutions and groups to participate at a global and local level by participating in the in-put of innovation and sharing the output. Open Source is a creative, imaginative leap that led to a key innovation to encourage creativity and innovation itself. The result is maximized by including a diverse range of "creative agents" in the development process. It is decentralized and seeks to match global and local needs. By contrast, traditional creativity and innovation systems are too hierarchically driven from the top by corporations, institutional policy makers and formal R&D institutions. The new organization ethos is by contrast porous and permeable. This method is especially significant in generating service innovations that require co-development and co-creation with intended audiences.

How Open Source Is Organized

The patent driven approach, still dominant in industry and elsewhere, is seen as a "closed creativity or innovation system" where exploration and research operates in silos of activity in research entities with little or no connection to outside worlds. The results are expected to "trickle down" in time. There is then a bias to being self-referential without relating to audiences and broader communities. The skills and knowledge generated from the creative processes are restricted to a limited group, thus contributing little to the overall creative consciousness and experience of an institution, a community, or a city. Most closed innovation tends to funnel down narrow tunnels linked to business, funding, or policy requirements. The unlimited potential of spin-off ideas that are not part of the brief are reduced, even though they may interest other creatives "out there," who represent a broad diversity from potential customers and stakeholders with varied backgrounds and interests. These are the complex cluster of interdependent organizations that comprise a system or a city from small and large business to community groups to the public domain who might want innovations but do not have the knowledge or resources to carry them out. Open systems of innovation or a creative milieu are more able to leverage creative capacities by being a network of nodes or entities that are bound together by pursuing common and individual goals simultaneously. Furthermore, such innovation can be cheap because it spreads the experience of innovating and creating knowledge as well as spin-offs. The traditional approach does not necessarily provide an optimal return on investment. By encouraging experiment, the open approach allows inventions to "leak to the market," as happened with texting, not foreseen initially as a central component of mobile phones. This development in turn has encouraged a whole series of innovative spin-offs downstream.

As Charles Leadbeater (1999) notes, Linux represents a new model for open, networked innovation. Linux is perhaps best known for being a leader in launching the open innovation movement. Linux started when a Finnish programmer Linus Torvalds put the core of his proposed computer operating system on the internet. He invited other programmers to make improvements and literally hundreds began to join and proposed changes, additions and deletions. An innovation community then developed around the Linux programme, which operates so strongly that business uses it widely.

Linux combines many elements which traditional organizations usually keep separate. The open innovation community is competitive but also cooperates and shares and has a sense of the public good. It flourishes because many small initiatives are aggregated. The source code is available to all on the internet and the participants can be distributed across the globe and they operate virtually. At the same time though it is managed and integrated from a centre. As Leadbeater notes: "A cacophony of localised innovation cumulates to create system wide change."

This innovation approach allows stakeholders to engage in defining the goals by interacting, encouraging feed-in and feedback in a dynamic manner. It provides an umbrella within which commonly rules guide development and direction setting. It is a living system. According to Leadbeater, Linux combines three ingredients:

Modularity: The programme could be broken down into sub-systems, so that innovators could focus on particular modules without having to change the entire system. That allowed multiple and parallel efforts at innovation to take place simultaneously.

Open Standards: Torvalds, the inventor of Linux, set clear standards against which proposed innovation could be judged. The source code for the programme was left open for innovators to examine and modify. This meant that information could be readily shared among many developers, but their efforts could be fairly judged against a publicly visible yardstick.

Central Design Authority: Torvalds kept control of the kernel of the Linux programme and was the final arbiter of changes to the kernel. The community had an authoritative leader.

In a closed system the life journey of ideas whether to affect public policy change or to get to market is highly "sealed," tending to narrow down possibilities, which then have to "squeeze" through a series of hurdles (feasibilities, criteria for investment) to then push on to the public domain. Open Source, by contrast, leaks out and opens possibilities. In comparing the two, one narrows down options and the other expands them and is dynamic. This is the creative milieu.

Who Owns the Creativity and Who the Innovation?

This different view changes how we think of intellectual property (IP). The debate over creative control has tended to go to the extremes. At one end total control, where every use of a work is regulated and the phrase "all rights reserved" is the norm. At the other end, anarchy rules where creators enjoy freedom but are left vulnerable to exploitation.

"Creative Commons" is working to revive the initial driving forces of a copyright system that valued innovation and protection equally based on balance, compromise, and moderation. They create private rights to create public goods and set free for certain uses. Like the free software and open-source movements, their ends are cooperative and community based. They try to offer creators a best-of-both-worlds way to protect their works while encouraging certain uses of them—to declare "some rights reserved." Effectively, there are three possibilities: free to use, shared use amongst a small group, or owned and accessed through rights, licences, partnering agreements, and contracts.

The Creative Milieu

Open Source creativity and networking provide a specific template to think through how a similar logic might work on a city level. Conventionally when we think of a creative milieu, we imagine an area of a city usually just off the center. The center is seen as staid, but in fact it may be the financial powerhouse of a city where prices are far too high for innovators and especially the young to base themselves. The off center site is where prices are somewhat lower and older former industrial buildings can be recycled into incubators and creative economy or artists studios that over time become desirable as the edgy crowd make the place safe and hip.

Then over time, so the logic goes, those that give the flair, attract the restaurants, bars, and alternative shops that then slowly go mainstream are pushed out. Then in move the urban professionals who hike up the prices, and so the process moves on with artists in search of new cheap space that in spite of themselves gets gentrified again. This has been repeated now several hundred times over the last 20 years from Granville Island in Vancouver to Newtown in Johannesburg to Fitzroy in Melbourne.

But that was then—there are now scarcely enough old buildings left to recycle and one cannot imagine the chic of the future trying to convert the new sheds, built with a shelf life of 20 years, into hip living space.

The time has come to rethink the spatial geography of creativity, remembering that Silicon Valley broke the stereotype of where innovation occurs. A soulless, spread-out place of multiple road networks of no distinction and no heart created a frenzy of innovation that has shaped how we live. With San Francisco nearby acting as a safety valve, the setting worked.

The old more technologically driven innovation of the Valley required computer wizards and scientists, not renowned for their artistic and sensory sensibilities. This model, though, is not working anymore. The beginning of the twenty-first century has finally seen a rapprochement between the two great ways of exploration, discovery, and knowing: art and science. The concept of Sci-Art encapsulates the desire to bring artists and scientists of all kinds together to work in structured environments on projects of mutual discovery from which it is predicted a raft of new inventions will occur. The Sci-Art concept is based on the premise that the most fruitful developments in human thinking frequently take place at those points where different lines of creativity meet. However, how does one attract artists to live in corporate sheds? What then are the places of creativity of the future and what do they look and feel like?

The Creative Economy

Most observers take for granted the increasingly well-researched and known facts about the growth in the creative economy over the last two decades. They do not need repeating here in detail. They generate jobs, income, and economic value, foster image, attract tourism, and help physical regeneration. They work in several arenas and have several roles. The three main domains are the arts and cultural heritage, the media and entertainment industries, and perhaps most importantly creative business-to-business services and the way they can add value to every product or service. In the latter sphere, especially in design, advertising, and entertainment the creative industries act as drivers of innovation in the broader economy and shape the so-called "experience economy." Technological excellence in the "experience economy" is no longer the only important feature in competition between goods and services. Increasingly, experience, feel, and lifestyle value decide about market success of goods and services. One important factor that I have discussed extensively in a publication called *Culture @ the Crossroads* (the short book is available for free download on www.charleslandry.com) is how the commercial sector is finding it important to bring the artistic imagination into experience based retailing and attractions as a lot of leisure based activity is seen as shallow. This collaboration can range from creating hybrid forms of retailing and in reverse new kinds of interpretation centers.

More broadly, the cultural environment is a magnet for the creative talent needed to generate innovations and for a region to be competitive. In the battle between hard and soft location factors, cultural climate, especially an open and tolerant one, is increasingly emerging as vital for the so-called creative class. This means that urban identity and reputation act as pre-condition for the creative economy and culture, art, and heritage provide cities with a stronger identity.

John Howkins and Richard Florida are two key protagonists in the creativity debate. They are in agreement with most of the discussion on creative cities, but each has his own focus. Their views reinforce thinking about the city, its economy, and spatial patterns in new ways. More specifically, Howkins examines how we make money from ideas and personal expression highlighting how being a creative society results in the greatly increased importance and value of Intellectual Property (IP) expressed through patents, trademarks, copyright, and design—in other words, variations of brands. Howkins looks at industrial sectors in his analysis of the creative economy, whereas Florida bases his definition and measurements on individual occupations. If creativity is to accrue commercial value it has to take a shape through an innovation and be embodied in a tradeable product. For Howkins the Creative Economy involves transactions in the creative output of the four "Creative Industries," namely:

> *Copyright Industries:* industries that create copyright as their primary product, e.g., advertising, computer software, photography, books, film.
>
> *Patent Industries:* industries that produce or trade in patents, e.,g, pharmaceuticals, electronics, information technology, industrial design, and engineering.
>
> *Trademark Industries:* widespread and diverse types of creative enterprises that rely on the protection of trademarks.
>
> *Design Industries:* widespread and diverse creative enterprises that rely on individuality in designs.

This depiction differs from how the British Department of Culture, Media and Sport categorizes the industries. According to this department, the creative industries are those that have their "origin in individual creativity, skill and talent and have a potential for wealth and job creation through the generation and exploitation of intellectual property. This includes advertising, architecture, the art and antiques market, crafts, design, designer fashion, film and video, interactive leisure software, music, the performing arts, publishing, software and computer games, television and radio."

These definitional distinctions are important because they shape and determine ideas, policy, and strategy. For example, the focus on trademark industries draws far greater attention to branding issues than does defining sectors in terms of radio or crafts. Importantly, the "industrial economy" and the "creative economy" differ in relation to *diminishing returns*: In the old economy, returns start to diminish particularly in three scenarios: when the scale of production increases; when the cost of inputs goes up; when returns become scarcer. In the creative economy, though, there are no limits: endless increasing returns are possible from production of more ideas and the subsequent innovation that generates more transactions. What types of places encourage such ideas? This question is a central concern of Richard Florida who focuses on "quality of place," when he asks rhetorically:

> *What's there?:* the combination of the built environment and the natural environment; a proper setting for pursuit of creative lives.
>
> *Who's there?:* the diverse kinds of people, interacting and providing cues that anyone can plug into and make a life in that community.
>
> *What's going on?:* the vibrancy of street life, café culture, arts, music and people engaging in outdoor activities—altogether a lot of active, exciting, creative endeavours.

The Creative Class

Richard Florida makes an important conceptual shift by focusing on the creative role of people in the "creative age." Companies now move to people and not people to jobs, he asserts. He is not discounting the importance of the company, but he argues that the economy is moving from a corporate-centerd system to a people-driven one. Cities need a people climate as well as a business climate. He notes the emergence of a new social group or "class"—the creative class, which are demographic segments of the population, and he develops indicators to measure the attributes of places that attract and retain the creative class, which in turn attracts companies.

The old way of thinking projected that people moved to jobs and that regions adopted investment attraction policies to entice companies to relocate to their region. Florida, though, reverses the argument, highlighting how companies and especially innovative companies offering high-value employment are attracted to regions where the creatives like to live. From this idea follows the question: Who are these people and what are the key attributes of a region or place that attracts the creative class?

The "core" of his creative class are people in science, engineering, architecture and design, education, arts, music, and entertainment. Perhaps 12% of the population in the United States comprise this group. Around them is a broader group of *creative professionals* in business and finance, law, health care, and related fields. This group accounts for roughly another 18% of the U.S. citizenry. Since 1980 this total group has doubled.

Florida then found strong correlations between tolerant and diverse places and economic growth leading him to conclude that development is driven by lifestyle factors. Florida developed a series of interesting and contentious indices to compare regions and cities, such as the creative class index, which measures the percentage of people employed in creative class positions; the High Tech Index—based on the percentage of national high-tech output and percentage of a region's output that comes from high-tech; the Innovation Index measured as patents granted per capita; the Talent Index measured as percentage of people with a higher degree or above; the Gay Index, a measure of over- or under-representation of coupled gay people relative to the nation as a whole; the Bohemian Index calculated similarly to the Gay Index based on occupations such as authors, designers, musicians, composers, actors, directors, painters, sculptors, artist printmakers, photographers, dancers, artists, and performers; the Melting Pot Index, which measures the relative percentage of foreign-born people in a region; the Composite Diversity Index, which combines the Gay, Bohemian; and Melting Pot Index; finally there is the Creativity Index with a composite measure based on the Innovation, High-Tech, Gay Indexes, and the creative class.

Urban development, then, should be seen as much more of an organic process that is never finished as the creatives want to feel they are part of the making, shaping, and co-creating of the unfolding setting. Creative places have potential, but they are not complete. The dynamic is less from top to bottom and less based on pre-digested and pre-chewed experiences that MacDonald's, Subway, or Disney, and other chains may offer. Similarly, traditional entertainment centers, regional shopping malls, or stadiums with acres of parking lots that neutralize the surrounding space are out. The creatives yearn for "authenticity," the new mantra of the current age. *Authenticity* and uniqueness are difficult words, but in essence each refers to self-definition, thus suggesting that each place needs to develop its own capabilities, ideas, culture, and traditions. This movement empowers the locals, who may of course be global in outlook. Kjell Nordström and Jonas Ridderstråle (1999) have beautifully articulated this scenario in *Funky Business:* "The 'surplus society' has a surplus of *similar* companies, employing *similar* people, with *similar* educational backgrounds, coming up with *similar* ideas, producing *similar* things, with *similar* prices and *similar* quality."

Art and the Artist

The dominant values and attributes that are responsible for the malaise of the modern world and narrow conceptions of efficiency and rationality are almost *diametrically opposed* to the values promoted by artistic creativity. The prevailing worldview is characterized by words such as goal, objective, focus, strategy, outcome, calculation, measurable, quantifiable, logical, solution, efficient, effective, economic sense, profitable, rational, and linear. These words reflect the contrast between this approach and the alternative worldview that gives artistic creativity its power.

The quintessence of the arts is artistic creativity. What is unique about artistic creativity or its distinct attributes in the context of the city? What human values does it embody and share with others so that it is capable of having deep significance for individuals, communities, cities, and even, over time, for history? Do the arts shape the creativity of cities? Can the arts "re-anchor" humankind, "knit together what has been ripped apart"?

At its best artistic creativity involves a journey not knowing where it will lead or if and how one will arrive; it involves truth-searching; it has no calculated purpose; it is not goal-oriented, not measurable in easy ways, nor fully explicable rationally—its outcome can be mysterious; it has no quick or easy solutions; it denies instant gratification; it accepts ambiguity, uncertainty, and paradox; it endures the tedious and repetitious so as to reach mastery; it contains loneliness and the potential for failure; it recognizes that something beyond the rational such as the arational or a soul exists; it can offer glimpses of the (non-supernatural) sacred; it gives the spirit a connection outside itself; it originates in the self but aims to create work that enters the common space of humanity; it proclaims that humans have the right to pursue freedom and urges confidence in exercising that right; it inspires others to be brave and to risk failure; it generates openness to new ideas and new ways of doing; it lives in the "now"—it takes place in the moment; it is transgressive and disruptive of the existing order; it is often uncomfortable, even frightening.

What is it that is special about artistic activities: singing, acting, writing, dancing, performing music, sculpting, painting, designing or drawing, in relation to developing cities? The arts use the imaginary realm to a degree that other disciplines, such as sports or most of science, do not. Those are more rule-bound and precise. The distinction between involvement in arts and writing a computer program, engineering, or sports is that the latter are ends in themselves in that they do not, or very rarely, change the way one perceives society, and they tend to teach specific outcomes.

The arts have wider impacts by focusing on reflection and original thought; they pose challenges, and on occasion they want to communicate. Turning imagination into reality or a tangible object is a creative act, so the arts more than most activities are concerned with creativity, invention, and innovation. Reinventing a city or nursing it through transition is a creative act, and thus engaging with or through the arts helps this endeavor. In other words, participating in the arts uses the imaginative realm to a heightened degree.

Engagement with arts combines stretching oneself and focusing, feeling the senses, expressing emotion, self-reflection, and original thought. The result can be too, if co-ordinated well, for the arts to help build a story of a place. It does this by broadening horizons to convey complex ideas and emotions iconically with immediacy and depth, to nurture memory, to see the unseen, to learn, to uplift, to anchor identity and community or by contrast to stun, to shock by depicting terrible images for social, moral, or thought-provoking reasons, to criticize, to create joy, to entertain, to be beautiful. The arts can even soothe the soul and promote popular morale. At its best, art on occasion can lift us into a higher plane beyond the daily grind, to a realm that people call spiritual. These activities highlight the role of the arts in tapping potential.

The arts help cities in a variety of ways. First, with their aesthetic focus, they draw attention to quality and beauty. This feature affects the evolution of urban design and architecture. By using imagination and lateral thinking the arts can help invigorate other disciplines such as planning, engineering, social services, or the business community, especially if allied with other emphases such as a focus on local distinctiveness. The arts challenge us to ask questions about ourselves as a place. "What kind of place do we want to be and how should we get there?" The arts ask us to create a story about the places we live in. They challenge us to describe where we are going and how we might get there. Second, arts programs can challenge decision makers by undertaking uncomfortable projects that force leaders to debate and take a stand. For example, an arts project about or with migrants might make us look at our prejudices and find ways of living together better. Arts projects can empower people who have previously not expressed their views, so artists working with communities can help consult people. For example, a community play devised by a local group can tell us much more than a typical political process. Lastly, arts projects can simply facilitate joy and enjoyment.

Artistic Thinking

More important than perhaps the artist or the arts is encouraging artistic thinking in individuals. It draws on understanding culture and the importance of art and is the ability to focus on the elemental factors of life: what we see, what we hear, what we smell, to listen to deep emotions, fears and delights, to sense and understand materials and the material world around us and the feelings they engender. At its best, it is then able to pull out the important and interesting ideas and to make lateral connections, thus building a mosaic of ideas or projects that capture the quintessential and significant for people. It can even plunge us deeper into a realm where we feel more in touch and in tune with larger life forces, where the "me" and the "we" somehow merge and individual wants and collective needs coalesce. Aligning these two elements is perhaps the central challenge of creating human settlements. The solution or challenge might be physically expressed as in how a building looks, or in an activity like a performance, or how a process was undertaken like a consultation exercise to unleash the creative contribution of local inhabitants.

This way of thinking is not the exclusive domain of artists, although it is legitimized in that field and therefore more common. At its core, it involves an attitude and openness to all the phenomena in existence, an openness anyone can have. Looking at the world in this manner is particularly important for highway engineers or planners, for example, because it directs them on the human dimension and impact of what they do. It is different from the technocratic approach to place making, which has many virtues, but which breaks down the world into ever smaller specialist functions, disciplines, and compartments that lead us to lose appreciation of how things hang together.

In summary, creative people are integral to the creative organisation, which is in turn a pre-condition for the creative economy, then the creative community, and ultimately the creative city, all of which are driven by creative learning and education rather than technology. Education is central to the creative milieu. In other words, creativity must be treated as a form of capital and needs to be fostered, measured, and managed like any other asset. In addition, creativity needs embedding into the genetic code of how places and organizations operate—it needs to be systemic. This concept involves a change in organizational cultures where unnecessary controls are reduced; reward systems are introduced that promote creativity and new perspectives on success, efficiency, and failure are instituted. As Florida and others note the public sector and companies are set up to "squelch" creativity, given their command and control model. Instead decision-making should be decentralized so people's potential is unleashed and harnessed.

Finally, as a consequence organizations can become more flexible, shifting in part from hierarchies to networks that operate on trust and so can respond faster to change, including rethinking the organization from its budgetary to administrative controls. All these elements may facilitate the emergence of a creative milieu. This way of thinking is different from looking at creativity or innovation as if it were a pipeline that starts with research in large laboratories and is fed through to manufacturing and sales, with an emphasis on hiring bright people, investing in research, keeping knowledge within the company, and protecting intellectual property through patents.

History shows economic power translates itself into cultural power in very complicated routes, and we suddenly find things interesting that may have bored us before. This phenomenon occurred, for instance, in the 1960s when U.S. art was suddenly deemed to be "it," with Jackson Pollock, Rauschenberg, Oldenburg, or Warhol as its exponents. Were they intrinsically better than other artists around the globe? As the world turns its axis eastwards, the same is happening with Fang Lijun, Cai Guo-Qiang, Zhang Xiaogang, or Zhang Huan. How then do we distinguish what is compellingly good from what we think should capture our interest?

Clustering and More

We know the arguments for clusters: mutual financial, technical, and psychological support, increasing the efficiency of markets, bringing together buyers and sellers, creating overlaps between adjacent disciplines or accessible centres of excellence and stimulating competition, so generating "multiplier" effects, synergy complementary interchanges, and swapping of resources.

Clustering of talent, skill, and support infrastructures is central to the creative economy and creative milieu. Clustering as a concept is not new—the convenience derived from being close to labor, expertise, suppliers, and being able to share information has been obvious since trading began. Early examples of clustering were the industrial revolution and the development of towns specializing in a product. However, with real and virtual worlds coalescing, the spatial geography of creativity and clustering are changing. Two events are happening simultaneously. Yes, people need other people to bounce ideas off, to generate enthusiasm, and to form teams to tackle projects especially, in areas where continuous innovation is important. They need to connect, disseminate, collaborate, and have ready access to creative skills, suppliers, and customers. Spatial clustering reduces barriers to entry for start-ups because there are so many opportunities nearby, and word of mouth and viral communication proliferate.

When we communicate electronically, the sheer mass of communication is huge, yet the small proportion of people we actually want to meet is greater than the number of persons we met in the pre-electronic age. However, digital clustering whilst widespread in a number of communities of practice still has a long way to go before it becomes compelling.

A Transition Point . . . .

Behind this seemingly crushing inevitability toward becoming the creative age lurk dangers and dilemmas, and these need to be precisely understood in order to make decisions about what might be right, relevant, and full of impact. Before focusing downward, we need to understand the bigger picture within which they are enfolded.

Coping with an Old Intellectual Architecture

The key texts on the emerging age agree, adapting Einstein's dictum, that if one looks at the world within the mindset that created the opportunities or problems we worry about, one will only replicate those solutions or problems: The mind that created the problem is unlikely to be the mind that solves them. Our intellectual architecture was constructed for the age of industrialism and has embedded itself into our minds like a cityscape of familiar streets and buildings which we simply take for granted. Since such mental architecture gets out of date, it causes a particular set of conundrums and strategic dilemmas when we try to apply it to the emerging world. Furthermore, we attribute incomprehension to "complexity" rather than revisiting and questioning the appropriateness of that mental architecture. However, each generation claims its age is more complex than the previous one. What we really mean is, "This is a pattern of events that I do not understand."

There are numerous projects around the world, for instance, in urban development that continue to be made with the mental landscape of a bygone industrial age, and inevitably the results have a factory feel. Consider, for example, the new workplaces in hospitals, schools, or universities. Most are still designed with long corridors with cells on each side, or there are communal work rooms where brain workers sit in their serried ranks—and this in a world of the ideas economy. There seems little recognition or understanding of how to cocreate environments where healing oneself or learning could occur. This requires completely new relationships between the expert and the citizen or user. The same applies to services that tend to infantilize individuals. All public services could be rethought, and the artistic imagination has a role to play. The British Design Council's Red Initiative of co-creating health services is an example of the way forward.

Another example is how we look at economics and its spatial implications in terms of retailing, streets, and urban environments. The "theory of the long tail" breaks up traditional economic theories of diminishing returns. It applies to companies like Amazon or Wikipedia. Chris Anderson, describing its effects on current and future business models in *The Long Tail: Why the Future of Business Is Selling Less of More* (2006), argues that products that are in low demand or have low sales volume can collectively make up a market share that rivals or exceeds the relatively few current bestsellers and blockbusters—if the store or distribution channel is large enough. As an Amazon employee noted: "We sold more books today that didn't sell at all yesterday than we sold today of all the books that did sell yesterday." In the similar sense, the user-edited Internet encyclopedia Wikipedia has many low popularity articles that, collectively, create a higher quantity of demand than a limited number of mainstream articles found in a conventional encyclopedia.

The Speed Regime

With any dynamic there are leads and lags, yet when change moves fast people, organizations, and governments can be many iterations behind and be lost. Indeed, speed is a key driver of the emerging world, and we need to make a brief detour as speed determines much of the way we do things and its resulting content. In comparison to 1974, 16 times as many products are produced and the population has only doubled, which means we need to consume eight times more. The consuming logic that is never fulfilled means people want to experience more, perhaps 30 hours of experience in a 24-hour day. More items are offered, but only the same amount of time is available. In our desire not to waste time, we are left with even less of it. Speeding things up means substituting quantity for quality, and along the way a certain depth to life is lost. Travel is faster, communicating electronically is faster, eating has become faster, lunch breaks are shortening, with little time for eating, let alone digesting, getting to know people, and relationships are speeded up through speed-dating, the length of time we keep clothes has shortened, and disposability is key, the shelf-life of buildings is shorter, room decorations can be bought off the peg, and discarded at will. This pattern signifies the throwaway city. With everything speeding up, the very fast instant response is needed to capture the blips of experience racing past. To grab attention, the power of advertising comes into its own, and increasingly the city is an interactive billboard that spectacularizes experience, just as we spectacularize products and services. The city then becomes more like a performance and art work using design, visuals, and theatrics along the way. Remember, too, that in urban settings today we see as many images in a day as our forebears did in a lifetime in the Middle Ages. This reality creates its counter-movements such as slow cities.

The Complicated and the Complex

A conceptual framework is proposed to unscramble this mush of trends, fads, fashions, ad-creep, and speed-experience that often fly in different directions. It seeks to distinguish the deep from the shallow, the superficial from the meaningful, the strategic from the trivial and to understand timelines and connections as well as the differential rates of change and the nature of their impact. This perspective, though, is far more extensively covered elsewhere.[2]

Escalating change is in evidence. The shift of the global axis toward the East is a primary example, changing global terms of trade another, and growing global disparities a third, not to mention climate change, pollution, and the growth of fear culture.

How do you unlock and understand the change elements that interweave, interlock, and reinforce? To begin, recognizing the distinction between the terms *complicated* and *complex* is useful.[3] The complicated is essentially mechanical, whereas the complex essentially relational. "Complicated" is about acting *on*. "Complex" is about acting *with*. "Complicated" is appropriate to a world of predictable outcomes. "Complex" must acknowledge and respond to uncertainty. Putting a rocket on the moon is complicated. Indeed, an enormous number of detailed steps are involved from engineering to navigation. Building a series of numbered housing blocks with units is complicated; building a road or motorway is, too. Anything to do with people is "complex" like raising a child. We learn and adapt from day-to-day experience, and we coevolve in relationship to one another. Understanding what it feels like to live in uniform housing blocks and to predict emotional outcomes and reactions of that experience is complex as is appreciating the effects of the dulling monotony of endless roads and asphalt. However, we think about the future with the mindset of the merely complicated.

Embedded Assumptions of the Past

Some trends are deeply etched. Say, the nexus of emancipation built around individuality, choice, and independence spilling out from the Enlightenment has been with us for some 250 years. Some feel this motor of change is at the edge of exhaustion: Its self-focused energy is causing more negative than positive effects, but it still has enough energy to shape everything from how politics appeals to constituents to how we customize products and services and generate wants: Who would have thought ten years ago that we deeply needed an iPod?

There is little doubt that a realignment of individual desires and a broader public purpose is in the offing. The environment is just one example. With an incentives framework in place, thousands of products and services wait to be invented at the right cost to wrench our habits and behavior in a more sustainable direction. We now know that individuals pursuing personal wants do not add up to a harmonious whole.

Another trend is the renewed vigor and degree of globalization facilitated by IT that both makes operating across boundaries easier and has helped shift global terms of trade. In the context of cities, it makes operating globally an imperative for success. Such deeply embedded trends are like the air we breathe, so we can take them for granted or forget they exist. However, it does not mean they will not have considerable further effects. They will continue to affect urban lifestyles, social and economic structures, policies and choices, and creativity potential. The significant issue is where the continuities and especially discontinuities are likely to fall.

The Crux of Everything

The central dilemma of our age is how we live together, and most of this dynamic plays itself out in cities. It is the goal of civilization. Avoiding the "clash of civilizations"[4] should be one overarching intent of politics.

It focuses us on considering boundaries, barriers, or borders within cities, such as ghettoes, voluntary or imposed, and between cities, cultures, and countries. The urge to identity, self-definition, and broader belonging is everywhere, even a punk wants to be the same and different simultaneously. It draws attention to our tribal tendencies and our insider/outsider instincts, as well as how we claim territory. This might be physically as when gangs occupy an area or by distinguishing ourselves from others through lifestyle choices or making people like the homeless feel like outsiders. It challenges us to ask how porous we are as individuals or cities whilst still feeling confident about who we are. In a world coming closer together virtually and in real time, it becomes crucial to assess more what we share as common citizens of the world rather than what divides us. This is the intercultural agenda, and it is not to claim some cosy togetherness, but rather to stress how we negotiate conflicts and be *together in difference.*

The Paradox of Risk and Creativity and More

A *paradox* is an incongruity that seems to be contradictory or an outcome that is different from that envisaged. In a world of paradox, the counterintuitive comes to the fore. Shaping everything we consider is the major paradox of our times: The simultaneous rise of a risk culture and the creativity agenda with risk avoidance strategies often cancelling out inventiveness. Creativity, openness, and risk taking are demanded of us to solve problems or to be competitive in a globalized world and to be inventive to adapt to twenty-first century needs. At the same time creativity is denied.

We are caught between a rock and a hard place. How in this context can we discuss creative cities or the role of artists or the creative economy with its tendencies to push at boundaries and into the unknown? The evaluation of everything from a perspective of risk is a defining characteristic of contemporary society. Risk is the managerial paradigm and default mechanism that has embedded itself into how companies, community organizations, the public sector, and most cities operate. Risk is a prism through which any activity is judged. Risk has its experts, consultants, interest groups, specialist literature, an associational structure, and lobbying bodies. A risk industry has formalized itself. It is a growth industry. Hardly a day passes without some new risk being noted. It is as if risk hovers over individuals like an independent force waiting to strike the unsuspecting citizen. The notion of *accident* seems to have gone from our vocabulary. Cleansing the world of accidents means scouring the world for someone to blame.

The notion of risk subtly encourages us to constrain aspirations, act with overcaution, avoid challenges, and be sceptical about innovation. It narrows our world into a defensive shell. The life of a community self-consciously concerned with risk and safety is different from one focused on discovery and exploration. The mood of the times becomes *averting the worst rather than creating the good* and guidelines are drawn up on worst case scenarios. The media shape perceptions of risk creating a climate that disposes us to expect bad outcomes. It heightens dangers like food poisoning, which is far less risky than the risks caused by sedentary lifestyles encouraged by urban planning that reduces walkability in cities. Other paradoxes worthy of note in the context of cities are listed below.

Calculating tangibles in a world of intangibles. We live in a "weightless economy" or an economy of ideas, where 80% of wealth is created through intangibles. However our systems of measurement and the calculation of value are out of step with realities. Accountancy systems remain largely focused on measuring assets as material entities. People, who as idea generators create most value, are by contrast treated as an accounting cost, even though in the sale of a company they can be part of its goodwill. How does one assign cost to the value of intangibles like insight generated through intuition?

Accessibility and isolation. Can there be too much access? Being swamped with cascades of uncontrolled information that are impossible to filter is a well-known problem. Accessibility is deemed an unquestioned good, yet too much accessibility can destroy what it sets out to do. Unfettered access can make goods too popular or bring matters into reach too easily. Places can be overwhelmed by popularity fired by accessibility and mass mobility. The critical mass tips from being "just right" to being "out of control." A heritage setting can inspire and generate welcome tourism. However, if too many visitors appear, they can drain the lifeblood out of local identity. The result may be that a city's future is determined by the nostalgic past that visitors want to see, but which residents do not need, with knick-knack shops, souvenir outlets, and interpretative centers that gel the past into aspic.

Porousness and identity. People and cities need to be porous to new influences to thrive as well as to retain their identities. Places need to be both local and global to survive, selectively open and closed at the same time: Boundaries and borders to ground and anchor identity as well as bridges to connect to the outside. Identity is shaped by factors from upbringing and friendship networks to work. Crucially, it is also rooted in geography and place. In spite of increased mobility, a sense of place remains a core value and often acts as the pivot point around which a person acts. Applying this concept to cities means that they need to balance being parochial and cosmopolitan.

Space and density. People want space and density at the same time. Some will want both; others one or the other. With space at a premium, it will become the benchmark of luxury. Perceived lack of space will drive location decisions, lifestyle choices, densities, and technological development. Systems to optimize and manage physical and virtual space from autonomous vehicle control devices involving smart card technology will emerge. Simultaneously, and in a seemingly contradictory way, densities will increase as the number of households rises, and urban vitality is deemed to come from close-knit mixed uses, so shaping the look and feel of cities.

City and country. The more we move to the country, the less like countryside it will become. The overwhelming majority of the British, for example, want to live in the countryside, exacerbating the intense pull out of urban areas, putting pressure on market towns and villages whose formal integrity will be blown apart by in-fill, edge developments, and rises in population. It will all merge into a built-up mass. The overall feeling will be of many highways connecting some settlements rather than many settlements connected by some roads. The battle between perceived urban and rural values will surely get worse.

Age and technology. For the first time in history the young become the translators of the modern world for the old. The capacity to handle technology is a form of power, and the young feel more comfortable with it than older generations. Technological change drives the economy, thus transforming the power relationships between generations. We already know that children teach parents how to use videos, e-mail, and the Internet: In a global culture where age has engendered respect, what will technology do to social relations when older people feel increasingly disenfranchised? For some older people, there is a growing sense of being an immigrant in their own country. East Asian cultures perhaps venerate the old more than Western ones, and what will be the consequence? These overarching premises then may trigger the thinking about fault lines, battlegrounds, drivers, and strategic dilemmas.

Fault Lines

Fault lines are change processes, global in scope and cross-culturally relevant, that are so deep-seated, intractable, and contentious that they shape our entire worldview. They determine our landscape of thinking and decisions across multiple dimensions and being global in scope, they affect our broadest purposes and ends. They may create insoluble problems and permanent ideological battlefields. Even if they eventually solve themselves, such problems are likely to linger for a very long time: 50 years, 100 years, or more, and then they become more a question of mediating and managing conflict. They affect the dynamics of places and what they feel and look like and the scope of creativity required.

The five most important fault lines are the battles between faith-based and secular worldviews, battles between the rational, irrational, and a-rational, between environmental ethics and economic rationality in running countries or cities, between the artificial and the organic and realigning individualism with collective good. These different forms of dynamic affect a mass of downstream decisions. Taking the first, the most obvious aspect at a global level are the varieties of religious fundamentalism. Fundamentalists are responding to disappointments with material progress and consumerism that neither makes us happy nor answers genuinely fundamental questions such as, "What is life for?" What in this context are the agreements that bond and anchor communities when fundamental views of the world are so different and people with diametrically opposing views now live in the same place—a city, a neighborhood, a street?

The rational, irrational, and arational. A big put-down occurs when a logical rationalist claims someone with whom he or she does not agree is irrational or arational. Being arational is not to be irrational (that is, to act without reason). It implies instead acknowledging that a narrow rationalist, linear approach is not the answer to inextricably interwoven issues where to untangle the threads involves thousands of variables. Being arational is being full of reason and openness. It implies the belief that an imaginative leap in thought can occur; that very deep instinct exists; that there are higher registers of understanding, knowledge and insight, some of which may remain intuitive for a long time. The arational mind sees matters less as a machine or a defined structure and more as an organism that evolves and is emergent, where the seeming randomness is not mindless. It can be intuited from within a higher pitch. The arational person understands the principles of connections and processes and is not scared of emotion, believing instead it is a source of great value. The narrow rationalist eschews emotion and so misses out and makes decisions without sufficient knowledge and insight. Artists more than many other professionals feel comfortable in the realm of the arational. Perhaps indeed this is the arena in which they can give most back to the city of creativity.

Environmental ethics and economic rationality. The rise of environmental ethics is a sustained challenge to an economic rationality increasingly regarded as an impoverished theory of choice-making as it implies "the invisible hand" in the longer run equates to public good. Its central fault is that it assumes the environment is a free exploitable resource. Eco-efficiency on its own is a small part of a richer web of ideas and solutions that require a fundamental rethinking of the structure and reward system of commerce. This rethinking implies combining creativity and innovations in business practice and public policy to develop a regulatory and incentives regime attuned to encouraging resource efficiency.

The more urban we become the more we hanker after the wild, the untamed and un-explored, the undisturbed. This desire mirrors the divide between culture and nature or that made by humans and matter that pre-exists. It mirrors too the urban/rural split. Put differently it is the clash between *the artificial and the organic.* The urban stands for the rational, the logical, the instrumental, and the constructed. Thinking driven by the urban mindset appears to those on the opposite side of the fence as lifeless, lacking in understanding of natural cycles, seasons, forces, and rhythms. The divide expresses itself in many manifestations contrasting *the fast and the frenzied* with *the simple and the slow.*

The final fault line is the struggle to *realign the individual* and *the collective* in twenty-first century terms. Many feel individualism has gone too far. Expressed differently, it is about how much we take or how much we give. The choice of individuals as trumpeted has largely reduced people to consumers with a parallel loss of what it might mean to be a citizen.

Are there fault lines missing? Is the difference between male thinking or female thinking one, especially in a world that probably requires much more soft creativity? Soft ceativity is the ability to understand and work with relationships and emotions and to appreciate the fragility of things.

The search for authenticity is perhaps another fault line, and, if so what does authenticity mean? Is the fate of the "fragile" and the weak in power-driven societies yet another one? Is the fight for intrinsic values a third in a world dominated by instrumental reasoning and by the ideal of measurement?

This combination of unresolved fault lines makes people experience a lack of meaning in their lives, an emptiness, a void, and a desire for fullness that is filled often materialistically but does not provide satisfaction. Private life becomes more important and civic life (and the city always stands for civic life) atrophies. When life is moving fast it "spins out to a rationalization that the average citizen is accomplishing a great deal simply by coping with or even surviving in this modern milieu never mind being expected to assume responsibility for civic engagement and concern."[5]

There is a yearning for completion, being at one, having a sense of wholeness that might result in fulfilment. That desire is filled in various ways: religion, ritual, spirituality, internal mediation, or meditation. Ultimately these seemingly abstract concepts are expressed in the city. It might be a place of worship, an urban festival, or the way a public space is laid out. Artists again more than others make it their business to know about the oscillations of emptiness and fullness. They remind us too of the human tendency to flip between needing anchorage and wanting exploration. Seemingly contradictory but sensible, it highlights the desire for stability and familiarity and the constant strive to experience the new. Resolving the contest between new experience, and familiar habit, and the fixed surroundings creates cultural identity.

Battlegrounds

Discussions and policy debates around the fault lines discussed above often become *battlegrounds* because the nature of debate is intense and contested. However, there are other battlegrounds less concerned with ultimate purposes, although at times they touch on them. They are usually about significant policy choices and thus more concerned with pragmatics. Each battleground has implications for the future of places. We elaborate briefly on a few. They include the following.

Creativity: Superficial or deep. With the conflict between the demand for "creativity" and how, consciously or unconsciously, we undermine creativity in the education system, in city planning, in economic decision-making, and elsewhere. Allied with this concept are the superficial ways in which society "values" creativity—as style, as fashion, as edge, as a controversy for its own sake: they are attributes without substance.

Authenticity* versus *global markets. The contrast between the real, the virtual and the fake will move into a new gear. The search for the authentic, distinctive, and the unique has become pervasive as our sense of the "real" and the local is dislocated by virtual or constructed worlds, such as those of cyberspace and theme parks and standardized, global mass products with little link to a particular place. Related to this dynamic is the battle between chain store power and its homogeneity and locally distinctive shopping. Once basic facilities exist, it is difference not sameness that contributes the most.

Multiculturalism* versus *interculturalism. In the multicultural city we acknowledge and ideally celebrate our differing cultures and entrenched differences. In the intercultural city, we move one step beyond and focus on what we can do together as diverse cultures sharing space. The latter leads to greater well-being and prosperity, yet funding structures are usually predicated on the first.

The technology fix or behaviour change. Will the regulatory and incentives regime at differing levels (city, state, nation) be constructed to encourage behavior change through recycling, using renewable energy resources, energy efficiency, or will it just be left to the market to produce new technologies?

Social equity* versus *disparity. Can cities bend markets to broader social needs given that the inclusion and empowerment agenda will remain with us?

Central* versus *local. The battle between central and more localized power is ever present. Can cities be creative when central governments hold power and often stifle creativity?

Compaction* versus *dispersal. Density creates a better urban fabric since it results in viable activities born of the increased vitality and economic efficiency that sprawl dissipates. Can cities counteract decades of city-building and habits that encourage sprawl? Is sprawl good for creativity or density?

There are other battlegrounds, such as: ***The holism* or *specialism focus.*** Do you see issues, such as urban decline or how cities as a whole operate, as being composed of interacting wholes that are more than simply the sum of the parts? By contrast do you look at the fragments within narrow forms of specialism? Is the creativity of cities measured by *hard or soft indicators.* Will hard indicators such as levels of employment, growth, income, or GDP suffice? What about soft factors of competitiveness such as a city's networking strengths, governance capacity, cultural depth, its creative milieus, and its atmospherics? How do we know where a city stands when these are not measured?

The fast and the slow. Competitive pressure with IT as an enabler speeds up life and makes it shrill to the extent that the slow is increasingly desirable. From the slow cities movement to the "clock of the long now" that will chime only once every 1,000 years, people are trying to avoid existence becoming a whistle stop tour through life and then you are dead. This connects to another battle line between always emphasising *the next or the past* with the futurists fighting the nostalgics with rarely anyone *living in the present.*

Drivers for Change

I have deliberately not discussed possible futures by using "drivers for change" as the template. The basic drivers who determine much of what will happen are known. However, they tell us little about the depth or severity of a change process, its possible timeline, and deeper movements. For this reason fault lines, battlegrounds, and paradoxes were discussed first.

The core drivers include *demographics* and especially an ageing population in the West with inward migration balancing out expected skills and job shortages; *globalisation,* which will move forward unabated; shifting *global terms of trade* inexorably to favor the East; technology transfer periods which will reduce dramatically; and *climate change,* which will hasten the end of the oil economy and speed up the search for energy alternatives.

New issues will rise to the fore and shape urban decision-making. These include the health and urban design agenda, which will push debate on city-making more toward the New Urbanism agenda, and in this process public health and urban design will come together.

Safety, surveillance, and a public realm. Safety and responses to terror will determine how cities are built and managed. The watchful eye of surveillance will be with us wherever we go in cities. People will choose to live in voluntary physical ghettoes, and gated communities will proliferate. Indeed, they parallel the mental ghettoes they create to block out a seemingly uncontainable world. Therefore, the fake experience is easier for many to cope with than reality.

Time and the spectacular. People increasingly perceive themselves to be "time poor," and yet dream of being time and experience rich. The commercial sector will respond and increasingly seek to make all experiences, especially leisure activities, more intense and spectacular in an attempt to give them greater impact and meaning. This trend will affect design, especially for shopping, culture, and education facilities. The same is true for public authorities who will increasingly feel they need to play the game of "urban iconics," throwing up ever more spectacular buildings to catch attention. Additionally new, more invasive "spectacularizing" technologies will emerge as knowledge from brain research cascades down into commercial applications, giving rise to neuro-marketing.

Crucially, *pre-existing decisions, dominant ideas, and mindsets* are the forgotten drivers, rarely if ever mentioned. What shapes present decisions more are the decisions that have preceded them and the intellectual architecture of those that make them. Pre-existing decisions, such as those that have resulted in the houses, shopping malls, roads, and industrial sheds already built, are significant determinants of the future look and feel of the city, narrowing the range of alternative choices. The shape, style, and form of the future city is in essence embedded in the laws, regulations, codes, guidelines, and plans of the present.

One dominant idea and mindset challenged by artistic creativity is the central idea of our civilization, which is the notion of the business logic, efficiency, and economic rationality. It has significant merits, but it does not tap the complexities of human behavior. Its ideas provide the warp on which the patterns of our behavior are encouraged to be woven. It affects the language we use and the discourse of public affairs. It entraps us, however much we talk of "thinking outside the box." Cold economic logic is coupled with the rise of managerialism with its colourless, grey, neutralised language of process that has little flavor or energy. Not surprisingly, civic engagement and connection are in decline.

The managerial logic spills over into other domains that traditionally worked on different principles such as ethics, morality, justice, voluntary work, and the idea of the public. However, discussion of such concepts is now shaped by the language of "efficiency." When efficiency is itself the end, it strips out other life values, creating as many problems as it solves by promoting short-term thinking.

Are we, some ask, in the dying days of the Enlightenment, which began by allowing a space for the spirit and imagination, but "rationality" and efficiency have now pushed out mystery, ambiguity, and paradox, treating these, if at all, as frivolous or disorderly.

References

Centre for Creative Communities. (2004). *Creative community building through cross-sector collaboration.* London: Centre for Creative Communities.

Chris Anderson The Long Tail: Why the Future of Business Is Selling Less of More, Pub: Hyperion (2006).

Coyle, D. (1998). *The weightless world: Strategies for managing the digital economy.* Cambridge, MA: MIT Press.

DCMS. (2004). Definition of creative industries. Department of Culture Media & Sport. UK: Retrieved from http://www.culture.gov.uk/creative_industries/default.htm

Florida, R. (2002). *The rise of the creative class: And how it is transforming leisure, community and everyday life.* New York: Basic Books.

Hargreaves, D. H. (2003). *Education epidemic: Transforming secondary schools through innovation networks.* London: Demos. Retrieved from http://www.demos.co.uk

Howkins, J. (2001). *The creative economy: How people make money from ideas.* London: Penguin Books.

Huntington, S. P. (1998). *The clash of civilizations and the remaking of world order.* New York: Simon & Schuster.

Kalantzis, M., & Cope, B. (2008). *New learning: Elements of a science of education.* Cambridge: Cambridge University Press.

Kelly, K. (1999). *New rules for the new economy.* London: Fourth Estate.

Landry, C. (2000). *The creative city: A toolkit for urban innovators.* London: Earthscan.

Landry, C. (2006). *The art of city-making.* London: Earthscan.

Landry, C., & Bianchini, F. (1995). *The creative city.* Bournes Green: Comedia/Demos.

Landry, C., & Wood, P. (2007). *The intercultural city: Planning for diversity advantage.* London: Earthscan.

Leadbeater, C. (1999). *Living on thin air: The new economy.* London: Penguin Group.

Mauzy, J., Harriman, R., & Arthur, K. A. (2003). *Creativity inc: Building an inventive organization.* Boston, MA: Harvard Business School Press.

Mayor's Commission. (2002). *Creativity—London's core business.* London: Greater London Authority, September 2002. Retrieve from http://www.creativelondon.org.uk/commission.

Nordström, K., & Ridderstråle, J., (1999). *Funky business: Talent makes capital dance.* London: FT Prentice Hall.

Porter, M. E . (1998). Clusters and competition: New agendas for companies, governments and institutions. In *On competition.* Boston: Harvard Business School Press.

Seltzer, K., & Bentley, T. (1999). *The creative age: Knowledge and skills for the new economy.* London: Demos.

The Adaptive State. Authors: James Wilsdon, Tom Bentley; Demos; Publication Date: 2003

Notes

1. Thanks to Uffe Elbaek from Kaos Pilots, who made the point about "for" and "in" in relation to his organisation's goals in education.
2. See Charles Landry, *The Art of City-Making* (London: Earthscan, 2006).

3. Thanks to Colin Jackson for pointing out Eric Young's speech at Policy Learning and Distributed Governance: Lessons from Canada and the UK 5th June 2003, quoting Brenda Zimmerman
4. Samuel P. Huntington, *The Clash of Civilizations and the Remaking of World Order* (New York: Simon & Schuster, 1998).
5. Joy Roberts, private note, Musagetes Foundation.

18

Innovate, Innovate! Here Comes American Rebirth

Jan Nederveen Pieterse

General problems that form the backdrop of the contemporary emphasis on innovation are postindustrial society, globalization, and the financial crisis. Postindustrial societies face the challenge of how to manage the application of labor-saving technologies and secure sufficient employment opportunities. With contemporary globalization comes the trend toward offshoring production and outsourcing services to low-wage zones and countries where levels of skill, education, and infrastructure are increasingly competitive with advanced countries. The financial crisis from 2007 to 2009, especially if it is understood as an economic crisis and the implosion of an accumulation model, carries broad implications. These problems are particularly salient in the case of the United States. Can contemporary innovation meet these challenges?

"Innovation can give America back its greatness," according to Jeff Immelt, CEO of General Electric. "This downturn is not simply another turning of the wheel but a fundamental transformation. We are, essentially, resetting the US economy. An American renewal must be built on technology" (Immelt 2009). If the United States is to recover from this crisis and regain its place as a leading world economy, it will be through new technologies, especially green technologies and smart solutions to contemporary problems. This assessment reflects an American consensus, shared by media and commentators, CEOs, business forums, and policy makers, and politicians from both parties. Specifically, this consensus comes at the confluence of several trends—a long-term commitment to innovation as part of modernity and part of the American self-image; and the role of information technologies, the green turn and the cultural turn, now in conjunction with financial risk and crisis. "Innovation economics" is in vogue. "Beneath the gloom, economists and business leaders across the political spectrum are slowly coming to an agreement: Innovation is the best—and maybe the only—way the U.S. can get out of its economic hole. New products, services, and ways of doing business can create enough growth to enable Americans to prosper over the long run" (Mandel 2009: 52).

Several innovation scripts are in vogue in the United States as ways out of the economic crisis and ways forward—in brief, engineering scenarios, going green, managerial innovations, the cultural economy, and various crisscross combinations. This treatment first reviews classic problems of innovation as an "applied Enlightenment" theme and then turns to contemporary American innovation scenarios. This is not a critique of innovation nor is it to detract from innovative proposals for technical, social, and institutional transformation and the forward thinking they represent. Rather, I distinguish between innovations and, on the other hand, innovation marketing and fluff. If we view innovation ideas and marketing as the "supply," we can measure it against the "demand" of economic, social, and institutional problems. So this is a critique of innovation rhetoric in relation to the challenges of American and global transformation. No doubt, innovation is a fruitful and valid theme; the question is valid for what purpose, on what terms, to what degree? Some problems are classic, some apply generally and widely, and some apply to the American situation in particular.

1. Innovation

The theme of innovation is a late-twentieth century extension of classic beliefs in progress, especially the Enlightenment theme of progress through reason and applied science (Nisbet 1980). In Europe this trope stretches from seventeenth-century scientific developments and eighteenth-century admiration for Chinese inventions, from Francis Bacon to Condorcet, Saint-Simon, Comte, and well beyond (Kumar 1987). Nineteenth-century scientific utopianism extended the Enlightenment faith in reason to social questions. The utopians of applied science and industry, such as Saint-Simon in France and Owen in England, believed that the application of new technologies could solve social problems. Amid the dislocations of early industrialism, this assumption triggered a sequence of problems.

"Scientific socialism" and historical materialism were responses to the scientific utopianism of the Saint-Simonians. Although there is a strong strain of technological determinism in Marxism, Marx held and Marxists generally hold that what matters is not technology per se (changes in forces of production) but the ownership and, accordingly, the uses of technology (relations of production). This issue remains the basic problem of the techno fix. Upheld by scientists, entrepreneurs, and policy makers, techno fixes claim or assume that technological change equals social progress, yet while claiming to address social questions, they tend to gloss over the relations of production and social dimensions.

In parentheses, broadly similar equations apply in development policies. Various forms of social engineering and "development from above" claim to achieve social development but in fact concern economic growth with "trickle down" factored in as an optimistic clause or an opportunistic assumption. Development without growth may be a difficult proposition, even so what matters is the *quality of growth* and experience shows that fast-lane growth may be detrimental to social development.

Technological determinism assumes that technologies steer society and culture; the constructivist view holds that technology is socially embedded and social forces steer and shape the application of technologies. Not technology rules but society rules, with its political and economic inequities. A key problem, also for ICT, is "disembedding technology from capital" (Nederveen Pieterse 2005: 26). In his probing critique of the "future industry," Rein de Wilde argues that "technological finalism" (the assumption that technologies point to and determine social outcomes) syncs with neoliberal market ideas (2000). Writing in 2000, this held true. However, what is striking overall is the flexibility of techno fixes, as if they served as an all-purpose elixir. Techno fixes offer to achieve mastery over nature and then promise to fix the problems caused by mastery over nature. Technical innovations produce and hone entrepreneurial competitiveness and sync with market forces, and when the market fails, techno fixes offer paths beyond the market. Innovations play multipurpose roles as crisis maker (as in financial engineering and quantitative investments) and as crisis breaker (as in "innovation economics").

This scenario illustrates the wide-ranging nature and applicability of human ingenuity and is as such unremarkable. Even so, in the process, techno fixes are large-scale instances of what economists call the "expert problem": experts often have a stake in the problem they diagnose and the solution they advocate. With some simplification, innovators are wont to plead innovation scripts in which their expertise holds trump cards. Urban analysts advocate urban solutions; university educators counsel the strategic importance of university education; asset managers offer financial solutions. Knowledge is power; expertise is not neutral, and independent advice is a rare bird.

Innovation talk is both stimulating and soothing, stimulating for it portends to offer something new and soothing because it fits the matrix of decades of discourse and exhortation—innovate, innovate! To achieve and maintain competitiveness nations must innovate. Entrepreneurs and companies must innovate or perish. Consumers must innovate and keep their lifestyle and gadgets up to date. To accumulate, innovate! To compete and improve, innovate! Innovation talk reproduces the enchantment of the "new," what Vattimo called modernity's "tradition of the new," the charm of novelty and the fascination with newness and its identification with "better, improved, efficient," an association that dates from the late nineteenth century (Williams 1976) and defines modern times.

Innovation often combines with "leadership," a cherished trope in American-style business and management studies, which from there seeps into general culture. In the context of business, leadership originally refers to market share (as in a leading company that sets standards for a sector), but translated into corporate governance (as in the "imperial CEO") and as it seeps into public life, it inevitably carries authoritarian connotations and, of course, feeds into the growing remuneration gap between top managers and employees. The distinction between these two registers of "leadership" is rarely made. Cambridge University offers a master's degree in Sustainability Leadership and notes "Extraordinary times need Extraordinary Leadership."[1] Maastricht University (2009) opened its 2009/10 academic year with speeches devoted to "Innovation and Leadership" and a keynote address by an alumnus who is a social media entrepreneur who founded a large cyber company in China. He tells the audience, "To stay competitive CEOs should not only read blogs, but also actively write them." They should also Twitter and do "crowd sourcing" (Chijs 2009: 18).[2]

These ideas are extensions of the idea of postindustrial society, Toffler's "third wave" and the knowledge economy. Technological transformation is widely viewed as a major driver of economic change; in Schumpeterian perspectives, innovation is viewed as key to the business cycle and to the fifty-year long wave or Kondratieff cycle. Thus, science and technology policies are central to economic policy. Universities play a strategic role in knowledge and science and technology upgrading. Research parks and partnerships between universities and corporations embody this approach. Patent and licensing lawyers are to convert innovations into intellectual property. It is interesting that when it comes to science and technology and economics the critiques of postmodernism hardly seem to matter.[3]

This thinking is en vogue worldwide. In emerging societies it is "the race to the intelligent state" (Connors 1997), from Japan and the East Asian developmental state to the Singapore model of a highly educated, multilingual populace, smart and wired, with governance geared to promoting education, infrastructure, and technological change in a society open for business. The ideas of the smart developmental state, in parentheses, parallel the classic French idea of the revolutionary state as an intelligent "educator state," a state that attracts society's best educated elite. They are reflected in the growing role of higher education in emerging and developing societies. They are interwoven with the general recognition of human capital as the key ingredient in development, which is a keynote in the human development approach (Haq 1995; Sen 1995) as well as at the World Bank and its aspirations to be a "knowledge bank."

2. Innovation in the United States

Innovation matters and matters particularly in the United States. According to many accounts, it is the master key to whether or not the United States will recover and regain its global lead. In view of the depth of American economic malaise and its levels of indebtedness, three future scenarios for the United States are the *Titanic*, or a complete crash, the Phoenix, or a comeback, and a twenty-first century New Deal, or a social turn in American capitalism (discussed in Nederveen Pieterse 2008). Joachim Rennstich (2004) makes a case for an American Phoenix scenario on the model of the British experience. Britain "ruled the waves" during the commercial-maritime era, then declined, and in the course of the nineteenth century made a comeback as the industrial "workshop of the world." The United States, too, could have two shots at global hegemony, first through its lead in mass production and Fordism, which has now come to an end in the bankruptcy of the Detroit automobile industry, and then through a lead in high-tech products and services. In this and other scripts, the decisive component is innovation. With the wakeup call of crisis, the United States now experiences a scramble for innovation with media, magazines and books offering a steady stream of innovation ideas and exhortations. As White House chief of staff Rahm Emanuel noted, "you don't ever want a crisis to go to waste," and this sentiment runs through many reflections.

Ralf Dahrendorf referred to the United States as the country of the "angewandte Aufklärung" (applied Enlightenment). Industrial innovation, in particular mass production and Taylorism, automobiles, highway construction, aircraft, military industries, and space travel, and engineering feats such as the TVA, exemplify American technological and engineering prowess. Postwar innovation policy in the United States has been driven by two missions, to "fight communism and cancer" and focused on military industries and health care. Both have become leading hi-tech sectors. While this focus has led to major advances, it has also produced white elephants. It has produced an expensive health care system that caters to elite needs and seeks return on fancy technologies with costly interventions and at times questionable medical necessity, alongside a byzantine insurance system that leaves many Americans (46.3 million in 2009) uninsured. It has made the military-industrial complex and the health care sector luxury liners of American society, well connected and politically powerful. In some respects, they have become worlds in themselves, functionally autonomous in relation to their original mission, with limited capacity for self-correction and virtually impervious to outside correction. The steadily growing military budget and the tremendous difficulties in reforming the American health care system signal the resilience of these institutional complexes. They have drained resources and talent away from other fields. With their growing cost they have become millstones around the neck of American society. Arguably, the United States has become a victim of its specialization, has overspecialized, in part as the price of American hegemony. Innovation leads, but it does not always lead forward; it can also lead sideways, or into a cul de sac.

Conventional approaches to socioeconomic change in the United States have been the techno fix and the spatial fix. Spatial fixes include suburbanization and the highway system, industrial zones, research parks, special zones, and tax incentives for corporations and, more recently, gentrification, gated communities and the new urbanism. A major spatial fix has been to move manufacturing production to low-wage, low-tax, low-service, no-union zones, initially in the American South and Southwest (moving industries from the Frost Belt to the Sunbelt), which I refer to as Dixie capitalism (Nederveen Pieterse 2004). In time it has extended to Mexico's maquiladores and to overseas special economic zones and low-wage zones from the Caribbean to Asia. Spatial fixes typically ignore and circumvent social questions by bypassing or deftly maneuvering around them while using public incentives to generate "spaces of capital" and fund private gains (Harvey 2001).

A recent techno fix, the new economy boom of the nineties, ended in the NASDAQ crash of 2000. Techno fixes serve as a circulation mechanism for excess liquid capital, for there is nothing like the lure of new technologies and the promise of new products (a better mouse trap) to attract and capture venture capital. "Internet economics," cyber-utopia scripts, and green tech are in vogue worldwide. Going digital is now the way forward from France to Malaysia. For every cyber utopia, of course, there is a dystopia (e.g., Harkin 2009). After the high-tech bubble of the nineties and the dotcom crash in 2000, this theme is less salient in the United States. Companies that one might expect to lead the recovery and a new wave of innovation—such as Microsoft and Dell—posted significant losses over the first quarter of 2009.[4] In contrast to the nineties, Silicon Valley now experiences a credit squeeze; in Palo Alto the talk now is of lack of venture capital—along with speculation whether it might come from China. Unemployment in the San Francisco Bay Area stands at 11.8%, higher than the national average.[5]

An ambiguity in the American situation is that government contracts occupy vast swaths of the economy, notably military industries, and pharmaceuticals through the Medicare prescription drugs program, yet the dominant ideology is "free enterprise." Successive waves of deregulation, particularly since the Reagan era, of financial services, telecoms and energy create corporate oligopolies, yet this unfolded under the banner of the "free market." So, the commitment to "free enterprise" and the belief in the efficiency of market forces combine with massive government intervention. However, at the same time industrial policy has traditionally been anathema in the United States. While government policies and subsidies play a key part in innovation strategies worldwide, this is a hard sell in the United States, aside from select sectors. The 2009 stimulus measures and the $787 billion American Recovery and Reinvestment Act (ARRA) change this only marginally. The Act and White House policy "envisage a knowledge-based, 'green' economy, jumpstarted by a serious ramping up of science, technology and education expenditure" (Hayden and Basset 2009: 1).

Jeff Immelt of GE departs from orthodoxy when he argues that "the US government can play a catalytic role. . . . Today, my country needs an industrial strategy built around helping companies to succeed with investment that will drive innovation and support high-technology manufacturing and exports." Globalization and a "robust trade policy" are part of his proposal (Immelt 2009). However, Republicans in Congress, media and think tanks continue to blame government ("bureaucracy and red tape") for economic ills, oblivious to private sector excesses and government-led industrial strategies worldwide. This stalemate is difficult to overcome even amid a crisis.

Against this backdrop let's review the main American innovation scripts. As mentioned before, innovation scripts that rank as ways forward in the United States are engineering technologies, going green, managerial innovations, and the cultural economy.

1. *Engineering scripts* and the call to new technology figure in most innovation scenarios. The basic problem of current innovation scripts is simple: if it's feasible why hasn't it been done already? Besides a few sectors—such as military industries, pharmaceuticals, agricultural machinery, biotech—American corporations have not produced major new engineering products for some time. A case in point is the Detroit car industry. Why innovate when established

product lines offer steady profits? Rather than venturing new products such as the electric car, although the technology was available, GM and Ford opted to continue established value chains (pickup trucks, minivans, SUVs) and, instead, Toyota and Honda led the development of technologies such as the hybrid engine. According to Michael Mandel,

> Innovation has fallen short of its promise in recent years. While some info tech corporations are still thriving, other sectors that were supposed to drive growth have faltered. Biotech companies have produced new drugs, but so far no real breakthroughs. And nanotechnology has been slow to generate commercial products. Worse, the historic link between jobs and innovation seems to have vanished, at least for now. In the past, pioneering industries such as automobile manufacturing and aerospace were big job creators. Today, jobs in cutting-edge sectors are down 12% since their 2001 peak. (Those industries include computer and communications hardware, software and computer-systems design, aircraft, drugs and medical devices, telecom, and Internet outfits such as Google and Yahoo!). (2009: 54)

Innovation has precisely *not* been the common trend in American corporations for some time, with some exceptions such as drugs, military industries, aircraft, software and ICT. The problem of innovation in the American case is the radical disproportion between innovation rhetoric—pervasive, habitual and part of common sense—and the meager record of industrial innovation, particularly since the 1970s when offshoring and outsourcing became standard. Why innovate when low wages and special conditions overseas offer ample profit margins? The dearth of domestic investment in plants and technology is noticeable broadly since the 1970s and 80s when the trend of relocation to the Sunbelt and overseas took hold. Major industrial sectors—such as automobiles, consumer electronics, machine tools, computer chips—have been taken over by overseas producers that *have* continued to invest and innovate. "Instead of investing in new technologies to spawn further productivity gains corporate managers overpaid themselves, doled out cash to investors, consumed luxury items, and engaged in corporate takeovers and mergers and acquisitions" (Leicht and Fitzgerald 2007: 66).

Although engineering is the dominant model of innovation, most actual innovations in the United States in recent decades have been in services, in management and business processes, in health care (14 percent of the GNP) and financial services (20 percent of the GNP). Deregulation achieved major innovations; the deregulation of financial services, telecoms, and energy created vast new market opportunities (Schiller 1999) and set the stage for the Enron and WorldCom series of corporate scandals (Nederveen Pieterse 2004). Mergers and acquisitions generate revenue for executives and financial intermediaries with a meager record in improving products or productivity. Special interest arrangements involving lobbyists, lawyers, lawmakers, and corporations; creative accounting, as in the Enron case; tax evasion and offshore tax havens; patents and licensing all represent innovations without necessarily adding value. Mathematics applied in quantitative investments (the 'quants') and hedge funds and financial products such as fancy futures and derivatives paved the way for the 1998 crash of the fancy Long Term Capital Management hedge fund, and CDOs (collateralized debt obligations, credit packages passed on to other banks) and sub-prime loans and set the stage for the sub-prime mortgage crisis of 2007.

2. *Going green* is a major discourse of economic revitalization in the United States and worldwide. Going green is an extension of engineering scripts, sometimes cast as a "Green New Deal" (Dickey and McNicoll 2008). For the United States, the problems are glaring. The US has been the world's guzzler of energy and other resources, has long kept aloof from international environmental agreements, such as the Kyoto Protocol, and has been a laggard in energy saving technologies. This was the point of President G.H.W. Bush's statement, "the American way of life is not negotiable." No wonder the United States lags in these technologies. Others lead in key technologies—China in solar panels, wind turbines and "clean coal," Germany in solar energy as well, Japanese companies in hybrid engines, France in nuclear technology, and so forth.[6] Red China is becoming "Green China" with large-scale investments in wind turbines and solar panels in the Gobi desert and has overtaken the U.S. as the largest market for wind energy (Garschagen 2009).

No doubt green tech is a major way forward, globally and for the United States. It plays a major role in government stimulus funding. However, it is unlikely that the U.S. can obtain a lead in these technologies, and it is more likely that, in the medium term, it will be an importer of green tech.

3. *Management innovations* have long been a major strand of American innovation. Business analysts distinguish several types of innovation—companies known for innovative products (e.g., Apple, Microsoft, Samsung), for innovative processes (e.g., Toyota, Wal-Mart), for innovative business models (e.g., Goldman Sachs, HP, Reliance) and for innovative customer experiences (e.g., Google, Amazon). According to the *Business Week* Innovation Index, "the companies with innovative business models tend to have the highest average stock returns and highest average revenue growth of all the companies in the index" (Jana 2008: 48). Thus, by this account, launching new products generates less revenue than innovating business models. No wonder that in many perspectives managerial innovations and new management methods in recruit-

> ing, deploying talent, producing, and valuing services, take precedence over product innovations. Interestingly, in *Business Week*'s Innovation Index, Goldman Sachs comes out as the most innovative and the most revenue and shareholder value generating company. Placing the leading and politically best connected Wall Street investment bank on top suggests a rank order of priorities with financial engineering in the lead, precisely at a time when the social value of financial innovations is being widely questioned.

Placing Toyota and Wal-Mart in one category is odd as well. Toyota has been a production process innovator, pioneering the flexible production techniques of lean manufacturing or just-in-time production, which is also known as Toyotism, whereas Wal-Mart's contribution is minor and in logistics, not in production (cf. Friedman 2005).

One type of managerial innovation seeks to help companies recover from previous innovations and reorganizations—by going back to core business. The corporate pendulum swings from innovation and expansion to revamping oneself back to basics and implementing innovation to overcome innovation (cf. Collins 2009).

With managerialism as a cultural ethos comes recurrent reorganization in institutions subject to managerial innovation, including public services, hospitals, and universities. Part of the innovation experience is that regardless of the efficacy of reorganization, invariably the upshot is that the influence and remuneration of administrators are vastly increased. The heading is innovation, but the outcome is the steady increase of managerialism.

Given the emphasis on generating revenue and shareholder value, wave upon wave of MBAs inflict innovations on new and existing product lines and services to cut cost and enhance revenue. Airlines squeeze seating space by inches, no longer serve peanuts (pretzels are cheaper), require payment for checked luggage; service personnel are scarce on the floors of big box stores, and so forth; countless cost cutting and revenue enhancing measures, large and small, shape our lives. We happen to inhabit the world of corporate revenue generation and have no choice but to volunteer as extras in their scripts. Besides, business models seep into public life and general culture. Low-tax and low-service conditions in most American states have long privatized many services such as waste collection. British conservatives take the "no frills" airline business model as a template for council public services, so "residents pay extra for service above basics" as part of a "relentless drive for efficiency" to cut cost in a time of economic crisis.[7]

So the question is innovation for what purpose? Innovation in the shareholder model of capitalism yields different criteria of success and different outcomes than innovation in stakeholder capitalism. Arguably, one set of innovations limits options and reduces the quality and well-being in the other set; shareholder and stakeholder interests are not generally a win-win equation. Hence the generalizing, multi-purpose talk of innovation is misleading—innovation for what purpose is the question. To foster economic growth is the standard answer. However, this approach only shifts the problem. Economic growth during past decades has come with sharply increasing social inequality. Thus, *what kind of growth* and what kind of innovation are the real questions.

Business organization outside the U.S. has been drawing attention too. Prahalad and Krishnan (2009) examine business models of companies in India and Asia. Work on business innovations that cater to the poorest consumers (Prahalad 2006) also breaks the mold. Other work focuses on the growing international competitiveness of companies in emerging societies (e.g., Sirkin et al. 2008). Here science and technology policies are as salient as in the OECD economies, but because of the development problems that these societies also face, there is often greater attentiveness to the general economic policy that innovation is embedded in.

4. A further script centers on the strategic importance of the *cultural economy* or the creative economy in economic growth and recovery. Variants of this script include the creative class and Richard Florida's urban revitalization and renewal perspectives. Florida's argument is essentially a cultural variant of the economic geographers' spatial fix—with urban spaces remapped as cultural spaces and culture redefined as human capital and redeployed as a growth engine. Thus, key to the revitalization of American cities is to attract the "creative class" of "scientists, engineers, managers, and professionals," as a recent article declares: "The spillovers in knowledge that result from talent-clustering are the main cause of economic growth. Well-educated professionals and creative workers who live together in dense ecosystems, interacting directly, generate ideas and turn them into products and services faster than talented people in other places can . . . Big, talent-attracting places benefit from accelerated rates of 'urban metabolism' . . ." (Florida 2009: 50).

Florida notes, too, "it's not that 'fast' cities are immune to the failure of business, large or small," and he refers to the 1873 crisis and credit freezing up. However, "unlike many other places, they can overcome business failure with relative ease, reabsorbing their talented workers, growing nascent businesses, founding new ones" (2009: 51). There are various slips in this argument. Yes, education, talent, and infrastructure are resilient. The most famous case is the success of the Marshall Plan, though it is worth noting that it is the *only* success of major foreign aid. The key problem is that the clustering argument applies if and as long as certain conditions at the margin are met—in particular, access to credit and capital and an institutionally supportive environment. Without credit, clustering is powerless. This dilemma now prevails in Silicon Valley. The talent is there, but where is the money? Without venture capital, the Valley is dry. A similar conundrum was faced when "social capital" was sold as the solution for poor urban neighborhoods and as a strategic ingredient in empowerment and enterprise zones; as Portes and Landolt noted (1996), social capital is powerless without jobs. Thus, in these instances the expert advice focuses on the necessary but not on the sufficient conditions for recovery and glosses over the margin conditions for clustering to deliver.

Florida sells the same product twice—once as an elite project and again as an egalitarian project. Thus, he argues, rightly, that the world isn't flat but spiky (Florida 2008). Throughout his article on "how the crash will reshape America," he argues for "elite cities" and the clustering of talent; yet, at the end, as an afterthought, he observes "we need to make elite cities and key mega regions more attractive for all of America's classes, not just the upper crust" (2009: 56). Since he does not give specific reasons why this should be done, it can be read as a social whitewash of what is essentially an elite project.

Florida notes that not "every factory town is locked into decline. You need only look at the geographic pattern of December's Senate vote on the auto bailout to realize that some places, mostly in the South, would benefit directly from the bankruptcy of GM or Chrysler and the closure of auto plants in the Rust Belt. Georgetown, Kentucky; Smyrna, Tennessee; Canton, Mississippi: these are a few of the many small cities, stretching from South Carolina and Georgia all the way to Texas, that have benefitted from the establishment, over the years, of plants that manufacture foreign cars" (2009: 52).

There is a remarkable silence in this argument. That the Sun Belt benefits from the decline of the Rust Belt is the cliché of the great American shift to the South, which dates back to the seventies and eighties. This is hardly news. It is quite odd that this trend should form part of a 2009 post-crisis feel-good narrative, for it is rather a manifestation of and contributing factor to the crisis. What is not mentioned is the rationale and downside of this shift; it is a shift to low wage, low tax, low service, no union states—a turn to Dixie capitalism (Nederveen Pieterse 2004, 2008). The American South and Southwest represented and continue to represent as it were a vast special economic zone where access to cheap labor reduced the incentive to innovate. So this is not a recovery scenario but a high-exploitation capitalism script with steep social inequality built in. With ample irony, it may be termed the revenge of the Confederacy. The banner success companies of the South, such as Wal-Mart, Enron, WorldCom, HealthSouth, have typically not contributed new products but have thrived on business process and financial innovations, often of a questionable nature. The current fiscal crisis and state of financial emergency of California is another manifestation of the limits of the low-tax model. Here fiscal crisis also affects the knowledge sector such as public education and the University of California system.

At any rate, the down-turn throughout the United States is of little comfort to Detroit and Miami. Current American recovery is hampered and mortgaged by the previous recovery from the slump of the seventies and the "second slump" of 1987. Then the way out was to recover profitability by moving plants to low-wage, low-tax zones in the American South and overseas. As American median wages stagnated in line with the shift to the low-wage model, consumption levels continued to rise. This conundrum was papered over by vast credit expansion—household credit card debt, home equity financing, adjustable rate mortgages and sub-prime mortgages were enabled by a Federal Reserve low-interest regime and gargantuan borrowing on a world scale, which absorbed 70 to 80 percent of world net savings. The American pattern of low wages and high consumption has been papered over by a vast debt expansion of which the bill is now coming due. These recovery solutions now limit the available choices. The low-tax, low-wage, high-profits, and high-social cost constellation is not a way forward. Low tax revenues and high debt, external and domestic, constrain state and federal government capacities. It does not work to offer the script that has precipitated crisis as a way out of crisis now.

If we interpret the cultural economy as a sector (including, e.g., Hollywood, television, the arts, design, fashion), it is vibrant and significant, but not nearly significant enough in job creation to make up for the millions of jobs lost in manufacturing and through outsourcing. As a sector, the cultural economy also faces a credit squeeze, and foreign ownership has been rising (for instance in the Hollywood studio system). The cultural economy, though surely significant, is simply not large and substantial enough to employ enough American workers; just as software, high-tech, and back office services in India will never employ enough of India's workforce. India needs a vibrant agriculture and a manufacturing sector. The United States, too, needs an industrial sector.[8] A related problem is that when manufacturing goes offshore, service jobs in design, research and development, transport, insurance, in other words the infrastructure of manufacturing, are also lost. If we interpret the cultural economy as a slice and dimension of production and services generally—as in Florida's "creative class" of "scientists, engineers, managers, and professionals"—it is certainly a key dimension, but precisely because it is interwoven with the economy generally it cannot also serve as a master key to renewal or as an economy rebirth snake oil.

3 Rebirth Bottlenecks

Part of the backdrop of this discussion is the gradual "decoupling" of the world economy from the American economy.[9] Reports by the CIA and the U.S. National Intelligence Council (2008) anticipate a drastically reduced global role of the United States by 2025. This issue is not in dispute; the "rise of the rest" is here to stay. Meanwhile in the wake of the crisis, innovation talk has gone in overdrive in the United States and Europe. "How innovation can fight the downturn," "Hard times can drive innovation," and "Why an economic crisis could be the right time for companies to engage in 'disruptive innovation'" are common headlines on both sides of the Atlantic.[10] Crisis is a rupture with old paths and stimulus funding opens new windows. Some innovation talk, of course, reads like advertisements and funding solicitations.

The credit squeeze will pass, yet the horizon is dark. Bailouts, stimulus spending, and fiscal pressure from aging baby boomers add to the American debt overhang. The external account deficit is at 13 percent of GNP. The status of the U.S. dollar as world reserve currency has been slipping. If the U.S. loses its AAA credit rating, interest rates will rise and will burden recovery. However, in the end, it is not clear whether the main bottleneck for American renewal is finance or lies deeper. Michael Mandel questions the importance of funds:

> If money alone were enough to guarantee successful innovation, the US would be in much better shape than it is today. Since 2000, the nation's public and private sectors have poured almost $5 trillion into research and development and higher education, the key contributors to innovation. Nevertheless, employment in most technologically advanced industries has stagnated or even fallen. The number of domestic jobs in the computer and electronics sector continues to plunge while pharmaceutical and biotech companies lay off as many workers as they hire. And even the industry category that includes Google (GOOG)—Internet publishing and Web search portals—has added only 15,000 jobs since 2003. (2009: 52–53)

Indeed, money does not explain the lack of domestic investment. After all, where have all the corporate profits from offshoring and outsourcing gone? The larger problem is *profitability* and the circumstance that American corporations have become habituated to operating in low wage, low tax, and low regulation environments, at home and abroad.

The key comparison is between the United States and other advanced societies. All have offshored production to low-wage zones, but companies in Europe and

Japan have generally balanced this outsourcing with domestic investments in new plants and technology, whereas most American companies have not, so American deindustrialization has been far more drastic and far-reaching.[11] There are three sets of hypotheses that may explain this difference. First, the availability of vast low-wage, low tax-zones within the United States—the American South and Southwest, of which there is no equivalent in other advanced countries (in many countries there are poor or backward areas, but not with a different legal and institutional structure). Essentially this is a legacy of slavery, Reconstruction, and Jim Crow. Indeed, there is domestic investment in the United States and also foreign direct investment, but the bulk is in the low-wage zones.

The second hypothesis is the perks and the price of American hegemony. The commerce department and the Export-Import Bank have long facilitated and supported the outward investment of U.S. companies as part of American outward expansion, going back to the Cold War era. Outward investment meant doing a service to the American cause when the cause was global expansion. No such mission or comparable support existed for European or Japanese companies. Part of hegemony, too, was the military-industrial bias in American innovation policies. The laissez faire, free enterprise philosophy further meant, unlike other advanced nations, no industrial policy and no national economic strategy (cf. Prestowitz 2005).

The third hypothesis concerns the overall character of American society. Unlike other advanced nations, American modernity is not a post-feudal modernity but a late-start, historically thin modernity. One of the implications is: no feudalism, no noblesse oblige. As an immigrant society, the United States is the envy of many other societies and rightly so. Almost nowhere else can (some) new immigrants rise to prosperity, status, and high office. However, taking a step back, this also has a dark side. Why invest domestically when the society is heterogeneous from the outset (Native peoples, slavery, indentured labor) and is an immigrant society in which ethnic and racial prejudice are rife, social solidarity is thin, inequality is high and growing, and assorted spatial fixes shelter the rich from the less well-off and their problems of crime, violence, drugs, unhealthy lifestyles, and obesity? The United States is a mixed society but also a fractured and class segregated society in which elites generally display less social solidarity and domestic allegiance than in other advanced societies. By world standards, American elites are deviant (Robinson and Murphy 2009). Patriotism (which is exceptionally high in the United States) is not the equivalent of social cohesion.

These factors together have prompted a greater disposition toward outward investment for American companies than for European and Japanese companies.

With this commitment comes path dependence. It is also an illusion that basic industry and research and development can be separated. With offshoring traditional industries, increasingly research and development move offshore, too, as do corporate profits.

The main bottleneck in the American "reset economy," then, is that corporations have become habituated to low-wage, high-profit investment. If this indeed is the main explanation for the relative lack of U.S. domestic investment, then stimulus funds and innovations will make little difference. I do not share the views of the "deficit hawks" and think government deficit spending is the right way to go. Even so, a Keynesian approach works in relation to a Keynesian problem and does not work when investors seek profits in a globalized economy; in fact, stimulus spending may increase imports into the United States. When the key problem is not innovation but profit margins, government policies will have a limited impact.

References

Chijs, Marc van der. 2009. Entrepreneurship and internet innovation: Lessons and experiences, in Maastricht University. *Innovation and leadership: Opening academic year 2009/2010.* Maastricht University Office.

Collins, Jim. 2009. *How the mighty fall and why some companies never give in.* New York: HarperCollins.

Connors, M. 1997. *The race to the intelligent state: Charting the global information economy in the 21st century.* Oxford: Capstone.

Conor, B. 1991. *Japan's new colony—America.* Greenwich, CT: Perkins Press.

de Wilde, Rein. 2000. *De voorspellers: een kritiek op de toekomstindustrie.* Amsterdam: De Balie.

Dickey, C., and T. McNicoll. 2008. Why it's time for a "Green New Deal." *Newsweek*, November 10: 49.

Florida, Richard. 2004. *The rise of the creative class.* New York: Basic Books.

Florida, Richard. 2008. *Who's your city?* New York: Basic Books.

Florida, Richard. 2009. How the crash will reshape America. *The Atlantic*, March: 44–56.

Friedman, Thomas L. 2005. *The world is flat.* New York: Farrar Straus and Giroux.

Garschagen, O. 2009. Windmolens in de Gobiwoestijn, *NRC Handelsblad*, September 12–13: 18.

Gross, Daniel. 2007. The U.S. is losing market share, so what? *New York Times*, January 28: BU5.

Haq, M. ul. 1995. *Reflections on human development.* New York: Oxford University Press.

Harkin, James. 2009. *Cyburbia: The dangerous idea that's changing how we live and who we are.* London: Little, Brown.

Harvey, David. 2001. *Spaces of capital.* New York: Routledge.

Hayden, Jennifer, and Julie Basset. 2009. Innovation for recovery: recovering innovation. *The Inno-Grips Newsletter*, February: 1–3.

Immelt, Jeff. 2009. Innovation can give America back its greatness. *Financial Times*, July 9: 9.

Jana, Reena. 2008. Indata. *Business Week*, September 22: 48–50. http://www.businessweek.com/innovate/global_index/

Kumar, Krishan. 1987. *Utopia and anti-utopia in modern times*. New York: Oxford University Press.

Leicht, Kevin T., and Scott T. Fitzgerald. 2007. *Postindustrial peasants: The illusion of middle class prosperity*. New York: Worth Publishers.

Lyotard, J-F. 1979. *La condition postmoderne: Rapport sur le savoir*. Paris: Minuit. Trans.: *The postmodern condition: A report on knowledge*. Manchester: Manchester University Press, 1986.

Maastricht University. 2008–2009. *Leading in learning: Innovation is our focus*. Maastricht University Communication and Relations Office.

Maastricht University. 2009. *Innovation and leadership: Opening Academic Year 2009/2010, articles, speeches, interviews*. Maastricht University Office.

Mandel, Michael. 2009. Can America invent its way back? *Business Week*, September 22: 52–60.

National Intelligence Council. 2008. *Global Trends 2025*. Washington, DC. Report released 20 November 2008. http://www.acus.org/publication/global-trends-2025-transformed-world-.

Nederveen Pieterse, Jan. 2004. *Globalization or empire?* New York: Routledge.

Nederveen Pieterse, Jan. 2005. Digital capitalism and development: The unbearable lightness of ICT4D. Iin Geert Lovink and Soenke Zehle, eds. *Incommunicado reader*. Amsterdam: Institute of Network Culture.

Nederveen Pieterse, Jan. 2008. *Is there hope for Uncle Sam? Beyond the American bubble*. London: Zed.

Nisbet, R. A. 1980. *History of the idea of progress*. New York: Basic Books.

Portes, A., and P. Landolt. 1996. The downside of social capital, *American Prospect*, 26: 18–21, 94.

Prahalad, C. K. 2006 *The fortune at the bottom of the pyramid*. Upper Saddle River, NJ: Wharton School Publishing.

Prahalad, C. K., and M. S. Krishnan. 2008. *The new age of innovation*. New York: McGraw-Hill.

Prestowitz, Clyde. 2005. *Three billion new capitalists: The great shift of wealth and power to the East*. New York: Basic Books.

Rennstich, Joachim Karl. 2004. The Phoenix cycle: Global leadership transition in a long-wave perspective. In Thomas E. Reifer, ed. *Globalization, hegemony and power: Antisystemic movements and the global system*. Boulder, CO: Paradigm.

Robinson, M., and D. Murphy. 2009. *Greed is good: Maximization and elite deviance in America.* Lanham, MD: Rowman and Littlefield.

Schiller, Dan. 1999. *Digital capitalism: Networking the global market system.* Cambridge, MA, MIT Press.

Sen, Amartya. 1999. *Development as freedom.* New York: Knopf.

Sirkin, H. L., J. W. Hemerling, A. K. Bhattacharya. 2008. *Globality: Competing with everyone from everywhere for everything.* New York: Business Plus.

Williams, R. 1976. *Keywords.* Glasgow: Fontana and Croom Helm.

Zysman, John, and Stephen Cohen. 1987. *Manufacturing matters: The myth of the post-industrial economy.* New York: Basic Books.

Notes

1. Advertisement in *Financial Times,* September 21, 2009: 13.
2. A motto of Maastricht University is "Innovation is our focus." Its innovation perspective is social: "Innovation is most effective when it is anchored in society." "It is innovation that holds the key to a wide range of social problems." Through it the university seeks to "contribute to a sustainable form of globalization" (Maastricht University 2008–2009: 5). The implication is that the university adopts a stakeholder perspective on innovation.
3. E.g., Lyotard 1979, 1986.
4. A report notes, "Microsoft revenues down 17%," "Results damp hopes of broad tech recovery." Results for Apple, Intel and IBM have been better. F. Waters, "Microsoft revenues down 17%," *Financial Times,* July 24, 2009, 12.
5. The report is mixed. B. Johnson, "Gloom of recession can't cloud over Bay Area's spirit of hi-tech optimism," *Guardian Weekly,* September 11, 2009: 9.
6. A specific consideration is that "China produces more than 99 percent of the world's supply of dysprosium and terbium, two rare minerals essential to recent breakthroughs in high-technology industries," in particular wind turbines and hybrid engines. K. Bradsher, "Backpedaling, China eases proposal to ban exports of some vital minerals," *New York Times,* September 3 2009.
7. "A leading Conservative council is using the business model of budget airlines Ryan Air and easyJet to inspire a radical reform of public service provision that is being seen as a blueprint for Tory government." R. Booth, G. Hinsliff, "Tories take budget airline route with 'no fringe' council cutbacks," *Guardian Weekly,* September 4, 2009, 13.
8. Cf. Zysman and Cohen 1987.
9. "The US' share of global GDP fell to 27.7 percent in 2006 from 31 percent in 2000 . . .the share of the BRICs rose to 11 percent from 7.8 percent. China alone accounts for 5.4 percent. . .in 2007 the BRICs' contribution to global growth was slightly greater than that of the US for the first time. In 2007. . .the US will account for 20 percent of global growth, compared with about 30 percent for the BRICs" (Gross 2007).

10. The *Inno-Grips Newsletter* has done a literature review of "innovation in times of crisis," February 2009: 8, including "Hard times can drive innovation," *Wall Street Journal*/Business Insight, December 15, 2008. "Why an economic crisis could be the right time for companies to engage in 'disruptive innovation," Knowledge@Wharton, November 12, 2008. C. Leadbetter, J. Meadway, "Attacking the recession—How innovation can fight the downturn," NESTA Discussion paper, December 2008. Euractiv, "Investing in innovation 'key to economic recovery,'" January 29, 2009. A *Financial Times* article notes "The weak economy is forcing companies to innovate" (September 21, 2009: 13).
11. This aspect is discussed more extensively in Nederveen Pieterse, 2008.

Part Three: Culture and Curriculum

19

Working the Paradigm Shift: Educating the Technological Imagination

Anne Balsamo

Converging on the Singularity

Shift work is a fact of life in a 24/7 age. Unlike shifts that start and end with the punch clock, the hours of the paradigm shift are not determined. Much of the significant shift work is going on around the edges of the academy; some of it is going on in ad hoc cultural institutions; most of it involves the use of digital networks and new media technologies. The challenge is how to coordinate these efforts such that the distinctive paradigm shifting projects contribute to significant and beneficial social changes. Our futures depend on it. The broader paradigm shift has been characterized in different ways. Educational theorists describe it as a transition from a paradigm of "teaching" to one of "learning" (Barr and Tagg, 1995). Scholars and historians who work with digital archives speak of a shift from a paradigm of scarcity to a paradigm of abundance.[1] Media theorist Henry Jenkins expands on both of these notions to suggest that the current paradigm shift has inaugurated a new cultural logic based on a reconfigured relationship between people and media; no longer simply spectators of media productions, we have become media creators ourselves.[2] As he elaborates:

> Convergence does not occur through media appliances, however sophisticated they may become. Convergence occurs within the brains of individual con-

> sumers and through their social interactions with others. Each of us constructs our own personal mythology from bits and fragments of information extracted from the media flow and transformed into resources through which we make sense of our everyday lives. Because there is more information on any given topic than anyone can store in their head, there is added incentive for us to talk among ourselves about the media we consume. . . . Consumption has become a collective process. . . . None of us can know everything; each of us knows something; and we can put the pieces together if we pool our resources and combine our skills. (2006: 3–4)

This shift in our relationship to media—from passive spectators to active participants—involves the process of productive consumption. Just as we harvest bits and fragments of information from various media flows, so too are we called to actively contribute to the information streams we fish in. This is one of the connotations of the use of the term "prosumer" as the name for a person who is as much a *pro*ducer as con*sumer* of media experiences.[3] For Jenkins, this paradigm shift is marked by a transition from individualized media consumption to the formation of "consumption communities" that enable new forms of participation and collaboration. Convergence for Jenkins lays the groundwork for the creation of a *culture of participation.* He goes on to suggest that one of the key consequences of this paradigm shift is the rise of a collective intelligence that derives from collective practices of information exchange and meaning making. The hallmark of this paradigm shift is its foundational assumption that intelligence is a distributed, multimodal ability that is developed, practiced, and expressed through the use of technologically mediated informational and social networks.

The specter of a superlative collective intelligence fires the imagination of science fiction writers as well as posthumanists. For these thinkers, the logical next phase of convergence culture will result in *The Singularity*—a science fictional concept now adopted by some futurists to describe the time when the acceleration of technological and social change exceeds humans' ability to keep pace with the changes. The Singularity is a moment of significant discontinuity with what has come before; it is brought about by the acceleration of technological progress, as defined by increases in artificial intelligence and the speed of computer networks. In many science fiction tales, *The Singularity* results in the birth of a superhuman AI species that transcends and, in some cases, annihilates the remnants of humanity. In the darkest posthumanist scenarios, human beings are at least an endangered species, if not outright extinct. In the milder post-singularity story worlds, humans mutate and accommodate themselves to a lower berth on the intelligence chain-of-being; life after the Singularity, in this version of the post-paradigm shift future, represents the next stage of human evolution in which we are no longer agents of our own destinies.[4]

For mathematician and science fiction author Vernor Vinge, *The Singularity* manifests as superhuman intelligence.[5] Vinge believes that although such a development is inevitable, its disposition is not. He takes issue with the posthumanists who celebrate the decorporealization of the human being and the victory of mind over matter. As a self-professed "technological optimist," he asserts that human beings are still the initiators who have the "freedom to establish initial conditions, [and] make things happen in ways that are less inimical than others" (Vinge, 1993). He proposes that we focus attention on an alternative paradigm of superhuman intelligence he calls "Intelligence Amplification" (IA). This approach differs from artificial intelligence (AI) programs in that it concentrates more stridently on the creation of more robust human-computer interfaces. The nuanced difference between the two approaches (AI versus IA), according to Vinge, is that the IA approach keeps the human being positioned as an empowered agent within the unfolding technological project rather than (as he claims for AI) at the periphery as an entity to be modeled (human intelligence) and, as the stories would have it, eventually discarded.

While the apocalyptic rhetoric of the science fictional accounts has certainly fomented lively debates about the nature of the relationship between human beings and the technologies we spawn, my interest here is not to wade through these debates again, but rather to focus attention on a near-term set of issues that arise on this side of our ascension (or collapse) into *The Singularity*.[6] Like Jenkins and Vinge, I believe that this paradigm shift proffers important opportunities to positively shape the conditions of human life in the future. In this chapter, I am specifically interested in the implications of this paradigm shift for the cultural work of the university. I argue that if it is to remain an important site for the creation of knowledge in a networked digital age, the university must embrace new lessons and reconfigure its practices.[7] While there are many issues to consider in thinking about how the university might reconfigure its practices, in this chapter I discuss the nature of the identity of those who are poised to participate in these new learning efforts, the students and teachers, to examine how their sensibilities and dispositions are in flux as a consequence of paradigm shifts already in motion. This discussion is part of a broader effort to consider the process whereby technological innovation (within the university) contributes to cultural transformation through the development of new educational programs, transformative (technology-based) research, pedagogical experimentation, and new modes of outreach and dissemination. (Those are some of the issues I examine in my book called *Designing*

Culture: The Technological Imagination at Work [Duke]). For the purposes of this chapter, I focus attention on two specific paradigm shifting projects already underway at one particular university: the University of Southern California. I do so not because these are the most advanced projects, but because as a participant in these projects I can better describe how they illustrate elements of a wider paradigm shift. The shared backdrop of these discussions concerns the education of the technological imagination. I identify what I believe to be the key institutional elements necessary to support the cultivation of the technological imaginations of students and faculty who will be engaged in designing the cultures of our future by virtue of their involvement in technological innovation. My objective here is to sample the expanse of the paradigm shift work already underway and to suggest work yet to be started.

Learning in a Digital Age: New Habits, New Practices, and New Subjectivities

In Thomas Kuhn's famous account, paradigms take hold in the academy, where they are gradually institutionalized through the production of new experiments, theories, articles, books, methods, and educational programs.[8] When Kuhn was writing, the academy was at the center of a constellation of institutions and practices that served as the context for the production of scientific knowledge. While Kuhn's account included reference to the role of affiliated organizations—such as professional associations, government bodies, and publication institutions—the academy was taken-for-granted as the center for the production of knowledge. Indeed, while this may have been true for the great paradigm shifts of the past century (and certainly of the scientific paradigm shifts that Kuhn focused on), this is NOT so true of the current shift. What we now realize is that the academy is but one site among *many* where significant learning and knowledge production happen. This is one of the first lessons of the current paradigm shift.

The practices that define this current paradigm shift focus on new forms of knowledge production, dissemination, and learning: none of these are based on the traditional practices of formal educational institutions. This approach has two general implications. The first is that to understand the paradigm shift—characterized by a transition from teaching to learning, or by Jenkins's logic of convergence—we need to take seriously the learning practices that are going on in places other than in the academy. If it is true, as I asserted earlier, that the hallmark of this paradigm shift is its assumption that intelligence is developed, practiced, and expressed through the use of digital informational and social networks, then we need to examine how learning happens when people engage in these networks. The second implication, and for my purposes, the most relevant is that we need to then think about how these practices might be incorporated into the university such that the academy be reconfigured as a vital learning institution for the twenty-first century. The changes necessary for the new paradigm to take root within the academy will involve more than the introduction of new textbooks or new technologies into traditional educational settings. At the very base, we need to stop thinking about new digital technologies as the channels through which education is delivered and instead explore the ways in which these technologies are implicated in the reconfiguration of knowledge production across domains of human culture. The aim then is to take these insights as the basis for rethinking structures and pedagogies within formal educational institutions.

The young people who are now showing up in U.S. university classrooms are ranking members of the *born digital* generation. This moniker is used by researchers John Palfrey and Urs Gasser (2008) to characterize the cohort of young people who were born after 1987. The use of the term was influenced by the phrase "digital natives and digital immigrants" developed by Marc Prensky (which in turn was influenced by a phrase of Douglas Rushkoff's) that extended the metaphor of geographic immigration to the consideration of those who were born into a digital landscape.[9] Even though the degree of access to network information flows available in their school classrooms or in their homes differs according to family economic position and geographic location, these students, born between 1987 and 1990, are members of the first generation to grow up in a world that included portable computers (laptops, 1985), network communication applications (AOL, 1989), and ubiquitous graphics applications (Photoshop, 1990). By the time they reached school age, say between 1993 and 1997, the Web (WWW, 1991; Mosaic, 1993) was already being touted as an educational resource. Given this contour of the technocultural scene of their birth, it is reasonable to assert that their beliefs and assumptions about the way learning occurs would have been significantly shaped by their early encounters with pervasive digital worlds and network technologies, and the ubiquity of "smart" and responsive environments.[10]

Nevertheless, as David Buckingham rightly points out, it is important to be aware of the significant differences within the group who comprise the "born digital" generational cohort.[11] Teasing out these differences is a complex process. For example, students of certain social classes may not have had access to computers in the home, but they could have encountered computers in schools. Other groups of young people within this generational cohort have not had direct access to computers and networks at all, but still have been audience members for advertisements, consumer products, and media messages that referenced these digital technologies. Moreover, access, in homes or in classrooms, does not always translate into use and mastery. For example, research suggests that girls and boys growing up with similar access to computers show different levels of interest in computing, gaming, and other technology-based practices.[12] However, even for those without access to computers in the home, the ubiquity of computers and digital technologies elsewhere would have, inadvertently if subtly, influenced the contexts within which they grew up, their basic understandings about the power of technology and of the media, and their imaginations about what is possible in the future. They, too, are members of the born digital generation, because even with differential experiences with digital technologies, these youths inhabit cultural landscapes marked by the pervasiveness of computers and digital applications.

As members of the born digital generation move through different kinds of networks (social, institutional, technological, peer-based, as well as familial), their imaginations are restructured and shaped through their encounters in different kinds of mediated worlds. When they learn to use new instant messaging programs, text messaging, and on-line chat spaces, and new practices such as searching, surfing, browsing, bookmarking, flaming, and tweeting, they are learning not only new communication habits but also new cultural *rituals* of self-fashioning. As they engage in repeated experimentation with new digital technologies, they literally create elements of the "self" through text, avatars, and online representations. The "self" is literally distributed throughout various digital networks: in the form of online accounts in different social spaces (i.e., *MySpace*, *Facebook*, and *LookBook*), through photographs and videos stored and shared through different sites (i.e., Flickr and YouTube), through game play and virtual world participation (i.e., *World of Warcraft* and *Second Life*), and through creative portals (i.e., *Scratch*, *Garry's Mod*, and *Outer Post*). Identity for these network denizens is best understood as a fluid construct that emerges through the performance and staging of multiple personae. In this sense, they can be understood as quintessential decentered postmodern subjects with identities that are marked by differing intensity flows and shifting affinities. While it has been more than a decade since researchers first took up the question of "life on the screen" and the implications for the creation of identity, what we have come to appreciate is how the acquisition of new habits, practices, and notions affects the formation of subjectivity and new subject positions.[13] It is these new subject positions that must be considered in the reconfiguration of formal learning environments.

From Just-In-Time Learners to Original Synners

It is already evident that many members of the born digital generation understand themselves as just-in-time learners, confident that when they need to know something, they'll know where to find it.[14] These young people understand how to mine their networks (both digital and social) for their information needs. Many of them treat their affiliation networks as informal Delphi groups. The statistical phenomenon of Delphi groups demonstrates that even when a factual piece of information is not known to each person, the aggregate mapping of responses from group members tends to cluster around the correct answer. For these youths, the process of thinking now routinely (and in some cases, exclusively) relies on social network navigation. Just as the creation of a self emerges through the navigation of different information flows, so too does the creation of knowledge. As they navigate intersecting digital networks, they are exposed to different knowledge communities: those of peers, of popular pundits, of parents, of media shills, of informal teachers, as well as of formal educators. This fact of network life has several implications. For one it means that the notion of learning is itself a distributed experience. Learning can no longer (if indeed it ever was) be understood as happening only in formal institutions such as schools or universities. While this insight may not be news to those community-based educators who were responsible for the creation of media literacy programs throughout the 1990s, it is an insight that has been gaining popularity more recently.[15]

A second implication suggests that the skills of creative and critical *synthesis* remain a vitally important cognitive capacity for those whose learning happens across digital media flows, among distributed learning sites, and in dialogue with contradictory sources of cultural authority. In this sense, the ability to "read" critically and integrate information is not an outmoded text-based literacy.[16] In an effort to shift my own thinking about the identity of the members of the born digital generation who are or soon will be students in my university classes, I have started to think of them as "Original Synners."[17] In tagging them with this identity (borrowed from science fiction author Pat Cadigan), I see them as fluid subjects-in-formation that create knowledge through their travels in different media flows. It helps me understand that my role as a teacher is to help them become "original synthesizers" of information that is expressed in different modalities and that is produced within different knowledge ecologies. Their media prosumption habits already incorporate creative synthesis practices such as remixing, modding and data mining. However, in order to engage in the reflexive production of *knowledge* they must also learn how to *incisively critique* the information flows they remix. From this critique emerges a set of questions about what is not yet known. Data = information = knowledge may be their taken-for-granted epistemology, but it is exactly this epistemological position that needs to be refined. Data mining does not necessarily yield valid information. In addition information requires a context to count as "knowledge."

Even though it is not the only knowledge ecologies that matter for *Original Synners*, for many of them, the university will be one that they have to learn to navigate. The university as a knowledge ecology has specific structures, conventions, and rituals of knowledge production. Some of these elements will and should persist in a digital age; others will definitely need to be transformed. In this ecology, *Original Synners* must learn how to integrate information that comes from different sources, critical frameworks, and academic disciplines. They need to understand how knowledge is produced in the dialogue among disciplines, through practices of social negotiation, and in creative collaboration with peers and experts. Thus, they will need to understand the structural function of disciplinarity as an institutionalized practice of knowledge verification. They must learn how to engage in conversations with those who do not hold the same cultural values or intellectual commitments as they do. However, equally importantly, this notion also suggests other considerations they may not yet fully appreciate and embrace. For example, although they are already global citizens by virtue of their consumption habits and residence in particular national contexts, they need to understand how the global flows of information and capital affect people in other geographic and cultural contexts. In this sense, they need to become deeply multilingual, not only in the use of languages but also in their understanding of different cultural logics and global politics. In this task, the born digital generation has a daunting learning agenda: they must acquire appreciation for the depths of disciplinary knowledge, but not get mired in the merely academic, so that they can forge connections across disciplinary contexts in the service of creating new understandings and formulating new questions to guide their educational pursuits. Learning is a practice; knowledge requires contextualization. They will have to learn the value of both.

The connotation of the term "synners" also suggests another important element of their "born digital" generational identity: a basic sensibility of transgressiveness that emerges from repeated experience of fluid boundaries. Their successful navigation of media flows and distributed learning and social environments requires the fluid mutation of interests, identities, and affiliations. While this may vex those born before the ubiquity of digital networks, this *mutability* is exactly what we (collectively) need them to foster, because it is the foundation for a life-time of learning and success in a rapidly changing knowledge-based world. The challenge for them and for those of us involved in the design of learning practices for the future is how to support and channel this mutability in creative and productive ways. More specifically, we must take seriously their experiences with the use of new social networks and digital media technologies. As such teachers and educators need to develop new pedagogical protocols that acknowledge and embrace the essential mutability of the subjects of digital technologies. But at the same time we must also imagine ways to offset the intense *presentism* of their digital experiences. As they navigate different media flows, there is a persistent sense of the now and the next, but very few markers of the past and of history. As was true in the analysis of postmodern logics of cultural bricolage, history is often treated as a mere repository of images, slogans, and remixable clips. Issues of historical context and the specificity of media are difficult to convey in the wash of images that is the vernacular of remix culture. While we might be right to be wary of the ideological work performed through the dutiful rehearsal of grand narratives of history—of empire or of progress, for example—there remains a challenge of how to engender the development of a historical imagination for those whose learning happens through surfing media flows. Is it useful to return to the work of John Dewey (or Paulo Freire and Lev Vygotsky for that matter) for inspiration in helping us revitalize our thinking about education in a digital age that is grounded in an appreciation of the subjective experiences of the born digital generation that also recognizes the value of the past for the purposes of cultural reproduction?[18] While a fuller discussion of their insights about the relationship between learning, experience, and the development of the historical imagination is beyond the scope of this chapter, recall the question that Dewey (1938/1997) framed as the core problematic of education: "How shall the young become acquainted with the past in such a way that the acquaintance is a potent agent in appreciation of the living present"? (23) Dewey was adamant that a robust educational theory had to be grounded in a notion of experience, and yet, he was also careful to clarify that: "while all genuine education comes about through experience," it does not follow that "all experiences are genuinely or equally educative" (25). The project for educators, he argued, is to explicate the relationship between experience and education.

Indeed, this is some of the important work going on at the edges of the learning sciences, not by learning theorists per se but by those who understand the importance of examining the experiences of those young people who participate in online communities and multiple knowledge creation ecologies. Of particular interest is the work by Douglas Thomas and John Seely Brown that analyzes the emergent epistemological sensibilities of young people who live, learn, and play among different kinds of digital and embodied networks. Thomas and Brown (2008) identify these emergent sensibilities as a *disposition*:

> Dispositions are not consciously chosen and enacted (any more than a glass chooses to break when dropped on a hard surface). Dispositions are triggered because they are deeply embodied states of comprehension that we act upon at a tacit level. Dispositions are the means by which we make sense of our experiences, a "grasping of disjointed parts into a comprehensive whole." (Thomas and Brown, 2008, no page)

Inspired by Dewey and Vygotsky, Thomas and Brown (2007) examine the implications of shifting and mutable subjectivities of online game participants. In particular, they analyze the epistemological contexts that constitute massively multiple online games (MMOGs) to describe the dispositions that enable participants to function in these elaborate information environments. Borrowing the notion of "conceptual blending" from Gilles Fauconnier and Mark Turner (2002), Thomas and Brown describe the epistemological experience that unfolds for a player in a MMOG wherein the imagination is engaged in the creation of a complex sense of reality that is simultaneously physical and virtual, real and imagined, social and private, digital and corporeal. Conceptual blending is a nuanced take on the cognitive process of synthesis—what Thomas and Brown describe as the fundamental process of "grasping of disjointed parts into a comprehensive whole." Bringing this insight to bear on the experiences of *Original Synners*, we could say that when they participate in online network cultures, especially those that involve complex virtual environments such as online games, they are not merely synthesizing *information* from disparate sources but rather creating entirely new experiences of "knowing and being." In reflecting on the specific epistemological quality of conceptual blending, Thomas and Brown argue, "whether it be in worldwide multiplayer games like *World of Warcraft* or new online tools such as *Wikipedia* or *Facebook*, the similarity among all of these spaces is the ability of people to create, shape and produce knowledge *on a constant basis*" (emphasis added, Thomas and Brown, 2008, no page). As an insight into the disposition of *Original Synners*, this understanding focuses our attention on their experiences as active participants who engage in the constant creation of knowledge as they navigate diverse network cultures and move through different kinds of virtual environments. While this notion of conceptual blending is only one of the dispositions created through the experience of online game playing, it is one that is particularly well attuned to the cognitive demands of a rapidly changing world. For this reason, it is one of the key dispositions that we must creatively engage in designing learning environments for the twenty-first century.

Learning Lessons from the Edge

Educators have the responsibility to design learning environments and institutional practices that foster the acquisition of dispositions that students will need for a lifetime of network navigation, information synthesis, social participation, and creative living. To do so, we must become the change that we advocate. We, the teachers, must make a commitment to becoming lifelong learners. Just as the notion of *Original Synners* helps to shift our ideas about the identity and dispositions of students, so too should the notion of teachers as lifelong learners inspire a reconsideration of the effort required of teachers to develop new dispositions and knowledge making practices. The language of learner-centered pedagogy recasts both students and teachers as collaborative participants in learning activities. To be responsible collaborators, faculty members, instructors, and researchers must develop their skills in the use of new digital applications and tools. The next two sections describe two paradigm shifting projects going on at the edges of one formal educational institution in the United States: the University of Southern California. These projects share a set of cultural values and philosophical commitments to take seriously the dispositions of *Original Synners* and the educational needs of Life-Long Learners. For the most part these efforts employ existing applications, information networks, and emergent social practices that themselves are borrowed from the edges of digital culture. Both projects contribute to the development of what Henry Jenkins described as a model of participatory learning. At the base, they illustrate three key characteristics required for the creation of new educational programs, pedagogies, and institutional structures:

- *Media rich*: Incorporating the use of multiple modalities of expression, including textual, visual, audio, dynamic, interactive, and simulated
- *Hybrid*: Combining networked and physical spaces, blurring lines between academic and everyday social, creative, and expressive practices; crossing traditional generational and cultural boundaries
- *Open*: extensible, participatory, non-proprietary, collaborative, distributed, many-to-many, multi-institutional, global

Neither of the projects discussed below proposes a transcendental model of digital learning; their innovations are context specific, mutable, and recombinant. This is as it should be: for in a digital age, it is unimaginable to think that any single set of best practices will address the varied learning needs of all members of the born digital generation. Our shiftwork is a remix project in its own right: where we strategically select and combine elements from a range of theoretically grounded innovations for the purposes of developing new literacies, pedagogies, and programs.

Multimedia Across the Curriculum: The Institute for Multimedia Literacy

Because literacy is a community based metric, the way in which new literacy programs will take shape will be influenced by different institutional contexts. For example, the Institute for Multimedia Literacy (IML) has, for more than a decade, served as a test bed for the development of new curricula and new pedagogies that involve the production of multimedia literacy educational activities and projects. Because it is housed within the USC School of Cinematic Arts, IML's approach to multimedia literacy draws deeply on traditions of visual expression, narrative, and sound, as well as on the emerging use of interactive media, ranging from games to immersive and mobile experience design. IML undergraduate programs are built on the assertion that in order to be fully literate in today's world, students should be able to read and write using the languages of multimedia as readily as they read and write using text.[19] The learning objectives that guide IML courses engage multiple intelligences: the social, cultural, and emotional, as well as the cognitive and the technical, in the service of exploring new modes of digital authoring.[20] The goal of the IML programs and research is to explore the full range of expressive potentials offered by moving images, sound, and interactive media, with a continuing emphasis on the integration of text as part of the expressive palette of multimedia.

Participants in IML programs learn to *write* multimedia by first learning to critically *read* it. Students develop proficiency with the modes of formal analysis required for the critical evaluation of a wide range of multimedia artifacts—including images, video, sound design, information visualization, typography, interface design, and interactivity. In addition, students become familiar with the major theoretical frameworks guiding the development of contemporary multimedia applications and interactive experiences. One of the key concerns of multimedia pedagogy is ensuring that students avoid the uncritical adoption of conventions of commercial or entertainment media. IML instructors introduce students to a broad range of multimedia genres—such as argumentative, documentary, essayistic, experiential, game-based, narrative, and archival forms—by performing expert critical readings of these media-rich cultural works. Based on these demonstrations/performances/explications, students are taught to evaluate the rhetorical strengths and weaknesses of different genres to serve different communicative objectives. In their own projects, students are required to justify their authoring and design decisions to demonstrate that their use of media and techniques are appropriate to their overall communicative goals.

As students become critical readers of multimedia, they also learn to produce multimedia in a scholarly way. Students gain experience in both individual and collaborative forms of multimedia authorship. In IML classes, students learn to choose appropriate media platforms for their projects, including video and audio productions: interactive DVDs, Web sites, games, exhibitions, and installations. Thus, students learn not only how media rich documents and modes of expression are constructed, they learn how to create them using a range of media making tools (cameras, sound devices, animation programs, game engines, storyboarding, and presentation applications). This wide range of authoring modes necessitates a highly skilled and diverse support structure, which includes teaching assistants, technical support staff, and student mentors, in addition to full-time faculty.

Rather than positioning "multimedia literacy" or scholarly multimedia as an emerging field, the IML developed strategies to integrate multimedia literacy efforts within existing disciplines and academic practices. IML classes are routinely taught within disciplines as diverse as history, philosophy, religious studies, geography, linguistics, and anthropology as well as more traditionally visually oriented fields such as cinema, communications, visual arts, and art history. The methods used by IML instructors, which draw from the fields of cinema studies and communication, are readily adaptable to fields within the humanities and social sciences, many of which are in the process of adapting to accommodate or experiment with audio/visual expression and different forms of electronic publication and technologically enhanced teaching. Multimedia, in these contexts function essentially to catalyze and promote innovations in research and pedagogy that are already emerging organically within various fields. In this sense, IML programs are explicitly designed to be transformative of the educational experience of students across the university. These programs seek to educate a new generation of students *and faculty* in strategies to enhance traditional academic practices through the use of media-rich modes of expression. For this reason, the IML also developed training programs for faculty such as in-service seminars and workshops on "Transforming Teaching Using Multimedia." This effort supported the integration of multimedia literacy into the disciplines because it served as a bridge to integrate faculty members' teaching interests with their research interests. As faculty learned new literacy skills, they were encouraged to rethink their scholarly publication and authoring practices using the affordances of new media. In this way, the time they spent learning new skills (literacies) benefited both their pedagogical practice as well as their scholarship. In the process, the modes of scholarly authorship within the disciplines were also subtly transformed. In the end, students as well as faculty who participate in these programs are expected not only to be multimedia literate, but also to be critically aware of the embedded social, political, and cultural values surrounding the uses of media, and ultimately to use this set of new communication tools in both creative and scholarly ways. The long-term goals of the Institute are to define and expand emerging scholarly vernaculars at the levels of undergraduate, graduate, and faculty publication and pedagogy.

Backchanneling in the Classroom: Opening the Classroom to Multiple Media Flows

For *Original Synners*, the process of knowledge creation happens across diverse settings, in formal institutions as well as through informal social and technological practices. Teaching and learning already occur in different kinds of informational spaces—distributed communities linked by wireless networks and mobile devices as well as on remote campuses, in smart classrooms, and in virtual environments. The multiplication of learning spaces is facilitated in part by increased access to high-speed data networks, but perhaps more important are the increasing familiarity and ubiquity of collaborative online activities as a part of many people's daily lives. Just as Thomas and Brown (2008) turned their attention to a consideration of the dispositions created in the course of online game playing, other researchers are also investigating the particular experiences developed during game playing that may have a bearing on learning in complex media environments.[21] An analysis of the game playing experience of young people suggests that in addition to providing dynamic visual, auditory, and sometimes bodily stimulation, interactions within game worlds also offer opportunities for players to express emotions, to engage in structured goal setting, and to gain a sense of accomplishment and social belonging. Higher order competencies include learning the ability to strategize, to interact tactically, to problem solve, to interpret, to use hints and aides, and to abstract general principles from specific situations. Moreover, for youths of school age, the bounded nature of a game world holds their attention in a way that traditional classroom educational activities may not. Many games require participants to move between multiple planes of reality: the world of the game, of the strategy, of the goal, of other players, and of the real world. Gamers learn by cycling through information spaces; they learn to iteratively scan multiple spaces and to adjust their activities in line with new information. In this way, games teach and condition a disposition of partial attending. In the process, the performance and temporality of "attending" are transformed. This type of attending is not easily accommodated by traditional classroom practices. This inherent complication suggests that although there is great interest among educators across the spectrum in the development of educational games, also now referred to as "serious games" or learning games, the deployment of games as an educational activity is going to require a broader refashioning of expectations of students' classroom behavior.

The pedagogical task here is to design classroom activities that take advantage of the new habits and modes of attending that are learned through the use of new digital media. Indeed, Linda Stone, former Vice President of Microsoft's Virtual Worlds research group, argues that the new disposition of attending common among gamers—a disposition she refers to as "continuous partial attention"—can be an extremely powerful mode of engagement.[22] This notion of "continuous partial attending" has stimulated experiments with the "opening up" of the classroom space to multiple information flows. For example, classes offered by the Interactive Media Division at the University of Southern California use multiple wall-sized screens as "windows" onto different kinds of media flows. Some screens might display Web pages; other screens present a text messaging "backchannel" space; other screens display prepared image and slide sets. Rather than wrestle with students to command their attention to a single-channel of communication (the "sage on the stage" classroom model), by incorporating publicly viewable multiple screens for the display of material, the pedagogical structure of the classroom seeks to enjoin the experience of "continuous partial attending." A typical IMD class session involves a speaker, presenter, or discussion leader who may display a set of digital slides or bookmarked urls on several screens, a "Google jockey" who commandeers another set of screens for the purposes of annotating and questioning the emergent classroom conversation by surfing through Web sites and online resources, and an IM backchannel (displayed on another set of screens). The viewable backchannel subtly changes the rituals of classroom communication by enabling multiple conversations to happen simultaneously.[23] In a similar way the display of Web searches (through the use of Google or other search engines) is a pedagogical technique that embraces the web as an important media flow that has something of value to add to the classroom based learning experience.[24]

The architecture of the IMD classroom embraces the hybrid nature of digital learning activities. The physical classroom space includes fourteen screens that create a panoramic room environment where any one of the screens or all of them can be accessed by a cluster of personal computers, commercial grade DVD players, or individual laptops. The room accommodates up to 30 participants (students, faculty, outside speakers), who sit at a large conference table or at individual workstations arrayed around the perimeter of the room. Some set of screens is visible from every seat in the room. The creation of this classroom, referred to as a "digital atelier," was the result of the applied research projects of several IMD faculty on the development of environmental and immersive panoramic displays. Special software enables the panoramic display of an image across all fourteen screens. Thus, the screens can be used as individual displays of images or digital content, or as tiled parts of a single panoramic image.[25] In this way, the research interests of IMD faculty result in the development of innovative technologies that in turn serve as the platform for pedagogical experiments, not only in the use of new modes of digital display but also more importantly in the integration of digital media flows into the classroom.

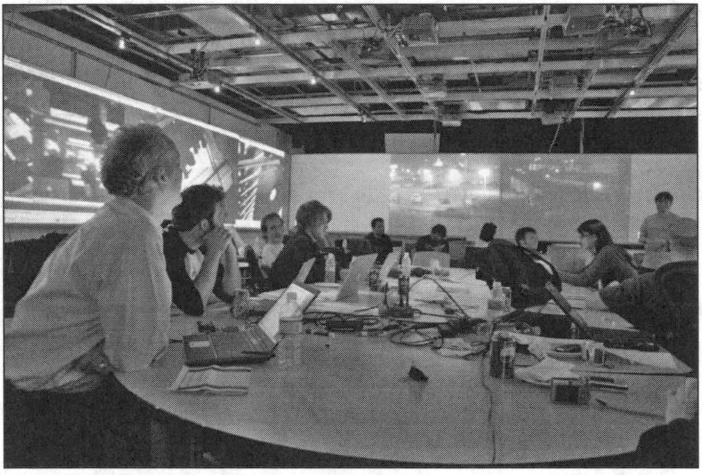

Figure 1: The main IMD classroom consists of 14 screens that cover three walls. Users can project onto any one of the screens or select multiple screens to display the same image.

As Stone (2006) points out, when individuals participate in multiple information streams, they learn to reinvent themselves as nodes within networks, who are capable of contributing to information flows as well as receiving them. Through interactions in a backchannel, an individual's agency in the classroom expands in interesting ways. Simultaneously interpellated as "listener," "audience member," and "peer," the student oscillates between technologically mediated subject positions. None of these positions is *purer* than the other; in the oscillating among them emerges the opportunity for the creation of new insights as one set of cognitive skills (of the listener, for example) interferes and collides with another set of cognitive practices (of texting). As a wild departure from the staging of communication in a traditional classroom, the early IMD back channel experiments suggest that the multiplication of information flows can productively stimulate meaningful interactions among students-as-peers in a classroom space. This classroom ecology of multiple media flows explores the epistemological experience of conceptual blending. All participants (students, presenters, teachers, visitors) have to find ways to meaningfully engage the multiple levels of reality and mediated information. Indeed, some participants find the backchanneling practice extremely distracting. In fact, when the conversation on the backchannel devolves into sniping, in-jokes and tangents, it is difficult to understand what "whole" can be usefully created from the disjointed flows and conversation streams. Howard Rheingold has argued that, although they are extremely promising, the existing cultural vernaculars that emerge in peer-to-peer social networking practices (such as IM-ing) are not always applicable to academic contexts. To use these tools effectively, teachers must not only understand the technological potential but also the kinds of structures needed to focus the energies these tools unleash. In particular, they need to introduce into these open classroom spaces also critical methods of reflection and assessment that consider the nature of interactions unfolding across multiple reality planes. In this version of the open classroom, the role of the educator is more like an orchestra maestro whose artistry is to *orchestrate* the creation of a meaningful experience. The Interactive Media Division's experiment in the creation of a digitally augmented learning space explores new conditions of possibility that take inspiration from the experiences of *Original Synners* who live in intersecting networks of information and media flows. It rests on a fundamental rethinking of the classroom, not as a bounded physical space within a fixed institutional geography, but rather as a mixed-reality learning ecology that is malleable, plastic, and usually able to open onto digital knowledge networks.

As much as these projects—the programs developed at the Institute for Multimedia Literacy and the backchannel experiments at the Interactive Media Division—are influenced by the particular institutional context of the University of Southern California, they are also part of a broader set of edge initiatives that define the current paradigm shift to a more far-reaching participatory culture. Several questions follow from this discussion, some of them are institution specific (important and of interest perhaps only to those at USC), others are more far-reaching. The reexamination of the role of the university in a digital age involves a wider set of issues than the ones I focus on here. For example, I do not explicitly discuss related issues such as the waning cultural authority of the professoriate, the notion of education versus credentialing, or the organization of formal education within a specific national context. Instead, I focus on the paradigm shift work that is already underway at the edges of the academy to argue that the university should explicitly embrace the mission to cultivate the technological imagination. By doing so, I believe that the university would be poised to make a significant contribution to the process of cultural transformation that is at the heart of the current paradigm shift. From the development of new identities and dispositions, to the transformation of the way we think about technology, to the labor required for specific projects of institutional reformation, the range of shift work we need to engage in and support is magnificently and dauntingly far-reaching. This chapter argues that this is a thoroughly worthy ambition for the exercise of the technological imagination in a digital age.

References

Adler, Mortimer, and Charles Van Doren. 1940/1972. *How to read a book.* New York: Touchstone.

Anderson, Steve, and Anne Balsamo. 2007. A pedagogy for original synners. In *Digital youth, innovation, and the unexpected,* ed. Tara McPherson, 241–259. Boston: MIT Press.

Balsamo, Anne. 1996. *Technologies of the gendered body: Reading Cyborg women.* Durham, NC: Duke UP.

———. 2000. Engineering cultural studies: The postdisciplinary adventures of mindplayers, fools, and others. In *Science + Culture: Doing cultural studies of science, technology and medicine,* ed. Roddey Reid and Sharon Traweek, 259–274. New York: Routledge.

Barr, R. B., and J. Tagg. 1995. From teaching to learning—A new paradigm for undergraduate education. *Change Magazine,* November–December, 12–25.

Bernstein, Danielle. 1991. Comfort and experience with computing: Are they the same for women and men? *SIGCSE (Special Interest Group in Computer Science Education) Bulletin,* 23.3: 57–60.

Boellstorff, Tom. 2008. *Coming of age in Second Life: An anthropologist explores the virtually human.* Princeton, N.J.: Princeton UP.

Buckingham, David. 2002. The electronic generation? Children and new media. In *Handbook of New Media: Social shaping and consequences of ICT,* eds. Leah A. Lievrouw, and Sonia Livingstone, 77–89. London: Sage Publications.

Cadigan, Pat. 1991. *Synners.* New York: HarperCollins.

Carr, Nicholas. 2008. Is Google making us stupid? *The Atlantic,* July/August, 56–63.

Cassell, Justine, and Henry Jenkins. 1998. *From Barbie to Mortal Kombat: Gender and computer games.* Cambridge, MA: MIT Press.

Cole, Michael, Vera John-Steiner, Sylvia Scribner, and Ellen Souberman. 1978. *L. S. Vygotsky, mind in society: The development of higher psychological processes.* Cambridge, MA: Harvard UP.

Dewey, John.1916. *Democracy and education.* New York: Macmillan, 1916.

———. 1938/1997. *Experience and education.* New York: Touchstone.

Fauconnier, Gilles, and Mark Turner. 2002. *The way we think: Conceptual blending and the mind's hidden complexities.* New York: Basic Books.

Fisher, Scott S., Steve Anderson, Susana Ruiz, Michael Naimark, Perry Hoberman, and Richard Weinberg. 2005. Experiments in interactive panoramic cinema. In *Stereoscopic displays and virtual reality systems XI.* ed. A. Woods, M. T. Bolas, J. O. Merritt and I. McDowell. Proceedings of SPIE, vol. 5664: 626–632.

Freire, Paulo. 1970/1995. *The pedagogy of the oppressed.* New York: Continuum.

Frenkel, James, ed. 2001. *True names and the opening of the cyberspace frontier.* New York: Tor.

Gardner, Howard. 2006a. *Multiple intelligences: New horizons.* New York: Basic Books.

———. 2006b. *Five minds for the future.* 2006b. Boston, MA: Harvard Business School Press.

Gee, James Paul. 2003. *What video games have to teach us about learning and literacy.* New York: Palgrave Macmillan, 2004.

Greenfield, Patricia Marks. 1984. *Media and the Mind of the Child: From Print to Television, Video Games and Computers.* Cambridge, MA: Harvard UP.

Hall, Justin, and Scott Fisher. 2006, April. Experiments in backchannel: Collaborative presentations using social software, Google jockeys and immersive environments. Computer and Human Iinteraction (CHI) Conference. (April 2006) Accessed on October 27, 2007. <http://nvac.pnl.gov/ivitcmd_chi06>.

Hayles, N. Katherine. 2005. *My mother was a computer: Digital subjects and literary texts.* Chicago: U of Chicago Press.

Herz, J. C. 1997. *Joystick nation: How videogames ate our quarters, won our hearts, and rewired our brains.* Boston, MA: Little, Brown.

Jenkins, Henry. 2006. *Convergence culture: Where old and new media collide.* New York: New York UP.

Kafai, Yasmin, Carrie Heeter, Jill Denner, and Jennifer Sun, eds. 2008. *Beyond Barbie and Mortal Kombat: New perspectives on gender and gaming.* Cambridge, MA: MIT Press.

Kozulin, Alex, Boris Gindis, Vladimir S. Ageyev, and Suzanne M. Miller. 2003. *Vygotsky's educational theory in cultural context.* New York: Cambridge UP.

Kuhn, Thomas. 1962/1970. *The structure of scientific revolutions.* 2nd ed. Chicago: U of Chicago P.

Margolis, Jane, and Allan Fisher. 2002. *Unlocking the computer clubhouse: Women in computing.* Cambridge, MA: MIT Press.

Meadows, Mark Stephen. 2008. *I, Avatar: The culture and consequences of having a Second Life.* Berkeley, CA: New Riders Press.

Palfrey, John, and Urs Gasser. 2008. *Born digital: Understanding the first generation of digital natives.* New York: Basic Books.

Prensky, Marc. 2001a, October. Digital natives, digital immigrants. *On the Horizon MCB University Press, vol.* 9, no. 5 (October 2001). Accessed on January 16, 2010 from: http://www.marcprensky.com/writing/Prensky%20%20Digital%20Natives,%20Digital%20Immigrants%20-%20Part1.pdf

Prensky, Marc. 2001b. *Digital game-based learning.* New York: McGraw-Hill.

Rosensweig, Roy. 2003, June. Scarcity or abundance? Reserving the past in a digital age. *American Historical Review* 108 (3): 735–762.

Salen, Katie. 2008. *The ecology of games: Connecting youth, games and learning.* John D. and Catherine T. MacArthur Foundaiton Series on Digital Media and Learning. Cambridge, MA: MIT Press.

Seely Brown, John. 2000, March/April. Growing up digital: The Web and a new learning ecology. *Change,* 10–20.

———. 2002. Learning in the digital age. *The Internet & the university: Forum 2001,* ed. Maureen Devlin, Richard Larson, and Joel Meyerson. Published as a joint project of the Forum for the Future of Higher Education and EDUCAUSE.

Seely Brown, John, and Paul Duguid. 1998. Universities in the digital age. *The mirage of continuity: Reconfiguring academic information resources for the 21st century,* ed. Brian L. Hawkins and Patricia Battin. Washington, DC: Council on Library and Information Resources.

———. 2000. *The social life of information.* Boston, MA: Harvard Business School Press.

———. 2002. Local knowledge: Innovation in the networked age. *Management Learning* 33 (4): 427–37.

Selfe, Cynthia, and Gail Hawisher. 2007. *Gaming lives in the twenty-first century: Literate connections.* New York: Palgrave Macmillan.

Shaviro, Steven. 2003. *Connected, or what it means to live in the network society.* Minneapolis, MN: U of Minnesota P.

Stone, Linda. 2006. Attention: The *real* aphrodisiac. Talk delivered at the 2006 Emerging Technology Conference. Accessed on April 1, 2007.

Stross, Charles. 2005. *Accelerando.* New York: Ace Books.

Thomas, Douglas, and John Seely Brown. 2007. The play of imagination: Extending the literary mind. *Games and Culture* 2: 149. Accessed on April 5, 2009. <-Gac.sagepub.com/cgi/content/abstract/2/2/149>.

Thomas, Douglas, and John Seely Brown. 2008, Nov. 25-Dec. 1. The power of dispositions. *Ubiquity* 9 (43). Accessed on April 1, 2009. <http://www.acm.org/ubiquity/volume_9/v9i43_thomas.html>.

Toffler, Alvin. 1981. *The third wave.* New York: Bantam.

Turkle, Sherry. 1984. *The second self.* New York: Simon and Schuster.

———. 1997. *Life on the screen: Identity in the age of the Internet.* New York: Simon and Schuster.

Vinge, Vernor. 1993. The singularity. Talk presented at VISION-21 Symposium sponsored by NASA Lewis Research Center and the Ohio Aerospace Institute, March 30–31, 1993. Accessed on October 27, 2007. <http://mindstalk.net/vinge/vinge-sing.html>.

Notes

1. As Roy Rosensweig (2003), a prominent theorist of the digital humanities, claims, "historians, in fact, may be facing a fundamental paradigm shift from a culture of scarcity to a culture of abundance" (2003: 739).
2. Jenkins (2006) characterizes this paradigm shift in the following way: "[C]onvergence represents a paradigm shift—a move from medium-specific content toward content that flows across multiple media channels, toward the increased interdependence of communication systems, toward multiple ways of accessing media content, and toward ever more complex relations between two-down corporate media and bottom-up participatory culture" (2006: 243).
3. Alvin Toffler (1981) first coined the term "prosumer" in his book, *The Third Wave.*
4. Charles Stross's (2005) science fiction novel, *Accelerando,* explores the contours of post-Singularity human existence. Critics as well as fans identify it as one of the most elaborated thought-experiments about human life after the Singularity.
5. I remind readers that Vernor Vinge was one of the first to imagine (in his fictional story "True Names") the network of computers that we would now recognize as the Internet. See: James Frenkel, ed. *True Names and the Opening of the Cyberspace Frontier* (2001).
6. Indeed much of my early work on technologies of the gendered body focused on the cultural implications of the decorporealization of post-humanist fantasies of human evolution (Balsamo, 1996).
7. In taking on this topic, I am greatly inspired by conversations and writings by John Seely Brown: see Seeley Brown, 2002; Seeley Brown and Duguid, 1998, 2000, and 2002.
8. Thomas Kuhn (1962, 1970) was clear to assert that new paradigms are not adopted, initially, because of the presentation of rational evidence. If that were the case, there would be no need for a NEW paradigm; the evidence and the construction of "rationality" would have been sufficiently provided by the previous paradigm. Instead, Kuhn explains:

   > The man [sic] who embraces a new paradigm at an early stage must often do so in defiance of the evidence provided by problem solving. He must, that is, have faith that the new paradigm will succeed with the many large problems that confront it, knowing only that the older paradigm has failed with a few. . . . Something must make a least a few scientists feel that the new proposal is on the right track, and sometimes it is only personal and inarticulate aesthetic considerations that can do that. . . . Rather than a single group conversion, what occurs is an increasing shift in the distribution of professional allegiances. . . . Nevertheless if they are competent, they will improve it, explore its possibilities, and show what it would be like to belong to the community guided by it. (1970: 158–159)

9. Even though Marc Prensky (2001a) is the one who first popularized the terms "digital natives and digital immigrants," he acknowledges the influence of Douglas Rushkoff's statement about kids as natives and adults as immigrants in a digital landscape. For Prensky's acknowledgment of Rushkoff's influence, see the footnote on page 414 of Prensky (2001b).
10. John Seely Brown (2000) develops the notion of "growing up digital" to describe the experiences of young people who grew up with the Web. Brown makes the important observation that the Web for these young people is not a discrete technology but part of an emergent multimodal ecology of learning.
11. David Buckingham (2002) cautions against making grand claims about the digital literacies and experiences of children born within a particular time period, such as the dates that mark the diffusion of new media technologies. He calls for the careful investigation of the differences that play out among groups of children in their access, use, and mastery of particular new media applications (such as games) and devices (such as mobile phones).
12. See especially Dannielle Bernstein, "Comfort and Experience with Computing: Are They the Same for Women and Men?" 1991. There is a significant amount of feminist research that investigates the differences in experiences with technology. This was the subject of my first book (Balsamo, 1996). For more recent work on gender and technology use, see: Jane Margolis and Allan Fisher, *Unlocking the Computer Clubhouse: Women in Computing* (2002); On the topic of gender and game playing, the first major serious treatment was Cassell's and Jenkins' edited collection (1998) *From Barbie to Mortal Kombat: Gender and Computer Games.* Continuing the discussion and reflecting on the rapid changes in gender and gaming experiences the volume edited by Yasmin Kafai, Carrie Heeter, Jill Denner, and Jennifer Sun, *Beyond Barbie and Mortal Kombat: New Perspectives on Gender and Gaming* (2008) focused attention on the broader media context that serve as the ecologies within which gaming experiences unfold.
13. Sherry Turkle's work is foundational on this topic: *The Second Self* (1984) and *Life on the Screen: Identity in the Age of the Internet* (1997). See also Steven Shaviro (2003), *Connected, or What It Means to Live in the Network Society;* N. Katherine Hayles (2005) *My Mother Was a Computer: Digital Subjects and Literary Texts;* Mark Stephen Meadows (2008) *I, Avatar: The Culture and Consequences of Having a Second Life;* and, Tom Boellstorff (2008) *Coming of Age in Second Life: An Anthropologist Explores the Virtually Human.*
14. This section is influenced by the article "A Pedagogy for Original Synners," Steve Anderson and Anne Balsamo in Tara McPherson's (2007) edited volume *Digital Youth, Innovation, and the Unexpected* that is part of the John D. and Catherine T. MacArthur Foundation Series on Digital Media and Learning.
15. Although it is difficult to determine when and where the very first community media literacy program was started, some noteworthy early efforts include the San Francisco-

based program called TILT. Started in 1995 by video artist Lise Swenson, TILT's mission is to "teach young people who are typically underrepresented and misrepresented in the mainstream media, the fundamentals of movie-making and media literacy through hands-on training in media production." www.tiltmedia.org

16. The importance of learning skills of synthesis and of integration has been part of the discussion about the development of critical reading techniques for more than fifty years. Mortimer Adler discussed this notion in the first edition of his book, *How to Read a Book* published in 1940 (Adler and Van Doren, revised edition, 1972).
17. This is an explicit reference to Pat Cadigan's novel, *Synners* (1991). For another discussion of the education of original synners in the context of creating a cultural studies curriculum within an engineering institute see Balsamo (2000) "Engineering Cultural Studies: The Postdisciplinary Adventures of Mindplayers, Fools, and Others."
18 . While it is not the purpose of this chapter to delve into a discussion of a philosophy of learning for the digital age, suffice it to say that the time is ripe for revisiting the constructivist (or instrumentalist) learning theories of John Dewey, Paulo Freire, and Lev Vygotsky. While John Dewey's classic work, *Democracy and Education* (1916) is one of his best known, perhaps his relevant work on this topic is found in the later book called *Experience and Education* (1938, 1997) where he reflects on two decades of experimentation in the creation of progressive educational programs. Paulo Freire was one who strongly advocated for the dissolution of the dichotomy of teacher and student, in favor of a more fluid exchange between who teaches and who learns in the educational encounter. See Freire, *The Pedagogy of the Oppressed* (1970/1998). The classic publication by Vygotsky is his 1926 volume, *Educational Psychology.* For a useful introduction to Vygotsky's important essays see the volume edited by Michael Cole, Vera John-Steiner, Sylvia Scribner, and Ellen Souberman, (1978) *L.S. Vygotsky, Mind in Society: The Development of Higher Psychological Processes.* For a more recent discussion of Vygotsky's learning theory see the volume edited by Alex Kozulin, Boris Gindis, Vladimir S. Ageyev, and Suzanne M. Miller (2003) *Vygotsky's Educational Theory in Cultural Context.*
19. The IML administers two educational programs: the Honors in Multimedia Scholarship Program is a four-year, undergraduate program open to students across the university, while the Multimedia in the Core Program introduces multimedia authoring into the University's General Education program via single-semester classes designed to reach a broad segment of the USC undergraduate population. iml.usc.edu
20. Howard Gardner's (2006a and 2006b) work explicitly informs the development of IML's programs and pedagogical frameworks.
21. Jim Gee's (2003) has established the key terms of the conversation about games and learning. Cynthia Selfe's and Gail Hawisher's (2007) edited volume expands the topics under discussion by including important essays on gaming and difference that consider issues such as girls and gaming, racial representations in games, sexual identity and game play, and the experience of gray (mature) game players. Early work that con-

tributed to the discussion on game playing and literacy include: Patricia Marks Greenfield (1984) *Media and the Mind of the Child: From Print to Television, Video Games and Computers,* and J. C. Herz, *Joystick Nation: How Videogames Ate Our Quarters, Won our Hearts, and Rewired Our Brains,* 1997. Justine Cassell's and Henry Jenkins' important edited volume *From Barbie to Mortal Kombat: Gender and Computer Games* (1998) carefully sorted through the facts and fictions surrounding the issue of girls and games (among other topics). Marc Prensky, *Digital Game-Based Learning* (2001) considers the use of video games for teaching adults in corporations and the military. More recent work that explores the relationship between gaming and the design of educational experiences include Katie Salen's edited volume in the John D. and Catherine T. MacArthur series on Digital Media and Learning called *The Ecology of Games: Connecting Youth, Games and Learning* (2008).

22. Stone discussed the concept of "continuous partial attention" in her talk at the 2006 Emerging Technology Conference.
23. See Justin Hall and Scott Fisher (2006) "Experiments in Backchannel: Collaborative Presentations Using Social Software, Google Jockeys and Immersive Environments."
24. The use of a google jockey refutes the polemic media blitz that posited the question: "Is Google Making Us Stupid?" (Carr, 2008).
25. The creation of this classroom space was only one of the research efforts conducted by IMD faculty to invent and implement new forms of display technologies for immersive and semi-immersive applications (Fisher, et al. 2005).

20

Learning in the Creative Economy

Patrick Whitney

Model T Education in a Twitter World

If Henry Ford could return this afternoon by time machine, he would be utterly confused by modern enterprises. Production, communication, corporate organization, work, and leisure would be unrecognizable to him. However, there would be one place that Mr. Ford would find familiar and unchanged—the public school.

Henry Ford would recognize the public schools of today precisely because they are structured in the same way his world was—that is, in the technological, organizational, and socio-economic context of the industrial age. Ford revolutionized mass production with his River Rouge plant. Standardization and economy of scale drove essentially all industries then, whether in manufacturing or media. In Ford's day and after, leading businesses tended to be organized vertically and were self-contained; GM and AT&T seemed like unassailable institutions. For working men (few women worked, after all) the archetypes were the "organization man," the union job for life, and the hierarchy of the established order. Of course, none of this is true in the business world anymore, but schools still operate as though it were true. Schools are like production lines with children processed in the same way at each station, or, in this case, classes and grade levels. Can such a system help children become the adults they need to be to succeed, thrive, and contribute to a creative economy characterized by speed, flexibility, uncertainty, change, and shared knowledge?

Changing the education system to incorporate all that digital technology has to offer and to cultivate the citizens and workers we will need in the future is no small task. Digital technology is a disruptive force, but we know enough about similarly disruptive forces to discern a pattern. Adopting a new and disruptive technology happens in two stages.

First, the technology enters an industry by providing a better and faster way of performing a function; second, the technology transforms the fundamental way organizations work. For example, in the 1950s, computers entered organizations as a faster and more accurate way of doing accounting. Only in the second stage, as people gained experience and the technology matured, did they alter the way organizations were structured and how they worked.

The outcomes of disruptive technology can be seen in a country store in Arkansas called Wal-Mart that created a supply network, enabling it to become the largest company in the world; an underdog computer company named Apple reinventing the music industry; a digital search company—Google—simultaneously disrupting the industries of publishing and advertising; and a community bulletin board—Craigslist—replacing the most profitable source of revenue for the local daily newspaper. However, just as the second stage of transformative powers in retail, music, advertising, and publishing sectors only revealed themselves over time and through arduous creative efforts, it is not yet obvious what long-term fundamental changes will occur in schools as a result of the digital revolution.

The end result of this disruptive transformation for schools remains unknowable, and there are too many variables changing too quickly for classic research and planning methods to accurately predict what the best result might be. An example of this dilemma can be found among business leaders who increasingly find themselves in uncertain and fast-changing times. The rapid growth in their ability to make anything and the increased number of choices available to consumers leaves corporate leaders in a situation in which they know how to make almost any product but are uncertain about what to make. As a way of coping businesses are rapidly adopting design methods to gain a new understanding of their position and to explore alternate futures, so they can still make reasoned decisions with incomplete information. Why would these businessmen look to design? Design is normally thought of as the skillful drawing and styling of a product, message, or building. However, behind most important design initiatives in history, there was a process invisible to the public preceding any drawings. This process involved, one, not accepting a problem as it was given and, two, redefining or reframing the problem in order to understand it in new ways. Prototypes would be used as a concrete way of exploring potential answers, rather than as test to arrive at one single correct answer. The part of the design process concentrating on reframing the problem and exploring options was not particularly relevant in the stable world of Henry Ford. However, now, in an era of uncertainty, it is key to help organization leaders decide the best course of action when faced with the combination of uncertain outcomes and an immense number of possible solutions,

Traditional development processes taught in engineering and businesses accept problems as given, research what customers say they like and dislike, choose a direction, and work toward an optimal solution.

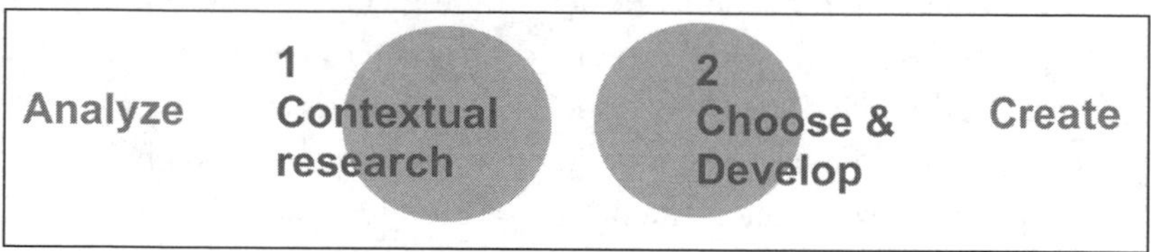

Organizations that seek innovative methods to accomplish the respective company's goals are using advanced design processes. More specifically, they do not accept a problem as given in and of itself, but instead they look at it in a more abstract manner and take into account normally external—but nevertheless important—factors. This approach facilitates the reframing of the problems and the creation of a portfolio of options that offer choices for both near-term and long-term innovation.

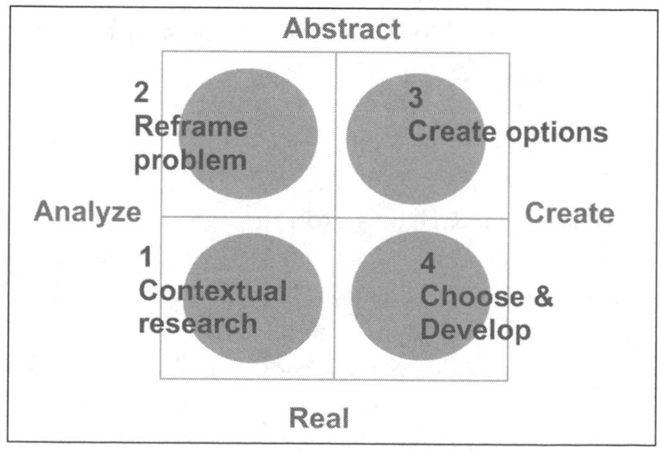

While schools' long-term response to digital technology cannot be known, in the short term, they are following the pattern of other organizations faced with a disruptive technology. Time and again, the standard pattern is for organizations to initially ignore disruptive technology, claiming it is not relevant to their core needs. Then they adopt it, using it at first as a faster and better way of doing an existing necessary task. Schools are in the middle of this shift, with some banning cell phones and others creating electronic textbooks.

Ultimately, however, schools will not be improved if they only transform the medium of delivering content while ignoring both the changes in the way organizations in the larger world work and what society needs from its students-turned-adults. If schools only learn to deliver content digitally, it would be analogous to a company limiting its use of computing to faster accounting.

It will not be easy for schools to determine a correct course of action. The changes in technology and related experiences happen too quickly for the standard research methods used in the field of education. Kids' lives are changing faster than anyone can gather and analyze data. What good is it to conduct a two-year study of how people use new media if the nature of the media and their use are changing every three months?

To overcome the problems of standard education research practices, the John D. and Catherine T. MacArthur Foundation has supported design research at the IIT Institute of Design where teams including faculty, PhD candidates and graduate students have been exploring what learning tools might be like in the near future. The projects move quickly by developing multiple assumptions early on that cannot be proven in the academic sense, but are believed to be self-evidently correct and fundamentally sound. For example, two of the core principles are: innovations should be learner-centered and make use of their intrinsic interests; and innovations should be designed for learners to work on their own and in groups.

Rather than taking time to prove that these principles are true, we accept the principles because they are applicable in the rest of the world. These principles are used as guides in the creation of several rough prototypes to discover how to increase kids' active engagement in learning processes. This type of research does not limit variables but looks for all the factors that seem to influence how a group of kids learn. It involves users with iterative prototypes as part of the discovery process rather than for testing or validation. In addition, it alters the principles on the fly, as we learn more from working with the kids.

We give up on certainty about a small set of information at the outset, but we acquire a broader look at the problem and gain experience grounded in the complexities of what kids actually do as they use our prototypes. Compared to traditional research, this type of design research leads to bolder, more comprehensive results that are both logical and obviously engaging. Furthermore, our "experiments" involving kids using prototypes bring us much closer to final innovations that allow the ideas to be deployed to kids in much less time than data derived from traditional research.

Schools in the Creative Economy

Networks are making the world evolve faster, and cities, communities, and schools in the twenty-first century will thrive because of their connectedness. Ideas, capital, and talent can play together and recombine in unplanned ways in this new environment. Digital technology can facilitate different exponents of talent to play and work together through its ability to support different types of organizations, learning styles, and access to content. Having the ability to network with others, share original thoughts and creations, and work together will only become more important in the future, as will the tools to leverage the power of community members, students, and others to contribute within schools.

In this context, the focus of our research has been to help transform schools from the last major living industrial age institution to an organization that belongs to an information-intense world. Because this new world is not standing still, the design processes related to reframing problems and creating multiple options were at the core of our work.

Reframing

Reframing involves not accepting the problem as defined, bringing in factors normally thought of as external to the problem, and seeing the problem in a new and expanded way. We observe, for example, user work-arounds, often-unconscious techniques people employ to overcome deficiencies in processes or products. These observations provide us with invaluable insights that drive us toward more thorough, expansive, and radical solutions to complex problems.

The traditional way of looking at education has three components: teachers, schools, and students. Our reframing of the education problem fell into four clusters: the context of education, the learner-school disconnect, learning and technology, and organizational structure and bureaucracy. These clusters are described more fully below together with the supporting circumstances we observed.

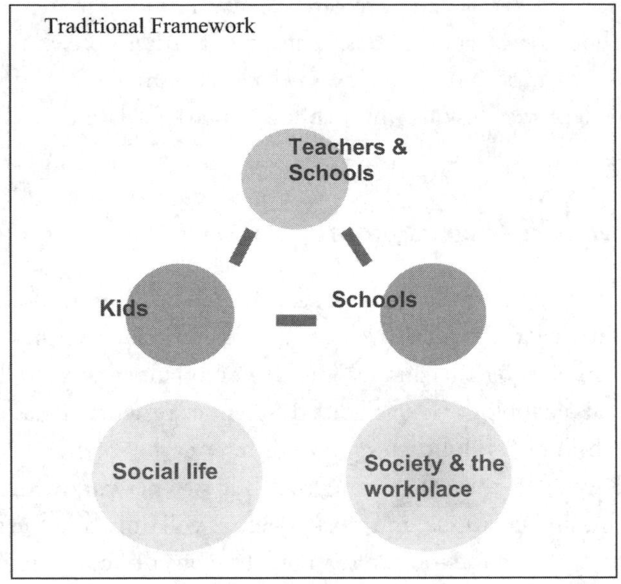

Traditional Framework

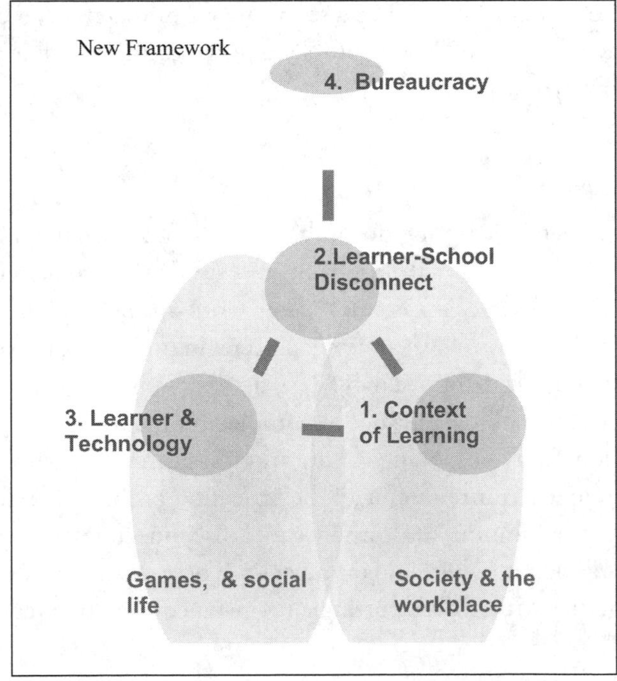

New Framework

1. Context of learning

Schools have failed to respond to change in education's social and economic context.

Undervaluation of education

To attract vital funding and talent, a mechanism for properly valuing education as a critical factor of production in our society must be found. Unlike companies, schools do not go out of business, so they do not feel the threat of demise pushing them to adapt to a changing competitive landscape. Because of the lack of competition in the marketplace for public schools and teachers, teaching is undervalued, generally underpaid, and sometimes attracts dilettantes. Teaching must be repositioned as a respected profession in American society, so the best and brightest teachers can be recruited.

Commercialization of education

Rather than ignoring organizations that treat learning as a business, schools should leverage the innovations happening in commercialized educational settings and understand the impact and value of innovations happening in the for-profit educational market. Whether it be the structure of the University of Phoenix or the market demand shown by the millions of parents choosing home-schooling, these activities on the edge are the source of many for presenting information, assessing what is being learned, structuring the time of the student, and linking the learning experiences to the rest of the students' lives.

Overcoming dogma

Strongly held educational belief structures limit large-scale innovation efforts. People have deeply ingrained ideas about what school is and how it should be organized, and changing these belief structures is key to innovation in education. Our educational beliefs have not adapted rapidly enough to keep pace with the changes in the structure of our economy. Educational dogma is inhibiting schools from changing in ways that will prepare kids for the knowledge economy.

Smart education systems

Schools need a way to assess the impact of smart education systems, and they need guidance on implementing system-wide solutions. There is a large and growing market for digital educational services like online tutors, curriculum development services, and professional development for teachers. The problem is that there is no integrated, system-wide solution. Affordable online tutoring in places like India is now available at 10 to 40 times less than similar services in America. Vendors of educational materials often provide a product solution for discrete districts in need of a system solution (e.g., something combining curriculum programs, integrated assessments, and professional development materials), but these vendors could better serve the market by selling data-integration services that would support broader solutions.

Out-of-school learning

There should be an institutional structure that accepts and values both out-of-school experiences and formal learning in schools. Kids' learning experiences outside school are increasingly more relevant to modern life than what they learn inside school. Innovation in schooling is happening at the edges of the educational institutions, but there is no mechanism for moving these innovations toward the center. Kids are increasingly motivated and engaged by what they learn in out-of-school programs and in their virtual online lives. A mechanism for capturing and integrating these online innovations must be found instead of resisting them.

As the nature of work changes, we need a new way to describe what children have learned and track their accomplishments through portfolios containing work for hire, school work (homework), volunteer work, and study work outside of school as well as accomplishments in the classroom.

Personal digital interaction

Schools should integrate personal digital interaction into their curricula. Currently they do not and, therefore, are unable to take advantage of its benefits, including peer assessment and information sharing. Digital technology can facilitate many interesting new kinds of interactions among teachers, students, and the external world. Embracing interactivity will require schools to surrender some of their authority for the benefit of their students.

Learner-School Disconnect

The culture and experiences of kids and parents are unrelated to school.

Connectedness and co-creation

Instead of acting as isolated factories, schools need to become nodes on a network connecting with other institutions. Schools see themselves as separate institutions. Each school has a principal and teachers, each of whom perform the same role as principal and teachers in every other school. Schools are not using digital tools to network kids with teachers, other educational resources, or kids in other schools, nor are they facilitating co-creation in ways that fit with modern ways of working and communicating.

The relationship between production and consumption is changing with the rise of co-creation tools like Wikipedia, user-controlled distribution channels like blogs, aggregation and sharing of information like Flickr, and Craigslist and the growth of peer-to-peer social networks like Facebook. As these new ways of communicating and collaborating pervade society and culture, the gulf between a kid's use of these tools outside school and the lack of them inside school widens the existing gulf between school and the outside world. These out-of-school communication channels diminish the relevance of learning inside school.

Schools need to become network organizations, establishing themselves as hubs at the center of diverse, overlapping networks of learning that reach out to the fullest possible range of institutions, sources of information, social groups, and physical facilities.

This generation's technology

Digital networking inside schools must accommodate problem solving, collaboration, storage of knowledge, and collection of information. Kids lead high-tech lives outside school and decidedly low-tech lives inside school. This digital divide is making the activities inside school have less real-world relevance to kids. A blend of intellectual discipline within a real-world context can make learning more relevant, and online technology can bridge the gap between the two. Kids prefer interactive products because they can be incorporated into their network of friends and provide a basis for play and interaction. Today's kids are skilled multitaskers, using multiple types of technology at one time to communicate. Kids growing up today live in a world where digital communication devices are ubiquitous, and they have fully embraced digital interaction in their lives.

Responsible collaboration

Schools need to create a method for assessing teamwork. The way schools teach today focuses the individual on improving and learning for his or her individual benefit, while the focus in industry, culture, and society is toward learning for the benefit of the group or community. Success in online games such as World of Warcraft requires maturity, responsibility, and a vested interest in the welfare of other players. This dynamic is similar to modern work environments, which are increasingly project focused and where success or failure is determined by group results rather than individual results. Collaboration favors community success but is also self-interested.

Technologically enhanced learning

Schools need a framework for understanding the impact of and determining how to implement new technologies. Kids are using technology outside school in increasingly sophisticated ways—using online technology such as universal access, location of related information, and communication with others—yet schools are failing to provide the tools and techniques that enable kids to take advantage of the digital media services available to them. One of technology's best uses for learning is as a platform for collaboration and feedback, but schools are currently unable to effectively leverage technology this way in the classroom. A range of technologically enhanced learning products is available, such as MIT's OpenCourseWare or Berkeley's podcast project, but most schools are unable to determine how to use these tools. The more kids connect using digital networks and the learning tools available to them outside school, the weaker the schools' institutional pull on children and their community will be.

Support structures

The support structures for learning are changing as the structure of family life and the nature of work and employment change.

Schools have been asked to take on new responsibilities as some traditional family and community-learning systems have deteriorated. Several of our experts decried the increasing responsibility of the schools to administer driver's education, sex education, parenting, child-care, and even basic civics, but part of the problem is the schools' inability to look at the whole picture.

Real-world relevance

Schools need to discover ways to bring real-world relevance into the classroom to engage children at higher levels. If the school experience is unconnected to the world outside the classroom, it will fail to engage a substantial percentage of capable learners. The problem is that the connection to the real world has been severed, a connection that in some fields does not return until college—a place beyond the reach of those who most need to see the practicality of what they learn. To keep students connected to education, education has to connect with students' lives and the workplaces they will someday inhabit. This involvement on the part of the schools can be accomplished by systematizing the involvement of professionals through mentoring Web sites and virtual field trips, utilizing the developing technology of high-definition television.

Learning and Technology

Change in the knowledge and learning environment is brought on by advances in digital technology, especially communication technology.

Personalized learning

For learning to be relevant to the future, it must be personalized, and only digital media provide the means to bring personalization to the classroom. Most public schools have a rigid structure of grades, grade levels, curricula, and credits that stifle creativity and move high-achieving kids toward the lowest common denominator. By creating gifted programs—now under attack as elitist, and they often follow the interests of the teachers rather than those of the students—and promoting a few problem achievers to a higher grade ("I can't give her what she needs in this classroom"), learning may be achieved, but personalized learning—the key to real achievement and leadership—is ignored.

In their lives outside of school, kids increasingly interact in a digital meritocracy that allows them the opportunity to push themselves to solve complex problems or explore personal passions for subject matters beyond the mandated curriculum of schools. To serve the future, schools need to learn how to teach, use, and assess each child's need to excel not just on the linear path laid down in state guidelines, but also through the now impossibly large body of knowledge in which each child can find for himself or herself information on the joys of chopper repair, calligraphy, or string theory.

Age-agnostic performance

For schools to allow all learners to advance at the pace best suited to them, a means of education that eliminates age barriers must be found. Schools are organized by age, but age and achievement are often not equivalent. Although some students in primary grades are "promoted" or sent to more advanced (older, including the local college) classrooms for a particular subject, most educators believe that a large difference in age between learners is unworkable due to interpersonal issues of physical maturity and status. In the K-12 demographic, individuals from widely different age groups will not accept each other as equals.

The social barriers among mixed age groups leave both younger and older learners at a disadvantage. Younger students are effectively held back, prevented from advancing according to their abilities. Older students are encouraged in their belief that age equals achievement. In the context of multiplayer online gaming, on the other hand, mixed age groups work very well because the virtualization of individuals through avatars allows for interaction without the interference of physical and social conventions. In such virtual meritocracies as World of Warcraft (WoW), eleven-year-olds can instruct and lead eighteen-year-olds without the difficulties they might encounter in a real social setting. Digital learning is part of the solution, but what we learn from WoW is not just that virtualization overcomes some problems, but also that the partition of learning into discreet, bounded units with defined goals—like games—facilitates a limited but rich instantiation of learning in which age is irrelevant.

Scope and sequence

Schools are not particularly good at identifying a child who is out of phase with the curriculum. There is an essential order to learning, and if it gets out of order, comprehension is more difficult. Since learners are able to name their needs only to a point, they need a smart system to push relevant collateral materials to them at the right time. It is not enough to turn a child loose in a library, not even an interactive, media-rich digital library of unimaginable size. The role of teacher as librarian, guide, coach, and scorekeeper needs to be emphasized over the role of presenter and authority figure. Furthermore, a quick and easy means of measuring the physical and social readiness of learners and correlating these to cognitive learning goals must be created, a means that is accessible to both students and teachers.

Organization Bureaucracy

Schools' bureaucratic structures encourage educational protectionism that allows only incremental improvements and thwarts all real innovation.

Educational protectionism

Schools need to find ways to circumvent strongly held outdated beliefs in order to move forward. Educational protectionism grows out of a deeply ingrained belief structure about what education should be. Our findings indicate that educational protectionism is a key factor in preventing personal digital interaction, connectedness, and co-creation from being integrated in today's classrooms.

Creating Options

Reframing the education problem led to three high-level criteria that helped us generate ideas for options that could be piloted in fast, small-scale trials.

- Innovation will come from the edges of the field, not the center.
- Innovations should assume schools are nodes on a network, not stand-alone institutions.
- Innovations should be kid-centered, not test-centered.

Using details from the reframed problem and applying it to our three general criteria, we developed a set of potential pilot projects. Through these we would gain actual experience with kids, teachers, and parents that would tell us more than traditional research could about the right answer to a tightly defined question.

Potential Topics for Prototypes and Trials

Electronic learning record (ELR)

The ELR is the electronic toolkit we conceived for managing a kid's educational history, goals, aptitude, and aspirations. The ELR is the interface that identifies each kid as a node on the network and allows other nodes (e.g., parents, teachers, other kids) to discover and convey the kid's achievement and needs based on learning style, aptitude, history, and personalized curriculum. The ELR also allows kids to interact on projects and to record their individual portfolio of learning and work accumulated inside and outside of school.

Cultural studies

Kids' lives outside school are very different from their lives inside school, due mostly to digital media. Kids today have access to many digital tools for socializing, playing, learning, and communicating. Contrast that with a kid's day inside school, which is mostly devoid of digital media. Understanding the culture of kids through ethnographic research into their daily lives is important in understanding the future of digital media in learning.

Little red schoolhouse

Some of the trends seen in learning and digital communications today are closely analogous to the world of the little red schoolhouse (LRSH). The LRSH is a metaphor for a system of learning where the student is both the teacher and the learner, and the teacher is the orchestrator and inspirer. A potential research project might map the activities, hierarchy, and structure of a traditional little red schoolhouse to an imagined virtual little red schoolhouse.

Role of the teacher

The role of the teacher in the twenty-first century is one of the most important and expansive research opportunity areas. The teacher's role today is much the same as it was 50 years ago and stands as a vital link to integrating digital media into classrooms and learning. We must first understand the role of the teacher in a rich media environment in order to craft the learning environment. An important component in this process is determining a teacher's value and the mechanisms that will see teachers compensated based on their economic worth.

Tools for educators

For educators today and in the future, the pace of change will continue to increase, and with change comes uncertainty about how to react. Today's educators suffer from a tyranny of choice in new digital teaching tools. Educators need to be able to analyze the new teaching tools at their disposal. More importantly, educators need to understand the potential impact of changes that technologies, laws, societal changes, community dynamics, domains of knowledge, and more have on the educational system and the kids they teach.

Smart transparent learning systems

Schools should use the transparency enabled by digital media as a means of uncovering problems in schools and in kids' different learning styles. Digital learning systems created by schools and researchers could discover individual and group mastery among students and convey the information back to them in ways that give kids additional control over their learning. Researchers could create smart digital learning systems that integrate play with learning and incorporate mechanisms giving parents and teachers a view into the system. Researchers could also focus on the power of connected digital systems to enhance students' learning.

Use real-world relevance to inspire

Inspiration is a critical component of an effective learning system. Schools could incorporate outside world experience in any number of ways like vocational training, connecting kids to social causes, leveraging gaming as a means of productive inquiry and so on, reframing schools as places of inspiration.

Home-school network

Many parents involved in home schooling are forced, due to lack of infrastructure, to be extremely innovative in finding and developing learning experiences for their kids. This research pilot would create a communications network to help parents aggregate their experiences with other parents, creating a rich and large laboratory generating and testing ideas. In some ways it would be like the open-source communities developing software, except in this case parents would be collecting and trying learning experiences instead of writing and testing code. Unlike wikis, it would not be open to everyone. Parents would have to apply and agree to participate, following the rules and structures set up by the research team. Parents would be motivated to participate in the community by the resources efficiently made available to them from other parents. The research team would have access to millions of examples of work contributed by hundreds of thousands of parents. Because the examples contributed by the parents would be described in a structured way, the researchers could find patterns of innovation that are now invisible and share and expand the best examples.

This home-schooling pilot follows the open-innovation model increasingly used by companies. Business leaders have discovered that their corporate research labs can never come close to matching the volume of relevant research being conducted in the rest of the world and have found means to tap into that larger pool of global information.

Arcade of learning nodes

The nature of new schools should respond to the new learning styles.

As structures, most school buildings reflect the current curriculum construct and the way each school day is organized around teachers delivering specific content to their students. Emulating the industrial-age thinking on which education is organized, one could see hallways as conveyor belts, classrooms as various production departments, and the sound of the bell as the factory whistle. However, what if the buildings were organized around the way kids learn? Rather than hallways connecting classrooms and people moving at the sound of the bell, there could be permeable environments that made it easy for kids to collaborate online or in groups and to be exposed to visitors from outside institutions. The school could become more like a village center than the River Rouge factory.

Building displaying student work

As belief systems regarding schools change, the buildings should change too, reflecting the guiding principles of being nodes on networks, innovations at the edge, and kid-centered. The building could have numerous large interactive screens displaying the student work, for example. This would allow the school to convey a sense of the value and the accomplishments happening inside. Imagine a place where a student approaching the school is recognized via his or her ELR and his schedule and most recent files were displayed.

Collaborative space

The classroom of the future should enable kids to collaborate and co-create in an environment that facilitates creativity and engenders inspiration. We envision a collaborative space where kids work on projects with real-world relevance, with the teacher acting as inspirer and orchestrator, where kids can connect to the outside world because all spaces, places, and objects are nodes on the network. We envision teacher studio spaces, individual reflection spaces, virtual desks, global connections, project-based learning, and real-world projects to inspire and connect to the outside world.

Teacher studio node

The changing role of the teacher and the changing nature of authority in the educational system require a new kind of space for teachers. A studio space would allow teachers to orchestrate and manage kids' activities as well as act as a collaborative workspace where teachers can inspire and direct kids in their more personalized learning spaces. The teacher studio would enable interaction between teachers and kids using displays that facilitate transparency of data at all levels.

Inventory space

Adjacent to the collaborative space would be an inventory space filled with objects, both virtual and real, that students could use for their projects. Imagine a group of seventh-graders working on a project measuring the amount of landfill space taken up by one year's worth of discarded plastic water bottles. The students would have access to water-bottle sales, to scales allowing them to measure volume, to virtual connections providing landfill data, as well as virtual connections to another seventh-grade class working on a related project in another state. All of these things would be available to participating students in an inventory space that included docking spaces for their ELRs.

Mobile genius lab

The mobile genius lab is a place we conceived where kids can learn from real-world experts to leverage real-world expertise into their school projects. A genius could be a pre-eminent physicist who would set up office for one week in a city and be accessible to all the students in a school district in his mobile genius lab. The mobile lab could be moved from school to school and would act as any other node in the arcade of learning nodes.

Prototyping as a Way to Explore Alternative Futures Electronic Learning Record

The electronic learning record (ELR), named BettrAt, is a Web application in the alpha stage of development. It helps kids and adults alike get better at their interests by placing emphasis on learning with peers and the support of mentors.

Main activities include:

- form and join groups of people with similar interests
- be mentored and act as a mentor to others in the group
- keep track of information related to accomplishments related to main interests
- plan how to accomplish goals related to their interests
- reflect upon what they learn and how they learn.

Thinkering Space

We believe that the act of making material goods promotes learning. Sometimes one needs to know before making something, and at other times one needs to make something in order to know. Our sense is that a particular kind of making, tinkering with physical objects, is in decline as kids spend more time in the digital world. In response, we conducted a project that builds interactive physical spaces and objects that kids can use to tinker as a way of knowing. We call this "Thinkering." We believe this process promotes the development of critical thinking skills that prepare kids when they encounter future scenarios of greater complexity. In particular, the following list of competencies, formerly seen as niche skills sets, is forecast to be of major importance for today's kids in their adult futures:

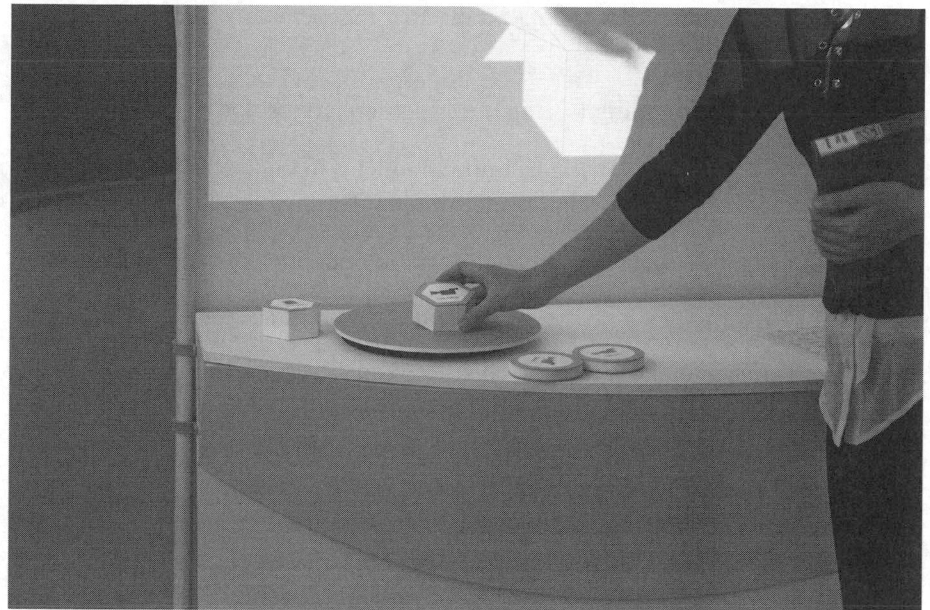

In Thinkering Spaces kids can manipulate tangible objects to explore and manage linked information drawn from various sources including archives of creative work, contents of books and data pulled down from the Internet.

Environmentally scaled interfaces display work as it is being created for an animated story being built frame by frame.

- creative thinking (developing intellectual independence and multiple perspectives)
- systems understanding (seeing meaningful relationships in complexity)
- innovative problem solving (framing problems in unconventional ways and connecting ideas through lateral thinking)
- information management (knowing how to find, organize, and use resources)
- interdisciplinary teamwork (collaborating effectively across disciplines)

The experience of Thinkering, self-directed discovery, and peer engagement within both physical and digital environments can help kids to develop these competencies.

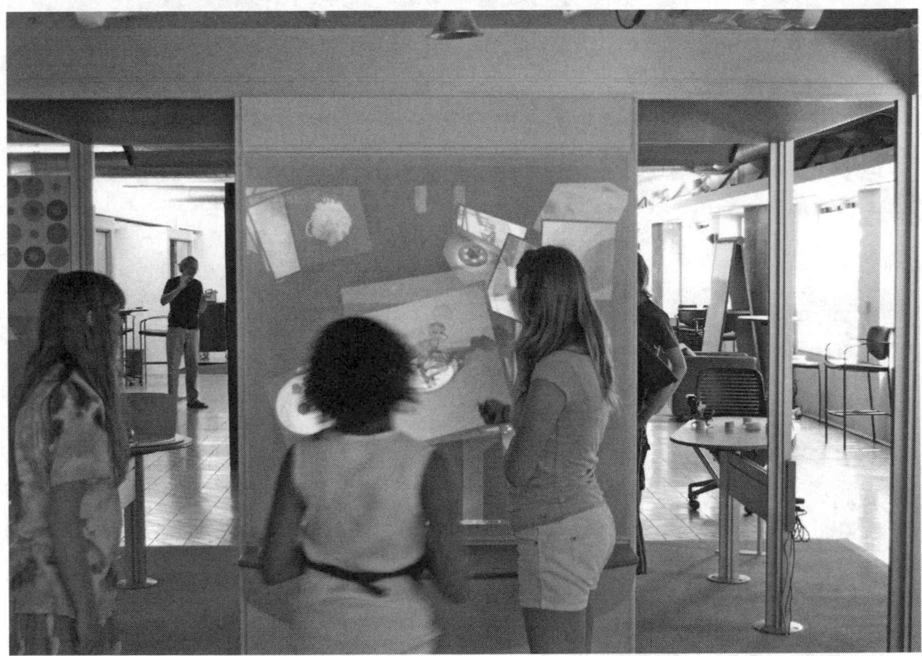

Collecting, sorting and arranging images together, kids co-construct a shared story.

Kids collaborate as they simultaneously draw on a circular table, each one using an individual controller.

Wrong Questions Answered with the Wrong Methods

Failure to respond to fundamental shifts in the needs and goals of society is fatal. This conundrum is facing schools now. The declining quality of K-12 education in the United States has created a political environment favorable to school reform, yet the schools are not reformed.

Federal and state governments have funded enough educational research to stock entire libraries. Phenomenal experiments have succeeded while the research is going on. However, even on the few occasions when the success finds its way into the classroom, it usually fails when the leader of the initiative is no longer involved. In fact, it is difficult to think of any successful research related to schools that has made a lasting improvement. After decades of investigations costing hundreds of millions of dollars, schools are still not the schools we would wish for our children. They do not serve the needs of the recent past, much less the present or future. Even so, in the face of the persistent lack of improvement, our society keeps funding more detailed research about how to improve schools and help kids learn.

Further, there is a growing gulf between the way children live and learn outside of school and the way they learn inside of school. Their experiences with Google, iTunes, Facebook, Twitter, Pandora, and other flexible customizable services are divorced from their experiences in school. Matters are getting worse, not better.

Most research in education is trying to answer the wrong questions in the wrong way. Our society cannot improve the ways kids are prepared for the twenty-first century by making incremental modifications to solutions designed for the early twentieth century. Today, it is ridiculous to imagine that a funding organization would support research about how to shuffle file cards faster to improve the way people search for information. No matter how rigorous the planned methodology, no one would fund research about applying robotics and modern supply chain theory to Ford's first assembly line. In addition, no researcher would think of asking for support for these projects. However, when it comes to learning, researchers write such proposals, and agencies fund them.

The answer to the problem is not obvious, but it is clear what the answer is not. To keep doing detailed large-scale research about improving schools meets Einstein's definition of insanity. However, there is hope and a sweet irony—the nature of the emerging creative society may contain exactly what is needed to overcome the industrial-age school system. Organizations in the creative society, when compared to those of the industrial age, are characterized by smaller size, employees who work collaboratively on projects instead of functions, people who are more comfortable making decisions with incomplete information, who are more tolerant of failures as long as they learn from them, and expect their offerings, technologies, business models and competition to be different in a few years from what they are now.

When these organizations notice that what used to work is no longer working, and their future is uncertain because of too many unknown variables, their answer is not to limit the variables and do another deep study about their old way of working. Instead, they set in motion fast trials, using information their intuition tells them is useful, even though they know it is unproven and it is incomplete. They try to get things approximately right, at first not knowing why it is working better than before. They fail early, often and on a small scale as a way of learning. They do not conduct a large study that forces one big bet that would likely lead to a failed, tardy solution offered on a large scale.

On the other hand, there are some industrial age companies and even whole industries that are unable to reframe what they do to fit into the new world. Our industry sectors of consumer electronics, music publishing, and automobiles are all a shadow of their former selves. As troublesome as their decline is, our society has found other sources. We cannot afford to give up our kids as easily as we abandon outdated technology.

Acknowledgment

The ideas and projects discussed in this paper are from three projects at the Institute of Design, Illinois Institute of Technology, supported by the Digital Media and Learning program at the John D. and Catherine T. MacArthur Foundation. The program is both marshaling what is already known about the field and seeding innovation for continued growth.

For more information, see: www.digitallearning.macfound.org.

The three research projects being done in the Institute of Design are *Schools in the Digital Age*, led by John Grimes, Patrick Whitney, and Kevin Denny; *Thinkering Space*, let by Dale Fahnstrom, Greg Prygrocki and TJ McLeish; and the *Electronic Learning Record*, led by Patrick Whitney, Kevin Denny, Ash Boohpathy, and John Grimes. All of the projects benefited from the involvement of numerous graduate students from the Institute of Design and numerous thought leaders in the area of learning and digital media. Much of this paper is based on and taken from various reports written to the Foundation by these people.

For more information, see:

ThinkeringSpace: www.thinkerinigspace.org

Schools in the Digital Age and the Electronic Learning Record: https://blog.id.iit.edu/wpmu/portfolio/

IIT Institute of Design: www.id.iit.edu

21

Creativity, Digitality, and Twenty-First-Century Schooling

Erica McWilliam,
Jennifer Pei-Ling Tan,
& Shane Dawson

Small 'c' creativity

In the twentieth century, it was characteristic of educational discourse to relegate *creativity* to the high ground of imaginative genius and the swampland of serendipity. In schools, this approach usually meant that creativity was thought to be the province of music or art or drama. Thus, *creativity* as a learning outcome generally found its way from being *de rigueur* for tiny tots to the flabby end of the secondary curriculum, far from the high status of the "hard" sciences. While this arts-bound understanding of big 'C' creativity still resonates in some quarters, there has been a profound shift in the ways that creative capacity is being valued and harnessed by global leaders of enterprise in this century. It is now understood to be an observable and valuable component of social and economic enterprise (see Haring-Smith, 2006; Cunningham, 2005; 2006; Hartley, 2004a). As small 'c' creativity, it is now a core attribute of an innovative work culture, "no longer a luxury for the few, but . . . a necessity for all" (Csikszentmihalyi, 2006, p. xviii).

According to John Hartley (2004b), the sort of creativity that drives enterprise may be more broadly understood as "the process through which new ideas are produced" while innovation is "the process through which they are implemented" (p. xi). Those with the capacity to generate such value-adding ideas are very much in demand for their ability to transcend routine thinking and doing. In his book, *The Rise of the Creative Class* (2002), Richard Florida paints a picture of the dispositions and the *modus operandi* of such workers, portraying them not as solitary individuals of artistic genius nor as a collective of computer nerds: rather, he names the creative class as comprised of people from all socio-economic and ethnic backgrounds who use their skills in a broad terrain of commercial settings.

Creativity, Digitality

While it takes more than a mere aptitude for being "digitally savvy" to make a claim to be a "creative" in Florida's terms, there is no doubt that technology and its affordances play a significant role in shaping the professional and social interests of Florida's "creatives." He describes creatives as individuals who are closely and constantly connected to their informational and relational networks, working in a diversity of fields such as business, finance, law, education, and health care but more likely they are based in digital/creative, lifestyle industries such as computer graphics and entertainment, many of which remain buoyant industries despite recessional economic times. While debates continue to rage around Florida's thesis—he has been accused of over-blowing the centrality, size, and significance of the creative class, the significance of urban amenities and their consumption, and the classism and populism of his focus—Florida has nevertheless drawn much-needed attention to unprecedented changes occurring in twenty-first century living, learning and earning, many of which are driven by new technological affordances. Indeed, computer-centered network technologies and their capabilities have impacted so powerfully on social systems and social relationships that, as Manuel Castells (2001) understands it, we cannot speak of the *social* in this century without speaking of the *technological*.

It is little wonder, then, that the nexus between creativity and digitality is demanding closer attention from within the educational sector. While schools are being asked to do much more than prepare future workers with necessary skills and capacities, the entire sector is expected to produce work-ready graduates with the kinds of literacies, numeracies, and dispositions that make for ready employability. Increasingly, employability has to do with creative and relational capacities rather than narrow instrumental skills, as employers seek more "multi-competent graduates" (Yorke, 2006, p. 2), who have "high level expertise emphasising discovery and exploiting the discoveries of others through market related intelligence and the application of personal skills" (p. 5). All this is driven, in turn, by the more fundamental recognition from the economic sector that productivity in the twenty-first century is inextricably bound up with "a deep vein of creativity that is constantly renewing itself" (NCEE, 2007, p. 3).

From Rhetoric to Reality

However, while there is more interest in creativity as a graduate attribute worthy of development through educational programs, and while digitality is acknowledged as a key means through which design dispositions can be built, mainstream schooling has yet to make the deep and sustained pedagogical shifts that are commensurate with in-school creative use of digital tools and technologies. For the education sector the picture is one of technology underutilisation, even when these resources are in plentiful supply (e.g., Becta, 2007; Kennedy et al., 2006; Tan, 2009; Tan & McWilliam, 2009; Warschauer, 2008a).

A number of reasons have been offered for this less than optimal scenario, not the least of which is the fact that fiscal constraints make it very difficult indeed for publicly funded schools to provide cutting edge, or even basic, digital tools and resources. While some schools have been able to access corporate funding to augment what governments are willing or able to provide (see, for example, FutureSchools@Singapore), most struggle in the absence of any targeted collaboration between government and the commercial sector. The current financial downturn does not augur well for any sudden change to this scenario, as the corporate sector hunkers down in these leaner and meaner times.

However, even where governments like the Australian Federal Government express a strong intention to go it alone in funding a "digital education revolution," progress can often be stuck in an unforeseen political mire. The Australian news headlines (summarised in six-month diary entry form below) provide a succinct case study of how the intention to digitalize schools threatens to collapse when it comes to spending:

> *Teachers told: Cash in or lose chance at laptops*
>
> Minister for Education Julia Gillard has written to secondary school principals across Australia inviting them to participate in the Labor government's so-called "digital education revolution."
>
> *Gillard opens coffers for tech-starved schools*
>
> Deputy Prime Minister and Minister for Education, Julia Gillard, has announced that the first round of funding for Labor's digital education revolution has begun and urged priority listed schools to apply for grants under the AU$1 billion initiative.

> *Digital education revolution: who's got the tab?*
>
> Deputy Prime Minister Julia Gillard revealed this week that the onus for funding federal Labor's digital education revolution will fall more heavily on the states than first expected, prompting raised eyebrows from some and the ire of the Opposition.
>
> *States speechless on digital education funding*
>
> After the Federal government was forced over the weekend to fend off claims that its digital education revolution is already coming unstuck, the offices of a number of the country's state education ministers have maintained a steady silence ahead of an intergovernmental meeting to discuss the next round of funding.
>
> *COAG hits green light for Labor school laptop plan*
>
> The Federal Labor government's digital education revolution received its final rubber stamp at yesterday's Council of Australian Governments (COAG) meeting, but one industry observer has advised education administrators to "take their money and put it elsewhere".(http://www.zdnet.com.au/tag/funding-school.htm, accessed 22 January, 2009)

As the above sequence of stuttering steps makes clear, 'going digital' takes more than a shift in policy direction, whether or not that shift is accompanied by adequate funding. The matter of digital up-take is more complex because it demands integration into a schooling culture where high degrees of regulation can have a paralysing effect on any such innovation. Educationist Ana Cristina Sousa, writing on the Wikinomics blog-site, underlines this fact:

> The biggest problems that we face in California are, on the one hand, poor wireless connectivity that can fail during class, meetings, presentations (namely in the Central Valley); on the other hand, school network systems that are managed like fortresses—neither students, nor teachers, or even administrators have permission to do basic things like creating and sharing a Google presentation. Any exchange and collaborative project may be stalled by a sudden change in security and access levels dictated by someone in the district office. (http://www.wikinomics.com/blog/index.php/2009/06/24/obama-should-lookto-portugal-on-how-to-fix-schools/, accessed 3 July 2009)

Moreover, even where schools claim to have managed to reconcile industrial model schooling and twenty-first-century learning needs, there remains the very practical problem of how to measure accurately the return on the financial investment in 'going digital.' As Mark Warschauer (2008a) makes clear, the business bottom line benchmark does not translate well into the schooling sector and is therefore not easily discerned in schools.

In the 2008 special issue of *Pedagogies: An International Journal* (vol. 3, no.1), Warschauer and other researchers of digital practices in schools indicate that little seems to have changed since Larry Cuban's finding nearly a decade ago that "computers most frequently remain underused or figuratively in the closet" (Cuban, 2001, p. 41). While it is possible to point to a growing number of innovations here and there, there is scant evidence of any significant cultural shift in the pedagogical work of schools. When contemporary schoolchildren speak of their experience of school, the picture they are most likely to paint is still 'blah blah blah' from a teacher instructing from the front. Whether or not the blackboard has been replaced by a whiteboard, a transmission culture of traditional curriculum remains dominant, particularly in the later years of secondary and in tertiary education, where disciplinarity wins out over digitality. Thus, even where technologically-mediated tools are more conspicuously employed, the 'lesson/lecture' may well continue to be designed along the lines of pre-existing transmission based models of pedagogy. When this happens, as it often does, laptops, data projectors, and interactive whiteboards stand in for pens and paper, blackboard diagrams and print-based worksheets. In other words, everything new becomes old again!

New Technology, Old Pedagogy

It is worth noting the counter-intuitive point that digital technologies may in fact exacerbate regression to transmission pedagogy. For example, Lane (2008) maintains that commercial learning management systems implemented by educational institutions are designed in such a manner that they adversely "impose limitations on instructional creativity and approach" (p. 5) for teaching staff. A study of the uptake and usage of digital technology among UK undergraduates further highlights the regression of digital-based pedagogy:

> "It's quite tedious. A lot of the teachers tend to just get the stuff off the internet and read it straight to you. Whereas before they might have explained it more, and they would have to have used their own words, rather than the internet's words."

> "I mean with my business teachers, they just go and visit, and copy stuff there, and they read it to you. Whereas I can just go on the internet and do exactly the same. I'm not particularly learning much from them." (Ipsos MORI, 2007)

In pure cost-benefit terms then, it is difficult to justify increased spending on digital tools in schools where they are essentially used as a stand-in for traditional modes of teaching. However, the decreasing hardware costs (e.g., the $100 laptop or PC Tablet) and the related improvements in software, alongside a continued growth in the societal importance for digital media, may result in a stronger case for relating a cost-benefit analysis to the education context (Warschauer, 2008b).

Digital Kids, Analogue Students

A recent empirical study by one of the authors (Tan, 2009) indicates that digital use and perceived usefulness in schools is a deeply complicated matter, involving student subcultures, school priorities, parent expectations, and the affordances of the innovation itself, as well as the extent to which it is normalised in the pedagogy of school classrooms. In her evaluation of a Web 2.0 digital innovation (a student-led in-school media centre) twelve months into its implementation, she found widespread ambivalence around the value of the initiative on the part of the hundreds of senior school students, for whom the centre had been established at no little expense to the school.

In brief, the evaluative study combined an extensive quantitative student questionnaire administered to a senior school student population of approximately 600 students with in-depth qualitative student focus groups. The numeric data from the questionnaire provided insights into the individual, technological, social and institutional factors that predicted the students' usage of and engagement with the media center. This set-up was complemented by the textual analysis of student focus groups' transcript data, which sought to uncover the students' shared 'common sense' and underlying reasoning practices about digitality and schooling.

Put simply, the student-led media centre, though equipped with cutting-edge digital tools, was being underutilised; thus Tan inquired further into why these students were rarely using the cutting-edge facility. She found a pattern of use and perceived usefulness that is much more complex than the "naming, blaming, shaming" literature can explain. While some students saw its possibility and endorsed its potential for their learning, the majority were less than convinced of its direct applicability or relevance to their performance in the high-stakes standardised tests that would determine their future academic pathways and hence professional success. Tan's study also made the paradoxical finding that, however 'cool' digital may be for Generation C in their own personal time and place, in-school digital use was not perceived as 'cool' for a range of reasons to do with the perceived identity of student users and non-users. In the students' own terms, if it's "nerdy," then it's "no go." Tan's study, therefore, takes us beyond the familiar terrain of deficit discourses that tend to blame institutional conservatism, lack of resourcing, and teacher resistance for low uptake of digital technologies in schools. It provides an empirical base for theorising school-based "techno-pedagogical innovation" (Tan, 2009) in a way that is more relevant to the lived culture within the school and its complex relationship to students' social identities and lives outside of school.

Teacher as Lead Learner

Because social identity work is involved in 'going digital in schools' (as pre-empted by Castells and found by Tan), the teacher-student relationship is in for a make-over if and when digital is to be normalised in mainstream education. The identity work will bring an end to pedagogical passivity, either the student passivity that accompanies the all-knowing Sage-on-the-Stage or the teacher passivity that is often excused as Guide-on-the-Side. The new relationship will reframe teachers and students as "meddlers-in-the-middle" (McWilliam, 2008), i.e., co-editors, assemblers, and dis-assemblers of knowledge and cultural products.

This shift is not easily achieved, given the expectation of many parents and employer groups that a compliant and risk-minimising model of classroom behavior and teacher-directed activity should be the pedagogical norm. However, this need not mean an 'anything goes' classroom—indeed, it must not. The twenty-first-century teacher still needs disciplinary and procedural expertise to be equipped as a "usefully ignorant" (Leadbeater, 2000) co-inquirer and designer of 'remix' curriculum. In other words, the teacher is not merely a content-free facilitator but a very smart designer of learning experiences. It is pedagogy that makes for optimal learning outcomes when students and digital technologies come together.

In formal education, the pedagogical approach needs to build a 30-to-one relationship (or even a 500-to-one relationship) as well as a one-to-one student-to-teacher relationship. It is too easy for teachers to sit down when students are anchored to computer terminals. Then, when teachers sit down, figuratively speaking, students lie down. Already we are hearing anecdotal evidence of teachers' wandering aimlessly along rows of students at banks of computers where everything is quiet, and the students seem to be doing something, but it is unclear what value is being added in terms of the learning. Where digital tools let students off the hook of rigorous inquiry and let teachers off the hook of support and direction, then they will be counter-productive for learning. If digital use enables more student passivity, then its potential is wasted.

Access to Learning Communities

The pedagogical challenge is not met simply by 'putting everything on-line' as many program innovators have found to their cost. Thus cutting back on instruction does not have to expunge useful lecturer talk. Face-to-face engagement, even 500-to-1, can create a wave that students can ride online once they have experienced the passion of an expert inquirer 'in the flesh.' Of course, this may also be experienced in a virtual environment if the program gives students access to professional practitioner communities who know how to communicate their thinking and doing to outsiders. The issue is giving access, and digital deployment is a further and underused means of giving access at this moment of time.

In a recent sequel to *Growing Up Digital,* Don Tapscott (2009) expresses a strong view that teachers ought to be "encourag[ing] [students] . . . to work with each other and show[ing] them how to access [expertise] . . . on the Web" (p. 148). This role reflects an important teacher function, but achieving access is not as simple as it sounds. Teachers are *in loco parentis* and therefore cannot easily open up the same invitations to limitless exploration that are possible in other social contexts. Schools have custody over children, and this duty of care militates against unfettered movement, physical or virtual. Moreover, collaboration is not necessarily something that all students are good at, given the extent to which we individualize them for sorting and credentialing purposes from an early age, and given the trend to single-child families, in which a child may experience more difficulty in sharing tasks with others. There is no doubting the desirability of collaborative team work, but the development of team dynamics is a complex matter (see McWilliam & Dawson, 2008). 'Sort yourselves into groups' or 'this is your allotted group' are not pedagogical strategies that make for productive collaboration of themselves.

Shared passions are more likely to produce strong collaborations, though passions have their pitfalls. Shared passion can be counterproductive if it means thinking with one brain. According to Richard Seel (2006, p. 3), "too much connectivity . . . can inhibit emergence . . . [in that] . . . diversity is excluded and groupthink is a very likely outcome." Young people need the capacity to be both separate and together, both leaders and followers, able to work across diverse and unified cultures, environments and purposes. This takes interactive agility, a capacity to network and broker across domains of knowledge production, to move where the creative action is. This sort of agility can be acquired in well-designed learning environments that are challenge-rich and respect-rich, as well as information-, conversation-, and structure-rich. Thus, the challenge of empowering students' is a complex pedagogical one, not amenable to achievement merely through paying lip-service to the rhetoric of student-centeredness. Indeed, empowering students to collaborate within poorly defined structures may well lead to unanticipated problems such as the cyber-bullying, which is now being commented on as the downside of unmediated social engagement.[1]

Testing Times

A further key reason for the underutilisation of digital tools is the disjunction between what is worth learning and what we measure in high stakes tests. While standardised assessment remains largely fixated on the memorisation of content, "coverage" of the prescribed curriculum content will continue to drive what happens in mainstream classrooms. These established patterns of instruction make it difficult to move teachers from standing and delivering, despite all the evidence that students may well learn better by other means. This practice is not merely the fault of the teacher, although good teachers will appreciate the falsity of the proposition that 'if we tell somebody something, then they know it.' As the Tan study shows (Tan & McWilliam, 2008, 2009), many good teachers and good students alike see the main game as achievement on standardised high stakes tests, and the point is to get the highest grades regardless of the relevance of the test. If asked how to bleach calico (something most grandmothers would have given up on long ago), a child is to remember the answer and write it down come test time. Teachers themselves, no matter how progressive, are not at liberty to step around high stakes assessment. Indeed, it is remarkable that some manage to do such exciting and stimulating work in classrooms despite this fact.

Fortunately, we are beginning to see experimentation with large-scale tests that inquire into students' capacity to think and do, rather than memorize. For example, Kay Stables and Richard Kimbell were assigned by the Assessment of Performance Unit of the UK's Department of Education and Science to develop an innovative approach to the assessment of design capacities. The specific challenge they faced was "to assess the design and technological capabilities of a 2% sample of the 15 year old population of England, Wales and Northern Ireland (about 10,000 learners)." They developed an assessment approach that made explicit how task-focused creative thinking occurs in the context of an explicit design task and how such thinking can be further facilitated through strategic use of evidence prompts (see Stables & Kimbell, 2007). However, such work as this is in its infancy and needs a great deal more attention from policy makers and from international bodies responsible for standardised testing (e.g., the OECD) than is currently the case. The pedagogical imperative is to test smarter, rather than stepping around the test.

The underutilisation problem is also borne out by the low uptake of e-portfolios as evaluative tools. Despite their increasing presence in the digital toolbox, e-portfolios are still rarely used, and this deficiency may well be related to instructor workload. The grading and ranking of one thousand students on a multiple choice exam is likely to be equivalent in terms of time to the marking of one individual e-portfolio. These obvious workload implications have effectively neutralised any potential motivation for extensive e-portfolio adoption.

Agile Cognitive Habits

What is at stake here is not simply the underutilisation of a particular digital affordance as a cost-benefit problem or a missed opportunity problem but more profound problem about the sort of learning that is needed to solve thorny twenty-first century problems, and what digital affordances can add to pedagogical imagination in the service of the sort of education that enhances small 'c' creativity. In his seminal work *The Act of Creation*, Koestler (1964) identified the decisive phase of creativity as the capacity to "perceive . . . a situation or event in two habitually incompatible associative contexts" (p. 95). Following Koestler, the capacity to select, re-shuffle, combine, or synthesize already existing facts, ideas, faculties, and skills in original ways may be understood to be evidence of creativity at work. David Perkins (1981) makes a similar point. In *The Mind's Best Work*, he insists that skills like pattern recognition, creation of analogies, and mental models, the ability to cross domains, exploration of alternatives, knowledge of schema for problem-solving, and fluency of thought are indicators of creativity as a set of learning dispositions or cognitive habits. To transcend the barriers of habitual thinking, an individual does not have to be a genius but does need a nimble or agile disposition to information processing. It is not about being able to flatten information out quickly in order to provide a simple story or explanation. It is about exercising a gymnastic capacity to work against the grain of sense-making. In the same way that a hurdler has to be agile enough to blur running and jumping, so, too, are creatives cognitively agile or dexterous enough to thrive in environments where unpredictability and complexity are the norm. Agility is a dynamic ability to create and utilize options at speed and thereby allow organisations to "gain competitive advantage by intelligently, rapidly and proactively seizing opportunities and reacting to threats" (Meredith & Francis, 2000, p. 138). Many digital tools can be harnessed for just this purpose.

Schools and universities are not well set up to build creative capacity for a range of reasons, not the least of which is the lock-step approach that is taken to progress through the system, and to the separation of knowledge into disciplinary silos. While the concepts of accelerated programming and interdisciplinarity have taken root over the last few decades, they remain the exception rather than the rule. Elective choice is fine but not when it comes to English, math, and science. Speed and fun with words are fine, but not at the expense of correct spelling. The idea that fast messaging capacity is enhanced through spelling short-cuts does not diminish community concerns that the spelling standards of Generation C are rapidly deteriorating. While these moral panics abound, and teachers feel obligated to address the concerns of parents, policymakers, and employers alike, the call to optimal digital use will be stifled by more long-term fears and attitudinal habits.

Self-Styling for Digital Times

The re-invention of a professional self is not an easy task, despite all the teacher-change rhetoric that garnishes educational policy documents. It is particularly difficult when a teacher, over a long career, has been rewarded for practices that are becoming increasingly obsolete. Many career rewards in education continue to reflect an outmoded educational world, in which academics are promoted for knowing more and more about less and less (at least in lay terms), and for doing research rather than teaching. The reward for being a strong and capable teacher in our schools may often be a promotion out of teaching altogether, into school management and administration.

There is a further and perhaps more personal issue to be grappled with—the extent to which a teacher is prepared to return to the status of a pedagogical beginner in a very different domain of information exchange and knowledge building. To become a beginner is to be able to tolerate the discomfort of not knowing, of learning from the instructive complications of pedagogical failure. It is one thing to tell educators that their role as beginners all over again is necessary and desirable. It is another for them to experience it firsthand. Teachers who are charismatic sages in the flesh may struggle when asked to customize their charisma for the chat-room. Their struggle is not just pedagogical but motivational. New pleasures need to be found that sustain them as well as their students. When change appears as 'the roar of the greasepaint' is replaced by an endless stream of emails, little wonder that teachers may feel nostalgia for a pedagogical past. In addition, as for changing from lecture to structured conversation, most teachers cannot bear even a few seconds of silence after they have asked a question. The struggle to achieve pedagogical meddling or co-inquiry might well mean waiting for much longer to insist on a shift from monologue to dialogue. With tens or even hundreds of eyes trained upon one sage, it is very hard indeed for the sage to refrain from providing the answer as well as the question.

Integrating Digital Dispositions

Digital integration in mainstream schooling will not be sudden, nor will it be a panacea for current educational ills. As Mark Warschauer (2007) has pointed out, we live in paradoxical times in which information literacy still depends to a large extent on print literacy. He asserts that "competence in traditional literacies is often a gateway to successful entry into the world of new literacies" (p. 43), citing American high school research into student use of computers and the Internet in defense of this claim. In doing so, he refutes "the romantic notion" that many reform advocates have of the "empowering potential of learning and new media" (p. 44) in and of itself. The ability to create multimedia presentations with the latest digital tools, or to fast text one's friends, or to play online games to a high level, can too easily be misrecognised as evidence of the sort of higher order thinking and doing that McWilliam calls "*epistemological agility*" (McWilliam, 2010). Agility in working across domains of knowledge does not come simply by spending more time spent at a computer screen. This fact is corroborated by the OECD (2006) finding that, while more experience with computer use is valuable, more frequent use does not necessarily lead to better performance. The 2003 PISA Study into computer use found that moderate users performed better than students who were either not using computers, using them rarely, or using computers very often. While the capacities associated with 'going digital' are useful and important, they are insufficient of themselves to build thoughtful action at its best.

Importantly, it should not be assumed that because a young person is highly digitally literate, he or she therefore knows how to optimise digital use for academic purposes. A key conclusion of the University of Melbourne Study (Kennedy *et al.*, 2006) concurs with the findings of a U.S. study (Kvavik & Caruso, 2005) of freshmen students that, in their first year, many students struggle, not to make technology work per se, but to make it work for their intellectual needs. Kennedy *et al.* further state:

> It is not that first year students are incapable of using technology for specialised, context-appropriate purposes; indeed many would have recently had these expe-

> riences at school. The critical point is that while first year students might use technology in a range of ways and may, apparently, be digitally literate, we cannot assume that being a member of the 'Net' Generation is synonymous with knowing how to employ technology-based tools strategically to optimise learning experiences and outcomes. (2006: p. 16)

Schools have a key role in sorting and credentialing young people in ways that can be validated by making them visible and calculable to parents, policymakers, and future employers. There is, in light of this complexity of roles, and the fact that the benefits of expensive technology are "difficult to realize and hard to measure" (Warschauer, 2008a, p. 3), an imperative from the research to be cautious about either extolling the virtues of digital tools, or seeking Canute-like to hold new waves of Web 2.0 digital tools out of mainstream schooling.

Concluding Remarks

Recent research distils a number of concerns about the digital/schooling nexus, including: a lack of clarity about the cost/benefit of expensive technology; underutilisation of technologies in classrooms; and, confusion over whether the main goal of education is improved performance in formal assessment or greater human capacity more broadly understood (Warschauer, 2008a; Ware, 2008; Cummins, 2001). Furthermore, there is burgeoning evidence that the twenty-first-century learning skills are being developed and utilised in the mainstream, outside formal schooling. These sites of cultural production and social interaction, like Facebook and YouTube, are spaces where young people can practice the forms of navigation, networking, and communication skills necessary to a 'creative worker' identity. Stephen Lunn (2007) highlights the fact that these are not command-and-control environments as many school classrooms are. Moreover, young people who are active users "tend to avoid dealing with anyone who could hold power over them on these sites, such as parents, teachers, bosses . . . [and] . . . those that want to prey on them" (p. 2). What drives these high levels of usage is the logic of "we-think" (Leadbeater, 2009), that is, the Web's culture of "lateral, semi-structured free association" (p. xxi).

All these different dynamics have profound implications for the nature of teaching and learning in schools. Aspirational students have to acquire formal academic skills (including a high level of print literacy), as well as skill sets relevant to a digital and conceptual age (Castells, 1996, 2000; Pink, 2005). At the same time, teachers are facing immense pressure to be relevant, in their content and pedagogy, to digital times, while ensuring that their students perform well in traditional academic tasks. Performing well in those traditional tasks, rightly or wrongly, is the passport and is an important first step to employablity. As futurist Sandra Welsman (2006) explains, formal educational qualifications will continue to be needed, but by the same token they will not be sufficient to the "edu-ventures" (p. 50) that young people will be undertaking in their future working lives. The logic then is 'both and' rather than 'either or' in terms of formal classroom-based learning experience and the sort of personalised, customised, and just-in-time learning that digital tools afford.

References

British Educational Communications and Technology Agency (Becta). (2007). *Becta: 2007 Annual Review.* Retrieved from http://publications.becta.org.uk/download.cfm?resID=33625

Castells, M. (1996). *The rise of the network society.* Oxford, UK: Blackwell.

Castells, M. (2000). *The rise of the network society* (2nd ed.). Oxford, UK: Blackwell.

Castells, M. (2001). *The Internet galaxy.* Oxford: Oxford University Press.

Csikszentmihalyi, M. (2006). Foreword: Developing creativity. In N. Jackson, M. Oliver, M. Shaw, & J. Wisdom (Eds.), *Developing creativity in higher education: An imaginative curriculum* (pp. xviii–xx). London: Routledge.

Cuban, L. (2001). *Oversold and under-used: Computers in classrooms, 1980–2000.* Cambridge, MA: Harvard University Press.

Cummins, J. (2001). *Negotiating identities: Education for empowerment in a diverse society* (2nd ed.). Los Angeles: California Association for Bilingual Education.

Cunningham, S. (2005; 2006, July). What price a creative economy? Platform Papers. *Quarterly Essay on the Performing Arts,* 9.

Dawson, S., & McWilliam, E. (2008). *Investigating the application of IT generated data as an indicator of learning and teaching performance.* A report produced for the Australian Learning and Teaching Council, Sydney, Australia.

Florida, R. (2002). *The rise of the creative class.* New York: Basic Books.

Haring-Smith, T. (2006). Creativity research review: Some lessons for higher education. *Peer Review, 8*(2), 23–27.

Hartley, J. (2004a). The value chain of meaning and the new economy. *International Journal of Cultural Studies, 7*(1), 129–141.

Hartley, J. (2004b). Preface in R. Wissler, B. Haseman, S. Wallace, & M. Keane (Eds.), *Innovation in Australian arts, media and design* (pp. xi-xxi). Flaxton: Postpressed.

Ipsos MORI. (2007). *Student expectations study: Key findings from online research and discussion evenings held in June 2007 for the Joint Information Systems Committee.* Retrieved from http://connect.educause.edu/Library/Abstract/StudentExpectationsStudy/45255

Kennedy, G., Krause, K., Judd, T., Churchward, A., & Gray, K. (2006). *First year students' experiences with technology: Are they really digital natives?* Preliminary Report of Findings, September, University of Melbourne: Centre for Study of Higher Education.

Koestler, A. (1964). *The act of creation.* New York: Macmillan.

Kvavik, R., & Caruso, J. (2005). *ECAR study of students and information technology, 2005: Convenience, connection, control and learning.* Retrieved from http://net.educause.edu/ir/library/pdf/ers0506/rs/ERS0506w.pdf, accessed February 16, 2009.

Lane, L. (2008). Toolbox or trap? Course management systems and pedagogy. *Educause Quarterly, 31*(2), 4–6.

Leadbeater, C. (1999). *Living on thin air: The new economy.* New York: Viking.

Leadbeater, C. (2009). *We-think: Mass innovation, not mass production.* London: Profile Books.

Lunn, S. (2007, July 27). Face to face in cyberspace. *The Australian.* Retrieved from http://www.australianit.news.com.au/story/0,24897,22143354–15302,00.html

McWilliam, E. (2008). *The creative workforce: How to launch young people into high flying futures.* Sydney: UNSW Press.

McWilliam, E. (2010). Learning culture, teaching economy. *Pedagogies: An International Journal, 5*(4).

McWilliam, E., & Dawson, S. (2008). Teaching for creativity: Towards sustainable and replicable pedagogical practice. *Higher Education, 56*(6), 633–643.

Meredith, S., & Francis, D. (2000). Journey towards agility: The agile wheel explored. *TQM Magazine, 12*(2), 137–143.

NCEE. (2007). *Tough choices or tough times: The report of the New Commission on the Skills of the American Workforce.* National Center on Education and the Economy. www.skillscommission.org

OECD. (2006). *Results from programme for international student assessment* (PISA) 2003, compiled by Andreas Schleicher OECD/Directorate for Education.

Perkins, D. (1981). *The mind's best work.* Cambridge MA: Harvard University Press

Pink, D. H. (2005). *A whole new mind: Why right-brainers will rule the future.* New York: Penguin.

Seel, R. (2006). *Emergence in Organisations.* Retrieved from http://www.new-paradigm.co.uk/emergence-2.htm, accessed 3 May 2008.

Seely Brown, J. (2006). New learning environments for the 21st Century. *Change, 65*(5), 2.

Stables, K., & Kimbell, R. (2007). *Evidence through the looking glass: Developing performance and assessing capability.* Retrieved from http://www:iteaconnectorg/Conference/PATT/PATT15/Stables/pdf

Tan, J. P-L. (2009). *Digital kids, analogue students: A mixed methods study of students' engagement with a school-based Web 2.0 learning innovation.* (Unpublished doctoral dissertation). Queensland University of Technology, Australia.

Tan, J., & McWilliam, E. (2008). Digital or diligent? Web 2.0's challenge to formal schooling In proceedings for the *2008 AARE Conference* Nov 29-Dec 4 2008, Brisbane, Australia.

Tan, J. P-L., & McWilliam, E. (2009). From literacy to multiliteracies: Diverse learners and pedagogical practice. *Pedagogies: An International Journal, 4*(3), 213–225.

Tapscott, D. (1997/2009). *Grown up digital.* New York: McGraw-Hill.

Ware, P. (2008). Language learners and multimedia literacy in and after school. *Pedagogies: An International Journal, 3*(1), 37–51.

Warschauer, M. (2007) The paradoxical future of digital learning. *Learning Inquiry, 1*, 1–49.

Warschauer, M. (2008a). Technology and literacy: Introduction to the special issue. *Pedagogies: An International Journal, 3*(1), 1–3.

Warschauer,M. (2008b). Laptops and literacy: A multi-site case study. *Pedagogies: An International Journal, 3*(1), 52–67.

Warschauer, M., & Grimes, D. (2008). Automated writing assessment in the classroom, *Pedagogies: An International Journal, 3*(1), 22–36.

Welsman, S. (2006, January 11). Enterprise is the key to the future. *The Australian Higher Education Supplement, 50.*

Yorke, M. (2006). Employability in higher education: What it is—What it is not. In M. Yorke (Ed.), *Learning and employability Series One.* The Higher Education Academy—employability@heacademy.ac.uk

Note

1. See http://www.news.com.au/story/0,,24450224–2,00.html

22

Digitized Youth: Constructing Identities in the Creative Knowledge Economy

Tina (A. C.) Besley

I don't feel that it is necessary to know exactly what I am. The main interest in life and work is to become someone else that you were not in the beginning. If you knew when you began a book what you would say at the end, do you think that you would have the courage to write it? What is true for writing and for a love relationship is true also for life. The game is worthwhile insofar as we don't know what will be the end.

—Michel Foucault (1982) *Truth, Power, Self: An Interview with Michel Foucault*

You must not for one instant give up the effort to build new lives for yourselves. Creativity means to push open the heavy, groaning doorway to life.

—Daisaku Ikeda

The opposite of creativity is cynicism.

—Esa Saarinen

Introduction

Foucault's comment alludes to how our identities are not fixed but can change and develop over time in a fluid, dynamic, and creative process. He notes the creative change from the past as we grow older, holding out a hopeful note that it is all worth it even if we don't know how it will end. Foucault demonstrates the understanding that identities or subjectivities are formed through language, signs, and discourse (Besley & Peters, 2007). This theoretical understanding of identity moves identity away from biological determinism to examining how identities are forged in relation to the global market. Many studies of the globalization of youth culture have focused on how children and teenagers, less influenced by the family, church or school, buy their values in the marketplace (Besley, 2003). More recent thinking has begun to apply social constructivist arguments to the economy, emphasizing its links to the cultural and creative components of the knowledge economy where young people now communicate with each other online and build their own personal profiles. Now that we are well into the twenty-first century, despite the global financial crisis, globalization continues apace, and new digital media abound. For the youth of today whose lives are being shaped by the rapidly changing digital world that includes mobile phones, iPods, MP3 players, social media,[1] social network sites[2] (SNS), online games, and video-sharing sites, their identities will be profoundly different from youth of the last century or earlier, different from those young people who are not engaged with the digital world for whatever reason—be it poverty, prohibition, or choice. Once in the workforce, youth will further develop their identities in the creative global knowledge economy, which is likely to paradoxically emphasize both a more individualistic, flexible, and meritocratic approach alongside small collaborative creative teams (Florida, 2002). As for anyone engaged with digital media, they will leave a digital imprint that maybe subject to data mining (for a critical example see http://personas.media.mit.edu/personasWeb.html). It is the youth who are engaged with the digitized creative economy who are the focus of this chapter.

The generation born after 1982 has been assigned various labels such as Generation Y, the Millennial Generation, Generation Next, Net Generation, and now Digital Natives, the first generation of young people born in the era of digital media (Prensky, 2001). They "share a common global culture that is defined not by age, strictly, but by certain attributes and experiences in part defined by their experience growing up immersed in digital technology, and the impact of this upon how they interact with information technologies, information itself, one another, and other people and institutions" (http://www.digitalnative.org/#about). The Digital Native Project "focuses on the key legal, social, and political implications of this group of youth." However, not all of today's young people can be considered digital natives since the label holds particular cultural biases toward advanced capitalist societies. Furthermore, the ease with which some older people use digital technologies mean that rather than an age-generational attribute/trait definition the psychographic based Generation C might now be a more useful term (see http://wiredworkplace.nextgov.com/2009/08/generation_c.php).

Identity is a complex, interlinked concept that holds both personal and public dimensions; the internal, psychological aspects that are displayed (consciously or not) with the public, interactive social side that we present to the world as actors in our own unique forms of social performance. Feedback about such performances from different audiences shapes our behavior. This dynamic constitutes an embodied performance unlike that of the digital world, which, despite all the possibilities of visual media such as videos and webcams, is basically still disembodied.

Digital identity is a set of characteristics asserted by one digital subject about itself or by another digital subject (human or otherwise) in a digital realm, i.e., it is what you publish about yourself and what others say about you and includes any electronic exchange with both human and with non-human digital agents. It comprises multiple pieces of formal and informal data, real or fictional/fantasized. Since digital identity can be broken up between several networks and Web sites, and these different pieces of identity might not be coherent, it is often considered to be fragmented. However, digital identity is not actually a fragmented identity since this is a social presentation that forms only one of many parts of identity that people hold in different contexts and at different times. The temporality of digital identity exists since it may or may not evolve over time especially if a comment or an old profile is not removed or updated with a digital imprint remaining forever. The temporality of digital identity means that we need to manage our digital identities by regularly updating profiles and information as far as possible and just as in person, the tendency is to put our best side forward. Consequently, creating digital identities and performance and understanding that of the Other are markedly different in the digital realm.

This chapter examines and elaborates on the notion of building knowledge cultures and the creative knowledge economy, referring largely to work jointly written with Michael A. Peters (Peters & Besley, 2006, 2009). The chapter then discusses some of the recent research findings about U.S. youth engagement and identities in the digital world that have become available since 2007. It examines the creativity of youth and the constructive means they use to develop new identities and subjectivities that resist the worst excesses of the market while engaging and negotiating the emergent social media and developing their own hybridized sense of style in music and culture. Finally, the chapter looks at youth and creativity—the implications for the creative knowledge economy with this new generation of digital natives and how education might finally take an active role rather than banning kids' participation.

Building Knowledge Cultures and the Creative Knowledge Economy

Building Knowledge Cultures

By 2009, following a series of policy reports by the Organization for Economic Cooperation and Development (OECD), the World Bank, and a burgeoning literature on the topic (e.g., OECD, 1996, 1999, 2001; World Bank, 2002), the notion of knowledge economy seems to be firmly established. In 2001 the OECD identifies a trend toward knowledge-based economies, where worker/student mobility and the rapid spread of information technology, particularly the Internet, are key features. The report identifies telecommunications, finance, insurance, education, and health industries as particularly strong knowledge based sectors of the economy and that knowledge based investments in these sectors are growing more rapidly than investments in fixed capital (OECD, 2001). The World Bank emphasizes the importance of the developmental aspects of the knowledge economy for countries to transition to a knowledge economy in the "Knowledge for Development Program (K4D)" which is based on a framework with four pillars:

- An **economic and institutional regime** that provides incentives for the efficient use of existing and new knowledge and the flourishing of entrepreneurship.
- An **educated and skilled population** that can create, share, and use knowledge well.
- An efficient **innovation system** of firms, research centers, universities, think tanks, consultants, and other organizations that can tap into the growing stock of global knowledge, assimilate and adapt it to local needs, and create new technology.
- **Information and Communication Technologies (ICT)** that can facilitate the effective communication, dissemination, and processing of information (http://web.worldbank.org/WBSITE/EXTERNAL/WBI/WBIPROGRAMS/KFDLP/0,,contentMDK:20269026~menuPK:461205~pagePK:64156158~piPK:64152884~theSitePK:461198,00.html) [bold in original]

In *Building Knowledge Cultures: Education and Development in the Age of Knowledge Capitalism*[3] (Peters & Besley, 2006), we coined the term "knowledge cultures" because we regarded the distinction and separation between the terms "knowledge society" and "knowledge economy"—an historical disciplinary division between sociology and economics—as unhelpful in understanding the present era. "Society" and "culture" can no longer be regarded as epiphenomena dependent or determined in some sense on or by the "economy." The distinction between knowledge economy and knowledge society is too dualistic and belongs to discourses that do not cross-thread: one term points to the economics of knowledge and information and of education; the other, to the concepts and rights of knowledge workers as citizens in the new economy, focusing on the subordination of economic means to social ends. Whereas the object of the economics of knowledge is knowledge as an economic good and the properties governing its reproduction, as well as its historical and institutional conditions determining its processing in the economy, the sociology of knowledge studies the social origin of ideas, and their effects on society. Until recently, economists have regarded knowledge as behaving like all other commodities, subject to the laws of supply and demand, and, in particular, to the law of scarcity. However, knowledge once discovered and made public operates expansively to defy the law of scarcity. Unlike most resources that become depleted when used, information and knowledge can be shared, developed and modified, and actually grow through application.

Simplified in the extreme, our argument is that knowledge production and dissemination requires the exchange of ideas and such exchanges, in turn, depend on certain cultural conditions, including trust, reciprocal rights, and responsibilities between different knowledge partners, institutional regimes and strategies, and the whole sociological baggage that comes with understanding institutions. We use the term "knowledge cultures" (in the plural) because there is not one prescription or formula that fits all institutions, societies, or knowledge traditions. In this situation, perhaps, we should talk of the ways in which knowledge capitalism rests on conditions of knowledge socialism, at least, upon the sharing and open exchange of ideas among knowledge workers, often called peer production (Peters & Besley, 2006, pp. 5–6).

We use the phrase "cultural knowledge economy" as a composite term trading on notions of "cultural knowledge," "knowledge as culture," and "knowledge cultures," as well as the now accepted term "knowledge economy" and the idea of "cultural economy" employed as an approach similar to political economy. Knowledge is now the dominant feature of the *social* transformations associated with globalization as the worldwide integration of economic activity (Peters & Besley, 2006, p. 26).

Today, more than any time in the past, the cultural has become the economic and the economic has become the cultural. "Culture" has now become as important as science to the extent that cultural policy is no longer seen as separate and divorced from science policy, competing for the same limited public purse, but rather complementary and overlapping in its interests. This is the basic insight of the knowledge economy that is based on the facility with signs and symbolic analysis and manipulation. The knowledge economy, then, rests on the production and use of knowledge and innovation, and communication through electronic networks that have become the global medium of *social* exchange. In this new configuration, the production of new meanings is central to the knowledge process and media or communication cultures once centered on literacy and printing, now are increasingly centered on the screen or image and the radical and dynamic concordance of image, text, and sound (Peters & Besley, 2006, p. 27).

The "culturalization" of the economy is clearly evident in a number of related developments: the creation, development, distribution, and production of both hardware and software as part of the information infrastructure for other knowledge and cultural industries; the growth of highly stylized consumer culture where ordinary products are increasingly aestheticized and imbued with cultural meaning in relation to questions of lifestyle and the 'fashioning' of personal identity; the convergence of telecommunications with enter- and edutainment media cultures based on radio, film, TV, the Internet, mobile phones with their assorted mixed media; the significance accorded to signifying and other cultural practices in the actual organizational life of firms as well as in the production, design, and marketing of products (Peters & Besley, 2006, p. 28).

Creative Knowledge Economy

Creativity seems to be most commonly identified with the arts and humanities, literature, music, high culture and popular culture, yet is in fact far broader than this domain, being almost ubiquitous. It occurs in the sciences, math, engineering, computing, all industries, such as business, medicine, building and construction, and so on. While the idea of a flash of insight, that genius eureka moment may occur, creativity is much more of a deliberate process. Creativity may have functionalist and technicist aspects and may involve a person's thoughts and or actions. It generally encompasses analyzing and synthesizing information, ideas, knowledge to develop or find some new item or ways of doing things that are markedly different from what went before. Creativity flourishes in a supportive environment and sets up new relationships with that environment. Foremost it implies change, something new and different, a rupture, challenge, even de-construction and destruction, and even conflict. Creativity is by no means always positive, a force for good, or neutral. Instead it may be profoundly negative, sweeping away vestiges of previous eras and cultures, or may involve creating destructive items such as guns, missiles, and weapons of mass destruction.

In "Academic Entrepreneurship and the Creative Economy" we discuss a number of recent theorists who examine the notion of creative economy since the 1990s, and we trace notions of entrepreneurship back to Richard Cantillon and Joseph Schumpeter (Peters & Besley, 2009).[4]

John Howkins (2001) defines the creative economy not simply in terms of the concepts of creativity, culture, heritage, knowledge, information, innovation, or in terms of the economic activities of arts, architecture, craft, design, fashion, music, performing arts, publishing, and so on, but rather more broadly as "an economy where a person's ideas, not land or capital, are the most important input and output (IP [Intellectual Property])."[5] He uses this broad definition because he understands that "All creativity—arts, sciences, whatever—involves using the brain's same physiological processes, the synapses fizz and splutter and make connections—or not—in the same way." Interestingly, Howkins argues that the Western paradigm of creativity is based on the idea that it is the preserve of a few talented people (artists and inventors) and a smaller number of investors. He argues that "we need a new paradigm for IP based on the public's demand for knowledge." Howkins's (2001) account of the creative economy follows a long line of development that emerges from different literatures—economics, sociology, and management theory, among others. The "creative destruction" of Schumpeter and his account of entrepreneurialism (1942/1976, 1951) is fundamental to the notion of the creative economy even though its individualistic bias is under review in contemporary networked environments (Peters & Besley, 2009).

Richard Florida in *The Rise of the Creative Class* (2002) draws on Howkins arguing that "[h]uman creativity is the ultimate economic resource" (p. xiii). Florida focuses on the institutions of the creative economy: new systems for technological creativity and entrepreneurship; new models for making things (including the creative factory and modular manufacturing); and the final element of the creative class, which he describes as the social milieu, an ecosystem within which creativity takes root. Florida's analysis plots the central importance of *place* as the key economic and social organizing unit of our time, as the ecosystem that harnesses human creativity and turns it into economic value. He also describes the emergence of a new social class as the fundamental source of innovation and economic growth. In addition, he charts the dimensions and new institutions of the creative economy, arguing that not only will it transform work, leisure, community, and everyday life but that it is both possible and desirable to build the creative community. Florida suggests that it is possible to build a creative community by creating appropriate working conditions, by managing creativity, by encouraging the shift from social to creative capital, and developing the city and university as creative hubs. While Florida's work is unabashedly utopian, we take from it both the geographical imperative and the notion that with appropriate investment and policies it is possible to build what we prefer to call *knowledge cultures*, from whence creativity can flow (Peters & Besley, 2009, pp. 4–5).

Youth Identities and Engagement with the Digital World

This section reviews some current research into the engagement of young people with the digital world and the impact on their identity. In particular it looks at a suite of reports produced by the Pew Internet & American Life Project: *Teens, Privacy and Online Social Networks* (Lenhart & Madden, 2007); *Teens and Social Media*, (Lenhart, Madden, Smith, & Macgill, 2007); *Teens, Video Games and Civics* (Lenhart, 2008); *Writing, Technology and Teens* (Lenhart, Arafeh, Smith & Macgill, 2008); *Teens and Mobile Phones over the Past Five Years: Pew Internet Looks Back* (Lenhart, 2009). It surveys work by danah boyd (2007); The Digital Youth Project (Ito et al., 2008, 2010); Henry Jenkins's (2006) work on participatory cultures and University of Minnesota research (2008).

The Pew Reports generally use telephone surveys and sometimes a series of focus groups and provide summary and full reports that provide methodology details and data sets on their Web sites (e.g., http://www.pewinternet.org/Shared-Content/Data-Sets/2006/November-2006—Parents-and-Teens.aspx; http://www.pewinternet.org/Shared-Content/Data-Sets/2008/February-2008—Teen-Gaming-and-Civic-Engagement.aspx).

The first Pew Report examined is *Teens, Privacy and Online Social Networks* (Lenhart & Madden, 2007) where Amanda Lenhart and Mary Madden look at teens' choices about sharing information online. Social networking has become a dominant force for teenagers, with 55% of online teens having online profiles, primarily using SNS to stay in touch with people they already know, either friends whom they see a lot (91%) or friends that they rarely see in person (82%). Approximately 49% use SNS to make new friends. One of the main concerns of parents and educators about teens using SNS is privacy and the dangers of sharing information online—potential harm to future college or job prospects or the risk of harm from sexual predators or from cyber-bullying. Some key findings are that teens "believe some information seems acceptable—even desirable—to share, while other information needs to be protected." To counter risk, UK police suggest that kids keep personal details to an absolute minimum. No phone numbers, no names, no school names and just keep that personal information to yourself or hidden behind that front page. Anyone with that intention or trying to identify you can piece all these little bits of personal information together and in a very short space of time can identify who you are, where you are, what school you go to, what your hobbies are, and who your friend are. (BBC, 2008, http://news.bbc.co.uk/2/hi/programmes/panorama/7180769.stm)

Teens hold diverse views about privacy and disclosure of personal information and say that much depends on the context to allow disclosure or not. A first name and photo are standard in teen SNS profiles. Girls are more likely to post photos of themselves and friends; girls are less likely than boys to post their physical location—perhaps a nod to girls being more safety conscious in this respect. Most teens (66%) take steps to protect themselves online from the most obvious areas of risk. To protect their privacy but also to be playful or silly, 46% of teens give some false information on their profiles, performing "keeping some important pieces of information confined to their network of trusted friends and, at the same time, participating in a new, exciting process of creating content for their profiles and making new friends." By deliberately creating fake identities and posting false information, youth treat SNS as a form of fictional creative writing. The problem is that the disembodied nature of digital identity means that others are not always aware that a fictional identity has been created, and it is by this same means predators can engage in digital media—especially in SNS (Lenhart & Madden, 2007).

The online risk is clear since "32% of online teenagers (and 43% of social-networking teens) have been contacted online by complete strangers and 17% of online teens (31% of social networking teens) have 'friends' on their social network profile who they have never personally met. 21% of teens who have been contacted by strangers have tried to find out more information about that person (7% of all online teens). 23% of teens who have been contacted by a stranger online say they felt scared or uncomfortable because of the online encounter" (Lenhart & Madden, 2007). Although many are aware of the risks and make careful choices, by no means all have such knowledge. This report highlights issues of identity for digitized youth: issues of friendship, trust, self-care, self-confidence, self-esteem, and risk-taking and how in turn these factors impact on youth engagement with digital media.

The second major report by Pew, *Teens and Social Media*, produced by Amanda Lenhart, Mary Madden, Aaron Smith and Alexandra Macgill (2007), replicates and compares results of research published in 2004. The report is based on a call-back telephone survey of a randomly generated sample of youth aged 12 to 17 and a parent/guardian. The sample size was 935 parent-child pairs.[6] The report looks at use of the Internet, social media, SNS, and blogging and considers the impact on educational activities. Here is a brief summary of the findings:

- The use of social media—from blogging to online social networking to creation of all kinds of digital material—is central to many teenagers' lives.
- The growth in blogs tracks with the growth in teens' use of social networking sites, but they do not completely overlap.
- Online boys are avid users of video-sharing Web sites such as YouTube, and are more likely than girls to upload.
- Digital images—stills and videos—have a big role in teen life. Posting them often starts a virtual conversation. Most teens receive some feedback on the content they post online.
- In the midst of the digital media mix, the landline is still a lifeline for teen social life. Multi-channel teens layer each new communications opportunity on top of pre-existing channels.

Email continues to lose its luster among teens as texting, instant messaging, and social networking sites facilitate more frequent contact with friends. Only 14% send daily emails to friends. MySpace is much quicker, and they like text messaging and sending pictures. (Lenhart, Madden, Smith & Macgill, 2007)

The data show changes since the 2004 survey. "Content creation" and participatory culture is blossoming; content creation increased from 57% of online teens in 2004 to 64% in 2007, i.e., 59% of *all* teens. "Content creators" are online teens who have created or worked on a blog or web page, shared original creative content, or remixed content they found online into a new creation. Contrary to how computers, gaming, software creation, and the Internet have seemed to be dominated by males (especially the specter of the nerdy computer geek), this report clearly indicates a shift: content creation in the social media realm is now dominated by girls, especially by the teen blogosphere. Indeed, 35% of girls blog compared with 20% of boys; 54% of girls upload photos compared with 40% of boys, but 19% of boys upload videos compared with 10% of girls. Content creation is not just about *sharing* creative output; it is also about *participating* in conversations fueled by that content. The report found that the most popular content-creating activities are self-authored artistic content and working on Web pages for others; many teens post comments on news sites, bulletin boards and group Web pages; and many create avatars to interact with others in games or Second Life. Content creators tend to be girls and older teens (55% girls, 45% boys); 55% are older teens aged 15 to17; 45% are 12 to14 years old. Surprisingly perhaps, lower family income did not seem to prevent teens from being content creators: 13% have a family income of less than $30,000; 21% from $30,000 to

$49,999; 19% from $50,000 to $74,999; 38% $75,000+; 23% are in urban areas, 52%, suburban, and 25% rural (Lenhart, Madden, Smith & Macgill, 2007).

SNS are used more heavily by girls, especially older girls, than by boys, with half visiting their sites every day primarily for communication (70% of older girls used SNS compared with 54% of older boys; 70% created online profile compared with 57% of older boys; 84% of teens post messages on their friends' profiles, 82% send private messages, 76% post comments and a third "poke" people). A subset of about 28% of all teens, primarily older girls were identified as "super-communicators" who use multiple technologies, including landline phones, cell phones, texting, social network sites, instant messaging, and email to relate to family and friends (Lenhart, Madden, Smith & Macgill, 2007).

In addition to SNS, blogs are central to the lives of many U.S. teenagers (19% of online teens in 2004, 28% in 2006). This number far surpasses the rate for adults of whom only 8% of Internet users have created a blog. Bloggers were likely to come from low-income and single-parent households and suburbs; 61% are older teens. The gender gap in blogging has grown since 2004, with girls dominating: 35% of online teens vs. 20% boys; 38% of older girls vs. 18% boys. Teens from lower income (less than $50k) and single-parent families are more likely to blog (35%) (Lenhart, Madden, Smith & Macgill, 2007)

This report is largely consistent with danah boyd's ethnographic study conducted from 2004 to 2006 of American youth aged 14 to 18 that focused on MySpace,[7] which at the time was the main SNS for this age group (boyd, 2007). boyd wanted to know: "Why do teenagers flock to these sites? What are they expressing on them? How do these sites fit into their lives? What are they learning from their participation? Are these online activities like face-to-face friendships—or are they different, or complementary?" She found that they use these spaces to "hang out amongst their friends and classmates . . . work out identity and status, make sense of cultural cues, and negotiate public life." Further, she states that "network sites are a type of networked public with four properties that are not typically present in face-to-face public life: persistence, searchability, exact copyability, and invisible audiences. These properties fundamentally alter social dynamics, complicating the ways in which people interact" (boyd, 2007). Interestingly, boyd found that for social networks, rather than an access digital divide along race or class lines where some teens were disenfranchised (denied access by parents, schools, or other public venues), there appeared to be a participatory divide where some teens, "conscientious objectors" chose to not engage for various reasons such as protesting against News Corp as owner of MySpace; agreeing with parental concerns about morals and safety; and those who were too "cool" and thought SNS were stupid.

Nevertheless, boyd "found that many of them actually do have profiles to which they log in occasionally" (boyd, 2007).

boyd's research on gender largely concurs with that of the Pew findings. She found that "younger boys are more likely to participate than younger girls (46% vs. 44%), but older girls are far more likely to participate than older boys (70% vs. 57%). Older boys are twice as likely to use the sites to flirt and slightly more likely to use the sites to meet new people than girls of their age. Older girls are far more likely to use these sites to communicate with friends they see in person than younger people or boys of their age" (boyd, 2007).

The third Pew Report, *Teens, Video Games and Civics* (Lenhart, 2008) based on a national U.S. survey of 1,102 youth aged 12 to 17 found "that virtually all American teens play computer, console, or cell phone games and that the gaming experience is rich and varied, with a significant amount of social interaction and potential for civic engagement," i.e., games are ubiquitous in contemporary U.S. youth culture—a contrast to the popular perception that it is mainly boys who play video games. The primary findings are that teens play diverse games, with "the most popular being racing, puzzle, sports, action and adventure categories.[8] Game playing is often social and can incorporate many aspects of civic and political life; game playing sometimes involves exposure to mature content, with almost a third of teens playing games that are listed as appropriate only for people older than they are" (Lenhart, 2008). Contrary to some adult perceptions, they "are not simply playing violent first-person shooters or action games. However, boys are more likely than girls to report playing these specific violent M-rated games." While most parents monitor their children's online activities to some degree, they tend to monitor their boys and younger children more closely, but, "monitoring . . . does not have an impact on whether or not teens are exposed to anti-social behavior or words in the gaming context." Many games involve playing with others and teens exert controls when negative behaviors occur:

> In multiplayer game play, different people control different characters in the game, and make individual choices about how to act and what to say in the context of the game. Nearly two-thirds (63%) of teens who play games report seeing or hearing "people being mean and overly aggressive while playing," and 49% report seeing or hearing "people being hateful, racist, or sexist" while playing. However, among these teens, nearly three-quarters report that another player responded by asking the aggressor to stop at least some of the time. Furthermore, 85% of teens who report seeing these behaviors also report seeing other players being generous or helpful while playing. (Lenhart, 2008)

An interesting aspect of this research was that there was little evidence linking gaming and civic action—either positively or negatively—i.e., it did not "promote behaviors or attitudes that undermine civic commitments" nor was it "associated with a vibrant civic or political life" (Lenhart, 2008).The fourth Pew Report is *Writing, Technology and Teens* (Lenhart, Arafeh, Smith, & Macgill, 2008). While e-communication, especially that related to texting, IMS, and SNS uses abbreviations, new acronyms, pays limited heed to traditional rules of spelling, grammar, and punctuation, and may arguably be a location for creative new forms of language, the concerns and debate about the effects of digital communication on formal writing continue. Some argue that it is or will have deleterious effects on formal writing. Others are aware that their children and teens write more than they ever did at a similar age with kids writing at school and informally for their own enjoyment in digital media via blogs, emails, SNS, texting, and wonder if such text-driven communication might inspire new appreciation for writing among teens. Paradoxically, though, they spend a "considerable amount of their life composing texts, but they do not think that a lot of the material they create electronically is ***real*** writing." "At the same time that teens disassociate e-communication with 'writing,' they also strongly believe that good writing is a critical skill to achieving success" (Lenhart, Arafeh, Smith, & Macgill, 2008).

The findings[9] indicate that in negotiating both the digitized world and the more formal world of school, young people are able to clearly discern and to maintain different stances in their usage of the new media. They, in effect, separate different aspects of their identities—the more formal related to school and education and the informal related to family and friends to e-communication, moving between these worlds with remarkable ease. While teens use computers for editing and revising work, surprisingly they "do not feel that use of computers makes their writing better or improves the quality of their ideas." Only 15% think that Internet based writing has improved their overall writing, 11% that it has harmed it, 73% see no difference to school writing, and 77% say there is no difference to their personal writing. Moreover, they mix longhand and computers in creating their written work (Lenhart, Arafeh, Smith, & Macgill, 2008).

The fifth Pew Report, *Teens and Mobile Phones over the Past Five Years: Pew Internet Looks Back* (Lenhart, 2009) mined several datasets including those mentioned above, but primarily used the Teens, Gaming, and Civics dataset. When Pew first surveyed teen mobile phone use in 2004, only 45% had one, but by 2008 the number had increased to 71% among young people and to 88% for parents. Clearly there was much higher uptake of the technology in this short time, with teens catching up with parents. Computer ownership has remained static over this time with 60% owning or accessing a computer. Other technologies that teens own/use are: a game console like an Xbox or a PlayStation (77%); iPod or mp3 player (73%); portable gaming device (55%)(Lenhart, 2009).

Landlines remain important means of communication with friends, with "87% of those with cell phones still using landlines." However, "for daily activities, cell phone-based communication is dominant, with nearly 2 in 5 teens sending text messages every day. Voice calling on cell phones is nearly as prevalent, as more than a third (36%) of all teens (and 51% of those with cell phones) talk to their friends on the cell phone every day." Sending text messages can be done via phone, computers or email with more girls than boys (42% of girls, 32% of boys) and older teens texting (25% aged 12 to 14; 51% aged 15 to 17). No ethnic differences were apparent but "teens from wealthier households are slightly more likely to text message frequently compared with teens from lower income households; 42% of teens from households earning more than $50,000 annually send texts daily, compared with 33% of teens from homes earning less than $50,000 per year. . . . Among social network users, 54% of teens on those sites send IMs or text messages to friends through the social networking system" (Lenhart, 2009)

The set of Pew Reports provides a nuanced and detailed picture of how young people are engaging with new digital media, negotiating these new spaces for themselves with little input or interference from parents and even less from schools, which mostly ignore or ban the new media. The opportunities arise for young people to forge new identities in this digital space, to become more creative and to develop the skills that will be necessary as the twenty-first century creative knowledge economy proceeds.

According to Henry Jenkins (2006), youth are engaging in "participatory culture," where they can develop their own styles, artistic and creative voices, and identities through multiple interactions with peers and others. Participatory culture has "relatively low barriers to artistic expression and civic engagement, strong support for creating and sharing one's creations, and some type of informal membership whereby what is known by the most experienced is passed along to novices." Participatory cultures comprise four forms: "affiliations" with informal and formal memberships built around various forms of media, e.g., SNS, message boards, and gaming communities; "expressions," which produce transformative types of creative expression, such as mash-ups, YouTube videos, fan fiction and fanzines; "collaborative problem solving" such as working in teams to contribute to a knowledge base using a wiki; and "circulations" which change the distribution and flow of media through such tools as blogging and podcasting (Jenkins, 2006).

The set of Pew reports links well with the next major set of research that the chapter outlines, the Digital Youth Project. November 2008 saw the completion and final report published of this three-year joint project by the University of Southern California and the University of California, Berkeley, funded by the John D. & Catherine T. MacArthur Foundation, "Kids' Informal Learning with Digital Media: An Ethnographic Investigation of Innovative Knowledge Cultures" (http://digitalyouth.ischool.berkeley.edu/node/1). Mizuko (Mimi) Ito[10] from the University of Southern California, led a team of researchers[11] who interviewed 800 youth over 5000 hours of online observations about how youth use digital media in their daily lives. The research focused on two major questions: "How are new media being integrated into youth practices and agendas? How do these practices change the dynamics of youth-adult negotiations over literacy, learning, and authoritative knowledge?"

The findings are available in several forms: as a two-page summary and a 30-page White Paper, Ito et al. (2008), "Living and Learning with New Media: Summary of Findings from the Digital Youth Project," http://digitalyouth.ischool.berkeley.edu/files/report/digitalyouth-TwoPageSummary.pdf); and a book by Ito et al. (2010), *Hanging out, Messing Around, Geeking Out: Living and Learning with New Media*, which is also available online. The title indicates different and deepening but fluid, non-sequential levels of participation that youth have with the new media—they may just "hang around" then "mess around" and at the highest level of participation may "geek out." Also available online at http://digitalyouth.ischool.berkeley.edu/projects, are full descriptions of individual research studies conducted by members of the Digital Youth Project.

In their chapter, "Creative Production," Lange and Ito draw from a number of case studies[12] and point out that they consider that creative production "includes imaginative and expressive forms that are also shaped by kids' individual choices and available media" (http://digitalyouth.ischool.berkeley.edu/book-creativeproduction).

They comment that a new media ecology for everyday sharing and creative production has emerged now that photos, videos, and music can be accessed by digital media-production tools that can modify, re-mix, and be posted and shared on social network sites.[13] Posting, commenting, and creating in the digital arena has changed young people from being simply media consumers to more active media producers or prosumers.

Web 2.0, user-generated content, modding, prosumer,[1] pro-am,[2] remix culture—these buzz words are all indicators of how creative production at the "consumer" layer is increasingly seen as a generative site of culture and knowledge. A decade ago, creating a personal webpage was considered an act of technical and creative virtuosity; today, the comparable practice of creating a MySpace profile is an unremarkable achievement for the majority of U.S. teens. As sites such as YouTube, Photobucket, and Flickr become established as fixtures of our media-viewing landscape, it is becoming commonplace for people to both post and view personal and amateur videos and photos online as part of their everyday media practice. In turn, these practices are reshaping our processes for self-expression, learning, and sociality. (http://digitalyouth.ischool.berkeley.edu/book-creativeproduction)

Lange and Ito note that while educational institutions have generally devalued popular culture, there is new work that challenges this position, for example, Buckingham et al. (2000). Similarly the New Media Literacy (NML) project headed by Henry Jenkins at MIT "explores how we might best equip young people with the social skills and cultural competencies required to become full participants in an emergent media landscape and raise public understanding about what it means to be literate in a globally interconnected, multicultural world" *(*http://newmedialiteracies.org/*).* Similarly, this project sets out a New Literacy Studies approach in the Introduction to the book:

> Seeing creativity as a process of not only creating original works but of recontexualizing and reinterpreting works in ways that are personally meaningful or meaningful in different social and cultural contexts. These approaches are efforts to bridge the more recreational practices and media literacy that kids are developing outside of school with more formal and reflective educational efforts that center on media production. As with all efforts to bridge the boundaries between instructional programs and everyday peer-based youth culture, these translations are fraught with challenges. Even in educational programs that recognize the importance of new media literacy, educators struggle to develop frameworks for assessing and giving appropriate feedback on student work. Teachers tend to assume the media are "doing the work" when kids engage in critical, remix, and parodic forms of production that use elements from other media (Sefton-Green, 2000). Teachers are also wary of media work that appears to be "too polished" or "suspiciously flashy," particularly those genres for which kids rather than teachers are more familiar (Buckingham et al., 2000).
>
> These difficulties in translating recreational media engagement into school-based forms point to persistent tensions between peer-based learning dynamics and genres and those embedded in formal education. Educators have examined a wide range of topics relating to the tension between in-school and out-of-school forms of literacy (Bekerman, Burbules, and Silberman-Keller, 2006; Hull and Shultz, 2002; Mahiri, 2004; Nunes, Schliemann, and Carraher, 1993); media literacy is somewhat unusual in that we are dealing with both an intergenerational tension

> (between adult authority and youth autonomy) and a tension between educational and entertainment content (Ito, 2007). This chapter, in order to inform educational efforts in media education, is an effort to describe the kind of new media literacies and creative production practices that youth are developing in their peer-based social and cultural ecologies. Any effort to translate popular and recreational social and cultural forms into educational efforts needs to be informed by these youth-centered frames of reference. The peer-based learning genres we see in youth online participation differ in some fundamental, structural ways from the social arrangements that kids find in schools. Simply mimicking genre or sharing and assessment dynamics are not sufficient to promote the forms of learning that youth are developing when they are given authority over their own learning and literacy in these domains. (http://digitalyouth.ischool.berkeley.edu/book-creativeproduction)

In June 2008, the University of Minnesota released a report "Educational Benefits of Social Networking Sites Uncovered" based on research led by Christine Greenhow (see: http://www.sciencedaily.com/releases/2008/06/080620133907.htm). The students were from families whose incomes were at or below the county median income of $25,000 and were taking part in an after-school program, Admission Possible, aimed at improving college access for low-income youth. Students aged 16 to 18 from thirteen Midwest urban high schools and a follow-up, randomly selected subset, were asked questions about their Internet activity as they navigated MySpace. Greenhow found that:

> low-income students are just as technologically proficient as their wealthier counterparts; 94% of the students observed used the Internet; 82% go online at home and 77% had a SNS profile; students said they learn technology skills , followed by creativity, being open to new or diverse views and communication skills—they are practicing the kinds of 21st century skills we want them to develop to be successful today developing a positive attitude towards using technology systems, editing and customizing content and thinking about online design and layout. They're also sharing creative original work like poetry and film and practicing safe and responsible use of information and technology. SNS offer more than just social fulfillment or professional networking, also have implications for educators, who now have a vast opportunity to support what students are learning on Websites. By understanding how students may be positively using these networking technologies in their daily lives and where the as yet unrecognized educational opportunities are, we can help make schools even more relevant, connected and meaningful to kids; very few students in the study were actually aware of the academic and professional networking opportunities that the Websites provide. Making this opportunity better known to students, is just one way that educa-

> tors can work with students and their experiences on social networking sites; educators can help students realize even more benefits from their social network site use by working to deepen students' still emerging ideas about what it means to be a good digital citizen and leader online (University of Minnesota, 2008)

With youth engaging so extensively with digital media in their lives outside educational environments, we need now to re-think youth subjectivity/identity, literacy, education, and attitudes to civics and democracy and work out how to link their interests into the creative knowledge economy.

Youth and Creativity: Some Implications for the Creative Knowledge Economy

One of the effects of the new digital media and programs that are so much more readily accessible to more than just a few aficionados and experts who were specially trained for programs that required high levels of software skills, has been a democratization of the potential for self-expression, to write, develop art, music and other creations—the creation of a new and larger 'casual creative' class, especially among creative youth. In order to better understand this evolving approach to media creation, Adobe recently worked with Create with Context to perform a series of in-depth, in-home, ethnographic interviews with 15-to 17-year-olds to understand how they express themselves digitally. The 2009 work of Evangeline Haughney and Bill Westerman led to a series of fresh insights into this market, including willingness to learn media creation software and websites "on the fly," a strong bias for learning by doing, collaboration and apprenticeship with an international pool of creatives, a desire for individuality leading to a rejection of standardized tools and templates, and publishing an ongoing stream of mashed-up, throw-away media. Insights from this research are driving the creation of next-generation products and services to target the creative youth today and as they enter the workforce (http://www.toccon.com/toc2009/public/schedule/detail/5177).

In researching youth understanding and attitudes to digital creativity and copyright law in students aged 12 to 22, John Palfrey, Urs Gasser, Miriam Simun, & Rosalie Fay Barnes (2009) found that young people who operate in the digital realm are overwhelmingly ignorant of the rights, and to a lesser degree the restrictions, established in copyright law. They often engage in unlawful behavior, such as illegal peer-to-peer music downloading, yet they nevertheless demonstrate an interest in the rights and livelihoods of creators. Building upon our findings of the disconnect between technical, legal, and social norms as pertaining to copyright law, we present the initial stages of the development of an educational intervention that posits students as creators: the Creative Rights copyright curriculum. Educating youth about copyright law is important for empowering young people as actors in society, both in terms of their ability to contribute to cultural knowledge with creative practices and to engage with the laws that govern society. (http://ijlm.net/fandf/doi/abs/10.1162/ijlm.2009.0022)

While many youth are not attempting to translate their skills in digital media into anything commercial, as Lange and Ito (2008) indicate, some youth have become entrepreneurial and have developed businesses. For example, Catherine Cook, the co-founder of SNS MyYearbook.com, started the site in 2005 when she was in high school. Together with her brothers she aimed to create an interactive digital yearbook to keep in touch after high school graduation. Now she has over 1.5 million members and has attracted over $4m in venture capital (see Stefanie Olsen, 2007, "Newsmaker: The Secrets of a Teen's Internet Success" www.news.com/the-secrets-0f-a/2008–1038_3–6202845.html). In 2004, Ashley Qualls, a 14-year-old girl interested in graphic design, created Whateverlife.com, a source for MySpace graphics and Web design tutorials. The site now includes free graphics and a magazine with teen-authored articles and reviews. It attracts more than seven million individuals and 60 million page views per month, according to Google (Chuck Salter, 2007). In Glasgow, Scotland, the startup WeeWorld with the avatar WeeMee was designed by girls for social networking sites (see: http://www.weeworld.com).

Many creative Web sites now exist for youth to develop their interest and talents, e.g., The Youth Creative Network (YCN), which posts portfolios, has blogs, links on Flickr, and Twitter and has now linked with Google in the eighth year of the YCN awards (http://www.ycnonline.com/home). There are also different types of Web sites set up by cultural and other organizations that now aim to enhance and develop youth creativity, e.g., with Toronto's goal to become a creative city, the Creative Youth Envoy program has been established (see http://creativecity.ca-/project-profiles/Toronto-CYE.html). Manchester Youth Arts in the United Kingdom targets 11-to 25-year-olds, aiming "to support young people, and those who work with young people, in gaining the knowledge to access what's available across the arts and also the means to develop new ideas or projects" (http://www.manchesteryoutharts.org.uk/). New digital technologies and graphics processing have already dramatically changed the face of drawing, design, photography, film, video, and music (e.g., Adobe Illustrator, Inkscape, and Nvidia, world leader in the development of graphics processing units and chipset technologies for workstations, personal computers, and mobile devices), and even how we interact with the world (e.g., Google-Earth, Maps, StreetView, SketchUp). Social networking is now changing how we relate to each other and is now being used by businesses as places to advertise and by consumers to report on customer service. A business ignores such feedback at its peril.

Regardless of concerns about copyright, the research clearly shows that youth are avid digital content creators and users—prosumers—they may create new knowledge and share knowledge (e.g., use Wikis & Wikipedia). They are part of or may form new networks and so interact with people in different and novel ways. This dynamic shifts traditional notions of identity away from that of passively consuming one's identity in the global marketplace. It emphasizes new aspects of techno-social relations, power relations, customization, and personalization in consumer culture. Many adults fear that youth do not have the critical facilities and maturity to be able to critically engage with the new media. However, they often seem critical or at least averse to the older style media and multinational corporations and pro the new advances in free, or at least cheap and/or open source systems (e.g., preferring Linux to Microsoft). With the new generation of mobile phones, the explosion of free and cheap apps has further altered the relationship between user and technology. The new trend for this generation is toward Open Source, Open Access Open Content, of peer-3-peer (p-2-p) systems towards notions of sharing and participation and collaboration rather than individualized competitive models of working and being.

Most companies and organizations need creative people so that their products and services and marketing of these remain competitive, yet schools and universities seem to not only dis-acknowledge this development, but many actually discourage creativity or any challenge to the existing order and perceived wisdom. For example, many schools (abetted by parents) seem fearful of student use of the Internet and social media and ban their use, citing multiple fears such as negative effects on learning, waste of time, simply absorbing all information and an inability to develop critical thinking, developing anti-social behaviors, cyber-bullying, exposure to undesirable items—especially pornography and violence. What they have not yet realized is that amidst some obviously negative aspects, the skills that young people learn as they become the first generation to grow up with the Internet in the Information or Post-industrial society is that they are learning the skills for the twenty-first century, rather than those for a Fordist, factory-model society. There are already new jobs in areas of the creative knowledge economy that were unknown until relatively recently such as positions in digital animation and digital games. These require not only the technical and creative skills but also the social skills of the world of today and the future.

References

BBC (2008). Panorama: 'One click from danger', Reporter: Jeremy Vine, recorded from transmission: BBC One, 7/01/08. Retrieved from: http://news.bbc.co.uk/2/hi/programmes/panorama/7180769.stm, accessed October 2009.

Bekerman, Z., Burbules, N.C., & Silberman-Keller, D. (Eds) (2006). *Learning in places: The informal education reader.* New York: Peter Lang.

Besley, A. C., & Peters, M. A. (2007). *Subjectivity and truth: Foucault, education and the culture of self.* New York: Peter Lang.

Besley, T. (2003). Hybridized and globalized: Youth cultures in the postmodern era. *The Review of Education/Pedagogy /Cultural Studies, 25*(2), 75–99.

boyd, d. (2007). Why do youth (heart) social network sites? The role of networked publics in teenage social life. In D. Buckingham (Ed.), MacArthur Foundation Series on Digital Learning: *Youth identity, and digital media.* Cambridge, MA: MIT Press. Retrieved from http://www.danah.org/papers/WhyYouthHeart.pdf

Buckingham, D. (2000). *After the Death of Childhood: Growing Up in the Age of Electronic Media.* Cambridge, UK: Polity.

Compete: Retrieved from http://www.compete.com/about/

Comscore: Retrieved from http://www.comscore.com/Press_Events/Press_Releases/2008/08/Social_Networking_World_Wide/(language)/eng-US

Florida, R. (2002). *The rise of the creative class.* New York: Basic Books.

Florida, R., Gates, G., Knudsen, B., & Stolarick, K. (2006). *The university and the creative economy.* Retrieved from http://www.creativeclass.org/rfcgdb/articles/univ_creative_economy082406.pdf Foucault, M. (1982). *Truth, power, self: An interview with Michel Foucault.* Retrieved from http://en.wikiquote.org/wiki/Michel_Foucault

Haughney, E. & Westerman, B. (2009). *Creative youth & digital publishing.* Retrieved from http://assets.en.oreilly.com/1/event/19/Youth%20and%20Creativity_%20Emerging%20Trends%20in%20Self-expression%20and%20Publishing%20Presentation.pdf Accessed October 2009.

Howkins, J. (2001). *The creative economy: How people make money from ideas.* London: Penguin.

Howkins. J. (2005). The creative economy: Knowledge-driven economic growth. Speech delivered at the Asia-Pacific Creative Communities: A Strategy for the 21st Century Conference. Retrieved from http://www.unescobkk.org/fileadmin/user_upload/culture/Cultural_Industries/presentations/Session_Two_-_John_Howkins.pdf

Hull, G., & Schultz, K. (2002). Negotiating boundaries between school and non-school literacies. In G. Hull and K. Schultz (Eds) *School's Out! Bridging out-of-school literacies with classroom practice.* New York: Teachers College Press; pp. 1–10.

Ikeda, Daisaku (n.d.) Retrieved from http://www.saidwhat.co.uk/topicquote/creativity

Ito, M. (2007). Education v. Entertainment: A cultural history of children's software. In K. Salen (Ed.). *The ecology of games: Connecting youth, games, and learning.* The John

D. and Catherine T. MacArthur Foundation Series on Digital Media and Learning. Cambridge, MA: MIT Press: pp. 89–116.

Ito, M., Baumer, S., Bittanti, M., boyd, d., Cody, R., Herr, B., Horst, H. A, Lange, P. G., Mahendran, D., Martinez, K., Pacsoe, C. J, Perkel, D., Robinson, L., Sims, C., & Tripp, L., with Antin, J., Finn, M. Law, A. Manion. A, Mitnick, S., & Schlossberg, D., & Yardi, S. (2010). *Hanging out, messing around, geeking out: Living and learning with new media.* Cambridge: MIT Press.http://digitalyouth.ischool.berkeley.edu/report. Accessed October 2009.

Ito, M., Horst, H., Bittanti, M., boyd, d., Herr-Stephenson, B., Lange, P. G., Pascoe, C. J., & Robinson, L., with Baumer, S., Cody, R., Mahendran, D., Martinez, K., Perkel, D., Sims, C., & Tripp, L. (2008). *Living and learning with new media: Summary of findings from the digital youth project.* http://digitalyouth.ischool.berkeley.edu/files/report-/digitalyouth-TwoPageSummary.pdf. Accessed October 2009.

Jenkins, H. (2006). Confronting the challenges of participatory culture: Media education for the 21st century (part one). Retrieved from www.henryjenkins.org/2006/10/confronting_the_challenges_of.html. Accessed May 2008.

Lange, P., & Ito, M. (2008). Creative production. In M. Ito et al., *Hanging out, messing around, geeking out: Living and learning with new media.* Retrieved from http://digitalyouth.ischool.berkeley.edu/book-creativeproduction.

Lenhart, A. (2008). *Teens, video games and civics.* Retrieved from http://www.pewinternet.org/Reports/2008/Teens-Video-Games-and-Civics.aspx?r=1.

Lenhart, A. (2009). *Teens and mobile phones over the past five years: Pew Internet looks back.* Retrieved from http://www.pewinternet.org/Reports/2009/14—Teens-and-Mobile-Phones-Data-Memo.aspx?r=1, accessed October 2009

Lenhart, A., Arafeh, S., Smith, A., & Magill, A. (2008) *Writing, technology and teens.* Retrieved from http://www.pewinternet.org/Reports/2008/Writing-Technology-and-Teens.aspx?r=1.Machlup. F. (1962). *The production and distribution of knowledge in the United States.* Princeton: Princeton University Press.

Lenhart, A., & Madden, M. (2007). *Teens, privacy and online social networks.* Retrieved from http://www.pewinternet.org/Reports/2007/Teens-Privacy-and-Online-Social-Networks.aspx?r=1, April 18, 2007. Accessed October 2009.

Lenhart, A., Madden, M., Smith, A., & Macgill, A. (2007). *Teens and social media.* Pew Internet & American Life Project. December 19, 2007. Retrieved from http://www.pewinternet.org/~/media//Files/Reports/2007/PIP_Teens_Social_Media_Final.pdf.pdf. Accessed May 2008.

Mahiri, J. (Ed). (2004). *What they don't learn in school: Literacy in the lives of urban youth.* New York: Peter Lang.

Manchester Youth Arts http://www.manchesteryoutharts.org.uk/

Nunes, T., Schliemann, a.d., & Carraher, D.W. (1993). *Street mathematics and school mathematics.* Cambridge, UK: Cambridge University Press.

Olsen, S. (2007). Newsmaker: The secrets of a teen's internet success. Retrieved from www.news.com/the-secrets-of-a/2008–1038_3–6202845.html.

Organization for Economic Co-Operation and Development OECD. (1996). *The knowledge-based economy.* Paris: Organization for Economic Co-Operation and Development.

Organization for Economic Co-Operation and Development OECD. (1999). *The knowledge-based economy: A set of facts and figures.* Paris: Organization for Economic Co-Operation and Development.

Organization for Economic Co-Operation and Development (OECD). (2001). *Scoreboard of science, technology, and industry indicators.* Paris: Organization for Economic Co-Operation and Development.

Palfrey, J., Gasser, U., Simun, M., Barnes, R. F. (2009, spring). Youth, creativity, and copyright in the digital age. *International Journal of Learning and Media, 1*(2), pp. 79–97. doi: 10.1162/ijlm.2009.0022. Retrieved fromhttp://ijlm.net/fandf/doi/abs/10.1162/ijlm.2009.0022.

Parr, B. (2009). MySpace's US traffic falls off a cliff. Retrieved from http://mashable.com/2009/10/12/myspace-traffic-plummets/

Peters, M. A., & Besley, Tina (A. C.). (2006). *Building knowledge cultures: Education and development in the age of knowledge capitalism.* Lanham, MD: Rowman & Littlefield.

Peters, M.A., & Besley, Tina (A. C.). (2009). Academic entrepreneurship and the creative economy. In M. A. Peters, S. Marginson, & P. Murphy (Eds.), *Creativity and the global knowledge economy,* New York: Peter Lang.

Prensky, M. (2001). *Digital natives, digital immigrants.* Retrieved from http://www.marcprensky.com/writing/Prensky%20%20Digital%20Natives,%20Digital%20Immigrants%20%20Part1.pdfSaarinen, Esa (n.d.) Retrieved from http://www.mycoted.com/Creativity_Quotes

Salter, C. (2007). Girl power. Retrieved from http://www.fastcompany.com/magazine/118-/girl-power.html

Schumpeter, J. (1942/1976). *Capitalism, socialism, and democracy.* New York: Harper and Brothers.

Schumpeter, J. (1951). Economic theory and entrepreneurial history. In R. V. Clemence (Ed.), *Essays on economic topics of Joseph Schumpeter.* Port Washington, NY: Kennikat Press.

Sefton-Green, J. (2000). Introduction: Evaluating creativity. In J. Sefton-Green and R. Sinker *(Eds) Evaluating creativity: Making and learning by young people.* London, UK, & New York: Routledge: pp 1–15.WeeWorld, http://www.weeworld.com

University of Minnesota, (2008) "Educational Benefits of Social Networking Sites Uncovered," Retrieved from http://www.sciencedaily.com/releases/2008/06/080620133907.htm, accessed October 2009.

World Bank . (2002). *Constructing knowledge societies: New challenges for tertiary education.* A World Bank Report & The International Bank for Reconstruction and Development, Washington, DC. Retrieved from http://www1.worldbank.org/education/pdf/Constructing%20Knowledge%20Societies.pdf

Youth Creative Network, http://www.ycnonline.com/home

Notes

1. According to Wikipedia, social media are: "designed to be disseminated through social interaction, created using highly accessible and scalable publishing techniques. Social media support the human need for social interaction, using Internet- and web-based technologies to transform broadcast media monologues (one to many) into social media dialogues (many to many). They support the democratization of knowledge and information, transforming people from content consumers into content producers. Businesses also refer to social media as user-generated content (UGC) or consumer-generated media (CGM). Social media can take many different forms, including Internet forums, weblogs, social blogs, wikis, podcasts, pictures, video, rating and bookmarking. Technologies include: blogs, picture-sharing, vlogs, wall-postings, email, instant messaging, music-sharing, crowdsourcing, and voice over IP, to name a few." (http://en.wikipedia.org/wiki/Social_media) accessed November 2009.
2. Social Networking Service definition at: http://en.wikipedia.org/wiki/Social_network_service. List of Social Networking Websites: http://en.wikipedia.org/wiki/List_of_social_networking_websites. e.g Facebook, MySpace, Bebo. There has been a huge expansion in use of social networking sites recently with such sites by no means only being used by young people—see "Social Networking Explodes Worldwide as Sites Increase their Focus on Cultural Relevance" RESTON, VA, August 12, 2008–comScore, Inc.—worldwide usage of social networking sites, indicating that while the growth in new users in North America is beginning to level off, it is burgeoning in other regions around the world. During the past year, the total North American audience of social networkers has grown 9 percent compared to a much larger 25 percent growth for the world at large. The Middle East-Africa region (up 66 percent), Europe (up 35 percent); Latin America (up 33 percent) each grown at well above average rates. (http://www.comscore.com/Press_Events/Press_Releases/2008/08/Social_Networking_World_Wide/(language)/eng-US)
3. This section is adapted from *Building Knowledge Cultures: Education and Development in the Age of Knowledge Capitalism* (Peters & Besley, 2006),
4. This section uses work jointly written with Michael A. Peters, "Academic Entrepreneurship and the Creative Economy" (Peters & Besley, 2009).
5. All references are to Howkins's (2005) seminar "The Creative Economy: Knowledge-Driven Economic Growth" delivered at the Asia-Pacific Creative

Communities: A Strategy for the 21st Century at http://www.unescobkk.org/fileadmin/user_upload/culture/Cultural_Industries/presentations/Session_Two_John_Howkins.pdf

6. Telephone interviews were conducted by Princeton Survey Research Associates International (PSRAI) between October 13 and November 19, 2006. PSRAI is an independent research company specializing in social & policy research. For results based on the total parent or teen sample, there is a 95% confidence level that the margin of error is +/- 3%. For results based on online teens or online parents, the margin of sampling error is +/ 4% (http://www.pewinternet.org/~/media//Files/Reports/2007/PIP_Teens_Social_Media_Final.pdf.pdf)
7. The digital media world epitomizes a fast paced environment such that in April 2008, Facebook, with over 300 million users became the world's leading social networking site. (http://www.comscore.com/Press_Events/Press_Releases/2008/08/Social_Networking_World_Wide/(language)/eng-US). MySpace, owned Rupert Murdoch's NewsCorp since 2005, has proven unable or unwilling to keep up with advances in social network technology, such that by November 2009, clients are either not using the site or are exiting in droves. According to Compete.com web analytics statistics (http://www.compete.com/about/), "MySpace's U.S. traffic dropped from 55.6 million unique visitors in August [2009] to 50.2 million in September nearly shed off 20% of its U.S. traffic since June" http://mashable.com/2009/10/12/myspace-traffic-plummets/).
8. The "five most popular games are Guitar Hero, Halo 3, Madden NFL, Solitaire, and Dance Revolution E rated as suitable for everyone."(http://www.pewinternet.org/Reports/2008/Teens-Video-Games-and Civics.aspx?r=1).
9. Some of the findings are: "teens are motivated to write when they can select topics that are relevant to their lives and interests, and report greater enjoyment of school writing when they have the opportunity to write creatively . . . writing for an audience motivates them to write and write well. Teens who enjoy their school writing more are more likely to engage in creative writing at school compared to teens who report very little enjoyment of school writing (81% vs. 69%). Non-school writing, while less common than school writing, is still widespread among teens and varies by gender and race/ethnicity. Boys are the least likely to write for personal enjoyment outside of school. Girls and black teens are more likely to keep a journal than other teens. Black teens are also more likely to write music or lyrics on their own time (47% of black teens write in a journal, compared with 31% of white teens; 37% of black teens write music or lyrics, while 23% of white teens do; 49% of girls keep a journal; 20% of boys do; 26% of boys say they never write for personal enjoyment outside of school). Multi-channel teens and gadget owners do not write any more—or less—than their counterparts, but bloggers are more prolific. Teens more often write by hand for both out-of-school writing and school work. As tech-savvy as they are, teens do not believe that writing with computers makes a big difference in the quality of their writing. Parents are general-

ly more positive than their teen children about the effect of computers and text-based communication tools on their child's writing. 27% of parents think the internet writing their teen does makes their teen child a better writer, and 27% think it makes the teen a poorer writer. Some 40% say it makes no difference." (http://www.pewinternet.org/Reports/2008/Writing-Technology-and-Teens.aspx?r=1).

10. Mimi Ito is now Associate Researcher at the Humanities Research Institute at the University of California, Irvine. In addition, she is a Visiting Associate Professor at the Keio University Graduate School of Media and Governance. (http://en.wikipedia.org/wiki/Mizuko_Ito).
11. "Led by [the late Peter Lyman], Mimi Ito, Barrie Thorne, and Michael Carter, the Digital Youth Project comprises an interdisciplinary team that draws upon expertise in fields varying from anthropology, communication, psychology, and sociology to computer science, engineering, and media studies. Many of our researchers also have experience working in industry and community organizations. In addition to our four PIs, the team for this project included seven postdoctoral researchers, six doctoral students, nine MA students, one JD student, one project assistant, seven undergraduate students, and four research collaborators who contributed fieldwork materials for the project." (http://digitalyouth.ischool.berkeley.edu/people).
12. "We draw primarily from our case studies on youth media production, MySpace Profile Creation (Dan Perkel), Hip-Hop Music Production (Dilan Mahendran), YouTube and Video Bloggers (Patricia G. Lange), Self-Production through YouTube (Sonja Baumer), Anime Fans (Mizuko Ito), and Harry Potter Fandom (Becky Herr-Stephenson)." (http://digitalyouth.ischool.berkeley.edu/book-creativeproduction)
13. For example, SNS sites such as MySpace and Facebook, blogs, online journals, and media-sharing sites such as YouTube, deviantART, and FanFiction.net.

23

Community as Curriculum

Dave Cormier

Mr. McGuire: I just want to say one word to you—just one word.
Ben: Yes sir.
Mr. McGuire: Are you listening?
Ben: Yes I am.
Mr. McGuire: 'Plastics.'
Ben: Exactly how do you mean?
Mr. McGuire: There's a great future in plastics. Think about it. Will you think about it?
Ben: Yes I will.
Mr. McGuire: Shh! Enough said. That's a deal.

—*The Graduate*, 1967

This classic scene from the 1967 blockbuster *The Graduate* illustrates the assumptions and premises of the traditional twentieth-century ontology of work and knowledge. The scene paints a world where access to knowledge is privileged, access to mentors and inside information key, and getting in on the inside of the next big thing a one-time stroke of luck on which an entire successful—and linear—future could be built. The film reflects the prevailing understandings of its era, including that a future could be chosen at the start of one's career.

In the scene, Benjamin Braddock, twenty-one years old, recent university graduate and star student, stands uncomfortably in his parents' living room being grilled by their peers about what he's going to do with the rest of his life. The enigmatic Mr. McGuire lures him aside in a semi-satirical performance that emphasizes both McGuire's delight in his own power and the bewildering weight of the choice ahead of Benjamin. The magic word that McGuire dangles promises Benjamin entry into a cabal of people who know who hold the future. As viewers, we can apprehend Ben's dilemma and the critique implied in McGuire's over-the-top delivery because the roles and the beliefs they were built on still carry power in our culture and particularly in our educational systems, where knowledge as a commodity with structured, gate-kept access paths to success is still a powerful narrative.

Benjamin Braddock has the package he needs to succeed in McGuire's world. He knows how to access the world of knowledge through the books and journals that live in the library at his university. He knows how to work with experts, receive the knowledge and expertise stored in their heads, and serve that back in a format that meets specific requirements. Ben has acquired an education, and now it can be put to use. He has a store of knowledge and the requisite skills to acquire more. Even as the final act of the film closes with Ben's rejection of convention and the world offered by Mr. McGuire, viewers are left with the assumption that he will be just fine. Benjamin Braddock is positioned not to fail.

This idea of learning as something that can be bought, acquired, and then completed is deeply ingrained in popular culture. It is a comforting model. If we are to believe that learning is as simple as this, then at any given time, we can simply find the "way to do something," pay the requisite fee to acquire the knowledge in question and then go about applying it to our lives. If we are in a field like, say, plastics, we need only be able to tell one plastics knowledge salesman from another. If we are clever or choose a reputable institution of plastic sales, we will have the item that we need and will be able to progress a little further down our chosen path. The model implies that there is a way to know what the answer might be, and that a person (our plastics expert) could have this knowledge, and that the knowledge can then be acquired by a learner. In addition, most importantly, once the knowledge is acquired, the learning is finished.

It is a simple model. Unfortunately, it is false. If it ever worked, it will not work anymore. The promise of the twentieth-century model of knowledge is an empty one today.

As the job descriptions of traditional professions change and diversify, people are realizing that the myth will no longer hold. There will be no more "plastics" salesmen. A manager might also now be a social media strategist; administrative assistants are becoming Web managers, and mechanics are plugging computers into cars. Where once universities and colleges would market to an established clientele and rely on word of mouth and prestige for attracting students, they now market to the globe and find themselves in need of types of literacy that were hitherto the province of embassies and multinational corporations.

In this climate of change and adaptability, where money is being spent on "making things digital" and "making people ready for the digital age," how does one know how to guide one's professional course? How can one choose learning opportunities that will contribute to success? How is the policy maker to choose between options, either of which are potentially equally likely to succeed, and both of which are being presented in terms that break current frameworks of judgment?

Often the answer to these questions has been, simply, "more." More training, more work, more hours spent in the office. More research, more money spent on research consultants. The training is too often too short, too boring, and too old-fashioned. The consultants are expensive, difficult to find, and come with no way to accredit their advice. People turn to passing fads in an attempt to stay current without the necessary experience or context to be able to make informed, professional decisions about them.

This is the climate that is now demanding a new way of learning and a new sense of what it means to learn. The implicit lessons of our educational system are still twentieth-century lessons: being on time, attentiveness, focus on single tasks, completion of tasks outside of context, and, perhaps most importantly, completion of tasks without any sense for why they are being done. These are the preparations for effective labor in the Dearborn factories of Henry Ford in the 1910s and 1920s. These are the skills for the members of a workforce who were not to sit down, not to stray from their task and earn their time in front of the fire from a hard day's work. They are preparations for the industrial revolution.

Most of us have, in spite of ourselves adjusted—at least incrementally—to this transmission-focused military model of education. There is a sense in many educators' minds that learners need to explore their way through their learning, and have the experience of learning, of searching out ideas and discovering them for themselves. This process, though, is usually bounded by the learning objectives laid out at the beginning of the course of study by the designer/instructor. There is still, implicit in most widely held conceptions of learning that the instructor, designer, or at least the institution knows what a learner should get out of a given course.

The problem, then, only comes into play when we are not sure what "people should be learning." What is the curriculum for innovation? How do we impart creativity? Where do students turn to be guaranteed that they are learning what is new and current? These are the questions that face us on a more or less regular basis now. As knowledge becomes a moving target and the canon starts becoming less reliable, we need a new—or in fact an old—model of education drawn out on a new canvas: community.

The answer is to stop trying so hard, to stop looking for a systemic solution, and to return to a human-based knowledge plan. We need to return to community as a valid repository for knowledge, and away from a packaged view of knowledge and expertise. Knowledge can be fluid; it can be in transition, and we can still use it. We need to tap into the strength provided by communities and see the various forms of community literacy as the skills we need to acquire in order to be effective members of those communities.

Community as curriculum is not meant as a simple alternative to the package version of learning. It is, rather, meant to point to the learning that takes place on top of that model and to point to the strategies for continuing learning throughout a career. There is a base amount of knowledge that is required to be able to enter a community, and there are methods for acquiring the specific kinds of literacy needed to learn within a specific community. A learner acquires basic forms of literacy and associates with different peer groups. Networks begin to form and, occasionally, communities develop. Knowledge is created and sometimes discarded as the community interacts. Knowledge does not develop and spread from and through concentric circles. There are no "plastics" to be learned and no canon to consult to ensure that a new skill has been acquired. Knowledge is a rhizome, a snapshot of interconnected ties in constant flux that is evaluated by its success in context.

We need a move toward a more practical, sustainable learning model that is less based on market-driven accreditation and more on the inevitable give and take that happens among people who engage in similar activities and share similar forms of literacy and worldviews.

Learning and Knowing

The rhizomatic view of learning reflects an organic, practical approach to thinking about learning and knowledge. It has a distinct connection to the traditional academic knowledge model, with its interlinking references and people. Each piece of information and knowledge is interlinked and supported by at least one other element, with no one place where knowledge about a matter begins or ends. The rhizomatic model, in contrast to the academic one, keeps the knowledge *in* the people and in the *community* rather than distilling it into a paper based product—be it the final publication of a journal, book or other 'changeless' medium. The problem with the paper publishing cycle is the time it takes to proceed through the entire cycle, and the constraints on time and space that go along with the medium place severe restrictions on the flexibility and applicability of the academic tradition. It is not to say that it is not valuable, just that it does not always—and cannot always, today—respond in ways that meet the needs of learners in a world where what is known in many fields changes from month to month.

If we are working in a field where what is new or current is continually in flux, then we need to have a way of keeping our knowledge up to date. With the huge increase of academic publications, the simple process of choosing has become more difficult, and the sifting through what is out there a significant task for any professional. Our ideas of learning and knowledge need to become more flexible to allow for this mutability. "The term [rhizomatic learning] encapsulates a sort of fluid, transitory concept; the dense, multi-dimensional development and integration of several different sets of tools and approaches, appearing in diverse forms under separate settings, using all the multidimensional networking information technology tools, the social web, etc." (Szucs, 2009, p. 4). Rhizomatic learning distributes the channels of knowing outside traditional hierarchical models and into the social realm, allowing for help in sifting through the flow of information and knowledge. These "social learning practices are allowing for a more discursive rhizomatic approach to knowledge discovery" (Cormier, 2008, p. 3). Rhizomatic knowers use a variety of approaches and tools to blend together bits of information and knowledge in order to form what they need. They especially need a learning community to help them test ideas, filter information and knowledge, and seek advice.

The skills that Benjamin Braddock—as a representative for his generation—learned for acquiring knowledge were listening, accepting hierarchy, and learning how to identify sanctioned bits of knowledge to apply to new situations. The traditional testing and identification model that Ben would have been accustomed to operated along principles of "here's what you need to know and here's how I'm going to know that you know it." These notions of hierarchy, sanction, passivity, and external validity are all further victims of the social shift we are currently experiencing. As learning becomes about participation and knowing becomes a negotiation (Cormier, 2008), it is no longer as practical to approach learning with a pre-existing notion of what we are trying to find out. Instead, knowledge production becomes a participatory process based in communities, with members trying to solve problems, tap into existing trends or simply exploring by helping someone else. The problem of how we know what we are taking out of a community environment might be true or useful or might, in some sense, be seen as knowledge that is at the heart of the change that is needed to cope with the accelerated speed of change.

The knowing, then, exists out in the networks and should be seen in this rhizomatic way. It should not be seen as a network in a digital sense but rather as a culmination of the connections between people. Knowledge becomes a snapshot in time of what is known in a community on a given issue. Publishing is done in order to crystallize and make knowledge about the community public—as in "public"-ation. Its value is in its ability to reach out from the community to others, not in its inherent knowledge.

In this rhizomatic model, the roles of the expert, authority, and reference in knowing all change considerably. A side benefit of this transition is the expulsion of the plastics salesmen, vendors of the "next best thing" that you simply *must* think about. Our plastics salesmen can no longer fall back on the old adage "research shows," or simply show us a major corporation using a given product or method and expect us to want to incorporate it into our work.

This being said, expertise and experience are still critical to the success of any community. Whether one believes in the 10-year-rule of expert-making (Ericsson, K.A., Prietula, M.J., Cokely, E.T., 2007), a healthy community will work faster and move toward goals quicker when the people living in that community immerse themselves in the specific goals of that community. Novices certainly can come together to learn without the presence of master-style members, but the progress is likely to be slower. Access to pre-existing work on a subject and trial and error experience can be of great value to novices trying things for the first time. The negotiation of knowledge is going to be more productive if the people involved have access to the basic literacies for that field.

Community

Community is a kind of network. A learning community is a specialized version that inverts the normal pattern of responsibility from being responsible to oneself to being responsible for the learning of the people with whom one is involved. More specifically, one is connected to these people in a variety of ways, be it professional lives, social networks, or other settings. There is a sense, however, that work goes better when people are working together. They are sharing information, working together to learn new things or sharing experiences. In a professional network, "taking care of oneself" is what is considered appropriate; in a community, lending a hand and helping to make sure that others in the same community are learning is the highest order. This idea of learning, that it happens between people on a relatively flat hierarchy, is the antithesis of the world that our Benjamin came from. In his education, having the right connection and having acquired the right information from the existing canon were critical. In the learning community view knowledge is rhizomatic and learning transactional.

Joining any kind of community can involve a fair amount of research and time. Communities are about commitment. They are about responsibility rather than goals described by others. There are many interpretations of the word, many of which attempt to call on a possible past full of barn-raising, communal sharing, and mutual survival. There is another version of the word community that calls more to mind the inequality of high school group work, the inefficiencies of the communist farms and the tedium of the weekly meeting. There is no guarantee that a given community is going to work in a way that will satisfy every prospective member. The definition of community used here differs from both of those archetypes in the sense that it is about choice. A choice to be responsible to a group of people and a choice to join in with a group of people who, while they might not share similar distinct goals share equivalent mores, skills, or worldviews. They are, on some axis, on the same part of the long tail.

Conventional instruction is based on a hierarchical model in which those who know teach those who do not know. As Cross notes, "Institutions of higher learning were set up with the express intent of attracting people who know and the tools of knowing into regional settings in order to allow for the building of centres of knowledge. Students would come from outlying towns to come to the locus of this knowledge in order to get access to 'those who know'" (Cross, 2005, p. 5). Communities, then, become this same kind of regional settings, allowing different kinds of members, some centrally invested and others peripherally so, to interact in centres of knowledge. The difference is, of course, that knowledge is negotiated in time. As this knowledge is negotiated, it is spontaneous rather than reified in books or articles. "Learning together depends on the quality of relationships of trust and mutual engagement that members develop with each other, a productive management of community boundaries, and the ability of some to take leadership and to play various roles in moving the inquiry forward" (Wenger, White, & Smith, 2009, p. 8).

The creation of a community in and of itself is not sufficient to guarantee the kind of life-long learning that promotes creativity in a professional context. Indeed, the creation of a community is often the easiest part of the process. It is important to understand, from each learner's point of view, the goals he or she is to achieve as well as the means available to obtain them.

The two types of learning communities presented here reflect two differing directions for learning. They should not be seen as mutually exclusive but rather as different in control. The first, the *guild model* of community, offers more control and better options for accreditation and verifiability and is also the easier one of the two options. If a guild style community can be established, then it gives a single locus of learning. The other, the *distributed model* of community, i.e., multiple membership roles in multiple communities, offers far greater flexibility, though less control. It would be difficult for an employer to track learning in this kind of environment, or for learners to take guidance on what they should learn. A combination of these two is certainly possible, or the path might lead from one end to the other.

Community as Guild

Creating communities for learning is the first path that many people take who begin to believe that the Benjamin Braddock model is failing them. They are looking for a connection between the organized world of learning and the new connected world of the Web. Indeed, there is a sense that a classroom can be this way. The community versions of these kinds of classrooms emerge around a particular topic or perspective and grow and adapt along non-institutional lines. We also see the same in different social communities that occur around different topics and fields online.

They do offer a number of very significant advantages—perhaps the most important being quality control. A look at traditional guild models and how they deal with issues of quality control offers an interesting perspective to the current open classrooms and social networks.

> It is clear from the records left by guilds that they [guild] were vitally interested in matters of "quality control" and quality assurance. The exclusive right of the guilds to sell certain goods in certain markets, coupled with quality standards written into the guild regulations, assured buyers that all goods under the guild's jurisdiction would be of a certain quality. The guild "imprimatur," in other words, took the place of the reputation of individual craftsman as a quality assurance device. (Merges, 2004, p. 7)

The members of the guild, be they a representative organization or a classroom, are constrained by a charter, social contract, or syllabus that defines the things that are done and known inside that community. It both allows for observers of a community member to know what that member is likely to know and allows new members a better sense of how they can get involved. There are observable dos and don'ts that a person can follow in order to be more successful. People need to know who they are in their community. They need to know how to succeed. They need to understand the roles that are available and what it means to participate. Guilds can work, particularly when they are open and people think of them as part of the whole knowledge building structure.

In the Classroom

> *"The community is not the path to understanding or accessing the curriculum; rather, the community is the curriculum."* (Cormier, 2008)

The ED366H Educational Technology and the Adult Learner classroom, a course I taught at the University of Prince Edward Island in the summer of 2008 was an attempt at putting the guild model community learning into practice. The goal was to create a sense of reliance and a sense of responsibility in each student toward the learning of their fellow students. The goal was not, however, to create some kind of community that would last past the time allowed for the course but rather to instill some of the literacies and demonstrate some of the advantages of community learning in the hope of fostering the desire to join or support community learning in their respective teaching environments.

The course is accredited by the university and designed to span 35 hours in a two-week period. The constraints of such a shallow time span had a considerable impact on the decision not to create a standing community. The course began with a day-by-day syllabus that suggested broad topics of research for week one and broad topics for student-led demonstrations for week two. It took a people-centred and, as far as possible within the required structure, a technologically neutral approach to introducing the idea of technology to the classroom.

One of the interesting features of teaching this kind of course is that any specific information, terminology, technology, or even approach is likely to be partially or totally outdated by the time the opportunity for the learner to actually use it in the classroom has happened. The focus of the syllabus was entirely on the students learning to rely on each other to find paths through the tasks that were set out. Competency for the course was simple, the students needed to teach something to the rest of the class in the second week that they had never heard of in the first. During this process, they were required to create a textbook, together, of the things that they learned and reflect on the process throughout.

The difficulty with the course scenario, however, is that the forced community tends to fall apart after the course. Many more formal attempts have been made to create this kind of community, but they face significant challenges. A case study made of the Education Network of Ontario illustrates many of the pitfalls of trying to sustain long-term interest and participation in an online community of practice. It was a 12-year project that began in the days of dial-up and, due to the inevitable challenges implicit in sustaining a long-term community, eventually ended in 2005.

Despite the obvious benefits of online networks, the complexities of forming and supporting online communities will need to be addressed if they are to be sustained. Designers will have to balance the needs of the community and the needs of individual members. The success of future online communities will be heavily dependent on:

- the level of information overload,
- the tone of the environment (including all of the community building practices needed for a healthy community), and
- outreach and marketing. (Riverin & Stacey, 2008, p. 55)

The guild style community, however, need not be an end in and of itself. It can be seen as a gateway to a more distributed, more flexible view of learning communities. It can be a safe place from which to set out on our own or simply a trusted node in a widening network of trust from which communities tend to form. What the leaders of these guild style communities need to teach people, then, has little to do with content and more to do with actually using communities to learn. The community is, in effect, the whole of the curriculum. Its members need to experience what it can be like to learn in a community mediated environment and take that away with them so that they can continue to be contributing members to their knowledge rhizomes.

Distributed Community

Seeing communities more broadly and taking multiple community and network membership online/offline might be a sustainable community learning model. In a sense, this is what academia has been, as opposed to a simple guild model. The guilds learned and worked mostly with people from their villages or the rare traveler; academics had the written word in books, and they traded with them. They developed methods by which, at a distance, the community could judge the applicability of a given bit of knowledge to the overall field. The methods of quality control by peer review and citation are the distributed community in paper form; adding the Internet, matters start to move faster at a fantastic rate.

The learner and the knowledge producer now can tap into a broad base of professionals in any field at the click of a button. There are professionals sharing their work at the time that it is happening, presentations being streamed online and data coming out from research long before papers are published. With connections through social networks it is possible to query the writers of professional works in order to get clarity, suggestions or confirmation about certain ideas or theories.

We are also able to find other people who have the same degree of interest in a given subject as we ourselves possess. This concept makes Chris Anderson's idea of the long tail so attractive. If we connect the whole world through social networks, then the people with very specific, very passionate interests will be able to collaborate. In Braddock's world they would have accomplished their goals—had they been able to find each other, but now even simple searches will reveal people with whom specialists can form strong connections.

The multiple memberships that make up online community participation can be overwhelming. Online participation can, from a technical perspective, include micro-blogging, bookmarking, blogging, Web cast memberships, and a host of other technical formats that require some degree of competency to participate. The key, however, is the varying layers of connection that they allow with the people who are actually present. Simply "using" the technology offers no particular benefit. Being able to participate in live knowledge building on a daily basis with a group of peers, on the other hand, is a privilege of the so-called digital age.

Having members of a community involved daily in activities of the digital age means teaching them to network first, to assess ways in which their networking can grow into relationships of trust that allow them to rely on people to care about their learning and about their success and where they learn how to judge people's opinion. They will, in a sense, have to become sensitive to the ways in which knowledge can be acquired, created and validated along the rhizomatic view.

Edtechtalk as a Model for Community Curriculum—A Piece of the Puzzle

Edtechtalk comprises a community of educators who come together in various ways on a regular basis to share their expertise and experience with each other. It is "a community of educators interested in discussing and learning about the uses of educational technology" (Edtechtalk, 2009).

My community learning experience started with Edtechtalk. My relationship to it and membership in it has changed and morphed over the last four years, but it has consistently occupied a very important place in my learning world. It is both a locus for the guild model of learning in the structured weekly programming and the inner core of producers of shows that works together and is also a distributed model of learning in the way that it becomes a locus for less central members to come together, meet others, and move on to other projects.

The problem, however, is that these are emergent communities. They are communities with no direct goal and have not been created by an entity with the specific goal of "learning." It is a learning community that was created out of an event as is sometimes created out of an existing event, as the Web heads were in 2002 and continue to be a strong supportive community. However, they are difficult to create on purpose to solve a particular problem.

The focus of this kind of community learning needs to be on the people and on the specific context of those people. There may be circumstances where much of this kind of learning would be face to face, where it might go in waves, or where it might be entirely online. This need not have a profound impact on the success of the learning experience. Each venue or platform, be it a Multi User Virtual Environment or a coffee shop offers different advantages and disadvantages. In either case an eventedness is offered to the student who can take them out of their context (Cormier, 2009, p. 545). Certain matters may be resolved easily in one situation, whereas in another situation, the same matter may prove to be a lot more complicated. More specifically, each matter needs to be seen in its individual context.

There are problems, necessarily, with this kind of approach. For example, if a number of practitioners, who are still in need of experience, attempt to band together without access to the "trails" of more experienced practitioners, they could be in for a very difficult and time consuming learning process.

Strengths and Weaknesses

This kind of approach is in no way going to appeal to all people and be ideal to all situations. There are some buttons that are blue that need to be pushed after the yellow one. While there are any number of potentially important community lessons to be learned around this concept (how to sit when doing so, what to do if the yellow button is missing), there are times when the training path is simply a question of acquiring certain simple facts. More broadly, there are objections that affect this kind of learning . . . as Mr. Szucs noted "We should notice that the strength and the weakness of this approach is at the same time, that the content and the competence are legitimated by the collaboration in the networked system" (Szucs, 2009, p. 4).

The content, in this case information and knowledge on how to accomplish tasks, e.g., grow professionally, choose a new course, whatever the case may be, is legitimated by collaboration and by its applicability to the context of the user. If that user does not return to the community with the results, is not able to assess those results, or is not honest about them, the system begins to get weaker. There is also the significant risk of groupthink. The more a community insulates itself from outside influences (a distinct risk in the guild model), the more it is likely to fall into established patterns and allow knowledge to become defined as "what we happen to be thinking right now."

Another place where online communities can get into trouble is the over-focus on the digital element. Much ink has been spilled in discussions of digital literacies and how they are critical to the survival of the twenty-first-century professional. Indeed, these literacies are important and are comparable to the effective use of the pen, train, or conference hall in the twentieth century. The digital landscape is yet to be painted, not necessarily in terms of contents but in terms of the medium that is used to "paint" digitally. The medium has a huge effect on what can be done and a more subtle influence on what does get done.

Conclusion

The world that rhizomatic learning commits itself to is a far less secure world than the one presented to Benjamin by Mr. McGuire. He would have us believe that the future is a novel, can be seen, learned about, acquired, and put in the bank against an uncertain future. What we are acknowledging here is that the future is uncertain that we do not know what we will need to know, or who will know it. We are committing ourselves to people, not to specific bits of knowledge or information and hoping that our commitment to those people will keep what we know relevant, and keep us above water.

We can create communities, out there in the online space. By far the easiest way of creating this kind of community is to find a Web based community that already exists. As the decision of "what" to learn, that is, the problem facing most professionals in the workplace today, it is probably easier to start one's learning inside a guild community model, where the literacies for community are easier to learn, and while the idea of "knowing you know" is a bit easier to manage.

References

Cormier, D. (2008). Rhizomatic education: Community as curriculum. *Innovate, 4* (5). Retrieved from http://www.innovateonline.info/index.php?view=article&id=550

Cormier, D. (2009). MUVE Eventedness—An experience like any other. *British Journal of Educational Technology, 40*(3), 543–546. DOI:10.1111/j.1467–8535.2009.00956.x

Cross, K. P. (2005). What do we know about students' learning and how do we know it? *Research & Occasional Paper Series, 7*(5). Retrieved from http://repositories.cdlib.org/cgi/viewcontent.cgi?article=1041&context=cshe

Edtechtalk. (2009). About Edtechtalk. Retrieved from http://edtechtalk.com/About_EdTechTalk

Ericsson, K. A., Prietula, M. J., & Cokely, E. T. (2007). The making of an expert. *Harvard Business Review*, (July-August), 5. Retrieved from http://www.washingtonbaseballinstruction.com/themakingofanexpert.pdf

Merges, R. P. (2004). From medieval guilds to open source software: Informal norms, appropriability institutions, and innovation. *Conference on the Legal History of Intellectual Property.* Retrieved from http://www.law.berkeley.edu/institutes/bclt/pubs/merges/From_Medieval_Guilds_to_Open_Source_Software.pdf

OJR: The Online Journalism Review. (2006, August 31). Five rules for building a successful online community [Web Log message]. Retrieved from http://www.ojr.org/ojr/stories/060831miller/

Riverin, S., & Stacey, E. (2008). Sustaining an online community of practice: A case study. *Journal of Distance Education,* 22(2). Retrieved from http://www.jofde.ca/index.php/jde/article/download/3/533

Szucs, A. (2009). New horizons for higher education through e-learning. *e-Learning Papers,* 14. Retrieved from http://www.elearningpapers.eu/index.php?page=doc&doc_id=14317&doclng=6

Wenger, E., White, N., & Smith, D. (2009). *Digital habitats: Stewarding technology for communities.* Portland, OR: CPsquare.

24

The Creative Campus: Practicing What We Teach

Ellen McCulloch-Lovell

Creative Capital

Colleges and universities are recognized as sources of intellectual capital for their regions and the nation, as their educational and research missions add to the storehouse of knowledge, and they train the next generation of scholars and creators. The service mission of educational institutions means that they interact with the larger community in many ways, building social capital by enriching many organizations with the talents of students, staff, and faculty members.

There is another key mission for us in higher education: as providers of creative capital. In recent years, our institutions have been recognized as educating for the innovations our society and economy will need to thrive as well as contributing ideas and jobs to the creative economies of cities and towns. Higher education is seen as perhaps the largest patron of the arts (The 104th American Assembly, 2004), employing artists; providing the creative spaces of museums, galleries, performing arts facilities, and film studios; presenting all the arts in series open to the other members of the educational institution and broader community; commissioning new works; and facilitating new innovators, who use materials of the past and synthesize knowledge across disciplines. Most importantly, colleges and universities are places where freedom of thought and expression is safe-guarded;

where new ideas can be recognized by a group of educated colleagues; where inquiry, risk, and learning from failure are said to be valued.

The key question of this chapter will be: even as colleges and universities focus on curriculum, teaching methods, campus activities, facilities and community partnerships on the arts and creativity, are we truly fostering the conditions in which innovation and creative thinking can occur? How can we model the creative thinking we aim to teach?

First I want to look at the "creative campus" movement that is gaining currency despite an economic downturn squeezing budgets of educational institutions. It is worthwhile to examine the idiosyncrasies of this movement and to examine why it captures both attention and imagination. Although the impetus came from the arts—the ways and areas of knowing that many of us most closely identify with creativity—I want to expand the "creative campus" to include other forms of inquiry and discovery, especially in the sciences.

The Creative Campus

The term "creative campus" emerged as the title of the 2004 conference held by the American Assembly of Columbia University, which brought together educators, performing artists, administrators, and policy thinkers to examine the role of higher education in the "training, sustaining, and presenting of the performing arts." The Assembly has a unique history of gathering informed individuals with differing positions to attain consensus and direction around a particular issue, whether in foreign or domestic policy, or in this Assembly, the arts. This conference concentrated more on defining and discussing the role of higher education in the performing arts than settling any controversy. It was directed by Alberta Arthurs, former director of arts and humanities at the Rockefeller Foundation and Sandra Gibson, president and CEO of the Association of Arts Presenters. The co-chairs were Lee Bollinger, president of Columbia University, and Nancy Cantor, then chancellor of the University of Illinois, Urbana-Champaign, who both had publicly supported the role of the arts in higher education. The days of discussion were greatly enriched by research compiled under the leadership of sociologist Steven J. Tepper, who is now the associate director of the Curb Center for Art, Enterprise and Public Policy at Vanderbilt University.[1]

The Assembly viewed colleges and universities as the greatest patrons of the arts in America. Douglas Dempster, now Dean of the College of Fine Arts at the University of Texas, Austin, made that case in a much-quoted paper written for the conference (Dempster, 2004). He and Tepper pointed out how little data there are that aggregate all the resources of higher education directed at the arts, collections, architecture, and other aspects of colleges' and universities' creative activities. At the same time, The Assembly convincingly noted the number of graduates in the arts, the wide variety of facilities and presentations, and the rich interaction with communities that support the arts in ways that amount to support beyond other more recognized sources.

The conference looked at how these resources are used or underused on campuses; at the need for better integration between presentation and curriculum; at the role of artists as teachers; at learning itself. I quote from the final report many of the findings that were articulated at the three-day meeting and published by The American Assembly. The complete text is available on its Web site (http://www.americanassembly.org).

The Assembly found that "Higher education and the arts . . . coincide as major arenas for education, experience and knowledge-building. They coincide . . . as builders, as makers, as shapers of society's values. . . . the real wonder is that higher education and the arts have persisted, in parallel and in partnership, all these years, in so many places, without articulating their relationship or taking full advantage of it" (The 104th American Assembly, 2004, p. 3). The writers point out that "the arts on campus have sustained in profound ways the academy's deep-seated, tripartite mission—to provide research, education and service to society" (The 104th American Assembly, 2004, p. 4).

Under the "research/creative activity" mission, the report affirms that artistic work is a form of knowledge valuable in its own right. The arts make discoveries and contribute to the ideas and understanding "that society draws on." As for the central mission of educating the populace, "the arts provide both subjects for learning and ways to learn . . . they can also illuminate other areas of the curriculum. . . . The arts enrich learning methodologies through their standards of observation, discernment and interpretation . . ." (The 104th American Assembly, 2004, pp. 4–5). In service and public engagement, "the performing arts are particularly successful at engaging communities on campus and off. . . . Through the diversity of their offerings, the performing arts encourage linkages between different cultures and different expressive traditions" (The 104th American Assembly, 2004, p. 5).

Much more could be said about the central role of the arts in higher education, expanded beyond the discussion of the performing arts. Furthermore, much of what can be said about creative inquiry and discovery must extend to other disciplines of knowledge as well.

There needs to be a national inquiry into the common elements of creativity in the arts and the sciences. The existing connections provide worthy subjects of research in themselves, and some inquiries show exciting promise. Just two examples show a scientist and an artist in collaboration with scientists who are discovering patterns of collective behavior that apply to the process of creation and to the social sciences as well.

Biologist Iain Couzin's research at Princeton University focuses on animal group behavior: "how biological patterns result from the actions and interactions of the individual components of the system" (http://www.princeton.edu/~icouzin/). Bennington College choreographer Susan Sgorbati is working with scientists at the Neurosciences Institute at La Jolla, California, "exploring the relationships between dance and music improvisation and complex systems," informed by Dr. Gerald Edelman and Dr. Stuart Kaufman (http://emergentimprovization.org).

Many conferences issue reports; the 2004 American Assembly was a rare one that stimulated action. The Doris Duke Foundation gave significant Creative Campus Innovations Program grants through the Association of Performing Arts Presenters (APAP) intended to "enmesh" performing arts programs on campuses in the activities of faculty, students, and community members. For example, Dartmouth College's Hopkins Center worked to illuminate the "Class Divide" through presentations, films, and discussion, while the faculty members were encouraged to develop new courses about the issues of class and culture in the United States. The APAP annual conference has become the place each year where ideas and projects exploring the creative campus are discussed and shared.

In 2006, with support from the Ford and Teagle Foundations, the Curb Center at Vanderbilt convened a national research conference on The Creative Campus to explore the creative experiences of students and the creative climate of campuses. Artists, educators and social scientists explored these topics in five working groups, looking at the terrain of such subjects as "Mapping the Creative Campus: Understanding Connections and Networks," and "Cultural Participation, Learning and Campus Engagement" (http://www.curbcentervanderbilt.org). The groups identified more research that is needed to fully understand what occurs in a truly creative campus.

At first look, it is most obvious to identify creativity on campus primarily with the arts. We see imaginations at work in all kinds of performances, exhibitions, media presentations, and we witness the innovations of the past in college and university museums and collections. In a 2006 article, Steven Tepper noted that the number of students who said they wanted to major in one of the arts had increased 44% in the past ten years (Tepper, 2006). I wonder, with today's increasing focus on earning a living and paying off student loans, whether that figure is increasing or stagnant.

Richard Florida's influential book, *The Rise of the Creative Class* (Florida, 2002) described creative workers who cluster in receptive communities to reinvent and revitalize the economy. Daniel Pink's book *A Whole New Mind* brought the idea of "left-brain" thinking further into the public mind, as he described the modes of thinking and skills that will be needed for the emerging economy. Pink states: "We are moving from an economy and a society built on the logical, linear, computerlike capabilities of the Information Age to an economy and a society built on the inventive, empathic, big-picture capabilities of what's rising in its place, the Conceptual Age" (Pink, 2005, pp. 1–2).

The research of Mt. Auburn Associates, an economic development and strategy firm, more carefully defined creative workers and showed how artists and arts-related jobs migrate across the private and nonprofit sectors, including periods of self-employment, to comprise a significant and growing segment of the economy. The first such influential study was conducted for the New England Council and the New England Foundation for the Arts and published in 2000. Mt. Auburn's expertise has now measured the creative sector for New York City, Louisiana, and the Berkshires region of Massachusetts (See their Web site at: http://www.mtauburnassociates.com). My article, "Colleges as Catalysts for the Creative Class" (McCulloch-Lovell, 2006), describes the crucial roles that colleges and universities play in building the creative economy in their communities. I claim that the creative economy will not thrive without the resources of "creative capital" found in nearby institutions of higher education.

In each of these studies, the arts-related employment of colleges and universities is seen as a key part of the creative economy. These are just a few sources for the growing attention to higher education as artistic patron, economic driver, as well as guardian of freedom of expression and stimulus to innovation.

Understanding Creativity

However, what is creativity and how does it get supported and recognized? Creativity is generally understood as the inherent ability to bring into existence new items or novel thoughts, in brief, as introducing innovation. In addition, it combines seemingly disparate parts into a new whole. It requires "fierce determination" and "unquenchable curiosity" says psychologist Mihaly Csikszentmihalyi in *Creativity: Flow and the Psychology of Discovery and Invention*, his study of innovative individuals. (Csikszentmihalyi, 1996). Creativity resembles the "synthesis" of scientists, which occurs "when disparate data, concepts, theories yield new knowledge, insights or explanations. Synthesis creates emergent knowledge in which the whole is greater than the sum of the parts. Synthesis takes stock of what we know and generates new knowledge from novel recombinations of existing information." (Carpenter et al., 2009).

As I noted in "A Vocation of the Imagination" (McCulloch-Lovell, 2005) ". . . every creative act overpasses the established order in some way," and "is likely at first to appear eccentric," quoting the poet and University of Utah professor Brewster Ghiselin from his 1952 introduction to *The Creative Process,* (Ghiselin, 1952, p. 3). Ghiselin observed that creativity takes acute attention, an openness of mind, and "a surrender" to the "widest and freest ranging of the mind" (Ghiselin, 1952, p. 14.). Like others, though, he reminds us that "what is needed is control and direction" (Ghiselin, 1952, p. 9).

Despite the myth of the lonely creator, Csikszentmihalyi asserts that creativity is a group activity. This observation is noted in science: "Teamwork speeds synthesis and thereby accelerates innovation" (Carpenter et al., 2009). Creativity thrives in collaboration, using the ideas and methods of the past as well as new expressions and practices. Without collaborators or witnesses, creativity rarely emerges or is recognized. The atmosphere in the college or university is crucial to such discoverers. "This is the period where they found their voice." According to Csikszentmihalyi, "College provided soul mates and teachers who were able to appreciate their uniqueness" (Csikszentmihalyi, 1996, p. 183).

The Conditions for Creativity in Higher Education

Most of us would agree that higher education is devoted to training the mind to serve society by producing new knowledge and connecting it to solving real-world problems; and to produce informed, humane citizens of this democracy. In each of these goals, creativity commands a central role. Do we agree that higher education should also train the imagination? Even as we face the overwhelming issues of climate change, providing effective health care for more people, sustaining an economy that includes more workers and is not dwarfed by other countries, and many other challenges one could identify, we rely on our capacity to invent, to solve problems, to discover remedies, to express the common threads of our humanity across differences.

Are our institutions of higher education supporting creativity, synthesis, and discovery? Are we living up to our mission to add to the storehouse of human knowledge, educate discerning citizens and creators, and thus serve society? The answer is not to be found in the number of arts graduates, square feet of arts facilities, or number of performances and exhibitions on campus. For the fundamental question is how are we creating the conditions in which synthesis and discovery can flourish?

In his article, "The Creative Campus, Who's No. 1?" (2004), Steven J. Tepper describes five "conditions for creative work." I will briefly cite and then add to them. Tepper asks whether an institution promotes "collaboration," whether students and faculty work together on research and papers and whether students are noted as co-authors. Do they work on class assignments in groups? He noted "cross-cultural exchange": the diversity of student and faculty, the amount of study abroad and foreign students. "Interdisciplinary exchange" is key in his thinking. Do students take a number of courses outside their majors? What is the distribution across majors? What is the number of research projects involving students and faculty from different disciplines? If creativity combines previously disparate or unrelated elements into something new, it is fed by the ability to work across disciplines. Tepper rightly points to the importance of "time and resources," involving rewards for new research, independent study, funding for performances and experiments, and new course development. "Tolerating failure" is another factor. Noting how hard this concept is to define, he nonetheless asks how we could see if faculty members encourage risk-taking. Thus, it would be interesting to examine how many drafts of papers are reviewed, whether students can repeat course work or exams, whether faculty are supported in "unconventional approaches"—these factors might reveal a tolerance for learning from failure. Some of these questions can be plumbed by the National Survey of Student Engagement (NSSE), a measure used by many colleges.

After thinking about Tepper's conditions, I want to expand on them. How easy or hard does the institution make it for a student to study across disciplines? Does she have to fit into the category of a "major," or is there a way to record and evaluate truly interdisciplinary work, where the inquiry is not just drawing on knowledge from different fields but whether the goal is approaching a synthesis between the findings from different areas of study? Does the institution recognize broad work that combines different data or ideas, rather than rewarding a narrow analysis of subfields? Is well-substantiated independence of thought recognized, or must the student conform to the faculty members' perspectives on their disciplines? How open and diverse is campus discourse; how civil?

Examining time and resources, how much support is there for both faculty and students to explore and create something that is different, even eccentric? Resources may not mean funding. Creative work needs other kinds of support, such as space, time outside class, collaboration of other students, faculty or staff, or the willingness to tap knowledge in other disciplines or departments.

Many institutions, especially liberal arts colleges, talk about the capacity to take risks and learn from failure. If there are no risks taken in combining previously disparate or uncombined elements, then there is no discovery. However, the squelching pressure on higher education is the movement toward measuring outcomes, toward the assessment that all of our accreditation associations now require. What are the expected outcomes of student learning, and how do we measure them? This question is reasonable, and certainly thoughtful faculty members might describe these outcomes in expansive ways. Is the student able to use the knowledge of the past? Can she make a well-reasoned argument? Can she critique it from another point of view? Can he pose an important question? Does he have the tools and skills to posit answers? Can she formulate a theorem and test it more than one way?

However, the pressure is to find quantifiable measures. The pressure is to test and grade. Grades can be a valuable form of assessment, but if that grade is not composed of carefully analyzed processes, if the institution is large and has to scan test answers by the thousands, how is the ability to learn from one's failures taken into account?

There is another factor: how creative and flexible is the administration? What do students hear from advisors or the registrar if they want to combine disciplines? How is independent study viewed? How broadly can a student study before choosing a major or area of concentration? We would gain insight from observing how faculty and administrators solve their own problems, learning from their mistakes.[2]

Next: how creative is the use of space? Is discovery confined to the studio or the lab, or does it happen in the campus center, the dorm room, or on the lawn? How many informal opportunities are there for students, staff, and faculty to participate in discovery? How many music groups, impromptu readings, group public art projects, student lectures are available to the students each semester? How hard is it to schedule and use campus facilities for rehearsals or experiments? What happens when art appears, unplanned, on campus? Is it "cleaned up?" Is there disciplinary action? Is it discussed in a public forum?

How do we rate our colleges and institutions as creative campuses? Do we enlist the staff in creating an atmosphere of open expression and experimentation, coupled with discipline and respect for others? What is our tolerance for the eccentricity of the new? Do we talk about the importance of creativity but stifle it with our administrative actions?

My questions raise other questions. Does fostering creativity take money and can smaller or poorer institutions be creative? Furthermore, is there a qualitative difference between smaller and larger institutions? I assert that money is not the primary element of support. At Marlboro College, one of the smallest liberal arts colleges in the United States with about 320 undergraduates and 38 full-time faculty members, there is a modest fund for faculty research and travel, some student grants for internships, and basic facilities in the arts and sciences. Three years ago we opened the Serkin Center for the Performing Arts, built for about $2M, providing a dance studio, lecture and concert hall, foyer gallery, music library, and new classrooms. I have joked that since the dancers moved out of their basement space, they got off the floor and added leaps to their choreography. It is a marvelous studio and testimony to the notion of "creative space." Even so, dance in the old place was also wildly inventive, just as theater has been without elaborate technical equipment. Perhaps it justifies our frugality, but some tell me that when students do all their own costumes, music, sets and tech, they are learning more and creating more than they do at richer facilities. Similarly, the science labs are not elaborate. However, in them and in the field, students learn to think like scientists, and they combine discoveries into the synthesis that propels them into graduate studies or medical school. The spaces and materials that money represents are important but are not the only or even the essential elements.

The key element is what we value and how we recognize it. If we value creativity, then we will organize our systems, measures, and even our budgets to encourage the conditions in which it flourishes.

The factors I have elaborated upon may be easier to develop in smaller institutions. Again, at Marlboro College, we do not have departments. All faculty members have the same rank and are granted the freedom to design their courses and practice the teaching methodologies they find effective. They teach in small classes, and a great amount of teaching and intellectual exchange with students occurs in group or one-on-one tutorials. In this intimate setting, they can prompt thoughtful evaluation that allows students to learn from what "didn't work."

These are conditions hard to reproduce, I admit, in larger institutions with departments, majors, and public funding that demands accountability. Nevertheless, I see large institutions committed to become creative campuses, such as Vanderbilt University with its new Creative Campus Initiative and Syracuse University with leadership from Chancellor Nancy Cantor, who spoke so compellingly at the 2008 APAP session. Princeton University and Rensselaer Polytechnic Institute have launched initiatives. Large universities can look at the ways in which they value or discourage creativity, examine class size and the conditions in their departments, and discuss how creativity is reflected in their administrative practices and budgets. It is my hope that this chapter stimulates debate, more examples and even more research.

Colleges and universities have to be orderly places, accountable to trustees, accreditation officials, parents, and the public. They and we education leaders want to know that the years spent at college assist students to flourish, to develop in positive ways so they will sustain themselves as they participate in a humane democracy. We want to be able to prove that indeed we do advance the quotient of knowledge and discovery that will solve the world's problems. Creativity requires skills and discipline to allow the expression or discovery to emerge and be shared with others. So too our institutions need careful guidance and sustenance. We provide the structure, the containers, in which the willing participants feel free, safe, and able to find the competencies they need to show us a new world.

References

American Assembly. (2004). *The creative campus: Training, sustaining, and presenting of the performing arts.* New York: The 104th American Assembly, Columbia University.

American Assembly. (2004). *The creative campus: Training, sustaining, and presenting of the performing arts.* Retrieved June 2006, from http://www.americanassembly.org.

Carpenter, S. R., Armbrust, E. V., Arzberger, P. W., Chapin III, F. S., Elser, J. J., Hackett, E. J., et al. (2009). Accelerate synthesis in ecology and environmental sciences. *BioScience,59*(8), 1.

Couzin, I. Retrieved from September 12, 2009: http://www.princeton,edu/~icouzin

Csikszentmihalyi, M. (1996). *Creativity: Flow and the psychology of discovery and invention.* New York: HarperCollins.

Curb Center for Art, Enterprise and Public Policy. (2006). *The creative campus—Higher education and the arts.* A report from the Creative Campus Research Conference at Vanderbilt University. Retrieved September 2009 from site http://www.curbcentervanderbilt.org

Dempster, D. (2004). *American Medicis.* Paper prepared for the 104th American Assembly on the Creative Campus, Harriman, NY.

Florida, R. L. (2002). *The rise of the creative class: And how it's transforming work, leisure, community and everyday life.* New York: Basic Books.

Ghiselin, B. (1952). *The creative process: A symposium.* Berkeley, CA: University of California Press.

McCulloch-Lovell, E. (2005, summer). A vocation of the imagination. New England Board of Higher Education. *Connections*, pp. 14–15.

McCulloch-Lovell, E. (2006). Colleges as catalysts for the creative class. *Chronicle of Higher Education,* p. B15.

Mt. Auburn Associates. (n.d.). Retrieved, September 12, 2009, from http://www.mtauburnassociates.com

New England Foundation for the Arts, Mt. Auburn Associates & New England Council. (2000). *The role of the arts and culture in New England's economic competitiveness.* Boston: Author.

Pink, D. H. (2005). *A whole new mind: Moving from the information age to the conceptual age.* New York: Riverhead Books.

Sgorbati, S. (n.d.). Retrieved, September 12, 2009, from http://www.emergentimproviza-tion.org

Tepper, S. J. (2004). The creative campus, who's no. 1? *Chronicle of Higher Education*, p. B6.

Tepper, S. J. (2006, July/August). Riding the train. Association of Performing Arts Presenters. *Inside Arts.*

Notes

1. This is one of the few remaining university centers for culture and policy, led by Bill Ivey, former National Endowment for the Arts chairman under President Bill Clinton, and supported by the university and record industry leader, Mike Curb.
2. Kristin Howrigan, a Marlboro College dance professor, offered this insight.

25

Creativity and Education: A View from Art Education

Michael Parsons

This chapter relates the current discourse of creativity and education and focuses on art education in particular. The importance of nurturing the creativity of students has become once more a popular topic for educators and politicians, mostly because of its perceived relation to innovation and the economy. My discussion of it asks whether art education, as distinct from, say, math or science education, has any special role to play in educating for creativity.

Of course, creativity is a desirable educational goal in all school subjects. One can be creative in any school subject (or in any discipline—for my purposes here school subjects are the same as intellectual disciplines), and one can teach to promote creativity in any subject. The current discourse of social and educational policy stresses the importance of nurturing creative scientists, mathematicians, and technologists. However, art and design have traditionally had a special relation to creativity, one that is somewhat different from that of other disciplines. Many people, including artists and art educators have thought for a long time that one can promote habits of creativity more easily and directly in the arts than in other school subjects. In fact, and in accordance with my own experience, most artists and art educators believe this to this day. I will briefly account for this sense of a special connection between teaching for creativity and the arts and then discuss whether there is any reasonable argument for holding it today. Because I am primarily a visual arts educator, my examples will come from the visual arts, but I believe the case is similar for dance, music, poetry, and the other arts. I include design as part of the visual arts as is taken for granted in the United Kingdom though not in the United States.

Part of the problem is deciding just what we mean by creativity. It has never been a very clear concept, related as it is to so many other concepts and assumptions of a psychological, philosophical, social, and political character. I am going to discuss rather rapidly several versions of creativity drawn from the history of art education in the United States. I will of course have to generalize, yet I hope this outline is useful because all of these versions, and perhaps more, exist today side by side in the popular mind, if not in the policy literature.

The first idea descends from the European romantic movement of the late eighteenth and nineteenth centuries. It was embodied in the idea of the lonely genius, usually an artist, who was driven to express his personal emotional life in highly expressive and creative ways. It was assumed that the creative person had some vision he must explore, some inner feelings he must express, an obsession of some kind deep inside the self that he must follow. The idea was associated with inspiration, often from God or the gods, sometimes from the unconscious mind, and with the idea of imagination, a faculty held to be different from and more insightful than the rational mind. It was wedded to the notion of an essential and individual self.

In art, a figure emblematic of this idea of the creative person might be Van Gogh. He was famous for his strikingly new and powerful paintings and, in the popular mind, for his unconventional personal, social, and emotional life, including the fact that he wound up in a mental hospital. His creativity was manifested especially in the new style of painting he developed—he painted with oils outdoors, often rapidly and without correction, used color more for emotional effect than for realism, and laid his paint on thickly with unrevised brush strokes. This style was held to express his inner turbulent states of mind.

The idea of the lonely genius persisted as the dominant image of creativity through the first half of the twentieth century and was associated with modernism in art, with psychoanalysis in theory, and with all kinds of unconventional behavior in society. Jackson Pollock is a good twentieth-century example of the stereotype. Pollock also had an unconventional personal life and created a new style of painting. "Autumn Mist" is one of Pollock's great works produced in this manner.

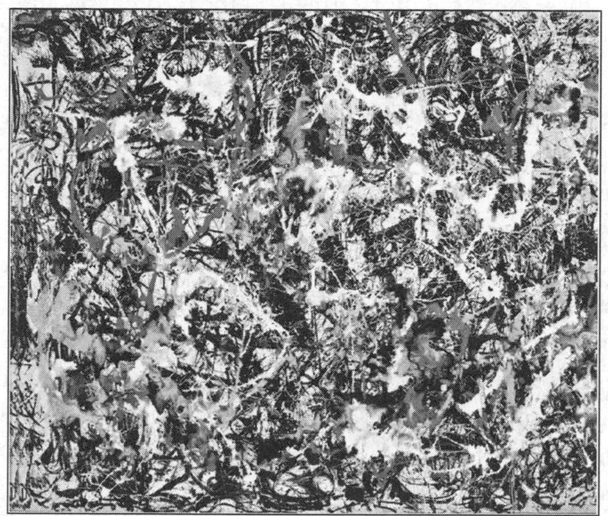

Figure 1: "Autumn Mist" *by Jackson Pollock*

An important part of this early idea of creativity is that it means ignoring, perhaps defying, the conventions of a discipline and the expectations of society, in general the authority of what I will call "rules," though "norms" might be another choice. A major theme here will be that our changing ideas of creativity are shaped in part by its assumed relationship to disciplinary conventions and social expectations, I will say a little more about what I mean by rules. The idea is very general. It refers to anything that is regularly expected by society: laws, moral expectations, spelling, dress codes, recipes, and especially the concepts and procedures of school subjects and academic disciplines; it is anything that carries a sense of the authority of society's expectations over individual actions. Such rules are experienced as both guides to behavior, what we should do in certain circumstances, and as restrictions, what we should not do.

One of the rules that Pollock defied was the convention that a painting should be done vertically on an easel and hence that it should have a top and a bottom that help define the work. Pollock famously laid his canvas on the floor and walked around it, working on it from each side, as young children sometimes do. He also used a brush in an unconventional manner: he threw the paint at the canvas instead of contacting the canvas with the brush. Laying the canvas on the floor meant that he did not need to decide which way was up and which was down until he was finished, nor did he have to decide where the edges of the painting were to be; only later would he cut the canvas where he thought best.

I mention these aspects of his treatment of the medium because they are a large part of where his creativity is thought to lie. The medium in which any artist works carries with it a set of rules about what should be done with it, rules which as I said are both guides and restrictions. I do not include as "rules" the restrictions that come from nature, though the natural/conventional distinction is a slippery one. For example, naturally you cannot do in two dimensions (painting) what you can do in three dimensions (sculpture). This is a time-honored distinction in the visual arts, but contemporary artists have certainly blurred it by using embedding various objects in their paint. However, there are many conventions built up within a tradition that govern the way artists treat their medium, such as the expectation that the difference between the top and the bottom of a painting be significant. In short, an artistic medium is one of the sites of "rules" that creative artists have to deal with. I will return to this topic later.

In art education in the first half of the twentieth century, Victor Lowenfeld was the dominant champion of this understanding of creativity as non-conformity to the socially established rules. (For a more careful and scholarly account of Lowenfeld's ideas about creativity, see Judith Burton, 2009). Non-conformity could be from defiance or ignorance. Lowenfeld argued that in the case of young children it was necessarily ignorance; they don't yet know society's rules and conventions. More specifically, when they expressed their thought, they were naturally honest and direct, and their expression was not yet filtered through the many screens of conventionality. Artistically, the eyes of young children, the slogan went, were innocent. Their art was creative by nature.

Lowenfeld wanted teachers to protect the innocence and creativity of children as long as possible by trying to prevent them from learning about adult artworks, from art history, aesthetic judgments, and adult talk about art in general because that would make their art imitative rather than creative. They were not to study artworks, copy models, or have coloring books as toys. In addition, he disapproved of any attempt to assess children's art. Assessment of children's artworks or abilities, whether about creativity or something else, was very much maligned because it represented the imposition of adult judgments on the children. Just as conformity was the opposite of creativity, so was assessment the opposite of good teaching. Art educators are still sometimes influenced by this idea.

At the same time, in the adult world there was a cult of children's art which can still be found in, for example, calendars and displays of children's work in airports and on refrigerators. Many modernist artists were affected by these ideas of innocence and creativity. They studied children's art and were often influenced by it. Dubuffet's "Cow with the Subtle Nose" is a good example.

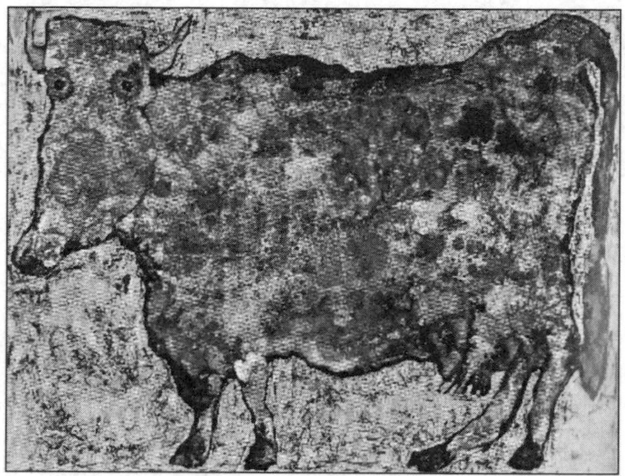

Figure 2: Dubuffet: "Cow with the Subtle Nose."

Artists like Dubuffet labored, paradoxically, to see the world with an "innocent eye" without the preconceptions of conventionality. Picasso famously said that he worked every day to unlearn all he had learned in order to paint like a child.

Of course, this understanding of creativity affected more than art. It was a part of the progressive movement and child-centered education in general, and it influenced scholars and teachers of all disciplines. Nevertheless it had a special association with the arts, and the teaching of the arts was much more affected than were other school subjects. One could use problem-solving methods rather than direct lecture in any subject and could teach for conceptual understanding rather than rote learning. One could regard learning as the gradual re-discovery by children of the significant concepts of any discipline. Nevertheless, in educational practice, most school subjects do not have the freedom of the arts; they have a sense of correctness to them and of the importance of the rules that is not present in the arts. One sign of this phenomenon is the fact that, prior to the "cognitive revolution" in psychology in the early 1960s, the arts were the only school subject classified as "non-cognitive." They were the product of the imagination, the others were the products of the rational mind, and the two were categorically different. The rational mind was full of rules and correctness, whereas the imagination was not.

There was an associated theory of child development in the arts that reveals this notion well. It was called the "U-shaped curve" and was popularized especially by early versions of Project Zero (Davis, 1991). The curve plotted creativity—which in art meant expressive quality—against age, resulting in a U-shaped curve. The idea was that children start out very high in creativity; then, as they go through elementary school, their creativity declines. This decline occurs because they are busy learning the rules and conventions of society and the disciplines, which in art have mostly to do with realism, perspective, proportion, correct colors, balance, design. They reach a low in creativity in the early secondary years, and most people do little better than that for the rest of their life. In late high school, however, as they become adults, some manage to regain their creativity, as Picasso did.

The point is that much of traditional schooling was considered a training in conformity, and behind this thought was a more general attitude toward society itself. It held society to be too often repressive of individuality and self-expression. The first half of the twentieth century was, in the West, an age of assembly lines, bureaucracy, industrial giants, and the use of ID numbers. It was felt to be a conformist society, emotionally cramped, in which the schools functioned to produce "the organization man." The identification of creativity with nonconformity was part of this larger formation. Society represented repression; creativity represented freedom.

Other versions of creativity also reflect attitudes to society, of course. A second version, which I turn to now, was associated with the "new disciplines" movement in U.S. education of the early 1960s. A major motivation for these new curricula is often said to have been the Cold War and the fear that America was falling behind the Soviet Union in a military, technological, and political competition. In 1959, the Soviet Union surprised most Americans by launching the first satellite into space, the *Sputnik*. The response included John Kennedy's project to put a man on the moon and a major attempt at educational reform that placed a huge emphasis on teaching the disciplines in greater depth.

Creativity in this version was associated with scientific innovation and development, especially in connection with space exploration and warfare and in general with the progress of society. This scenario called for a new understanding that associated creativity with mastering the academic disciplines. In addition, it was usually more about math, science, and technology than about the arts; the popular image of a creative genius during the Cold War was more likely to be Einstein than Jackson Pollock.

Creativity could no longer mean ignoring or defying the rules. One could be creative only if one knew the rules well—especially the rules of math and science. Mastery of a discipline meant mastery of its rules—its conceptual structure and traditional procedures (which constitute a complex combination of natural facts and social conventions). The characteristically creative task was not expressing oneself but solving a problem, usually one that arose from within the discipline. Creativity was possible within a discipline only after mastering its rules. Then one could be creative by applying the rules to new contexts or, in advanced cases, by changing them. The point is that you had to use or go beyond them, not ignore them.

In education, the change put a new emphasis on teaching the conceptual structures of the disciplines—starting with what was called the "new math" curriculum, then "new chemistry," and the other sciences, spreading to history, language, and eventually to the arts. As a consequence, teaching creativity became less important than teaching the discipline. Creativity was thought of as feature that followed mastery of the discipline, and mastery is rarely achieved while students are still in school. Thus, learning the rules often was of immediate importance, and creativity became an afterthought.

In art this dynamic meant a new attention to the work of adult artists and to art history. It also meant teaching art criticism and developing a capacity for artistic judgment. Art-making often became making art in the styles of particular artists.

Theories of child development changed too. An interesting, because complex, case is the work of Piaget. In general terms, Piaget thought of child development as showing the same three stages already discussed: the first where children learn through their own perceptions and their bodily interactions with the world; the second where they acquire the concepts and expectations of the disciplines and society; and the third where they have developed the full capacity for abstract thought. Piaget was interesting because of his conception of creativity at the second of these stages. Where others saw children as mastering the rules in preparation for possible later creative work, Piaget saw them as re-discovering the rules for themselves, in a gradual process of re-invention. According to Piaget, this process reflected a natural and individual evolution in which children re-created the disciplinary rules when they were ready for them. Indeed, re-creation followed the same process as creation. This observation led Piaget to believe, as Lowenfeld did, that schools could have no influence on creativity and only a superficial one on understanding.

Of course, we now have a social reconstructivist view of this process of re-creation. In this view, children re-create the rules not "naturally" and by themselves but with the assistance of adults. They still have to re-create the ideas and reasons of the disciplines but have the advantage of the active "scaffolding" that teachers and society in general can provide

In art, there have been similar adjustments. We realize now that children imitate styles they are exposed to—they never have an "innocent eye" but are always seeking to re-create what they see others doing. There is, for example, considerable documentation of young children's efforts to draw figures in the style of manga comics, a widespread phenomenon. This trend is a precursor to the current emphasis on the influence of visual culture in general on children, an influence that typically occurs out-of-school rather than in it.

Figure 3: Junior high school girl's manga drawing

Figure three is from a series of drawings that I collected from middle-school girls in Hong Kong. It is not strictly copied from a manga comic, though it is a copy of a manga style. These girls were in fact fervent and persistent in making these drawings—in re-inventing manga comics for themselves. They were trying to perfect their characters and were beginning to make up stories about them: a good case of creativity through mastery of the rules and of the scaffolding (and narrowing) of creativity by the social world. The general point is that creativity was no longer thought of as inherently nonconformist or critical of society. It required mastery of a discipline, was supportive of social progress, and a necessity for technological development.

The third version of creativity is the one we hear so much about in education and policy discourse today. It is more a gradual development of the previous one than a revolution of it. The typical creative figure in the public imagination this time is more likely to be a small group than an individual artist or scientist. The group is typically focused on solving a social problem or developing a new technology. The problem might come from any field. A quick list of examples might include: getting a political candidate elected, renovating an inner city district, controlling the spread of diseases, cleaning up toxic waste sites, designing a non-gasoline-burning automobile, a Web-based software program, a better way to store energy, or creating a public artwork.

These kinds of problems—the ones we care most about today—tend to be more complex and more interdisciplinary than the ones discussed earlier. They are more likely to require a combination of ideas from different disciplines. They, therefore, require a group of people, representing different disciplinary backgrounds, to work on them. Creativity in such cases is found as much in the interactions between members of such a group as in the individual members themselves; it is a product of their exchange of knowledge, ideas and points of view as much as of their individual abilities. Members of such a creative group must have a common purpose; focused by the problem at hand, they must each approach it with a different set of ideas and skills. They must, as before, each have conceptual mastery of a discipline, know its traditional ideas and skills, its strengths and its uses; but at the same time they should not be obsessed with their own discipline to the exclusion of an interest in others. They need to have an open mind to the possibilities of other approaches and a sense of the possible interactions of ideas from other disciplines. Creativity in these cases lies in the interaction of different approaches, the layering of one conceptual scheme on another, finding new uses for old concepts and skills.

Of course, in the public imaginary and in fact there is usually an acknowledged leader of the group, a figure who is some combination of organizer and visionary. This person is often an entrepreneur who wants to develop a new idea to meet a public need and at the same time make a financial profit. Such a figure often gets the credit, rightly or wrongly, for the work of the group.

This new socially interactive version of creativity means that members of a creative team must also have the skills and aptitude for social interaction. They include the ability to listen to others, to be able to present one's own ideas clearly, to be open to changing one's mind, to take risks, and to switch attention from ideas to people and back again. These are the kinds of habits of mind that we are encouraged to teach our students nowadays. They do not amount to social conformity, but they certainly require a social awareness and ability that the individual artist or scientist was not required to have.

Richard Florida has famously painted a portrait of the social traits of the people that he thinks are in the "creative class" (Florida, 2002). He argues that such people are well educated and are in a variety of professions—computer engineers, musicians, lawyers, teachers, designers, or medical workers to name a few. They are interested in new technology, use social networking sites frequently, care about design, and they are style conscious. They are socially tolerant of differences, be they of an ethnic, racial, cultural, or sexual-preference nature; indeed, they like difference in general, rather than avoiding it, and, importantly, they are very interested in the arts.

They tend to seek each other's company and to live in similar areas, which Florida calls "creative districts." These are usually found in inner cities where there are cafes, bookstores, concerts, health clubs, local festivals, and cheap rents. In addition, because they are associated with the cult of entrepreneurial activity, with start-up companies of all sorts, they are of economic interest to cities and policy makers, including educational policy makers.

This version of creativity involves a somewhat different attitude to disciplinary rules. A discipline this time is not understood as a structure that determines what a person should do, a set of rules that must be followed and determines what is correct. It is rather a flexible set of useful concepts and procedures that will be followed when relevant, but they can sometimes be ignored, overridden, or redesigned. Cognitively, creativity requires being able to take a distance from the rules, to be aware of alternatives, to critique them if necessary. It sometimes consists, as in the previous version, in applying the rules to new situations or in pushing them beyond their previous limits. However, it also includes choosing which rules to use, intuiting which ones will be useful or just interesting in particular situations. They are no longer felt primarily as restrictions imposed from outside but as tools that may and may not be fruitful in guiding one's thought. Having chosen them, one plays with them, exploring their consequences, to see where they lead.

This attitude shows up most clearly in contemporary art. Because the concept of art has widened to include work with so many kinds of materials and media, contemporary artists have much more choice of what they will work with. They make the choice, no doubt, for a variety of reasons, perhaps often intuitively. The most important reason is likely to be their sense of what device will be most useful to help them explore the issues of their concern. I will use as an example the work of Liu Bolin, a contemporary artist in Beijing. His work is distinctively creative because of the choices he has made about how to make them. Figure 4 is from his series *Hiding in the City.*

Figure 4: Liu Bolin: "Suojia Village"

His choices of media constitute a set of rules, which can be described in the subsequent manner: he stands very still in front of a significant background; he wears simple clothes of a neutral color; he and his clothing are painted by assistants; the painting is such as to make him almost disappear against the background; the result is photographed; the photograph is expressive of some quality ("Suojia Village" suggests, at least to me, a sense of the loss of individuality in a big city). These rules may appear rather constrictive, but they have resulted in a series of many works that are strongly expressive and creative. Indeed, one can see that they function to stimulate further creative works; he can maneuver within them—especially he can choose a different background to be partly seen against—to see what possibilities of meaning will occur.

Incidentally, the point about group work also applies here. Bolin needs a team to help make his artworks. He needs others to paint his body, an activity that calls for considerable technical skill, and to take and develop the photographs. Presumably there is discussion each time about which part of the background to select, where exactly Bolin is to stand, the best time of day for the light, where the camera will be, and so on. Much of the creativity of each piece will be the result of such a dialog, and the point applies generally. Many creative artists, though not all, have a team to work with and could not produce their work by themselves.

I believe one can analyze the work of most contemporary artists in terms of the rules they adopt. I will mention one more example: Cindy Sherman's series *100 Hollywood Stills.* Her self-imposed rules for these were, roughly: she dressed herself in a manner to suggest a type of woman (not an individual) recognizable from old Hollywood movies; she is not personally identifiable; she chooses a background situation to suggest a typical Hollywood situation—a kitchen, a beach, a street in New York; she adopts a relevant pose and expression against that background; she is photographed in that pose; there is no one else in the photo, though there is a strong narrative suggestiveness to it; and the result is a comment, sometimes strong, sometimes subtle, on the way Hollywood has typically portrayed women. These 100 photographs established her reputation as a major creative artist today.

John Baldessari articulated this idea in a recent interview with *ART21* (2009). He said about his own way of working:

> What's a system? I think my idea is this: not so much structure that it's inhibiting or that there's no wiggle room but not so loose that it could be anything. I guess it's like a corral around your idea, a corral that you can move around in—but not too much. And it's that limited movement that promotes creativity. (*ART 21,* 2009).

One might call this attitude play (in one or more of the many meanings of that word). There is work involved, too, of course; persistence is certainly a necessary part of it and so is close attention to detail. "Play," though, suggests the motivation—choosing some rules in order to see what happens when you follow them—and the freedom of choice behind it. The choices are less obligatory than usual, are more self-motivated, and are not required in detail by social expectation or the demands of the problem.

While this analysis applies rather obviously to contemporary artists, who have so much freedom to adopt new media for their work, it can be extended to artists in general. I have already mentioned Van Gogh. One could articulate in a similar way the rules he adopted that were the source of his creativity. They would be something like this (for much of his work): he painted outdoors; he did not revise; he made large impasto brushstrokes with thick paint; he used color for expression and not realism; his subjects were domestic or natural. I have just come across a short article by Stokes that makes a similar point about Monet: that Monet's creativity in the famous water lily series sprang from his decision to limit his vision strictly to the surface of the pool: no horizon, no banks, no solid objects, no perspectival rendering—only the surface of the water (Stokes, 2001).

No doubt one could find the same playful attitude to rules in the work of inventors and innovators in many fields. Creativity can happen anywhere. So why do many believe that art is particularly good at nurturing creativity in students, if it is taught in an appropriate way? Many of us still think so, without doubting that one can promote creativity in all school subjects.

One explanation has emerged in my discussion of attitudes to rules. If creativity requires a playful attitude to the rules of a discipline and of society, an understanding that they are not to be ignored or disrespected but can be chosen for their interest and utility in particular contexts, then that attitude is most apparent in the arts. This dynamic occurs for two slightly different reasons.

One reason is the psychological demands of the discipline on the learner. In the other school subjects, the conceptual structures are more elaborate and mastery of them is more time-consuming. A good deal of attention must be given to learning them, and they tend to be experienced more as demanding respect rather than as inviting playfulness. One cannot easily play with the rules of physics or math, especially at the level of beginning learners. Psychologically, they call for correctness rather than creativity. No doubt this is different at high levels of achievement but the point here is primarily about learners.

The conventional rules of art, on the other hand, do not necessarily have this character. Of course, one can teach art with an emphasis on correctness—emphasizing the demands of realism, for example, of perspective and proportion, of color and tone—or on art history. However, when art is taught well, the emphasis is on meanings and ways of expressing them. Art-making will give learners the direct experience of setting up some rules and of playing with them to see what meanings they suggest. Learners will both come to understand more about contemporary art and also cultivate the habits of creativity in general: free choice of rules, exploring them, looking for consequences, persistence in playing.

In addition, art is the only subject where creativity is an inherent value of the subject as a conceptual structure. It can therefore be an explicit target of teaching. In other subjects, the structure is determined by the answers to problems and, though responding to issues requires creativity, the focus is more often on the solution and not the process itself. The same is true with practical technologies. Artists, on the other hand, are often as much interested in their creative process as in their results. The art medium, understood in the way I have indicated, constitutes a corral that offers moving space for the artist. Most artists play with the shape of that corral, experimenting with the medium to see what it can do. The examples offered above—Van Gogh, Jackson Pollock, Liu Bolin—are exemplary of this process, and it is perhaps why we are often ambiguous about where their creativity lies: in their new ways of treating their medium or in the achievements so promoted. Of course, it lies in both, but the ambiguity reveals the tight connection of creativity with process.

This interest in process for its own sake distinguishes art from most other subjects. It is process that makes art unusually valuable educationally because it is the most direct way to promote attitudes of creativity that have a general application to all situations.

References

ART21 (2009). Fifth season of a television series by ART21, carried by PBS, October 2009. Materials available from ART21.com.

Burton, J (2009). Creative intelligence, creative practice: Lowenfeld redux. *Studies in Art Education, 50* (4), 323–337.

Davis, J. (1991). *Artistry lost: U-shaped development in graphic symbolization.* Unpublished doctoral dissertation, Harvard Graduate School of Education, Cambridge, MA.

Florida. R. (2002). *The rise of the creative class.* New York: Basic Books.

Stokes, P.D. (2001). Variability, constraints, and creativity: Shedding light on Claude Monet. *American Psychologist, 56* (4), 355–359.

26

Beyond the Academic "Iron Cage": Education and the Spirit of Aesthetic Capitalism

Eduardo de la Fuente

In my hometown of Melbourne, Australia, the closing date for *Liquid Desire*, an exhibition of art works by the surrealist Salvador Dali, looms. The National Gallery of Victoria (NGV), the premier art institution of the city, has decided to stay open all night to celebrate the final day of the exhibition. The NGV is planning to offer visitors food, drink, musical entertainment, and roving performers. The event will mark the NGV's entry into Melbourne's "night-time economy" with spillover effects for local restaurants, bars, nightclubs, and other venues usually open late. In a city where the creative edge is more often than not associated with gritty laneways, urban graffiti, hard to find bars and venues that simulate Weimar-era cabarets, the all-night opening of the NGV to end the Dali "blockbuster" exhibition taps into Melbourne's bohemian self-consciousness. It is a marketing strategy that echoes the rhetoric of the city's "Fringe Festival"—a piece of publicity designed to attract younger audiences, who usually bypass the NGV for the city's edgier art spaces.

So what kind of capitalism are we living through when staid, bourgeois art museums like the NGV, decide to rebrand themselves through association with bohemia and its nocturnal habits? "Aesthetic capitalism," the "experience economy," and the "creative city"—all these terms have been used to describe the patterns of social life associated with an increased role for culture, art, and aesthetic factors more generally (Lash & Urry, 1994; Boltanski & Chiapello, 2005; Pine & Gilmore, 1999; Florida, 2002; Landry, 2000). Art is now fully integrated with the economy; and the economy increasingly functions as if art were the model for all markets. As the sociologist of design, Harvey Molotch (2004) notes, under contemporary capitalism, "any industry is both a 'culture industry' and one that serves practical ends . . . all art does work and all markets are art markets" (p. 372).

One of the key institutional agents in the transformation of capitalism into its present aesthetic guise was the emergence of the art school as a specific breeding ground for the contemporary blurring of bohemianism and entrepreneurialism. As against those famous residents of Max Weber's (1976, p. 182) "iron cage" of modernity—"specialists without spirit, sensualists without heart"—today's bohemians like to combine work with play, "asceticism" with "aestheticism." As Elizabeth Currid (2007) has noted in her aptly titled *The Warhol Economy*, contemporary creatives do much of their conceptual work, conduct business deals, and market their products at nightclubs, art gallery openings, and other cultural spaces. For these folks, going to an all-night exhibition of Dali paintings to a backdrop of food, alcohol, and entertainment, would be less than novel. The question looms: how did creatives acquire such routines and habits? Some two decades ago, Simon Frith and Howard Horne (1987) noted how the British art school served as a training ground for a new marriage between art and commerce: "The art school experience is about commitment to a working practice, to a mode of learning which assumes the status of lifestyle . . . Art is everything. Art is life" (pp. 28–29). In fact, this educational experience foreshadowed the kind of the mixing of work and play, asceticism and aestheticism that is now widespread under the new aesthetic capitalism.

The argument I am advancing requires rethinking what we mean by terms such as "bohemian," "work ethic," and the "aesthetic." The literature on the new or "neo"bohemia has complicated our image of bohemianism. Elizabeth Wilson's (2000, p. 1) *Bohemians: The Glamorous Outcasts*, is reflective of the new scholarship announcing that the figure of the bohemian is a social construct and a complex one at that: "The bohemian is both a genius and a phony, a debauchee and a puritan, a workaholic and a wastrel." Echoing a theme present in the literature going back to Cesar Graña's (1964) *Bohemian versus Bourgeois*, Wilson (2000, p. 1) adds that the bohemian "identity" has always been "dependent on its opposite . . . [for example] the bohemian (good) is contrasted with the bourgeois (bad)." However, many of the characteristics of bohemia are open to revision. The columnist David Brooks's (2000, p. 10) depiction of a new hybrid class of "BoBos" (bourgeois bohemians) is predicated on the notion that it is "now impossible to tell an espresso-sipping artist from a cappuccino-gulping banker . . . [it is] getting harder and harder to separate the antiestablishment renegade from the pro-establishment company man." A similar sentiment is expressed by Richard Florida (2002) in the *Rise of the Creative Class*, where we are told that contemporary society and culture are experiencing a "Big Morph" and that at the "heart of the Big Morph is a new resolution of the centuries-old tension between two value systems: the Protestant work ethic and the bohemian ethic" (p. 192).

As a result, certain scholars have opted to speak of a "postmodern" or "neo-" bohemia. Richard Lloyd's (2005, p. 67) recent ethnography of the Chicago neighborhood of Wicker Park uses the term *neo-bohemia* to suggest a gentrification of bohemian districts by "affluent professionals" rather than "starving artists" and to indicate that "place" and "creativity" are strongly linked in the postindustrial economy. However, Lloyd (2005, p. 66) takes issue with what he sees as "Florida's relentlessly cheerful account of [the] creative economy." He suggests that "neo-bohemians" have been susceptible to the "fantasy that they could be creative, edgy, and rich all at once" when the reality has turned out to be "long hours, mediocre wages," and young creative workers having to service the restaurants, cafes, and bars frequented by the affluent residents who move to a neighborhood like Wicker Park in the expectation of living in a "bohemian-themed entertainment district" (Lloyd, 2005, p. 67).

However, even those critical of Florida's celebratory "creative class" thesis have had to admit that there is some validity in his claim that the creative class ethos is not some "nitro-burning strain of pure hedonism or narcissism" as some conservative defenders of the Protestant work ethic had suggested. Lloyd (2005, p. 50) shares the view that the bohemian rejection of the "utilitarianism" and "sobriety" of the Protestant ethic conceals the fact that "there is nurtured in bohemia the conviction that the artistic life constitutes a calling that encompasses the very soul of the producer." The notion that bohemians are "workaholics" as much as "wastrels"—to evoke Wilson's language—had already been argued by Frith and Horne in their study of art school students. They say: "contrary to popular stereotypes, most art students do work hard. In learning conditions other undergraduates would find adverse . . . [they] commonly work eight-hour days, often into the night" (Frith & Horne, 1987, p. 28).

Florida's (2002, p. 197) argument is that the problem with misrecognizing the kind of work ethic creatives subscribe to lies in "seeing work and life, or the economy and culture, as separate spheres with distinct value systems." The creative class simply does not subscribe to the view "work first, then live in your spare time" (Florida, 2003, p. 197). Contemporary creatives are at the forefront of a new work ethic by virtue of linking workplace to lifestyle and work to leisure. Florida's former student, Elizabeth Currid, has taken this argument the next logical step suggesting that creatives do not simply have the "lifestyles" or "tastes" indicative of their "creative class" status. For Currid, the new blurring of creative work and leisure entails much more than a preference—as it sometimes seems to do even with Florida (2002, pp. 173, 177)—for "active sports" such as "bicycling, jogging and kayaking" or an interest in "looking your best." In her schema, restaurants, bars, and nightclubs become an extension of creative work rather than a mere leisure activity:

> [W]hat I kept finding out when I talked to people was that it was not just about creating places for people to hang out or be entertained. It was not just that celebrities and creative people hung out in these places and got drunk, snorted coke, and danced all night. These places are also the sites of meaningful social and economic interaction—they are nodes of creative exchange . . . creative people use these places as ways to advance their own careers and the cultural economy more broadly. (Currid, 2007, p. 95)

Interestingly, given the anecdote with which I began this chapter, Currid (2007, pp. 94–99) has an entire section entitled "The Importance of Nightlife." The growing importance of the "night-time economy" reflects that conducting business in the contemporary cultural economy is not just about "social engagements like dinner or power lunches"—it is through "music venues, gallery openings, and DJ nights that real knowledge and collaborations and product review are occurring" (Currid, 2007, p. 95). Currid (2007, p. 87) refers to the social networks associated with the cultural venues of the creative economy as the "economics of the dance floor." This phenomenon is not restricted to her case study of the post-Warhol, New York creative scene. It is evident in any city where creative economic production is growing in significance. As Hemker and Koopman (2005: 105) note, with respect to creatives in Amsterdam, creative industries often contain "people who are opposed to the nine-to-five mentality":

> Meeting like-minded people, a characteristic of the night, is perhaps more important in the world of creativity than among other entrepreneurs. . . . They do not separate their private life from their work. Appointments overlap. Work-related meetings take place in the entertainment circuit, and friends drop in during working hours. The life-style is polychromic, i.e., diffuse and relatively undefined, rather than monochromic in which private and work activities are kept separate and juxtaposed. (Hemker & Koopman, 2005, p. 105)

It would seem then that terms such as *work* and the *work ethic* need altering in light of the social practices of creatives. What then of the terms *aesthetics* and *aestheticism?* Should these terms also be re-defined in light of the character of postindustrial capitalism?

One of the central understandings of aesthetics in Western modernity is that associated with Immanuel Kant's philosophy. It is based on an opposition between "utility" and "beauty." Such an opposition manifests itself in Kant's argument that the work of art is essentially "purposeless" and that the faculty of "aesthetic judgment" functions best when it is "disinterested." Leaving to one side the question of whether such an understanding of aesthetics is inherently ideological or has the capacity to deal with everyday objects and experiences rather than just "fine art" (Eagleton, 1990; Gronow, 1998), what interests me here is whether discussions of "aesthetic capitalism" and "creative economies" have forced scholars to rethink what is meant by aesthetics. There is some evidence of this in Wolfgang Welsch's (1997, p. 1) claim that we are "currently experiencing an aesthetics boom" that extends "from individual styling, urban planning and the economy through to theory." His notion that "aestheticization" is occurring both at the "surface" level (e.g., "hedonism," "aesthetically styled products" and a "society of leisure and experience") and is "deep-seated" (e.g., production processes involving "computer-simulation" and the aesthetic "constitution of reality through media"), suggests a rethinking of aesthetics. As against the Kantian concept of the aesthetic as "disinterested" or as lacking in "purpose," Welsch returns to the original Greek meaning of the term *aisthesis* as heightened "sensory perception." He adds that "*aesthesis* has two sides . . . *sensation* on the one side and *perception* on the other" (Welsch, 1997, p. 10). It is simultaneously "sensory" and "cognitive"; it is both about the "accentuation of pleasure" and "*theoristic*" in that it involves "perception's observational attitude" (Welsch, 1997, p. 10). The latter puts rest to the notion that the aesthetic is merely the irrational or the pre-cognitive. Indeed, the Greek word *theoria* from which we derive 'theory' is based on the kinds of observations that took place in the theatre. Theoretical knowledge was from the very beginning aesthetic in character.

The notion that "theoretical knowledge" might itself be aesthetic is not a bad place to start re-thinking Weber's metaphor of the "iron cage." In the *Protestant Ethic and the Spirit of Capitalism,* all theoretical knowledge seems to be rational and to be placed in the service of "instrumental reason": hence the author's fear that modern culture may be a breeding ground for "specialists without spirit, sensualists without heart." Indeed, the Protestant ethic is depicted as inherently hostile to aesthetic and other non-instrumental values. Weber (1976: 169) spends several pages cataloguing how the Protestant ascetic attitude favored "sober utility against any artistic tendencies. This was especially true in the case of decoration of the person, for instance clothing." In his view, the Protestant ethic involved the "repudiation of all idolatry of the flesh" or what we might see as aesthetic in relation to persons, objects and the domestic environment (Weber, 1976, p. 169). Weber (1976, p. 169) is willing to concede that "Puritanism included a world of contradictions" and that Rembrandt was very much a product of "his religious environment" but posits that the few instances of Protestant aestheticization did not "alter the picture as a whole."

The Weberian narrative has been modified by sympathetic critics in the following way: there is Colin Campbell's (1987, p. 99) suggestion that there was an "other Protestant ethic," namely the "Romantic ethic," whose celebration of "autonomous, pleasure-seeking," the cult of "sentimentality" and "hedonism" could be said have forged a link between Protestantism and capitalism through modern consumerism; and there is Daniel Bell's (1976, p. xxiv) claim that modern culture created the conditions under which lifestyle rather than work became the chief source of satisfaction and the "lifestyle that became the imago of the free self was not that of the businessman, expressing himself through his 'dynamic drive,' but that of the artist defying the conventions of society." While these accounts complicate Weber's causal links between Protestantism and capitalism they duplicate what Florida (2002, p. 197) sees as the tendency to deny the "possibility of synthesis between the bohemian ethic and the Protestant ethic, or of actually moving beyond these categories." We see an implicit bias against the bohemian ethic in how some neo-Weberian authors see the phenomenon of the 1960s. Bell (1976, pp. xxvi–xxvii) disparagingly describes the 1960s counter-culture as a "children's crusade that sought to eliminate the line between fantasy and reality . . . less a counterculture than a counterfeit culture." For him, the 1960s concern with "art," "imagination," and liberating "impulse," is nothing more than the expression of the "double contradiction of capitalism" between work and consumption (Bell, 1976: xxvii). There is little recognition that anything lasting could come out of the cultural impulse of the 1960s or, as Florida (2002: 207) puts it, that "Microsoft and Jimi Hendrix" may have more in common than first meets the eye. The figure of the geek or techno-savvy entrepreneur has altered Bell's opposition between the cultural types of the businessman and the artist:

> [N]ow the picture has been reversed. Business people are no longer vilified. Today, they and so-called bohemians not only get along, they often inhabit each other's worlds; they are often the same people. Jobs, Wozniak, Gates, Allen and others have inserted the idea of *Entrepreneur* into the fabric of popular mythology . . . Allen, cofounder of Microsoft . . . embodies this fusion of identity. [He]

> is the creator of Seattle's Experience Music Project, an interactive museum designed by Frank Gehry and initially created as a tribute to Jimi Hendrix. (Florida, 2002, p. 208)

For Florida (2002, p. 209), the geek or "engineer as pop-culture hero" is also a sign that a significant change in cultural types is taking place. He says of the geek: "Neither outsider nor insider, neither bohemian nor bourgeois, the geek is simply a technologically creative person" (Florida, 2002, p. 210).

I concur with Florida's claims that the opposition between a Protestant work ethic and a bohemian/artistic ethic is being superseded. However, what would happen if, instead of seeing these ethics as ideologies or fixed lifestyles, we saw them as akin to what Foucault termed "technologies of the self"? There is some evidence that this concept is precisely what Weber intended with his notion of an "ethic." In his essay "The Economic Ethic of the World Religions" (or, as Gerth and Mills translated the title, "The Social Psychology of the World Religions"), Weber (1948a, pp. 267–268) says that "similar forms of economic organization may agree with different economic ethics . . . [and an] economic ethic is not a simple 'function' of a form of economic organization." More important than the determination of conduct by a religious worldview is "man's attitudes toward the world" and the "inner factors" that give an "economic ethic . . . a high measure of autonomy" (Weber, 1948a, p. 268).

How can one put to work Weber's formulation that an ethic is identifiable by "man's attitudes to the world" and the "inner factors" that give an ethic a degree of autonomy? Fortunately, Weber (1948b) offered precisely such a framework in his "Religious Rejections of the World and Their Directions" essay. The essay operates with a formal typological distinction between "asceticism" and "mysticism" that significantly alters the discussions of religious ethics in *The Protestant Ethic.* In "Religious Rejections," Weber (1948b, p. 325) compares the "abnegations of the world" of an "active asceticism that is a God-willed *action* of the devout who are God's tools" to the "contemplative *possession* of the holy, as found in mysticism." The contrast is between "active mastering" of reality and the detached "contemplation" of reality:

> For the true mystic the principle continues to hold: the creature must be silent so that God may speak. He is "in" the world and externally "accommodates" to its orders, but . . . [there] is a minimization of action, a sort of religious incognito existence in the world . . . Inner-worldly asceticism, on the contrary, proves itself *through* action. To the inner-worldly asceticist the conduct of the mystic is an indolent enjoyment of the self; to the mystic the conduct of the . . . asceticist is an

> entanglement of the godless ways of the world combined with complacent self-righteousness. (Weber, 1948b, p. 326)

Weber (1948b, p. 326) adds that in "both cases the contrast can actually disappear in practice and some combination of both forms of the quest for salvation may occur." However, what is of interest here is how Weber situates art and an aesthetic outlook on the world within his schema. He suggests that to the "creative artist . . . as well as to the aesthetically excited and receptive mind" there are two forms of "coercion" that appear opposed to "their genuine creativeness and innermost selves": first, the necessity of "taking a stand on rational, ethical grounds"; and, second, the imposition of form from the outside. The rejection of externally imposed form implies that, for the creative or aesthetically minded person, there "is an indubitable psychological affinity of profoundly shaping experiences in art and religion" (Weber, 1948b, p. 342). It also means that the aesthetic attitude is closest to that of the religious mystic (Weber, 1948b, p. 342). This tendency is heightened in modernity, where under the conditions of "intellectualism and the rationalization of life . . . Art takes over the function of a this-worldly salvation . . . from the routines of everyday life" (Weber, 1948b, p. 342). However, art's fate is not set in stone. Weber (1948b, p. 343) states that in "empirical, historical reality" the "psychological affinity between art and religion" can lead to "ever-renewed alliances"; although, whenever the world religions have "emphasized either the supra-worldliness of its God or the other-worldliness of salvation, the more harshly has art been refuted." In short, art tends to lean toward the mystical attitude of the world, but it is an empirical question whether it is combined with religious or ethical asceticism or whether it is in competition with it.

Deferring the question of any potential synthesis between asceticism and aestheticism for the moment, what of Weber's equation of art with "mysticism"? This has been a recurring motif in Western modernity especially since the advent of Romanticism. Weber's characterization of the mystic as someone who in his "innermost being" is "hostile to all form" and believes "precisely in the experience of exploding all forms" resonates with the rhetoric surrounding art since about 1800. As Bernice Martin (1981, p. 79) writes, one of the central tasks assigned to art in modernity has been to attack "boundaries": "between public and private, between decent and indecent, tabooed and available, sacred and profane, between good taste and vulgarity, between creator and creation, artist and observer." She links this idea directly to the Romantic movement and its aesthetic appropriation of the religious themes of "offering alternative visions and intimations of transcendence" (Martin, 1981, p. 80). We are told that the "Holy Grail of Romanticism" is that art should be able "to grasp the ungraspable, to represent ecstasy, to transfigure mundane reality" (Martin, 1981, p. 81). In the process, the role of the artist as creator was transformed. With the onset of modernity, the myth of the artist as nonconformist was born:

> In medieval Europe the artist was a craftsman/artisan or, if he worked with words or music, a cleric . . . By the 18th century he may have been a gentleman . . . In any case . . . [the] artist himself might be a conformist or troublemaker in his personal capacity but no special *éclat* attached to nonconformity. Gesualdo was a murderer and Purcell is said to have died of a chill caught when his wife locked him out after one of the many nights out on the town . . . but Bach lived a model and orderly domestic and civic life . . . The idea that the true artist is a rebel and a sufferer or that insanity and wildness are inseparable from genius is a myth of Romanticism. (Martin, 1981, p. 85)

The artist as "genius" and nonconformist conforms to Weber's depiction of the creative personality as mystical in nature. Weber's schema, though, breaks with popular stereotype of the artist as mad genius in one important respect: it sees the "hostility to form" and the adoption of "silent contemplation" of the world as practical techniques for having a certain kind of experience. This point has not been lost on some of the literature on "being an artist." While on the whole more sympathetic to the artist-mystic figure than Weber ever was (i.e., he remained throughout his life wary of the seductions of artistic charisma and strongly resisted the cult of personality associated with figures such as Richard Wagner and Stefan George), Mark Levy's (1993) *Technicians of Ecstasy: Shamanism and the Modern Artist* makes some surprisingly similar claims. In a section entitled "Shamanistic Techniques to Stimulate Creativity," Levy (1993, pp. 311–313) includes the following practices: "isolation"; "not-doing"; "drugs"; "prolonged looking"; "the strong eye"; and "setting up dreaming." Reminiscent of some of the values of the 1960s counterculture, this handbook for enhancing creativity highlights the benefits of "meditation and occasional fasting," "doing practically nothing," "preventing the superimposition of personal projections on the object of attention," and "non-focused attention" (Levy, 1993, pp. 311–313). In Weber's typological distinction between "asceticism" and "mysticism," most of the techniques recommended by Levy tend toward the mystical end of the spectrum. Not for nothing, Levy's book is entitled *Technicians of Ecstasy* and includes the word *shamanism* in its subtitle.

How do the two ethical types come to be synthesized in the age of the creative economy? Borrowing from Weber we might say that all kinds of "alliances" are possible between "asceticism" and "mysticism" in "empirical, historical reality." If we stop seeing these as fixed ideologies or lifestyles and more as ethical orientations towards the world, we quickly see that both ascetic-mysticism and ascetic-aestheticism are possible. It is the responsibility of those studying the phenomenon to identity concrete socio-historical institutional or cultural structures where such "morphing" (to evoke Florida's phrasing) takes place.

In line with scholars who have written on recent syntheses between art and commerce, I argue that a key institutional setting for the emergence of "ascetic-aestheticism" has been the art school of the last few decades of the twentieth century. Frith and Horne's study of the British art school was driven by an interesting historical and empirical coincidence: a large number of creative individuals who were successful in the world of British rock and pop had studied at art colleges rather than at music conservatoriums. The list of people is impressive. It included, among others: John Lennon (Liverpool College of Art); Keith Richards (Sidcup College of Art); Malcolm McLaren (St Martin's College of Art and Harrow Art College); Pete Townshend (Ealing Art College); and Brian Eno (Ipswich Art School and Winchester School of Art). In other words, an entire generation of innovators in the world of rock and pop had acquired their aesthetic ethos at art colleges and/or used the opportunity to develop innovative musical styles.

The reasons for this historical and empirical relationship are many and varied. They range from meeting other like-minded individuals to having spaces to perform gigs (e.g., art college dances). However, in terms of a discussion of education and the emergence of aesthetic capitalism, what is interesting is the synthesis of work and leisure, art and entrepreneurialism that this milieu allowed to flourish. It was not so much that the British art school obliterated the opposition between the Protestant and bohemian ethics. Rather, it allowed a cultural space in which the two could be negotiated and managed in innovative ways:

> The modern British art school has evolved through a repeated series of attempts to gear its practice to trade and industry to which the schools themselves have responded with a dogged insistence on spontaneity, on artistic autonomy . . . Constant attempts to reduce the marginality of art education, to make art and design more "responsive" and "vocational" by gearing them towards industry and commerce, have confronted the ideology of 'being an artist' . . . Even as pop stars, art students celebrate the critical edge marginality allows, turning it into a sales technique, a source of celebrity. (Frith & Horne, 1987, pp. 29–30)

What does this "dogged insistence" on not letting the state or the market define what the art school experience should entail have to do with the milieu's nurturing of creativity, especially forms of creativity for which these institutions were not formally set up? For Frith and Horne, what marked the art schools apart from the rest of the higher education and vocational training sectors was the emphasis on experimentation and self-fulfillment. They found that even among the teachers at such colleges there was a tendency to downplay formal training and emphasize the cultivation of individual creativity. One art school teacher is quoted as saying: "True artists are in the end self-taught, for the work has its origins in the struggle with one's own being . . . an art school should provide the conditions in which the student is most likely to discover himself" (cited in Frith & Horne, 1987, p. 38). In other words, neither formal qualifications nor formal training were seen as ends in themselves. This observation is not to suggest that art education has resisted the process of academic professionalization (on artists and the contemporary American university, see Singerman, 1999). Rather, the question of "art as knowledge" and "art as vocation" can be answered in very different ways. Some responses are more conducive to generating a creative or entrepreneurial attitude than others. Arguably, the British art school with its emphasis on tolerating fun and play prevented participants from prematurely adopting a "silo" or "ivory tower" mentality. Participants also tended to see marginality as a blessing rather than a curse.

In some respects, the "learning"—if we can call it that—which took place in such art colleges did not actually require a formal institutional setting. We could well ask: what if some of these talented musicians had met each other in a café or at law school? The right degree of purpose and purposelessness does matter here. A chance encounter with a like-minded person at a café may result in pleasant and engaging conversation but without some sense that both persons share the burden of expected creativity, i.e., the responsibility of "being an artist," it may or may not translate into joint projects. Sitting in law lectures may create the fantasy that it would be nice to escape one's professional fate by running off to become a musician, but fantasies experienced in the context of a dry university lecture—even ones shared with others in the same predicament—are likely to remain just that: fantasies. By contrast, entering a milieu where creativity is possible but not expected "on tap" creates the conditions for some experimentation without abandoning the work-nature of experimentation. Once again, this meeting of the minds does not necessarily have to take place in a formal educational setting. Arguably, Andy Warhol's famous "Factory" simulated some of the conditions Frith and Horne attribute to the British art school. Currid explains how Warhol's studio helped to emphasize the collective nature of creativity and what resources were needed to be productive in a creative sense:

> Andy Warhol exemplified . . . in both his work and his Factory, the collective nature of creativity: that fashion, art, film, music, and design did not reside in separate spheres—that instead they were constantly engaging each other and shar-

> ing ideas and resources across creative sectors . . . Warhol also saw the significance of the social spaces in which these industries and creative people interacted—his Factory merged cultural production with a social scene. (Currid, 2007, pp. 15–16)

In short, the kind of cultural milieu that has promoted today's creatives is evident in the British art college of the 1960s and 1970s, as much as in Andy Warhol's Factory. Both settings merged art worlds with social worlds; scenes with networks of production. Just as central to participating in Warhol's Factory was being part of the scene. What seems to have mattered to the creatives in the Frith and Horne (1987: 29) study is that "British art schools *are* the art world."

However, can one learn in noneducational settings? The notion of learning implicit in some of the recent literature on creative economies is closer to the kind of acquisition of competencies Howard Becker described in his famous 1953 article, "Becoming a Marijuana User." The article uses the analogy of learning to suggest that a user will only "derive pleasure" from the drug once "he (1) learns to smoke it in a way that will produce real effects; (2) learns to recognize the effects and connect them with drug use; and (3) learns to enjoy the sensations he perceives" (Becker, 1953, p. 235). Now, while one hesitates to compare the knowledge acquired in initiations into the world of drug-taking to that received in formal education settings, there is enough evidence in the preceding discussion to suggest that creative work involves knowledge that is theoretical without necessarily being didactic, rational without necessarily being instrumental. How else can we understand the benefits of spending time in creative contexts for creative workers except that creativity is an activity that combines sociality with individual expression, practical application with leaps of the imagination, technical competence as a precondition for experiencing the pleasure of the task, and so forth. Indeed, the problem of correctly identifying the kind of learning processes that aesthetic creativity entails is partly language-based. The English verb "to know" lacks the subtleties of, for example, the Spanish verbs *conocer* and *saber*. That is, it fails to distinguish between knowledge acquired through direct and immediate experience as against knowledge acquired through formal and abstract means. However, it is precisely *conocimiento* or lived experience that creative work thrives on. In other words, despite the growing literature on creative work and the creative class, creativity cannot be bottled or learned through formulaic means.

In dealing with the paradox that creativity can be learned but not necessarily rote-learned, I would like to conclude by returning to the Weberian schema I outlined earlier on. Weber's contrast between the ascetic and the mystic is based on the notion that each set of attitudes to the world offers an implicit critique of its opposite. To the ascetic, the mystic seems indolent and too prepared to enjoy himself; to the mystic, the ascetic seems too complacent, too self-righteous and too closed off to the mysteries of the universe. However, the strength of each Weberian ethical type is underscored by a matching weakness: the ascetic yearns for control but lacks fulfillment and meaning; the mystic yearns for ecstatic transcendence of the self but is unsure whether a routine or practice can guarantee such an outcome. As both types take the "quest for salvation" seriously there is no room for fence sitting. How then is each ethical type to proceed? Weber's (1948b, pp. 323–324) suggestion is that the "motives" underpinning a "religious rejection of the world" and the "logical or teleological consistency" of the practical actions that spring from them are not one and the same. The *ratio* of an action and its ends are also not identical. Thus, in an essay devoted specifically to religion and salvation strategies, Weber surprises his reader by outlining the effects of worldly rejections in the spheres of economy, politics, aesthetics, the erotic, and the intellectual life. Each of these spheres produces tensions not unlike those we find in the religious sphere between asceticism and mysticism.

One of the imports of Weber's "Religious Rejections" essay is that no value-sphere is likely to be dominant over all others. Furthermore, spheres will borrow from each other in organizing their own inner logic. We are living at a time when the exponents of capitalism have realized they can learn from art, and the arts are seen as one of the drivers of economic activity. Thus, it is more or less inevitable that creativity becomes a buzzword, and its mysterious nature may generate theoretical speculation and attempts at empirical mapping. We may well be living in an age when the Protestant work ethic and the bohemian aesthetic ethic are being blurred and confused in all kinds of ways. However, if Weber is correct about the long-term tendencies present within capitalist modernity, then it probably still matters whether business people are trying to emulate artists or artists are trying to emulate businessmen. There are probably very different implications flowing from each of these scenarios for educational philosophies and institutions. Either way, a certain flexibility and room to experiment seem like desirable attributes of the education system in the age of aesthetic capitalism.

References

Becker, H. S. (1953). On becoming a marijuana user. *American Journal of Sociology, 59* (3), 235–242.

Bell, D. (1976). *The cultural contradictions of capitalism.* New York: Basic Books.

Boltanski, L., & Chiapello, E. (2005). *The new spirit of capitalism* (G. Elliott, Trans.). New York: Verso.

Brooks, D. (2000). *BoBos in paradise: The new upper class and how they got there.* New York: Simon and Schuster.

Campbell, C. (1987). *The romantic ethic and the spirit of capitalism.* Oxford: Blackwell.

Currid, E. (2007). *The Warhol economy: How fashion, art and music drive New York City.* Princeton, NJ: Princeton University Press.

Eagleton, T.(1990). *The ideology of the aesthetic.* Oxford: Blackwell.

Florida, R. (2002). *The rise of the creative class,* Melbourne: Pluto Press.

Frith, S., & Horne, H. (1987). *Art into pop.* London: Methuen.

Graña, C. (1964). *Bohemian versus bourgeois.* New York: Basic Books.

Gronow, J. (1998). *The sociology of taste.* London: Routledge.

Hemker, A., & Koopman, D. (2005). The creative class and nightlife. In S. Franke & E. Verhagen (Eds.), *Creativity and the city: How the creative economy changes the city* (pp. 100–107). Rotterdam: NAi Publishers.

Landry, C. (2000). *The creative city: A toolkit for urban innovators.* London: Earthscan.

Lash, S., & Urry, J. (1994). *Economies of signs and spaces.* London: Sage.

Levy, M. (1993). *Technicians of ecstasy: Shamanism and the modern artist.* Norfolk, CT: Bramble Books.

Lloyd, R. (2005). *Neo-Bohemia: Art and commerce in the post-industrial city.* New York: Routledge.

Martin, B. (1981). *A sociology of contemporary cultural change.* Oxford: Blackwell.

Molotch, H. (2004). How art works: Form and function in the stuff of life. In R. Friedland & J. Mohr (Eds.), *Matters of Culture: Cultural Sociology in Practice* (341–377). Cambridge: Cambridge University Press.

Pine, B. J., & Gilmore, J. H. (1999). *The experience economy: Work is theatre and every business a stage.* Boston: Harvard Business School Press.

Singerman, H. (1999). *Art subjects: Making artists in the American university.* Berkeley: University of California Press.

Weber, M. (1948a). The social psychology of the world religions. In H. H. Gerth & C. W. Mills (Eds.), *From Max Weber: Essays in sociology* (pp. 267–301). London: Routledge and Kegan Paul.

Weber, M. (1948b). Religious rejections of the world and their directions. In H. H. Gerth & C. W. Mills (Eds.), *From Max Weber: Essays in sociology* (pp. 323–359). London: Routledge and Kegan Paul.

Weber, M. (1976). *The Protestant ethic and the spirit of capitalism* (T. Parsons, Trans.). London: George Allen and Unwin.

Welsch, W. (1997). *Undoing aesthetics.* London: Sage.

Wilson, E. (2000). *Bohemians: The Glamorous outcasts.* London: Tauris.

27

Democratic Culture: Opening Up the Arts to Everyone

John Holden

This essay was written in a British context, but many of its arguments apply equally in the United States and elsewhere, despite differences both in our political systems and in the ways in which the arts are funded. There has been a recent trend for the U.K. and the U.S. to engage in a dialogue about the arts and culture because of a growing realization that we share many concerns, such as how to measure and account for the value of culture, and how to promote the arts in difficult economic times. I therefore offer this essay in the hope that it will add to that dialogue and that readers will forgive the references to practice in the U.K.

The aim of a democratic society is to release the talents of all its citizens, not just some of them. The arts and culture are good for people's general well-being, but they also help release the creative potential of individuals. A comparison of the arts and sports in schools shows that they can achieve many of the same aims in that they increase young people's confidence, their ability to work in teams, their communications skills, and so on. The difference between them lies in the fact that sports follow rules, whereas the creative arts encourage people to think afresh, to explore possibilities that have not been thought of previously. Indeed, this principle applies even when working within established art forms and traditions. Therefore, it is imperative that the arts are not the preserve of an elite or enjoyed by only a small part of society. Broadbased support for and engagement with the arts is an essential part both of a well functioning polity and a thriving modern economy.

Introduction

> *If art which is now sick is to live and not die, it must in the future be of the people, for the people, by the people; it must understand all and be understood by all*
> —William Morris. [1]

In a BBC Radio 3 programme, *Is This a Record?*,[2] the classical music critic Norman Lebrecht was in discussion with the record producer Tom Shepherd. This is an excerpt from their conversation:

Lebrecht: The pressure to record public celebrities originated in the 1970s . . .

Shepherd: Norman, every one of us at some point or other, I think, is tempted to ride the coat-tails of a success that may not have been there for purely musical reasons. I signed up Eugene Fodor, who was a perfectly decent violinist . . . but I didn't sign him up because he was good but because he was making headlines, because he was, like, playing the violin in a cowboy suit . . . Everybody in those times, in the mid 1970s, when everyone was trying to somehow democratise classical music, the person who wore the cowboy boots . . . was the person who you were going to go after because they were turning great art into what was considered popular taste, popular consumption, and I think we all fell victim to that. I mean . . . sometimes you do these things, which you do for purely mercenary reasons, and it pays off.

Lebrecht: The generic term for these compromises is "crossover," and, as times got tough in the 1990s when the compact disc boom ended, it spread across the record labels like a black death.

Shepherd's application of the verb *democratize* to the cynical use of gimmickry by a record company in pursuit of profit may seem surprising, but to bring the concept of democracy into the cultural discourse raises interesting questions: what might *democratise* mean in relation to culture? If democracy is desirable in the political system, why do some people consider it undesirable in the cultural world; why was democracy—the bedrock of American values—used here in a pejorative sense?

This chapter sets out to answer those questions. It asks what a "democratic culture" might look like, if we had it, and begins by addressing what "culture" means today.

Culture

The meaning of culture has always been difficult to pin down; the cultural critic Raymond Williams said it was "one of the two or three most complicated words in the English language."[3] In the late twentieth century, culture was principally used in two senses: to refer to the high arts of opera, ballet, poetry, literature, painting, sculpture, music, and drama; and in a more general, anthropological sense to encompass all of the practices and objects through which a society expresses and understands itself.

This dual meaning gave rise to much confusion. In the postwar world, culture in the sense of the high arts was defined and enjoyed by a mandarin class; in this context the phrase *democratic culture* becomes an oxymoron or a contradiction in terms—how can culture be democratic if it is confined to one small section of society? However, using *culture* in its other sense turns *democratic culture* into a tautology—how can culture be anything other than democratic if it is defined as the sum total of everything that people do?

T.S. Eliot implicitly acknowledged the two different uses of the word *culture* when he pointed out that individuals could adopt a "conscious aim to achieve culture,"[4] whereas at the level of society "culture is the one thing that we cannot deliberately aim at."[5] This assessment developed into an opposition between the individual and the mass, between what many saw as a debased popular culture and a refined higher culture. Eliot takes it for granted that there is a hierarchy, but when he concludes that "culture may even be described simply as that which makes life worth living,"[6] he sidesteps the question of who decides just what it is that "makes life worth living."

In our own time we need to look again at what we mean by culture because for practical purposes there are now three, deeply interrelated spheres: funded culture, commercial culture, and home-made culture. What counts as culture is decided by different groups in each of these cases, but the existence of a critical discourse, with arbitration of standards and quality, is a significant feature in all of them.

In funded culture, culture is not defined through theory but by practice: what gets funded becomes culture. This pragmatic approach has proved useful in allowing funding bodies like trusts and foundations to expand their definition of art over the last 50 years to include vaudeville such as circus, puppetry, and street art as well as high art such as opera and ballet, while still controlling what culture in this sense means. Similarly, *heritage* comes to be defined by decisions about what to protect and preserve. Who makes these decisions, and on what basis, is therefore a matter of considerable public interest.

Commercial culture is equally pragmatically defined: if someone thinks there is a chance that a song or a show will sell, it gets produced, but the consumer is the ultimate arbiter of commercial culture. Success or failure is market driven, but access to the market—the elusive "big bucks record deal" of Springsteen's *Rosalita,* the Broadway stage debut, the first novel—is controlled by a commercial mandarin class just as powerful as the bureaucrats of funded culture. So in publicly funded culture and commercial culture, there are gatekeepers who define the meaning of culture through their decisions.

Finally there is home-made culture, which extends from the historic objects and activities of folk art, through to the postmodern punk garage band and the YouTube upload. Here, the definition of what counts as culture is much broader; it is defined by an informal self-selecting peer group, and the barriers to entry are much lower. Knitting a sweater, inventing a new recipe, or writing a song and posting it on MySpace may take great skill, but they can be achieved without reference to anyone else—the decision about the quality of what is produced then lies in the hands of those who see, hear, or taste the finished article.

In all three of these spheres individuals take on positions as producers and consumers, authors and readers, performers and audiences. Each of us is able to move through different roles with increasing fluidity, creating and updating our identities as we go. Artists travel freely between the funded, commercial, and home-made sectors: publicly funded orchestras make commercial recordings that get sold in record shops and uploaded onto Web sites; street fashion inspires commercial fashion; an indie band may get a record deal and then play at a subsidized venue.

A recent Demos report, *Video Republic,* has pointed out that the internet has fuelled an "explosion in audio-video creativity."[7] In fact the rapid and enormous expansion of the Internet as a space for cultural interaction and as a facilitator of mass creativity has changed the possibilities for all types and all three spheres of culture, presenting, across the board, a wealth of new opportunities (such as audiences; art forms; distribution channels) and questions (for example, what to do about intellectual property; investment in technology; and censorship).

The Internet is credited with driving mass creativity, but in reality it is only one of the factors that explains it. Over the past fifty years musical instruments have become much more available. Cheap and easy-to-use digital technology means that one can make a film without needing the backing of a highly capitalized studio. Furthermore, investment in the infrastructure of performance spaces has made it possible for millions to participate in all sorts of cultural activities.

However, this upsurge in creative activity should not lead us to conclude that we have already achieved a "democratic culture" where everyone can enjoy culture equally. There are stark differences between individual capacities to make informed choices, and there are still parts of the cultural world where most people feel alienated: one of the main findings from Arts Council England's recent Arts Debate—a multifaceted inquiry into public attitudes to the arts—is their discovery of "a strong sense among many members of the public *of being excluded from something that they would like to be able to access*" (emphasis added).[8] Clearly, they are not referring to commercial or home-made culture—almost everyone reads books and listens to music—but to funded culture.

Excellence or Exclusivity?

Listening to Lebrecht, Shepherd, and other contributors to *Is This a Record?* reveals that there is a thin line between, on the one hand, mounting a defence of quality, and on the other, erecting barricades both against change and against outsiders. There is an urgent need to discuss where that thin line lies, because it is not always clear.

A report to government in the U.K., Sir Brian McMaster's *Supporting Excellence in the Arts,* stresses that the arts should be "excellent,"[9] but Sir Brian says that: "Excellence itself is sometimes dismissed as an exclusive, canonical and 'heritage' approach to cultural activity. I refute this."[10]

He is right to do so. There is no reason why excellence should imply a backward-looking culture and, equally, there is no reason why excellence should be conflated with exclusivity.[11] Conversely, we should be aware that appeals to excellence and quality can be used as a cover for maintaining social superiority. As John Seabrook, the author of *Nobrow,* has observed, in the cultural field, sometimes people are pretending to maintain standards but really just preserving status; we must beware of "taste as power pretending to be common sense."[12]

Three ideas are embedded in the conversation between Norman Lebrecht and Tom Shepherd: first, that collaboration between classical musicians and other musicians is always bad; second, that popular success is always bad, with *bad* here being used in the sense of "debased" or "of poor quality"; third, that there is some pure category of art that is polluted ("a black death") when it comes into contact with non-art.

All of those assumptions can be contested, but some of the program's contributors went on to say that these bad crossovers are tolerated because, in the words of one of them, "they fund Alban Berg.[13] In other words, the public and its constituents essentially exist to be exploited: crossovers sell, and however "compromised" they may be, they are a means to a higher end.

What is at work here is the belief that only a small minority can appreciate art, and that art of quality needs to be defended from the mob. If the mob gets its hands on the art, the art will be destroyed. Therefore, art must be kept as the preserve of the few, because only the few understand and value it. This attitude can be seen again in remarks by Francesco Corti, the music director of Scottish Opera, who says "We must have a faithful product, something true, not something cheap, just to catch an audience . . . I'm sorry; probably this is heretical, but I believe that opera is still something for the elite."[14] So while the members of the elite will enjoy "something true," the poor benighted public will be fobbed off with second-rate compromises in order "to fund Alban Berg." It is as if a baker justified adulterating his bread with sawdust so that he can make *madeleines* for those who have a taste for them.

Art and Life

It may be that there is a natural human tendency to protect those items we love by keeping them to ourselves. Some scientists put a cordon around their work and repulse public discussion of things such as nanotechnology and alternative medicine by saying that only they have the specialist knowledge that qualifies them to have an opinion.[15]

When it comes to the arts, there is indeed a sense in which they are "special." They are not the same as entertainment, and they take us beyond everyday life. Art's role as a substitute for religion,[16] its important social function in critiquing the status quo,[17] and its constant quest to explore new territory and to provide wonder, all militate in favor of seeing the arts as being a step away from mundane, diurnal experience. However, that step should not place them off limits to anyone because, while the arts are special, they are also simultaneously, inextricably, and healthily part of the everyday.

That may seem contradictory, but in fact, in most times and in most places, this synthetic point of view was commonplace. The Balinese dancer, or the medieval peasant standing in front of a fresco, or the Mughal prince admiring his collection of miniatures, could all appreciate both the normality and the extraordinariness of what they were doing. In our own time, public funding and tax-exempt private funding enable the arts to occupy a position that is neither divorced from nor wholly overwhelmed by the market. Funding for the arts provides one practical manifestation of our ability to live with these ambiguities.

As the three spheres of culture mentioned earlier—the public, commercial and homemade—become increasingly interconnected and networked, these ambiguities are likely to increase, and the reaction to that on the part of people who think that art is for an elite will be to try to maintain their power to define what art is by separating it from everyday life. For example, in an article titled "The Philistines Are upon Us," the think tank Civitas said that it was "vulgar" to talk about the economic effects of cultural activities. They concluded that "perhaps some of those people who are working in the 'creative industries' should go and stand in front of . . . *a fête champêtre* by Watteau, and ask themselves what it means to them."[18] But artists are part of the commercial world, even if their primary motivation is the creation of meaning and not financial gain: Watteau's final work was not a *fête galante,* but a shop sign, now in the Charlottenburg Palace in Berlin.

The arts and culture can be looked at in many ways, from many perspectives, and they are a legitimate concern for everyone. There should be no barriers to entry and no place for thinking that "opera is still something for the elite."[19]

Keeping the Mob at Bay

The defense of cultural exclusivity sits oddly with most other areas of life. No one, for example, would suggest that watching elite athletes competing at the Olympics should be the limited preserve of a group of cognoscenti well versed in the sport. In addition, it would be unreasonable to suggest that if the athletes were to lower their standards and give a worse performance, this behaviour would increase their popularity.

The notion that art needs to be defended from the mob also flies in the face of the economist's concept of art as a non-rival good, meaning that one person's enjoyment of a work of art should not, in theory, interfere with someone else's enjoyment. However, in practice, the very opposite seems to be true; indeed the cultural gatekeepers of the avant-garde go so far as to *define* art in terms of exclusivity. As Schoenberg put it: "If it is art, it is not for all. If it is for all, it is not art."[20] "Art" is another very difficult word to pin down, but surely, defining it in terms of its demographic reach is a poor place to start.

In addition, we should not tolerate the modernist artist's claim to authority over what counts as art and what does not. Duchamp said "I don't believe in art, I believe in artists."[21] When artists themselves exclusively determine what art is, no one else has a voice in deciding an important question in which everyone has an interest. Schoenberg and Duchamp are not alone in attempting to wrest questions about art away from the public. There are, in fact, three main categories of those who wish to erect keep out signs around culture. The first example is that of the malign—as distinct from the beneficent—expert. In culture, just as in every other part of human life, different people have different levels of expertise. It is a truism that a lifetime devoted to the study and practice of a subject should result in a greater degree of knowledge, expertise, and appreciation.

This concept applies equally to museum professionals, sculptors and dancers as well as plumbers or lawyers.[22] Expertise and informed judgment qualify people to speak with authority and to set critical standards. At its best, expert professionalism is used to educate, inform, serve, and enable the public; this approach might be termed benign, inclusive, socially useful, and democratic professionalism. At its worst it is used to bamboozle, patronize, and exclude the public; we might call this malign or antagonistic professionalism. This distinction between two types of professional attitude and practice applies well beyond the cultural world, but it is certainly apparent within it.

The second example of the gatekeepers is that of the cultural snob. The arts and culture have historically been used, and continue to be used, as a means of asserting social status. Some corporate sponsors align themselves with culture precisely because of the social status of the audience that it attracts, as this quote from a U.S. airline makes clear: "the huge success of the show and the strength of the brand association has helped us to raise our profile within the UK market, particularly to a core ABC1 audience."[23]

Again, a distinction must be made between *malign* and *benign* sponsors: those who delight in the exclusivity of an audience and those who wish to make that audience as wide as possible. As another sponsor, Ernst and Young, points out: "business sponsorship of the arts allows works to be shown to a wider audience than would otherwise have the chance to see them."[24]

From Veblen's[25] writing on conspicuous consumption, to Bourdieu's theories of cultural capital,[26] there are well-established models showing how processes of exclusion work. Although these theories are most often concerned with critiquing high culture, in the form of opera, ballet, music, drama, and the visual arts, the arguments are equally applicable in the context of contemporary subcultures, where knowledge for instance about hip-hop or reggae serves to define inclusion or exclusion in relation to a group. Pop music subcultures have their specialities, arcana, cults, and cliques just as much as the contemporary visual arts.

In the case of cultural snobbery, an individual's acceptance or rejection by a social group is based on the uninitiated accepting the norms and gaining access to the knowledge of the initiated. Paradoxically, the more people gain the knowledge, the more diluted the group becomes, leading to an ambivalent attitude on the part of members of the group to the education of the uninitiated. This is why some users of public libraries, some museum-goers and some classical music aficionados resist "new audiences" encroaching on their turf.

In the third example of exclusivity, that of the avant-garde, what started out as opposition to the Academy degenerated into antagonism toward the public, who are excluded not simply through preference, but by a process of logic. The avant-garde *defines* itself in oppositional terms: anything that is comprehensible by the mass is by definition excluded from the avant-garde. In order to maintain its own self-worth and status, the avant-garde *must* either alienate the public ("*épater le bourgeoisie,*" as the late-nineteenth-century French Decadent poets put it) or withdraw from contact with the public. The American composer Milton Babbitt, writing in 1958, put it this way:

> I dare suggest that the composer would do himself and his music an immediate and eventual service by total, resolute and voluntary withdrawal from this public world to one of private performance and electronic media, with its very real possibility of complete elimination of the public and social aspects of musical composition.[27]

All three positions—those of the malign expert, the cultural snob and the avant-gardiste—collectively the "cultural exclusivists"—rest on a belief that it is *they* who have a right to determine standards, that *they* know best, that *they* decide. Alliances form between malign experts and each of the other groups. Battle-lines (often but not always reflecting a right–left political divide) are drawn up between the cultural snobs and the avant-garde. Arguments and disputes frequently break out within the ranks. However, these are false battles, because the malign experts, the cultural snobs, and the negative avant-garde share a fundamental common purpose: to assert their exclusivity, to guard the territory that they have mapped together, in order to keep the public out.

The *real* battle is not between the different battalions of the cultural exclusivists, but between all of them, on the one hand, and, on the other, the fragmented, disputatious and heterogeneous public, plus the benign professionals who wish to enlarge the public franchise.

In fact, it is here that we find echoes of a parallel battle in politics. Whereas culture walls are built to defend the order of the canon, the discipline of practice, and the legitimacy of tradition against the disorder of popular culture and the threat of relativism, in politics there is a division between authority and anarchy.

The sociologist Bruno Latour, in his essay "The Invention of the Science Wars," makes the point that in Plato's *Gorgias,* where Socrates is debating with Callicles whether might or right should prevail in government, Socrates pulls off a neat intellectual sleight-of-hand. Socrates sets up an opposition between rule by reason and rule by a dominant nobility. Latour, though, points out: "What is beyond question for both Socrates and (Callicles) is that some expert knowledge is necessary, either to make the people of Athens behave in the right way or to keep them at bay and shut their mouths."[28] However, as Callicles and Socrates debate, "there is a second fight going on silently, offstage, pitting the people of Athens, the ten thousand fools, against Socrates and Callicles, allied buddies, who agree on everything and differ only about the fastest way to silence the crowd."[29] In other words, Socrates debates which type of expert should rule in order to avoid the question of whether the non-expert should rule.

In *Culture and Anarchy* (1869), the poet and cultural critic Matthew Arnold's position parallels that of Socrates—only he sees culture as the means to stifle the mob. What for him began as an argument about the civilizing effects of culture, its "sweetness and light," ultimately turned into a question of authority. Arnold explicitly asserts culture as a source of authority in the face of anarchy: "If we look at the world outside us we find a disturbing absence of sure authority; and culture brings us towards right reason."[30] In turn "right reason (is) the authority which we are seeking as a defence against anarchy."[31]

In exactly the same way, debates about dumbing down over the past half century pit a knowledgeable, self-defining, and self-regarding cultural *aristos* against the mob, which appears in various guises; in Postman's *Amusing Ourselves to Death*[32] as television, and in Lebrecht's *Is This a Record?*[33] as "crossover" music.

Letting the Public in

The point here is not to deny that popular literature, television, and crossovers on the whole produce rubbish (they do, although they occasionally produce brilliance); rather it is to assert, as does Latour, that there is another way out of this opposition between authority and anarchy, between cultural exclusivity and a debased, diluted, popular culture.

If we stop thinking of the *demos* as an anarchic mob and start thinking of them, of us, as a self-governing, enlightened citizenry, with the capacity to make judgments and decide questions, then a trialogue develops: "Instead of a dramatic opposition between force and reason, we will have to consider three different kinds of forces . . . the force of Socrates, the force of Callicles, and the force of the people."[34]

In culture, we will have to stop thinking of a dispute between high and popular culture and enter into public debate about cultural quality wherever it is manifested across all three spheres of publicly funded, commercial and home-made culture—in opera, crime writing, ballet, salsa, art galleries, TV, MySpace, and so on. We will ditch the sixth-form debates between the avant-garde and the cultural snobs, between one set of experts and another, between Robert Hughes and Damien Hirst, for example, and enter a much more interesting discussion that includes all of us. Furthermore by placing funded, commercial, and home-made culture together, they change from being, in their individual components, respectively marginal, entertaining and amateur, into a combined potent force of democratic expression

Questions of cultural quality and excellence cannot and should not be determined *solely* by a group of peers (who represent a producer interest), any more than questions relating to GM foods or nanotechnology or bioethics should be decided solely by scientists and big business (who equally represent a producer interest). It is essential that the many competing voices of the public are admitted into the debate as well. For we must recognise that the public is a collective term for what is in fact a multitude of different and sometimes opposing viewpoints.

We need to understand the public much better than we do. Our knowledge is increasing all the time, and we are getting a much more nuanced understanding of public attitudes and public segmentation in the U.K. through such endeavors as Arts Council England's Audience Insight research.[35] However, in funded culture, the public still tends to be seen in terms of audiences or attendees or non-attendees, whereas in contemporary society the individual is "the origin rather than the object of action."[36] As the Harvard Business School Professor Shoshana Zuboff and her co-author Jim Maxmin explain: "the new individuals seek true voice, direct participation, unmediated influence and identity-based community because they are comfortable using their own experience as the basis for making judgements."[37] If that is true in business and public services, why would it be different in the case of culture? As the cultural sector engages more with the public, it must of course be alive to the dangers of the public voice being captured by special-interest groups, and of what Raymond Williams called "administered consensus through co-option."[38]

Democratic Culture

Admitting the public into the cultural conversation will mean a battle. As we have seen, the cultural *aristoi* necessarily wish to exclude the public, the demos, from their ranks, because to admit the demos would undermine their own status. However, it is a battle worth fighting. In pursuit of the political ideal of self-government, and "active self-creation"[41] on the part of citizens, what better place to start than with cultural life? Moreover, if the *demos* were allowed in, what then would cultural democracy look like?

My argument is that it would look very much like political democracy and that it would ideally display characteristics of universalism, pluralism, equality, transparency, and freedom. It would be disputatious and contested, and it would develop representative institutions. Its professionalism would be rooted in public service, and it would have its basis in the rule of law.

The rule of law

In fact, the legal basis of cultural democracy already exists. It is underpinned by treaty obligations. Article 27 of the 1948 UN Declaration of Human Rights states that "Everyone has the right freely to participate in the cultural life of the community, and to enjoy the arts"; and article 31 of the 1989 UN Convention on the Rights of the Child says that every child "has the right . . . to participate freely in cultural life and the arts."

These treaty obligations form only the most basic building blocks of a legal framework for a democratic culture. There is little in domestic legislation to make the treaty obligations more than mere aspirations. To form a legal basis for cultural democracy, the treaty obligations mentioned above should be given force in domestic legislation that force central and local government to provide citizens with the tools and infrastructure to understand the cultures of the past and create the cultures of the present. This would include:

- commitments to cultural learning and cultural activity in the education system
- commitments to the arts and culture within public service broadcasting
- arts provision becoming a statutory obligation, with guaranteed access to, and animation of, cultural infrastructure, events, and participation

The goals would be for everyone to have physical, intellectual, and social access to cultural life and to have the ability and confidence to take part in and fashion the culture of today.

Representative institutions

When it comes to the institutions of culture, could they be said to be representative? In most cases the answer must be "no." Most cultural organizations are governed by nonrepresentative, self-perpetuating oligarchies. Governance arrangements are widely criticized by people within the sector and are hidden from the public gaze. There are pressing concerns within political democracy and across public services about how people can be represented and how they can be drawn into democratic participation. There are many suggestions about how to achieve this goal—making voting compulsory, proportional representation, financial rewards for voting, using new technology to make voting easier, reducing the voting age, to name a few. In the cultural world, there has been a long and healthy obsession with growing audiences by widening access and promoting diversity, but there has been little attention paid to what in the world of science is called "upstream engagement," in other words, having public representation to inform decision making about such things as the allocation of funding, choice of research goals and ethical questions.

In culture, consideration needs to be given to:

- the composition and appointment of boards
- public representation within the cultural funding system
- greater use of public consultation

Transparency

The cultural world is also poor at transparency. In the UK a recent report into Arts Council England's last allocation of grants confirms[43] it is often unclear how decisions about funding are taken. This dynamic is equally true when it comes to trusts and foundations and government grants. On top of that, many cultural organizations are opaque: the practice of making annual reports available to the public is not universal. In addition, the financial statistics are Byzantine: no one knows how much money is really being spent, and what effect it actually has.

Opacity is often used as an exclusionary tactic, and it is also dangerous; as recent events in the financial markets have shown, nondisclosure leads to malpractice at worst, and confusion at best. We need:

- full public disclosure of artistic policies and financial information
- clear criteria for funding decisions (which do not exclude the possibility of expert judgment but encourage clarity and explanation when that happens)
- public disclosure of how board appointments are made and by whom
- compulsory annual information from local authorities about expenditure on culture
- research into the influence of elected, non-elected and commercial interests on arts and cultural organisations
- clearly stated policies in relation to arts and cultural broadcasting on the part of broadcasters

Pluralism

A democratic culture necessarily implies plurality with competing ideas and multiple forms of highly developed critical approaches. Contemporary culture in Britain is strong in this area. Apart from a handful of voices on the political extremes, few would argue that our culture is anything other than eclectic and diverse, whether looked at in terms of art forms, content and practice, ethnicity or gender. Where there *is* a problem is with class, as discussed below—which reinforces the argument for expanding the cultural franchise.

Funders need to see their role as encouraging pluralism by nurturing a diverse (in every sense) cultural ecology. They must pay attention to the small-scale, the marginal, and the emergent, as well as the mainstream, and the established. Plurality can be encouraged through:

- commissioning and programming policies adopted by major cultural organizations
- devoting lottery funds to very small-scale grants to new artists and students
- closer collaboration between publicly funded and commercial subsectors to improve transition of work to larger and wider audiences

Freedom

Freedom of artistic and cultural expression is enshrined in the UN treaties referred to above. Limits are placed on that freedom in the form of censorship by legislation, self-censorship, and powers held by local licensing authorities, but as with plurality, in this area the UK's contemporary culture can be given a generally positive mark. Much more serious are the hidden barriers that limit participation in the arts.

Universalism

When the economist John Maynard Keynes set up the British system of arts funding he had the goal of making "the theatre and the concert hall and the art gallery a living element in everyone's upbringing."[44] That goal is very far from being met. Indeed, a recent paper has accused Arts Council England of more or less giving up on the job: "To quote an Arts Council England report of April 2008, 'even if we were able to eliminate the inequalities in arts attendance associated with education, social status, ethnicity, poor health and so on, a large proportion of the population would still choose not to engage in the arts. . . . Insofar as non-engagement with the arts is a matter of lifestyle choice, or "self-exclusion," should the state still intervene?' 'Self-exclusion'? Pardon? This takes the passivity of accessibility work to a new low. After sixty years of this work, not only is it the public's fault if they do not attend or participate in live arts events subsidised by an arts council. We now have a name for the condition they are suffering from."[45]

In a free society, no one should be obliged to enjoy the arts and culture any more than they should be forced to go to university, eat organic food, or exercise every day. However, the disproportionate adoption of all those things by economically privileged sections of society should be a source of concern to anyone who wishes to release the talents and increase the capital of the whole of society, which is surely what a democracy is trying to achieve.

People should have an equal capacity to make choices; otherwise they are not real choices at all. Celebrating people's inability to enjoy culture is one way in which cultural exclusivists have defended their territory; Lord Goodman's remark when he was Chairman of the Arts Council that "one of the most precious freedoms of the British is freedom from culture"[46] may be interpreted in this way.

Despite the UN treaties, most people have only limited and sporadic engagement with culture. The *aristos* has been a very effective gatekeeper. When asked how much they participate in culture, 84 percent of the British population say that they "do little if anything" or participate only "now and then"[47] A mere 12 percent count themselves as enthusiastic participants in the arts (which means they do something only three or four times a year, which is not much). And a tiny 4 percent can be described as voracious cultural participants, meaning that they go to all sorts of arts events frequently.

Importantly, the 84 percent of people—that's about 50 million—who rarely attend, are doing so not because of sheer indifference. On the contrary, as already noted, one of the main findings from Arts Council England's Arts Debate is their discovery of "a strong sense among many members of the public *of being excluded from something that they would like to be able to access*" (emphasis added). "They had a belief that certain kinds of arts experiences were not for people like them".[48] These findings are an indictment not only of the cultural exclusivists but of the entire education system.

Promoting universal access to culture means working with people where they are and not expecting them to come to the culture. In turn that means

- funders paying much more attention to demand-side factors, alongside nurturing the supply side
- more cooperation and coordination between publicly funded culture and mainstream broadcasters to reach wider audiences

- a greater understanding of how existing audiences resist new entrants and strategies to overcome such resistance
- working closely with schools and the education system

Equality

The regular enjoyment of culture is far from being universal, and it is far from equally distributed. Those who engage with the arts are still drawn overwhelmingly from educated and social elites.[49] Clive James recently said that "after World War II, the best of the Labour politicians knew what the gentry had but wanted the working class to have it too, and they were right. Any state that tries to eliminate the idea of gracious living will eventually impoverish everyone except pirates"[50]

He has a point. The full and free participation of an overwhelming majority of citizens in cultural life is attainable, but steps need to be taken to make it happen, and those measures are not all within the remit of cultural organizations. The achievement of cultural democracy lies as much within the education system as it does with arts organizations. We need:

- rapid universal adoption of schemes to put cultural and creative learning at the heart of every child's experience
- commitment to making arts and culture a part of primary education, to include compulsory annual visits to a museum, public library and performing arts venue, and opportunities to perform, write, and draw
- a review of the multitude of educational initiatives and programmes to ensure there is universal access to high-quality experiences in cultural education
- national reading programs in public libraries to develop high-level reading capacities in all young people; OECD research shows that a love of reading is more important for a child's success than his or her family's wealth or class.[51]

The role of the professional

In relation to cultural democracy, the role of the expert should be that of public educator and public servant. Experts should see themselves "as an agency of public education not of populist manipulation."[52] Expertise is hard-won and valuable, but everywhere from medicine to TV talent shows, the relationship between expert and non-expert is being renegotiated.

The concept of a vibrant democracy—whether political or cultural—rests on the existence of an informed, educated but not necessarily expert public. The development of such a citizenry rests on twin pillars of education[53] and a professional class intent on the creation of public value.[54] To quote Philippe de Montebello, the former Director of New York's Metropolitan Museum, "in the end, no-one appreciates being indulged or patronized and it is by treating our visitors with respect that we will gain theirs.[55]

Mutual respect is vital. Cultural democracy does not imply art by plebiscite, with artists, cultural experts and professionals being told what to do by whimsical public "input." On the contrary, it implies a mature relationship where the public recognizes, respects, and benefits from expertise, while simultaneously being alive to its dangers and able to question its credentials. It implies professionals recognizing that their role is to release the talents and potential of the whole community, not just one bit of it, and realizing that they are part of, not separate from, that community. Culture should be a feature of society that we all own and make, not something that is given, offered, or delivered by one section of society to another.

Defending democratic culture

Together with the positive features of transparency, universalism, and so on, a cultural democracy would also safeguard its integrity by adopting defensive measures similar to those taken by political democracy. Just as in politics, publicly funded cultural organizations and the funding system need to:

- guard against undue influence through donations and gifts
- fight against the tendency (inherent in all large systems) to become bureaucratized
- guarantee freedom of information
- engage in public debate

Conclusion

The artist and Turner Prize winner Grayson Perry said in an interview for the Royal Society of Arts: "Democracy has terrible taste. The public wants to bring back hanging as well don't they? The public is very unreliable. If you put it to a referendum, I think we'd have no immigration and no tax. In some ways you do have to be a bit dictatorial as an artist and say that sometimes the art person does know best."[56]

The fact, though, is democracy is *not* a referendum. We *do not* have hanging, and we *do* have immigration and taxation. Our political democracy works because we have developed, over many centuries, systems that benefit from expert opinion but that accommodate dispute, changing circumstances, media scrutiny, and populist sentiment.

We need to develop equally sophisticated approaches to cultural democracy. Arguably, culture has been at its most vibrant and most enduring when most exposed to the *demos*: think of Greek drama, the Elizabethan playhouse, and Italian opera in the nineteenth century—and increasingly perhaps also the art of today. Democratic culture is not an unattainable high ideal, nor is it, *pace* Tom Shepherd and Francesco Corti, synonymous with debased quality. Rather, it should be an essential part of a wider political democracy. A community of self-governing citizens, a *demos*, that understands, creates, and reinvigorates itself through culture. It is only when we have a cultural democracy, where everyone has the same capacity and opportunity to take part in cultural life, that we will have a chance of attaining a true political democracy.

References

ACE. *From Indifference to Enthusiasm.* London: Arts Council England, 2008.

Arnold, M. *Culture and Anarchy.* Ed. J. Dover Wilson.1869. Reprint, Cambridge: Cambridge University Press, 1961.

Bourdieu, P. *Distinction: A Social Critique of the Judgement of Taste.* Trans. R. Nice. London: Routledge, 1984.

Bunting, C. *Public Value and the Arts in England: Discussion and Conclusions of the Arts Debate.* London: Arts Council England, 2007.

Collings, M. *This Is Civilisation.* London: 21 Publishing, 2008.

Collini, S. *Common Reading.* Oxford: Oxford University Press, 2008.

Cook, M. "Review of *The March of Unreason*, by Dick Taverne. http://www.guardian.co.uk/books/2005/ apr/02/scienceandnature.highereducation (accessed 5 Dec 2008).

Corti, F. "Interview." *Herald,* 9 Oct 2008. http://www.theherald.co.uk/features/features/display.var.2458364.0.The_passion_of_Scotlands_new_maestro.php (accessed 5 Dec 2008).

de Montebello, P. "Art Museums, Inspiring Public Trust." *Whose Muse?* Ed. J. Cuno. Cambridge, MA, and Princeton NJ: Princeton University Press and Harvard University Art Museums, 2004.

Eliot, T.S. *Notes Towards the Definition of Culture.* London: Faber and Faber, 1948.

Hannon, C. et al. *Video Republic.* London: Demos, 2008.

Holden, J. *Publicly Funded Culture and the Creative Industries.* London: Arts Council England and Demos, 2007.

James, Clive. "A Point of View." BBC Radio 4, 2 Nov 2008.

Joss, T. *New Flow.*2008. http://www.missionmodelsmoney.org.uk/page.php?id=19 (accessed 23 Nov 2008).

Kovar, S. Review of *John Stuart Mill, Victorian Firebrand,* by R. Reeves. *The Liberal* (Summer 2008).

Latour, B. *Pandora's Hope: Essays on the Reality of Science Studies.*Cambridge, MA: Harvard University Press, 1999.

Lebrecht, N. *Is This a Record?* BBC Radio 3, 9 Aug 2008.

Levine, C, *Provoking Democracy: Why We Need the Arts.* Oxford: Blackwell, 2007.

Matarasso, F. "Whose Excellence?" *Arts Professional,* 2 Jun 2008.

McMaster, B. *Supporting Excellence in the Arts: From Measurement to Judgement.* London: Department for Culture, Media and Sport, 2008.

Moore, M. *Creating Public Value: Strategic Management in Government.* Cambridge, MA: Harvard University Press, 1995)

Perry, G. "I Want to Make a Temple." *RSA Journal,* (2008, Autumn).

Postman, N. *Amusing Ourselves to Death: Public Discourse in the Age of Show Business* (London: Methuen, 1986).

Robinson, K. *All Our Futures,* 1999. http://www.culture.gov.uk/PDF/naccce.PDF (accessed 5 Dec 2008).

Ross, A. *The Rest Is Noise.* London: Fourth Estate, 2008.

Seabrook, J. *Nobrow: The Culture of Marketing and the Marketing of Culture.* London: Methuen, 2000.

Sennett, R. *The Craftsman.* London: Allen Lane, 2007.

Veblen, T. *The Theory of the Leisure Class.* 1899. Reprint, Oxford: Oxford's World Classics, 2007.

Williams, R. "The Arts Council." *The Political Quarterly* 50, no 2, (1979, April).

Williams, R. *Keywords.* 3rd ed. London: Fontana, 1988.

Zuboff, S., and J. Maxmin. *The Support Economy: Why Corporations Are Failing Individuals, and the Next Episode of Capitalism.* London: Penguin, 2004.

Notes

1. "Hopes and Fears for Art," 1882, quoted in M. Arnold, *Culture and Anarchy,* ed. J. Dover Wilson (1869; rpt., Cambridge: Cambridge University Press, 1961).
2. N. Lebrecht, *Is This a Record?* BBC Radio 3, 9 August 2008.
3. R. Williams, *Keywords,* 3rd ed. (London, Fontana, 1988).
4. T. S. Eliot, *Notes Towards the Definition of Culture* (London: Faber and Faber, 1948).
5. Ibid.
6. Ibid.
7. C. Hannon et al., *Video Republic* (London: Demos, 2008).

8. See www.artscouncil.org.uk/publications/publication_detail.php?sid=4&id=609&page=4 (accessed 5 Dec 2008).
9. B. McMaster, B., *Supporting Excellence in the Arts: From Measurement to Judgement* (London: Department for Culture, Media and Sport, 2008).
10. Ibid.
11. Ibid.
12. J. Seabrook, *Nobrow: The Culture of Marketing and the Marketing of Culture* (London: Methuen, 2000).
13. Lebrecht, *Is This a Record?*
14. F. Corti, "Interview," *The Herald*, 9 Oct 2008. http://www.theherald.co.uk/features/features/display.var.2458364.0.The_passion_of_Scotlands_new_maestro.php (accessed 5 Dec 2008).
15. See, for example, Margaret Cook's review of Dick Taverne's book *The March of Unreason*, http://www.guardian.co.uk/books/2005/ apr/02/scienceandnature.highereducation (accessed 5 Dec 2008).
16. M. Collings, *This Is Civilisation* (London: 21 Publishing, 2008), p. 76: "Religion's depths, its beauty, its consolations, and answers to questions that can't be answered in any other way—this is now the realm of art. Art has that civilising possibility, it proposes the same resistance to chaos that religion used to have and was in fact invented for."
17. C. Levine, *Provoking Democracy: Why We Need the Arts* (Oxford: Blackwell, 2007).
18. See www.civitas.org.uk/blog/2007/07/the_philistines_are_upon_ us.html (accessed 5 Dec 2008).
19. Corti, "Interview."
20. Quoted in A. Ross, *The Rest Is Noise* (London: Fourth Estate, 2008).
21. See http://thinkexist.com/quotes/marcel_duchamp/ (accessed 5 Dec 2008).
22. See R. Sennett, *The Craftsman,* (London: Allen Lane, 2007).
23. See www.forum-arts.ch/page5.php (accessed 18 Nov 2008).
24. Ibid.
25. T. Veblen, *The Theory of the Leisure Class* . (1899; rpt., Oxford: Oxford's World Classics, 2007).
26. P. Bourdieu, *Distinction: A Social Critique of the Judgement of Taste,* trans. R. Nice (London: Routledge, 1984).
27. Ross, *The Rest Is Noise.*
28. Latour, *Pandora's Hope: Essays on the Reality of Science Studies* (Cambridge, MA: Harvard University Press, 1999).
29. Ibid.
30. Arnold, *Culture and Anarchy.*
31. Ibid.
32. Postman, *Amusing Ourselves to Death.*
33. Lebrecht, *Is This a Record?*

34. Latour, *Pandora's Hope.*
35. See www.artscouncil.org.uk/audienceinsight/ (accessed 5 Dec 2008).
36. S. Zuboff and J/ Maxmin, *The Support Economy: Why Corporations Are Failing Individuals, and the Next Episode of Capitalism* (London: Penguin, 2004).
37. Ibid.
38. R. Williams, "The Arts Council," *The Political Quarterly* 50, no 2 (1979, April).
39. See, for example, www.demos.co.uk/publications/ democratisingengagement (accessed 5 Dec 2008).
40. F. Matarasso, "Whose Excellence?" *Arts Professional,* 2 Jun 2008.
41. S. Kovar, review of *John Stuart Mill, Victorian Firebrand,* by R. Reeves, *The Liberal* (2008, summer).
42. See www.artscouncil.org.uk/publications/publication_detail.php?sid=4&id=609&page =4 (accessed 5 Dec 2008).
43. See www.artscouncil.org.uk/publications/publication_detail.php?rid=0&sid=&browse= recent&id=626 (accessed 5 Dec 2008).
44. "The Newly Established Arts Council of Great Britain (formerly c.e.M.A.) Its Policy and Hopes," broadcast on the BBC, 8 July 1945.
45. T. Joss, *New Flow,* 2008, available at http://www.missionmodelsmoney.org.uk/page.php ?id=19 (accessed 23 Nov 2008).
46. Quoted in S. Collini, *Common Reading* (Oxford: Oxford University Press, 2008).
47. ACE, *From Indifference to Enthusiasm* (London: Arts Council England, 2008).
48. C. Bunting, *Public Value and the Arts in England: Discussion and Conclusions of the Arts Debate* (London: Arts Council England, 2007).
49. See ACE, *From Indifference to Enthusiasm.*
50. Clive James, *A Point of View,* BBC Radio 4, 2 Nov 2008.
51. See www.pisa.oecd.org/knowledge/summary/e.htm.
52. Kovar, review of *John Stuart Mill.*
53. See K. Robinson, *All Our Futures,* 1999. http://www.culture.gov.uk/PDF/naccce.PDF (accessed 5 Dec 2008).
54. See M. Moore, *Creating Public Value: Strategic Management in Government* (Cambridge, MA: Harvard University Press, 1995).
55. P. de Montebello, "Art Museums, Inspiring Public Trust," *Whose Muse?,* ed. J. Cuno (Cambridge, MA, and Princeton NJ: Princeton University Press and Harvard University Art Museums, 2004).
56. G. Perry, "I Want to Make a Temple," *RSA Journal,* (2008, Autumn).

28

By Design

Bill Cope & Mary Kalantzis

We have in recent times heard much talk of the role of intangible cultural and epistemic factors in economic growth and social development—from the discourses of knowledge management and innovation at the micro-level to those of the creative economy and the knowledge society at the macro-level. This chapter explores the principles and practices of design pivotal to these arguments through twelve interconnected propositions.

The first third of this chapter (sections 1–4) examines the foundation of design as a process of meaning and making, as a process of human agency. It begins with definitions, then goes on to speak of the characteristics of our times that add particular loadings onto the notion of design-as-agency. The middle third (sections 5–8) explores the specifics of Design (with an upper-case 'D') as a discipline, and necessary tendencies toward interdisciplinary and even meta-disciplinary understandings of Design as a domain of knowledge and practice. The last third of the chapter (sections 9–12) develops a future-oriented agenda for the Design vocations, including a taxonomy of Design practices, elements of a Design grammar and, a method to read Design's programs.

1/ "Design"

What is this phenomenon, Design? "Design" has a fortuitous double meaning:

Morphology

On the one hand, design denotes something intrinsic to any found object—

inherent patterns and structures irrespective of that object's natural or human provenance. Things have designs. Design is morphology. This is design, the noun.

Agency

On the other hand, design is an act of conception and an agenda for construction. This meaning takes the word back to its root in the Latin word, *designare* or "to mark out" (Terzidis 2007). Design is a certain kind of agency. This is design, the verb.

We can make this duality of meaning work for us to highlight two integral and complementary aspects of design. In this paper, we are primarily interested in the second meaning, the stuff of human agency, how people "do design."

2/ Transformation

What do we do when we do design? The narrative of design-as-agency runs like this:

(Available) Designs

We live in a world of designs, available to us in the form of our cultural and technical heritage—found natural and human-made objects in our world of everyday experience, in the plans and interpretations of focused and specialized areas of knowledge, in situated actions and social processes. Designs are available to us as semantic resources, at once meanings in the world (intrinsic "sense") and meanings for the world (meanings we ascribe to the world in "sense making"). Meanings present themselves as if they were inherent to tangible objects, architectures, landscapes, social processes, human relationships, and cultural forms. We also give meanings to these things, varied according to the peculiarities of our life-formed perspectives, the focal points of our attention and our motivating interests.

Designing

Using the semantic resources of available designs, we engage in acts of designing. When we do, we never simply replicate available designs. We always rework and revoice the world as found. When language or imagery or space-making are understood to be design processes, each act of meaning merely reworks available design resources. However, in another sense, no two stretches of several hundred words, and no two photographs, no two built structures, even the seemingly most predicable or clichéd, are ever quite the same. Designing (of meanings, objects, spaces) always involves an injection of the designer's guiding interests and cultural experiences, their subjectivity and identity.

(The re-)Designed

The process designing, of making a meaning in the world, leaves tangible and intangible traces—a linguistic utterance, an image, a space, an object, a structure.

As the design narrative draws to a momentary close, the world has been transformed, in no matter how small a way. Indeed, for having been through this transformation, neither the designers nor their world will ever be quite the same again. The redesigned is returned to the world, and this return leaves a legacy of transformation. The redesigned joins the repertoire of available designs and so provides openings for new design narratives (Cope and Kalantzis 2000; Kress 2000, 2009).

We are not only talking about the design trades here, the architect or the advertisement creator or the fashion designer. We are also conceiving design as a foundational paradigm for representation and action. Let's consider a young child learning to write, also a design process: representing a meaning via the peculiar visual and linguistic modes of literacy. We could, and in fact in school we mostly do, present the child with available designs, not anticipating transformation—hence phonics drills, spelling lists, grammatical rules and the other formalisms of textual meaning. Take, however, the following example of what might be regarded as proto-writing. A child sits on his father's lap, and this is what he says: "Do you want to watch me? I'll make a car . . . got two wheels . . . and two wheels at the back . . . and two wheels here. . . . This is a car." Kress reads this double sign to consist of circles signifying "wheel" and wheels signifying "car": "Circles are apt forms for meaning wheels . . . and wheels here is a metaphor for car." These are what he calls "motivated signs": "it is the interest of the sign-maker at the moment of making the sign that leads to the selection of the criteria for representing . . .—'wheel-ness' and 'car-ness'" (Kress 2003: 42–43).

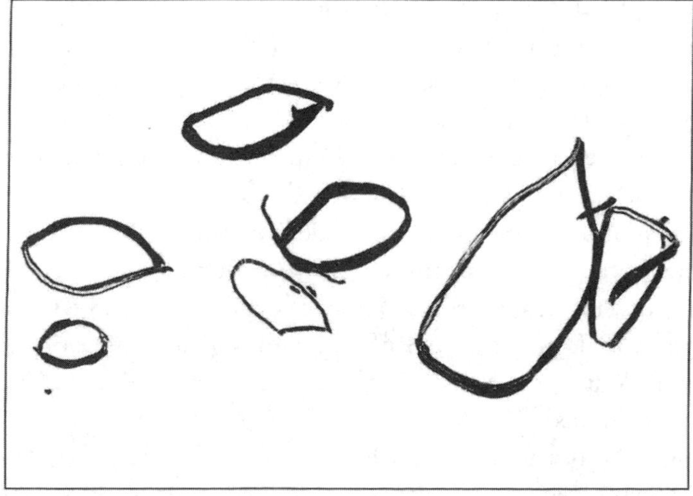

Figure 4.1 'This is a car'

By recognizing this method as a design process, we grant agency to a young sign-maker undertaking a piece of work. This work might not, however, be noticed as early writing in a literacy learning context where a teacher or a parent is anxious to tell young learners conventional formal literacy things that they do not yet know. However, in a designerly perspective, the available resources for meaning are to be traced to what the child does know and can do as a designer, grounded in their life experiences of cars, wheels, circles, and drawing. The child designs. The result is a proto-written, (re)designed meaning. Designing is this ordinary, and this extraordinary.

Design is never simply an instantiation of received "grammars," derived perhaps from what might at times seem to be the stable disciplinary rules of technology, semantics, or aesthetics. It is always and necessarily a process of transformation, and as such is an engine of change. Design is of course stabilized by the fact that we derive patterns of meaning and programs of action from structures of meaning that often appear rule-like in their persistent, at times insistent, presence in the world. It is also stabilized in the traces we leave in the redesigned. However, design is also and necessarily an act of re-voicing, re-working, re-meaning.

3/ Agency

Why should we conceive design this way? Why is agency so important now? More particularly, why now?

We are in a moment of radical reframing in the logic of modernity. In an earlier modernity, Fordist enterprises were run by line management. Bosses bossed, their orders passed down chains of command and control. Markets were sites of mass consumption of generic products. "Any colour you like as long as it is black," said Henry Ford, presuming to know what was best for all consumers, uniformly. Mass media provided a limited range of informational and cultural options through a few communications channels. We watched movies and sitcoms. We read novels, drawn vicariously into a voyeuristic relationship with narrative. Teachers taught and learners got their answers right (or failed). Government leaders commanded, reaching at worst for the governmentalities of communism and fascism (Adorno, Frenkel-Brunswik, Levinson, and Sanford 1950), and at best creating the "repressive tolerance" (Marcuse 1965/1969) of bureaucratic "welfare" states. Their citizens, by and large, complied.

We might call our times an epochal shift in the balance of agency. Here are some symptoms of change: In workplaces of the developed world at least, the command structures of Fordism are being replaced by the self-regulating compliance processes of post-Fordism: self-managing teams and the requirement that every worker personifies vision, mission, and culture of the organization (Cope and Kalantzis 1997). Mass consumer markets are being replaced by mass customization (Pine 1999) and the logic of niche marketing. Even the inner logic of the commodity is changing, now more open to variable designs in which "prosumers" (Toffler 1980) contribute as partners in the design process. Then there is the widespread appearance of that entirely new kind of artifact, the product with a configurable and reconfigurable interface (Krippendorf 2006: 9). As a consequence, no two computer desktops look the same and neither do two configurations of iPhone applications. Instead of the mass media, we have infinitely configurable new media. The top 40 have all but gone as a cultural force; people now make their own play lists for their iPods, and no two playlists are the same. Instead of the handful of network television channels, we have thousands of cable and satellite and millions of online video options—serving any number of interests and identities. We can create our own viewing programs, cut our own viewing angles on interactive TV, even make our own television show and broadcast it to an audience of maybe a handful or maybe millions. For narrative pleasure, we play video games, now a bigger industry than the movies, in which we are a character and can determine in part the ending of the narrative. Teachers, meanwhile, find themselves teaching the students of generation 'P'—for "participatory" (Jenkins 2006)—impatient with being fed someone's facts and theories, requiring instead engagement with their identities and experiences, and space to be knowledge makers themselves, observing facts, building theories, and connecting generalizations with the particularities of their own life worlds (Kalantzis and Cope 2008b). Furthermore, instead of powerful central states telling citizens what is good for them, we have increasingly participatory politics, which only works when built from the ground up: from within local communities, interest groups, professional organizations, workplaces, affinity groups, and knowledge communities. As for governance, who governs the World Wide Web? Nobody in the conventional spaces of government because it, like so many of today's emerging spaces, is self-governing—and, in any event, beyond the jurisdiction of any government as governments are conventionally understood. Even the heritage social patterns of design agency are deeply disrupted. These used to work across dichotomies of designer/consumer, actor-artist/audience, writer/reader. We are all users now. These are just some of the shifts indicative of what we are calling an epochal shift in the balance of agency.

We make this claim, for the moment at least, as somewhat disinterested historical observers. We might note, simply as an aside, that these changes are not always or necessarily for the good. For better but at times for worse, this shift in the balance of agency may underwrite a social order that remains rife with injustices. The new regime may at times be a site of post-Fordist hyperexploitation, rampant consumerism, narcissistic identity formation, and neoliberal renunciation of government and regulation (Harvey 2005). Nevertheless, whatever one's commitments, this much is clear: agency now counterbalances top-down power. It means that for every new development we might judge to be a travesty, we are also presented new openings for redress.

In the design vocations as conventionally understood, this broader drift in the balance of agency has been evidenced in recent times in a turn away from the heroic design personality of an earlier modernity. Take that archetypical command personality, Howard Roark, modern architect and towering individual in Ayn Rand's hyper-capitalist novel, *The Fountainhead* (Rand 1952/1996). At the vanguard of unadorned modernism, he stands alone against the world, unwilling to compromise his designs, and for his singularity of purpose, he triumphs. In almost the same moment, anti-capitalist Mexican artist Diego Rivera was painting the heroes of modernity into the murals of the Rockefeller Center in New York. Looking over the mighty works of modern man—the cities, the bridges, the industrial landscapes whose horizons are punctured by smokestacks—stand the heroic engineer, the heroic architect, the heroic intellectual, the heroic political leader, the heroic gang-supervisor, and his Rockefeller patrons also hoped, the heroic capitalist. Rivera was removed from the job when it became obvious that among the faces of the heroes was a likeness of Lenin. Notwithstanding twentieth-century sensitivities to their ideological differences, Roark and Lenin were equally command personalities, and in that sense at least substitutable in the tableau of modernism. Both left and right, in their time, lionized command personalities. Indeed, for every command personality, there had to be a multitude of unquestioning functionaries. Upon their compliance, the system depended. The ideal citizen of the central state was compliant; the ideal worker of the capitalist or communist industrial enterprise was compliant; the ideal learner in the classroom of disciplined knowledge was compliant; the ideal consumer impassively consumed generic products in an act of compliance, products designed by those who by virtue of their designing vocation, must know best. Command personalities like these are now an anachronism.

Today's design workers are more modest than Howard Roark in their aspirations. They are more respectful of users, more sensitive to their differences and more attentive to the knowledge their experiences may bring—hence the principles and practices of "participatory design" and "user-centred design" (Krippendorf 2006: 40). The archetypically insensitive technological know-all and the arrogantly superior artist are being progressively displaced by an emergent design democracy that turns the designer into conversationalist, facilitator, mentor, and pedagogue—so destabilizing a legacy of self-understanding whereby the designer presents himself (let's assume that our designer here is a man) as aesthete and technocrat. For sure, expertise and aesthetic commitment remain of enormous value, but the centre of gravity in an ideal relationship to users has shifted.

After this brief excursion, we now return to our foundational propositions about design. Design is inevitably transformative. Design cannot but change the world. However, in this moment of a general shift in the balance of agency, design has an even bigger role to play. Grasping its significance in this moment—its significance not only in the design trades as historically understood, but in all meaning-making—design could become a basis for new lines of enablement, even processes of emancipation. Upon this foundation we can perhaps begin to position ourselves as interested parties.

4/ Difference

However, before we express our interest, we need to observe one particularly important consequence of the shift in the balance of agency. By opening so much new scope for agency in spaces that were previously structured as sites of compliance, opportunities emerge for the flourishing of differences. Indeed, broadening the scope for agency fuels a dynamic of divergence, in which agents actively make themselves and their things more distinctly different. The convenient aspirations to sameness and the pressures to acquiesce and conform in an earlier era, suddenly become anachronistic—mass consumer uniformity gives way to a myriad of niche markets; nationalistic (and at times racist) identities give way to a necessary global-local cosmopolitanism; mass broadcast media give way to individuals who construct their own, invariably peculiar take on the world across an uncountable number of new media spaces. In other words, individuals use media according to their identity and according to their proclivity.

Here is a catalogue of differences that, in an earlier modernity, we tried to ignore, or if they could not be ignored, which we tried to put onto another side of a geographical border, or an institutional boundary, or a normative divide of "deviance": material differences (social class, locale, family circumstances); corporeal differences (age, race, sexual orientation, and physical, and mental abilities); and symbolic differences (culture, language, gender—an amalgam of gender and sexual identification, and identity). All of these differences in our late modernity present themselves as insistent demographic realities. They have become living and normative realities too, supported by an expanded conception of human rights (Kalantzis and Cope 2008b).

However, as soon as we begin to negotiate these differences in good faith, we find ourselves confounded by the categories. We discover that the gross demographic groupings used in the first instance to acknowledge differences are too simple for our needs. Instead, we find we are negotiating an inexhaustible range of intersectional possibilities—where gender and race and class meet, for instance. We face real-world specificities which confound generalizations about people who formally fit the ostensible categorical norm. In fact, if one takes any one of the categories, one finds that the variation within that group is greater than the average variation between groups. There are no norms. Rather, you find yourself in the presence of differences which can only be grasped at a level that defies categorization: different life narratives (experiences, places of belonging, networks), different personae (affinities, attachments, orientations, interests, stances, values, worldviews, dispositions, sensibilities); and different styles (epistemological, learning, discursive, interpersonal). The gross demographics might tell of larger historical forces, groupings and movements; they don't tell enough to provide a sufficiently subtle heuristic device or guide for our everyday interactions. For history's sake, we need to do the gross demographics, but also a lot more.

The rebalancing of agency in our epoch brings with it a shift away from a fundamental logic of uniformity in an earlier modernity to a logic of difference. Furthermore, we do not just have difference as a found object, legacies of lived experience that we can at last recognize. There is also today a tendency to divergence or to become more different. Here is one of the great paradoxes of what is also an era of globalization, when we are undoubtedly becoming more closely interconnected in many respects: communications, media, trade, travel, capital flows, ideas flows (Steger 2008). We also live in a time when the scope for agency allows us to make ourselves more different. What is more: because we can, we do. Take, for instance, the rainbow of gender identifications and expressions of sexuality in the newly plastic body; or the shades of ethnic identity and the juxtapositions of identity that challenge our inherited conceptions of neighborhood; or the locale that highlights its peculiarities to tourists; or the bewildering range of products anticipating any number of consumer identities and product reconfigurations by consumers themselves.

After the sociological description, we may choose to add a layer of agenda-implying interest. Differences have historically been overlaid by patterns of injustice (Fraser 2008; Fraser and Honneth 2003). Recognition of differences requires some acknowledgment of injustice, and acknowledgment suggests redress. Greater scope for agency suggests autonomous spaces to be different and to diverge. Agency plus difference/divergence prefigure a newly expanded conception of human rights.

For the design vocations, this new regime of difference and divergence means having to negotiate our differences in design teams and to recognize the virtue of difference in collaborative work. It means recognizing the differences amongst users as we create spaces, objects, and meanings that work the way they should for a particular kind of user or work differentially for more than one type of user. Inclusive design does not have to mean "universal design" as if one might magically find one size that finally fits all differences, one approach for all, a lowest common denominator, a design reduction. Moreover, we cannot allow that the hypothetical user is conceived in the singular, as if we could project a functionally determined and thus uniform relationship to an artifact (Suchman 2007: 188). Rather, our design work must set out to create many approaches for the many, replete with alternatives, layerings and infinities anticipating initially indeterminate possibility.

5/ Design Discipline

Having laid the ground of design, what of the design vocations? What discipline underlies these vocations? We use the word *discipline* advisedly here, to denote a focus of attention and accumulation of expertise that distinguishes activities that we do routinely because we are human, and because in our ordinary existences we mean, we make, we act. Design is in our natures. Now we mean "design" to denote a certain kind of additional work and extra effort. What happens when we discipline this aspect of our nature?

To make a necessary distinction, we will start speaking of two layers of meaning in the word design, design with lower case 'd' which we cannot help but do; and Design with a upper case 'D,' or disciplined design. What justifies the shift in case?

Disciplinarity is generally considered to be constituted: as contents (architecture, industrial design, graphic design, engineering . . . and this list could get quite long, getting even longer nowadays); as methods (of conception and planning, of graphic representation, of quantification and calculation and the like); as concepts (vector, consultation, algorithm, user-analysis); as sites of apprenticeship (school subjects, university departments, internships, first jobs); as peer communities (workplaces, professional organizations, relationships with colleagues); and as modes of public communication (conferences, Web sites, journals, magazines, books, blogs). The manifest doing of these disciplinary practices means that one can recognize a design professional when one encounters one. They are the visible elements of vocation.

However, more subtly and profoundly, discipline is constituted by sensibilities of practice: an epistemic frame (peculiar ways of knowing, deeper than everyday casual experience—an architect knows a building in different and significant respects on a deeper level than an inhabitant or visitor can); a mode of discourse (engineers know things differently because they speak about them differently, with the semantic precision of technicality not found in the everyday or "natural language" practices of crossing bridges or working at computers); a way of seeing (Web designers see screens in different and in some senses more perspicacious ways than regular readers, based on navigational logic, layout, and underlying code-functions from which variable renderings can be achieved across different Web browsers and reading devices); a way of acting (a professional stance, an orientation, a demeanor, an ethics); and a kind of person (a professional identity, a person who feels and thinks and sees some part of the world or some aspect of the world with a particularly studied focus, interest, responsibility, even obligation).

6/ Interdisciplinary Design

Where, then, does Design sit amongst the other disciplines, the other sites of knowledge, apprenticeship, and professional community that exist in the world?

Nigel Cross speaks of three cultures of human knowledge and ability. The sciences study the natural world; the humanities, human experience; and Design, the artificial world. Their methods are distinctive: the sciences use controlled experiment, classification, and analysis; the humanities, analogy, metaphor, and evaluation; and Design, modeling, pattern formation, and synthesis. The values of each culture also vary: the sciences, rationality, neutrality, and a concern for "truth"; the humanities, subjectivity, imagination, commitment, and a concern for "justice"; and Design, practicality, ingenuity, empathy and a concern for "appropriateness." Amongst the three, the Designer is characteristically the doer, the maker, the technologist (Cross 2007: 18).

However, we increasingly find that these traditional delineations are being blurred. There are new, hybrid professions that cross science, the humanities, and Design—people working in digitized communications, artificial intelligence, information architectures, design management or interface design, for instance (Krippendorf 2006: v). There are also new imperatives in every area of Design requiring ever stronger integration of sciences and humanities with Design. Two of the bigger imperatives of our time are sustainable Design and inclusive-equitable Design. We cannot achieve these objectives unless we have the capacities and the will to move beyond our discipline groupings, in other words, to be interdisciplinary. We have to bring humanities, science, and Design together.

Interdisciplinary work is grounded in the historical practices of more than one discipline, and consciously crosses disciplinary boundaries. We need to become interdisciplinary for pragmatic reasons, in order to see and do things that cannot be seen or done adequately within the substantive and methodological confines of a single discipline—things as big these days as "sustainability," or "globalization," or "inclusion."

The deeper perspectives of disciplinary work need to be balanced with and measured against the broader perspectives of interdisciplinarity. Even more finely grained within-discipline views may prove all-the-more powerful when contextualised broadly.

Interdisciplinary approaches need to be applied for reasons of principle, to disrupt the habitual narrowness of outlook of within-discipline work, to challenge the ingrained, discipline-bound ways of thinking and acting that produce occlusion as well as insight. If the knowable universe is a unity, disciplinarity is a loss as well as a gain, and interdisciplinarity may in part recover that loss.

Interdisciplinary approaches also thrive in the interface of disciplinary and lay understandings. They are needed for the practical application of disciplined understandings to the actually existing world. They are the raw material of dialogue between Designers and their clients. Robust applied knowledge demands an interdisciplinary holism, the broad epistemological engagement that is required simply to be able to deal with the complex contingencies of a really integrated universe.

7/ Metadisciplinary Design

Metadisciplinary Design, however, is still to think of design in some senses too narrowly, as a received catalogue of Design vocations, albeit stretched at times to address new exigencies that beg interdisciplinarity. However, if we return to our earlier definition, it becomes evident that 'd'esign is bigger than this narrow concept.

There is design in all human meaning, something that happens in all our spatial, tactile, gestural, visual, audible, and linguistic doings. Everything we mean and do is a re-presentation: (available) designs => designing => (the re)designed. These doings and their tangible traces at once have morphologies (design the noun) and are the stuff of human agency (design the verb). This is design with the lower case 'd.'

When we take design to be a fundamental category of meaning making, Design must conceive of itself as a metadiscipline. We need the principles and practices of Design to become a central concern of every discipline. Then we may also find more people being designers than we ever imagined was possible. There are many more people who are Designers by profession (instructional Designers, organizational Designers, labor process Designers, information Designers, communications Designers, artists and curators as Designers). Indeed, in an era of participatory culture (Haythornthwaite 2009), we will find growing numbers of people who are Designers by persuasion but not profession: amateurs with specialist interests; energetically self-defining homemakers; people offering their Design capacities in the digital "commons" (Benkler 2006); or participants in peer-to-peer production (Bauwens 2005).

8/ Undesign

Design with a lower case 'd' is all. However, if design is all, the word is shorn of its clarity, its useful specificity. This is why we need two understandings of design, the 'd'esign that is in our natures and the disciplined Design that is sufficiently focused to be deserving of our recognition as Design, proper. However, where does 'd'esign end and Design begin?

"I like boring things," said Andy Warhol. Henri Bergson called disorder an order we cannot see. Venturi, Brown, and Izenour quote Bergson and Warhol in support of their project to learn from the "Strip" in Las Vegas. "The emerging Strip is a complex order. It is not the easy, rigid order of the urban renewal project or the fashionable 'total design' of the megastructure." By this, they mean to unmask the pretenses and insensitivities of modernist design. They wish to acknowledge as design the "honky tonk improvisations" and what might be regarded as "commercial vulgarities" (Venturi, Scott Brown, and Izenour 1977: 87, 52, 72).

Learning from Las Vegas deconstructs with dazzling intellectual flair the vernacular and commercial grammars of space, sign, and structure, uncovering neglected design features of things found pervasively in our everyday lives. The modernist, by comparison, seems to have wished to impose abstract principles upon an at times unappreciative public.

However, in the rush to relativism, do we have to abandon all principle? "Las Vegas's values are not questioned here." (Venturi, Scott Brown, and Izenour 1977: 6) If there are peculiar virtues written into Design principles and forms of action characteristic of good Design practice, what is Design's other? If Design can no longer be located exclusively in formal places of work and institutionally accredited spaces of work; if there are now so many amateurs doing Design work; and if Design is being done in professions that are new, hybrid, and not classically understood to be design vocations—then where is it and what is it?

We can start to know Design by its counterpoints. What is undesign, something less-than-Design, understood normatively? Undesign is when things are made or done that are thoughtless or glib, unreflective or unreflexive, disrespectful or prejudicial, intolerant or obstinate, resource profligate or environmentally damaging, narrowly self-interested or insensitively opinionated, or which are non- or dysfunctional. (Of course, there is designing in all these things, with the lower-case "d," the stuff in our natures, but for the moment we are trying to determine what is not deserving of the label "Design.")

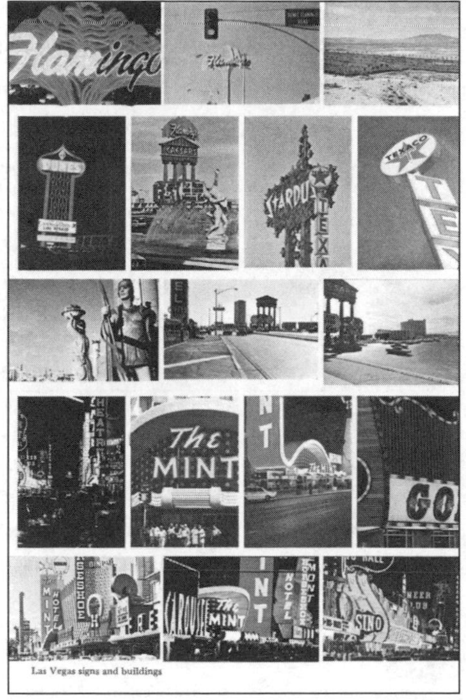

Las Vegas signs and buildings

(Venturi, Scott Brown, and Izenour, 1977: 62)

With their lower case 'd's, (available) designs, designing and the (re)designed are everything. Within this universe of design, we know undesign by its travesties; we know Design by its virtues.

9/ Design Virtues

To do something by Design, is to do it with a peculiar intensity of focus, in a designerly way. Design is premeditated, a series of extraordinarily focused stages of thinking and action: conceptualization, enactment, evaluation. Design is reflexive, aware of the range of its potential applications. Design is contextually aware—of its antecedents, of the scope of present needs, and of possible future consequences. Design is respectful, open to alternative perspectives and practices. Design is resource prudent. Design is functional, creating its objects for the world, and they are meant to be useable, useful, and enhance the quality of people's lives.

A lot of 'd'esign (meaning and making things), does not attain Design's ideals. With the lowercase "d," design is of our human natures; but like other traits in our natures, we can also develop a normative agenda by extrapolating from the ordinary. From "is," we can move to "ought."

<table>
<tr><td>Experiencing
Identity work (experiencing the known)
For instance:
- connecting with experience
- being explicit about perspective
- articulating interests, motivations, agendas, purposes
- being self-aware of representational modalities
- metacognizing, or thinking about one's thinking in order to think with greater acuity</td><td>Analyzing
Explanatory work (analysing functionally)
For instance:
- establishing cause and interpreting effect
- parsing structure and analyzing functions
- reasoning deductively and inductively
- specifying plans, projects, programs
- figuring solutions in relation to problems formulated</td></tr>
<tr><td>Empirical work (experiencing the new)
For instance:
- observing methodically
- measuring, recording, describing
- experimenting, testing
- consulting, interviewing, surveying
- researching similar or parallel cases</td><td>Critical work (analyzing critically)
For instance:
- interrogating goals, agendas, biases
- exploring scenarios and conjecturing options
- creating narratives and modeling alternative trajectories
- hypothesizing, conjecturing, predicting
- evaluating outcomes
- inferring and articulating ethics</td></tr>
<tr><td>Conceptualizing
Categorical work (conceptualizing by naming)
For instance:
- defining terms
- creating visual keys
- identifying physical elements
- classifying</td><td>Applying
Pragmatic work (applying appropriately)
For instance:
- implementing according to plan
- making things work, mechanically and humanly speaking
- engaging stakeholders
- realizing solutions</td></tr>
<tr><td>Theoretical work (conceptualizing with theory)
For instance:
- generalizing, linking concept to concept
- quantifying and calculating
- modeling, diagramming
- paradigm building
- ... and other abstracting</td><td>Transformative work (applying creatively)
For instance:
- creating hybrid, interdisciplinary solutions
- risk taking
- exploring hard-to-foresee, lateral transfers
- putting things to unanticipated use
- challenging paradigms</td></tr>
</table>

10/ Design Practices

So what do we do to instantiate Design virtues? How does one do more insightful and trustworthy, in short more virtuous, Design? Following is a Design schema, a taxonomy of Design processes. These (see p. 592) are some of the kinds of activities a Designer performs to do "Design" and to do it well:

This layout represents a repertoire of Design work practices and a set of pedagogical tags with which to mark up the range of learning engagements undertaken by Design initiates (Kalantzis and Cope 2005; Kalantzis and Cope 2008b). It suggests that improved Design may be achieved by expanding one's repertoire of Design practices and that better Design involves a balance of complementary Design practices, or a justifiable imbalance (related to specific defined purposes, specific agendas, or the subsequent integration of a narrowly focused practice into a wider program).

Each of these Design processes is a way of thinking and seeing, an orientation to the world, an epistemological take, a sensibility or way of feeling, and for shorter or longer moments in time, a way of living. These Design processes come in no necessary order. One may do some and not others in a particular Design practice. The distinctiveness of a Design practice may be identified by "marking up" or "tagging" the stages in the act of Design, thus bringing to explicit attention the weighting and sequence of Design moves. What is the mix and match? What are the transitions from one Design orientation to the next? These transitions might be likened to key shifts in music or mood swings in psychological affect. Indeed, those elusive concepts, innovation and creativity, may even occur in the moments of key change or mood swing, more so than in routine practice.

11/ Design Grammars

What, then, are the raw representational materials of Design? To answer this question, we will start with 'd'esign. These are the modes of meaning:

- *Written Language:* writing (representing meaning to another) and reading (representing meaning to oneself)—handwriting, the printed page, the screen.

- *Oral Language:* live or recorded speech (representing meaning to another); listening (representing meaning to oneself).

- *Visual Representation:* still or moving image, sculpture, craft (representing meaning to another); view, vista, scene, perspective (representing meaning to oneself) (McGinn 2004).

- *Audio Representation:* music, ambient sounds, noises, alerts (representing meaning to another); hearing, listening (representing meaning to oneself).
- *Tactile Representation:* touch, smell, and taste: the representation to oneself of bodily sensations and feelings or representations to others who "touch" them physically. Forms of tactile representation include kinaesthesia, physical contact, skin sensations (heat/cold, texture, pressure), grasp, manipulable objects, artifacts, cooking and eating, aromas.
- *Gestural Representation:* movements of the hands and arms, expressions of the face, eye movements and gaze, demeanors of the body, gait, clothing and fashion, hair style, dance, action sequences (Scollon 2001), timing, frequency, ceremony and ritual. Here gesture is understood broadly and metaphorically as a physical act of signing (as in "a gesture to . . ."), rather than the narrower literal meaning of hand and arm movement. Representation to oneself may take the form of feelings and emotions or rehearsing action sequences in one's mind's eye.
- *Spatial Representation:* proximity, spacing, layout, interpersonal distance, territoriality, architecture/building, streetscape, cityscape, landscape (Kalantzis and Cope 2008a).

In an earlier modernity, the modes drifted apart. At times they were even dragged apart. The radical iconoclasts of Protestantism tore the stained glass windows and the statues out of churches in order to force upon supplicants an unmediated relationship with the Word. The printing press required different processes for text (the offset letterpress) and image (engraving) (Cope 2001), and so, if image and text were to be in the same book, for the most pragmatic of manufacturing purposes they were best separated into different sections. In later modernity and in not dissimilar spirit, the theorists of the "language turn" assumed linguistic meaning was all, or at least primary. In addition, modernist architects, for their part, thought meaning could be expressed in a language of pure space. "Circulation in a big railroad station required little more than a simple axial system from taxi to train, by ticket window, stores, waiting room and platform—all virtually without signs. Architects object to signs in buildings: 'If the plan is clear, you can see where to go'"(Venturi, Scott Brown, and Izenour 1977: 9).

In our more recent modernity, the modes have been coming back together, merging into a renewed and at times new grammar of multimodality (Kress 2009). This trend can in part be attributed to the affordances of the new communications environment. As early as the mid-twentieth century, photolithography put image and text conveniently back onto the same page. Then, since the mid-1970s, digitized communications have brought image, text, and sound together into the same manufacturing processes and transmission media. The simple fact that these are all now made of the same stuff means we can more easily put them together, and because it is easier, we do, in complex overlays of text and image for instance—the magazine, the Web page, or the page of writing in which tools of spatial design that once were the exclusive preserve of typesetters have been made available to the masses. Anyone can make videos now and broadcast them to the world, bringing together image and gesture and sound and written-linguistic overlays (video has much more writing "over it" than was the case in the initial days of television). Meanwhile, architects have come to recognize that "complex programs and settings require complex combinations of media beyond the purer architectural triad of structure, form, and light at the service of space. [Airports, for instance] suggest an architecture of bold communication rather than one of subtle expression. . . . [T]he strip is virtually all signs" (Venturi, Scott Brown, and Izenour 1977:9). Our buildings have more and more writing on them.

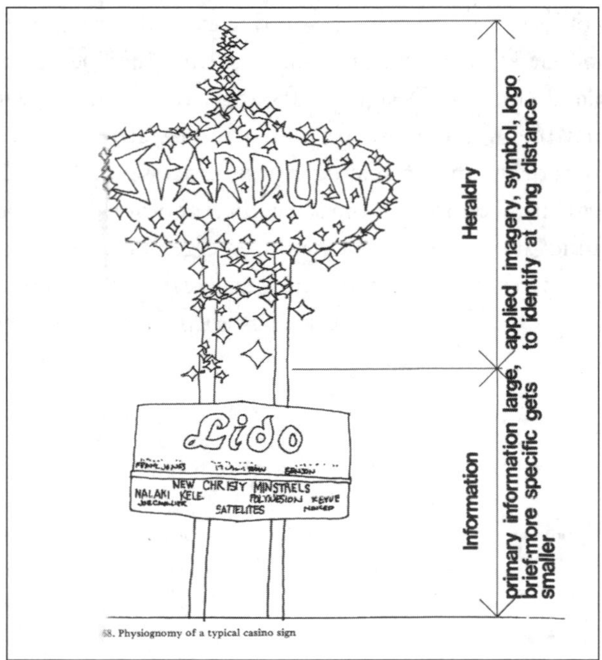

58. Physiognomy of a typical casino sign

(Venturi, Scott Brown, and Izenour 1977, p. 67)

Just as the objects of Design are in a state of flux, so are Design's modalities, its working tools of representation, communication, visualization, and imagination. Digitization of text, sound, and still and moving image has spawned new practices of modeling and simulation, of prefiguring the real in the virtual. It has also created the virtual as a Design end-in-itself.

The result is a new multimodality and an imperative to cross between modes. Designers need to be able to "do" professional Design discourse, as they speak and write their way through complex collaborations with co-designers. They need to be able to "do" visualization as they image design alternatives and picture them into reality. They need to be able to represent spatial realities, prefiguring the three-dimensional through the two-dimensional realm. They need to be able to turn plans into tactile artifacts, manipulable objects, architectural spaces, and navigable landscapes. Today's media inventions have become the mothers of Design necessity, and this innovation is not simply for innovation's sake. We need the new media for the most practical of reasons: the more frequent requirement to document for the purposes of planning and project management, regulation and compliance, risk assessment and risk management, project specification and contractual clarity. This dynamic reflects another aspect of the trend to interdisciplinarity.

This momentous shift toward multimodality in the world of 'd'esign as well as the practices of Design suggest that we need to expand our representational repertoires. Multimodality, though, is not to subtract from those legacy representational practices that define the Design professions. Rather it suggests an additive process in which the grammars of particular modes are integrated into a more expansive multimodal grammar. It also suggests processes of synaesthesia, or mode switching, representing designs in one way, then another. Once more, those elusive concepts, innovation and creativity, may emerge in the "key" or "mood shifts" from one mode of representation to another because, between and across the modes, there are profound parallels, as well as deep differences. Here are some of the parallels:

Dimensions of Meaning	*Modes of Meaning* >	**Linguistic**	**Visual**	**Spatial**
Representational: What do the meanings refer to?	**Participants:** Who and what is participating in the meanings being represented?	Naming words, which make sense in terms with their relationships with nearby words and contextual pointers.	Naturalistic and iconic representations, visibly distinguishable contrasts.	Objects in relation to nearby objects, part/whole relationships, contrasts.
	Being and Acting: What kinds of being and acting do the meanings represent?	Processes, attributes and circumstances.	Vectors, location, carriers.	Placement, topography, scale, boundaries, location.

Social: How do the meanings connect the persons they involve?	**The Roles of the Participants in the Communication of Meaning:** How does the speaker/writer mean to draw the listener/reader into their meaning?	Participant relationships and vicarious observer relationships.	Perspective, focal planes of attachment or involvement.	More and less negotiable spaces: e.g., parks versus prisons.
	Commitment: What kind of commitment does the producer have to the message?	The kind of affinity meaning makers have to the propositions they are making, and the degrees of certainty they express—"modality."	Contextualisation, depth, abstraction.	Emphatic (fences, barriers), or less insistent spatial designs.
	Interactivity: Who starts the interchange, and who determines its direction?	Agenda setting, turn taking, topic control.	Eye contact, response.	Spatially determined interchanges: audiences by a theater, students by a classroom.
	Relations between Participants and Processes: How are the participants connected to each other and with the actions and states of being that are represented?	Agency, or transitivity, "nominalisation."	Agency as represented through vectors, eyelines, perspective.	Principles of layout.
Organizational: How do the meanings hang together?	**Mode of Communication:** What is distinctive about the form of communication, and what conventions and practices are associated with this form of communication?	Spoken or written language; a part of what is going on or representing what is going on; monologic or dialogic.	Still or moving images, two or three dimensional representation, representational versus interactive.	Architecture topography geography.
	Medium: What is the communication medium, and how does it define the shape and the form of the representation?	Physical medium, such as recorded or ephemeral speech.	Different media, such as oil painting versus photography.	Natural environment, building, Web site.
	Delivery: How is the medium used?	Intonation, stress, rhythm, handwriting, typing.	Brushstrokes, photographic film.	Construction, landscape.
	Cohesion: How do the smaller information units hold together?	Information structure, reference, omission, conjunction, wording.	Left/right, top/bottom, center/margins, framing, salience/gravitational pull.	Structural, aesthetic.

	Composition: What are the overall organizational properties of the meaning making event?	Genre, such as romance novel or doctor-patient conversation.	Genre, such as landscape photography compared to photojournalism.	Building or environment types.
Contextual: How do the meanings fit into the larger world of meaning?	**Reference:** What how do meanings point to contexts and contexts point to meanings?	Frame of reference, pointers, metaphor.	Frame of reference, foregrounding/backgrounding resemblance/metaphor.	Location, prominence, metaphor.
	Cross-reference: How do meanings refer to other meanings?	Intertextuality, hybridity.	Pastiche, collage, icon.	Motifs.
	Discourse: How does the whole of what I communicate say something about who I am in a particular context?	Primary and secondary discourses, dialects, register, orders of discourse.	Imagery.	Topography, architectonics.
Ideological: Whose interests are the meanings skewed to serve?	**Indication of Interests:** How does the meaning maker declare their interests?	Authorship, context and purpose of meaning.	Naturalistic or stylized images.	Symbolism, facades, portals, signs.
	Attributions of Truth Value and Affinity: What status does the meaning maker attribute to their message?	Assertions as to the extent of the truth of a message, declaring one's own interest, representing agency.	Realistic (e.g., scientific diagrams), versus heavily authored (e.g., artistic) images.	Spatial arrangements, such as of a courtroom compared to a park.
	Space for Readership: What is the role of the reader?	Open and closed or directive texts, anticipated and unanticipated readings.	Highly detailed panoramas versus propaganda.	Alternative ways of using a space, directive or allowing alternatives.
	Deception by Omission if not Commission: What is not said and what's actively one-sided or deceptive—deliberately or unconsciously?	Selectiveness in foregrounding and backgrounding, non-declaration or obscuring of interests.	Foregrounding and backgrounding, distortion, perspective.	"Front" and "back" spaces, public and private.
	Types of Transformation: How is a new design of meaning created out of available designs of meaning?	Extent of creativity, degree of self-consciousness of representational resources and their sources.	Extent of creativity, degree of self-consciousness of representational resources and their sources.	New or hybrid forms of spatiality: e.g., Web sites, food courts (Cope and Kalantzis 2009).

These parallels are the reason why we can describe a picture in words, or turn a novel into a movie, or turn a plan into a building. A verb in a sentence is like a vector in an image. The "given" and "new" in a sentence are like the left and right in an image, at least if the language spoken is written from left to right (Kress and van Leeuwen 1996).

However, there are also deep and intrinsic differences between the modes, the linear temporality of writing delineating a logic of causality, and simultaneity of the image underwriting a logic of location (Kress 2003). Therefore, the movie is so much like the novel and at the same time irreconcilably so unlike it (Kalantzis and Cope 2008a).

12/ Design Programs

Thus, these are Design processes that deploy Design grammars. Design programs acquire their distinctiveness from identifiable patterns in their Design processes and grammars.

Le Corbusier's modernism speaks thus:

> The incredible industrial activity of today . . . puts before our eyes every hour, either directly or through newspapers and magazines, objects of arresting novelty whose whys and wherefores interest, delight and disconcert us. All these objects of modern life end up creating a certain modern state of mind. . . . If we set ourselves against the past, we see that new formulas have been found that only need to be exploited and that, if we can break with routine, will bring real liberation from the constraints hitherto endured. There has been a revolution in construction methods. . . . [W]e finally understand in a flash the true and profound laws of architecture based on volume, rhythm, and proportion; the styles no longer exist. . . . If we set ourselves against the past, we see that the old codification of architecture, weighed down by forty centuries worth of rules and regulations, ceases to interest us; it is no longer our concern; there has been a revision of values; there has been a revolution in the conception of architecture. (Le Corbusier 1924/2007: 297, 304–305)

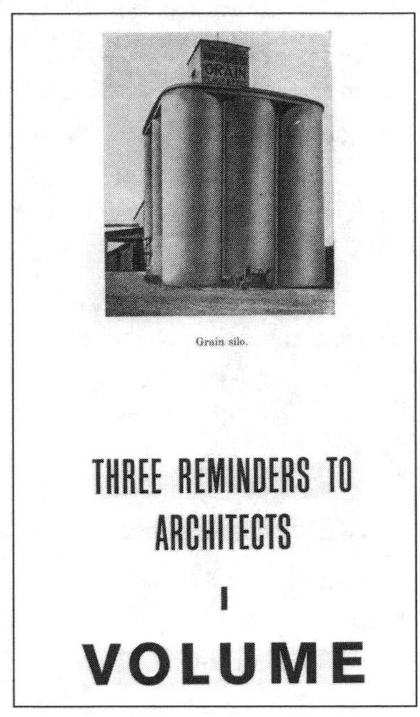

Grain silo.

THREE REMINDERS TO ARCHITECTS

I

VOLUME

(Le Corbusier 1924/(2007: 99)

Form follows function in the modernist lexicon, and the functions of industrial modernity are quintessentially expressed through the utilitarian architectonics of factories, grain elevators, airplanes, and automobiles. The logic of modernism's Design program, for better and at times for worse, uses some Design processes to unbalanced excess, skewing the overall program at times in such a way as to undermine the purported virtues of the program itself. For instance, using the terminology of Design processes that we introduced earlier in this chapter, modernism relies on categorical and theoretical (conceptualizing) work whose virtue is programmatic clarity, but which can also lead to dogmatism and doctrinaire insensitivity to context. It does a lot of analytical work (analyzing functionally), using the calculated engineering judgments to create new kinds of structure. Taken to excess, however, this kind of work relies on systems of formal reasoning disengaged from human and natural consequences; it creates systems of technical control without adequate ethical reflection; it elides means and ends; and it promotes a narrow functionalism, instrumentalism, or techno-rationalism. It also has a radical interest in transformation (application). Constructively, this approach may be viewed as an attempt to translate utopian agendas into practical reality. However, its occupational hazards are voluntaristic overconfidence that leads to a naive lack of pragmatism and a misreading of practical circumstances that produce failure.

In counterposition, Venturi, Brown, and Izenour's postmodernism speaks thus:

> Learning from the existing landscape is a way of being revolutionary for an architect. Not in the obvious way, which is to tear down Paris and begin again, as Le Corbusier suggested in the 1920s, but another, more tolerant way; that is, to question the way we look at things. The commercial strip, the Las Vegas Strip in particular . . . challenges the architect to take a positive, non-chip-on-the-shoulder view. Architects are out of the habit of looking nonjudgementally at the environment, because orthodox Modern architecture is progressive, if not revolutionary, utopian and puristic; it is dissatisfied with existing conditions. . . . But to gain insight from the commonplace is not nothing new: Fine art often follows folk art. . . . We look backward at history and tradition to go forward. . . . And withholding judgment may be used as a tool to make later judgment more sensitive. . . . Just as an analysis of the structure of a Gothic cathedral need not include a debate on the morality of medieval religion, so Las Vegas's values are not questioned here. (Venturi, Scott Brown, and Izenour 1977: 3, 6).

Postmodernism leans a different way to modernism, again for better as well as for worse, and again identifiable in its mix of the Design processes delineated earlier in this chapter. Postmodernism does a lot of identity work and empirical work (experiencing the now and the new). It is difficult to listen to voice, feel the sensual, recognize the embodied, frame the performative, account for the complex layers of the life world, explain the politics of identity, and understand the intuitive. For these virtues, postmodernism also encounters the occupational hazards of excessive subjectivism, the agnostic relativism of lived experience, and an impassive *c'est la vie* identity politics (Blackburn 2005). It also has at times a carefully empirical basis, reveling in the experience of moving into new and potentially strange terrains and deploying the processes of methodical observation, carefully regulated experimentation and cautiously sensitive reading of experience. Taken to one-sided excess, however, this dynamic leads to narrow empiricism devoid of big-picture analysis.

Thus, we can analyze the distinctiveness of Design programs by reading their characteristic moves, the Design processes they deploy, at times in an unjustifiably unbalanced way. Each Design process can be a site of occlusion as well as insight, particularly when isolated from other Design processes and taken to excess. Such an analysis may suggest at times that a more balanced program would serve us better for realizing Design virtues more completely.

Our Design projects today are more ambitious than ever. In fact, they are once again revolutionary because they must include universal agendas such as those of sustainability, agentive autonomy, and inclusion-in-difference. This inclusiveness demands that we develop an interdisciplinary repertoire of Design processes. It requires that we use an approach that moves comfortably between and across design grammars. It demands an expanding and balanced repertoire of Design processes. If we rise to these demands, for our troubles, might we be able to achieve a kinder, gentler modern?

References

Adorno, T. W., Else Frenkel-Brunswik, Daniel J. Levinson, and R. Nevitt Sanford. 1950. *The Authoritarian Personality.* New York: Harper and Brothers.

Bauwens, Michel. 2005. "The Political Economy of Peer Production." CTheory.

Benkler, Yochai. 2006. *The Wealth of Networks: How Social Production Transforms Markets and Freedom.* New Haven: Yale University Press.

Blackburn, Simon. 2005. *Truth: A Guide.* Oxford: Oxford University Press.

Cope, Bill. 2001. "New Ways with Words: Print and eText Convergence." Pp. 1–15 in *Print and Electronic Text Convergence,* vol. 2.1, *Technology Drivers Across the Book Production Supply Chain, from the Creator to the Consumer,* edited by B. Cope and D. Kalantzis. Melbourne: Common Ground.

Cope, Bill, and Mary Kalantzis. 1997. *Productive Diversity: A New Approach to Work and Management.* Sydney: Pluto Press.

———. 2000. "Designs for Social Futures." Pp. 203–234 in *Multiliteracies: Literacy Learning and the Design of Social Futures,* edited by B. Cope and M. Kalantzis. London: Routledge.

———. 2009. "A Grammar of Multimodality." *International Journal of Learning* 16: 361–425.

Cross, Nigel. 2007. *Designerly Ways of Knowing.* Basel: Birkhauser Verlag.

Fraser, Nancy. 2008. *Adding Insult to Injury: Nancy Fraser Debates Her Critics.* London: Verso.

Fraser, Nancy, and Axel Honneth. 2003. *Redistribution or Recognition: A Political-Philosophical Exchange.* London: Verso.

Harvey, David. 2005. A *Brief History of Neoliberalism.* Oxford: Oxford University Press.

Haythornthwaite, Caroline. 2009. "Participatory Transformations." In *Ubiquitous Learning,* edited by B. Cope and M. Kalantzis. Champaign, IL: University of Illinois Press.

Jenkins, Henry. 2006. *Fans, Bloggers and Gamers: Exploring Participatory Culture.* New York: New York University Press.

Kalantzis, Mary, and Bill Cope. 2005. *Learning by Design.* Melbourne: Victorian Schools Innovation Commission.

———. 2008a. "Digital Communications, Multimodality and Diversity: Towards a Pedagogy of Multiliteracies." *Scientia Paedagogica Experimentalis* XLV:15–50.

———. 2008b. *New Learning: Elements of a Science of Education.* Cambridge UK: Cambridge University Press.

Kress, Gunther. 2000. "Design and Transformation: New Theories of Meaning." Pp. 153–161 in *Multiliteracies: Literacy Learning and the Design of Social Futures,* edited by B. Cope and M. Kalantzis. London: Routledge.

———. 2003. *Literacy in the New Media Age.* London: Routledge.

———. 2009. *Multimodality: A Social Semiotic Approach to Contemporary Communication.* London: Routledge.

Kress, Gunther, and Theo van Leeuwen. 1996. *Reading Images: The Grammar of Visual Design.* London: Routledge.

Krippendorf, Klaus. 2006. *The Semantic Turn: A New Foundation for Design.* Boca Raton, FL: Taylor and Francis.

Le Corbusier. 1924 (2007). *Toward an Architecture.* Los Angeles, CA: Getty Publications.

Marcuse, Herbert. 1965 (1969). "Repressive Tolerance." In *A Critique of Pure Tolerance,* edited by R. P. Wolff, B. Moore, and H. Marcuse. Boston: Beacon Press.

McGinn, Colin. 2004. *Mindsight: Image, Dream, Meaning.* Cambridge, MA: Harvard University Press.

Pine, B. Joseph. 1999. *Mass Customization: The New Frontier in Business Competition.* Cambridge, MA: Harvard Business School Press.

Rand, Ayn. 1952 (1996). *The Fountainhead.* New York: New American Library.

Scollon, Ron. 2001. *Mediated Discourse: The Nexus of Practice.* London: Routledge.

Steger, Manfred B. 2008. *The Rise of the Global Imaginary: Political Ideologies from the French Revolution to the Global War on Terror.* Oxford UK: Oxford University Press.

Suchman, Lucy. 2007. *Human-Machine Reconfigurations.* Cambridge: Cambridge University Press.

Terzidis, Kostas. 2007. "The Etymology of Design: Pre-Socratic Perspective." *Design Issues* 23:69–78.

Toffler, Alvin. 1980. *The Third Wave.* New York: Bantam Books.

Venturi, Robert, Denise Scott Brown, and Steven Izenour. 1977. *Learning from Las Vegas: The Forgotten Symbolism of Architectural Form.* Cambridge, Mass.: MIT Press.

29

Beyond Education: Metaphors on Creative and Workplace Learning

Torill Strand

The urge to be first has as many forms of expression as society offers opportunity for it. The ways in which men compete for superiority are as various as the prizes at stake. Decisions may be left to chance, physical strength, dexterity, or bloody combat. Or there may be competitions in courage and endurance, skillfulness, knowledge, boasting and cunning [. . .] But in whatever shape it comes in it is always play, as it is from this point of view that we have to interpret its cultural function.

—Huizinga in *Homo Ludens* (1938)

Introduction

What are our images of creativity? How do these images relate to our ways of seeing workplace learning within the new and globalized symbolic economy? In this chapter, I address these questions through three philosophical discourses that metaphorize creativity as "expression," "production," and "reconstruction." A vital family resemblance is that they portray creativity as the *act* of creating something new. Hans Joas (1996) therefore holds that they "contain a potential for understanding human action" (p. 73). By implication, they may also carry a potential for understanding the shifting epistementalities—meaning the altering of epistemic cultures, practices and knowledge ties—now characterizing professional work and learning (Strand 2009; Strand et al. 2010).

Taking a bird's-eye-view of the philosophical discourses, I reveal their vital characteristics and distinct ways of portraying the relationships between creativity, educative experiences, and the new epistementalities occurring within and beyond the workplace. My ambition is to provide a better understanding of the contours of creativity in relation to productive workplace learning. The three metaphors, however, should be seen as ideal types, not mirroring any ongoing "epistemic war" or "historic progress," as their different ways of picturing creativity seem to be equally recognized and are used somewhat overlapping within the current literature (Banaji, Burn, and Buckingham 2006).

Philosophical discourses on creativity—old and new—demonstrate that creativity is hardly a new way of portraying the dynamics of knowledge and learning. However, since the contemporary economization of knowledge and ideas has contributed to a common call for a "creative ethos" (Florida 2002), it seems pertinent to explore the ways in which metaphors of creativity form our ways of seeing the dynamics of knowledge and learning within contemporary working life. I want to emphasize, however, that all three metaphors move beyond the current hype within business circles, in which creativity is paralleled with effectiveness and seen as economic imperative (Howkins 2007, 2009; Lewis 2004; Seltzer and Bentley 2002). Rather, this chapter may throw some light on this hype, since I here reveal the ways in which different ways of metaphorizing creativity offer distinct images, theoretical representations, or models of thought that not only mirror but also provide openings and limitations for our ways of thinking about workplace learning.

To avoid the pitfall of Thales, the ancient Greek philosopher who was so eager to observe the stars that he forgot to watch his steps and thus fell into a ditch, I illustrate my discussion with a few examples from an extensive ongoing study (Jensen et al. 2008; Strand et al. 2010). The study I am referring to is comparative and longitudinal and explores the epistemic trajectories of Norwegian nurses, teachers, auditors, and computer engineers over an eight-year span (2003–2010). "Epistemic trajectories" here denotes the discontinuous processes of epistemic change, transforming the knowledge ties, forms of practice, and epistemic cultures of the four professions. Epistemic trajectories thus differ from "learning trajectories" in that the epistemic shifts are generated by the knowledge-dependent practices and cultures of contemporary working life (Lahn 2009). Overall, the study reveals remarkable shifts in the epistementalities of these four professions. However, in which ways may these shifts relate to the intersection of global/local epistemologies? Furthermore, what may the three metaphors of creativity make us see and not see?

Reinvented Epistementalities

Before exploring how the three metaphors of creativity help to model workplace learning, it is pertinent to provide a somewhat better image of the Norwegian study. It is an extensive ongoing study documenting how nurses, teachers, auditors, and computer engineers are now geared toward lifelong learning, inclusion, information seeking, and knowledge production. The first part of the study (ProLearn, 2003–2008) disclosed a widespread willingness to learn due to some common expectations of epistemic change and an emergent epistemification of working life. A core challenge, described by all groups, was to navigate in the ocean of global information and epistemic networks available in a time-efficient and responsible manner. Despite the fact that the epistemic cultures of the four professions are permeated with differences, all groups have embraced the rhetoric of the knowledge society and seem to have developed new knowledge ties. In addition, they all underlined the values and necessities of updating their knowledge base, an exploratory attitude, and to link up to knowledge networks and epistemic communities outside their local environments (Jensen 2007; Jensen et al. 2008; Nerland 2008; Nerland & Jensen 2007).

The aim of the follow-up study (LiKE 2008–2010) is to further explore how and to what degree these aspects can be seen as signs of enduring epistemic transformations generated by the ways in which global/local epistemologies unavoidably interact, converge, convert, and offer new instruments of knowing, acting and constructing the world of objects in an increasingly dispersed knowledge world. The data material here consists of questionnaires, learning logs, individual interviews, and focus groups. The questionnaires were answered by all participants in the first and the final years of their initial professional education, as well as 2.5 and 5.5 years after graduation; the learning logs were written during the first year of the participants' professional life; individual interviews were performed in 2005 and 2009, and interviews with the focus groups were conducted in the early autumn of 2006. Focus group interviews were again performed in the spring of 2010 in order to validate the analysis. It is therefore necessary to stress that the preliminary findings of this follow-up study are by no way conclusive. Nevertheless, the preliminary findings from LiKE indicate shifts in the epistementalities of the four professions. The group of computer engineers, for example, describes how their Norwegian workplaces have turned into international settings, containing an international workforce, competing in a global market, cooperating with international research groups, not focusing on the new technologies but rather on the production of knowledge and ideas itself (Nerland 2010). Since 2006 they have adopted a project-organized and agile work-style, which contrasts with their earlier "torrent" style that demanded specific skills in the fast flow of problem-solving procedures. Now, the computer engineers stress the importance of sharing their expertise with colleagues in and beyond their workplace; they have access to, and use daily, a multitude of knowledge resources, databases, and centers; the access seems to have improved considerably since 2006. However, more than previously, members of this group stressed the importance of an oral exchange of ideas; they now attend more international forums and conferences than earlier; cooperate continuously with international research centers; and participate actively in knowledge centers beyond their workplace. Overall, we are now experiencing a new phase within the globalized knowledge economy, which again contests the very epistementalities of the professions (Strand et al. 2010).

Knorr-Cetina (2007) and Knorr-Cetina & Bruegger (2005) explain these shifts by the knowledge-related tasks of contemporary "macro-epistemics," meaning external agents and institutions monitoring, assessing, and validating the processes of knowledge production or verifying its outcome, for example, research policies, transnational institutions or NGOs. So-called "macro-epistemics" thus serve as an intermediary-level of arrangements between a larger knowledge culture (providing scaffolds) and the epistemic "micro-culture" studied (the profession). Despite taking "synthetic situations" into account (i.e., Knorr-Cetina 2009), Knorr-Cetina thus proposes an analytic split between the global and the local. By contrast, the Norwegian study implies that global/local epistemologies should be seen as inescapably intertwined (i.e., Nerland 2010).

In fact, several authors emphasize that a vital characteristic of the new era is the ways in which global/local epistemologies now interact and convert (Burawoy 2000; Castells 2001, 2004; Urry 2002). Beck and Sznaider (2006), for example, speak of a "globalization from within," characterized by a reflexive outlook carrying altered images and new habits of thought and action. Speaking of contemporary societies, they claim that the new " . . . is not globalization, but a global awareness of it, its self-conscious political affirmation, its reflection and recognition before a global public via mass media, in the news and in the global social movements" (Beck and Sznaider 2006, p. 10). Thus, on the one hand, shared images, worldviews, and habits of thought and action are now becoming more and more "global." On the other hand, there is also a growing *global awareness* of these new images, worldviews, and habits of thought. As this new awareness is generated by and in turn generates people's everyday life—including their working life—we may not only speak of a "globalization from within" but also of a "globalization from below." The reference does not necessarily relate to the global social movements and NGOs resisting the oppressive effects of a globalized economy and transnational policies but rather to the new *conscience collective*, collective habits of thoughts, images, or aspirations produced by—and producing—the current transformations of everyday life, including social institutions and societies. Following Castoriadis (1987), these collective images can be conceptualized as products and productive of an epistemic rupture: " . . . just as society cannot be thought of within any of the traditional schemes of coexistence, so history cannot be thought of within any of the traditional frameworks of succession. For what is given in and through history is not the determined sequence of the determined but the emergence of the radical otherness, immanent creation, non-trivial novelty. This dynamic manifests itself in the existence of history *in toto* as well as by the appearance of new societies (of new *types* of societies) and the incessant transformation of each society" (Castoriadis, 1984, pp. 184–185). The ways in which global/local epistemologies now unavoidably interact, converge, and convert can therefore be seen as an "immanent creation"; a "radical otherness"; a "non-trivial novelty" that "appears as a behavior that is not only 'unpredictable' but *creative* (on the level of individuals, groups, classes or entire societies)" (p. 44). By implication, the shifting epistementalities of the Norwegian professions can be seen as *creative* shifts parallel to the "non-trivial novelty"' of the ways in which global/local epistemologies unavoidably interact, converge, and convert. However, how do different metaphors of creativity help to model the epistemic ruptures now occurring within and beyond working life?

Metaphors of Creativity

Certainly, philosophical discourses—old and new—demonstrate that creativity is by no means a new way of metaphorizing the dynamics of knowledge and learning within a world of change. Rather, what is new is the current situation, a new phase of the global knowledge economy labelled as for example a "wave of creativity and innovations" (Landry 2000), "a creative economy" (Peters, Marginson, and Murphy 2009), a new "ecology of ideas" (Howkins 2009), or "an age of innovation" (Peters and Araya, this volume). Taking the new ways of the world, it is therefore pertinent to explore the ways in which, and to what extent, traditional metaphors of creativity may provide openings (or offer limitations) for our ways of thinking about the altering epistementalities now characterising professional work and learning.

The three metaphors discussed here—"expression," "production," and "reconstruction"—see creativity as the *act* of creating something new. Nevertheless, they offer different and somewhat contrasting perspectives on educative experiences, creative acts, and epistemic ruptures. By implication, they will also help to model contemporary workplace learning differently. Thus, while mapping their vital characteristics, I will here draw on data material from the ongoing Norwegian study to illustrate that point.

Creativity as Expression

Creativity can be seen as cultural forms of collective self-expression. In contrast to earlier ways of thinking, Johann Gottfried Herder (1744–1803) saw cultural forms as products of human activities, not of a divine action. In fact, he argued that creativity is innate to any form of human practice because of a creative predisposition. In his *Treatise on the Origin of Language* (1772) Herder rejected the divine origin of human speech, because—as he says—humankind could never have been granted the gift of speech unless they already were predisposed to discover language. Next, the acts of speech, in fact any cultural expression, are closely related to the origin and evolution of society: "A poet is a true creator of the nation around him, he gives them a world to see and has their souls in his hand to lead them to the world" (quoted from Bernard, 1983, p. 234). With his remarkable departure from earlier ways of perceiving creativity as a divine quality only, Herder is seen as the originator of the *Geisteswissenschaften* (the arts and humanities), which again gave rise to a cultural movement conceptualizing education as *Bildung* (Lovlie and Standish 2003).

At the heart of the idea of *Bildung* lies the conception of the vitality of human action and the dynamic transformation extending through culture and politics. *Bildung* therefore happens through a double movement; i.e., through our self-activities and through the cultural and political forms of any society. In short, *Bildung* is inseparable from the creative acts of self-activities and the collective self-expressions building social, political, and cultural communities. While creativity is innate in human action, it is mediated through the novelty of each newfound expression and articulated through cultural and political forms. In other words, we are educated concurrently through our self-activities and our social, political, and cultural interactions. Consequently, this way of metaphorizing creativity moves beyond a Kantian notion of the artistic genius giving rise to the fine arts, since the idea of creativity as a special quality for a few individuals is clearly dismissed. By contrast, creativity is here seen as a universal disposition, giving rise to the creative everyday practice of all human beings. Friedrich Schiller (1759–1805), for example, pictures play—*ludere*—as the very dynamic heart of *Bildung* (Schiller 1794). Friedrich Froebel (1782–1852), the father of the kindergarten movement, recognized the child's creative self-expression through free play: "Play is the purest, most spiritual activity of man at this stage, and, at the same time, typical of human life as a whole—of the inner hidden natural life in man and all things. . . . play at this time is not trivial, it is highly serious and of deep significance. Cultivate and foster it, O mother; protect and guard it O father!" (Froebel 1826/1987, § 30). Froebel, therefore, promoted an education through play, which he saw as the most creative self-activity at this stage of life. However, since he conceived of play as "typical of human life as a whole," he also paralleled play with the creativity innate to any form of human practice. To nurture the child's ability to play is therefore to nurture the child's future ability to create something new. In other words, play and creativity here go hand in hand.

Overall, creativity is here pictured as a collective form of self-expression happening in and through everyday work and play. Taking a somewhat naïve example from the LiKE study, the group of Norwegian teachers may illustrate the close connection between play, creativity, and workplace learning. For example, when they describe the introduction of smart boards to Norwegian schools as a positive event, bringing "the whole world into the classroom" and inviting them to play with the new technology in order to improve teaching and learning:

> Oh yes, there's a new thing I haven't tried before, wow,—seems interesting, look forward to trying it
>
> *Yes?*
>
> It's a new way of teaching, which I haven't thought of
>
> *Hmm*
>
> Overall, there are new ways of teaching all the time, aren't there? So it's . . . , you work . . . , so there's never *one* method. It's, I mean, the way you work, and it's never only *one* method. Because you continuously change the way you work

Describing the smart board as a fun thing, inviting her to play and trying out new teaching styles in order to "make the instruction interesting and engaging for the students," this teacher illustrates the fine line between play and work. At first glance, her playful approach to the smart board comes forward as fairly unserious. However, re-reading the whole interview it becomes clear that she is an experienced teacher, valuing her work ("I *do* love being a teacher"), and appreciating the fact that she is now a flexible expert able to interact with the smart board in order to create an engaging learning environment together with her students. The five years of working as a teacher have educated her and made her an expert. She now continues to educate herself by trying out new methods. Thus, she obviously enjoys being "educated" again through the new technology introduced to Norwegian classrooms. In short, we may say that, on the one hand, her playfulness appears as unserious. At the same time, though, there is a deep seriousness to her play, since it reveals her expertise, her flexibility, and her willingness and ability to initiate new teaching styles.

In sum, when following the tradition of Herder and education as *Bildung*, creativity is metaphorized as "expression," which means that it is seen as innate to all human activities and expressed through the cultural and political forms of society. Creativity, the act of creating something new, is thus inherent in all forms of practice, of working life, and of social, political, and cultural forms. On the one hand, education takes the form of self-education through our activities. On the other hand, we are continuously educated through our participation in and interaction with social, political, and cultural forms. Since creativity and play seem to go hand in hand, there is a fine line between play and work, and a deep seriousness to unserious playfulness within and beyond the workplace. Nevertheless, this way of metaphorizing creativity also carries some limitations: Since this metaphor only says that creation actually happens without further portraying the specific dynamics of the playful creation, it does not seem to provide additional openings for our ways of thinking about the altering epistementalities now characterizing professional work and learning.

Creativity as Production

The act of creating something new can also be seen as production, meaning the concrete act of bringing forward something quite new into the world through the object-related activities of human labor. This way of metaphorizing creativity thus goes beyond the aesthetic metaphor put forward by the tradition of the German *Geisteswissenschaften*. Karl Marx (1818–1883) critiques Hegel's metaphysical assumptions, but he recognizes his insight that labor is the essential spirit of human life. Human labor is not only the process of creating values and commodities but also the manifestation of man's vital powers: "The outstanding achievement of Hegel's *Phänomenologie* and of its final outcome, the dialectic of negativity as the moving and generating principle, is thus first that Hegel conceives the self-creation of man as a process, conceives objectification as loss of the object, as alienation and as transcendence of this alienation; that he thus grasps the essence of *labour* and comprehends objective man—true, because real man—as the outcome of man's *own labour*" (Marx 1844a, XXIII). In other words, human labor—as the concrete process of creating something new in the world—is a manifestation of human life.

We create ourselves through labor. In addition, the driving force—"the moving and generating principle"—is our desire to work, to labor, to learn, to create, which is generated by the dialectical movement between a loss (alienation) and the process of transcending this loss. A somewhat naïve example is the teacher who seems lost when the new technology is introduced, but through her labor—which is initiated and moved forward by her desire to learn and finding things out—she transcends the loss by creating a new situation; a renewed learning environment for herself and her students. Consequently, the object-related activities of human labor do not only produce values, goods, and the individual self. It also produces a productive and content community:

> Let us suppose that we had carried out production as human beings. Each of us would have *in two ways affirmed* himself and the other person. 1) In my *production* I would have objectified my *individuality, its specific character*, and therefore enjoyed not only an individual *manifestation of my life* during the activity, but also when looking at the object I would have the individual pleasure of knowing my personality to be *objective, visible to the senses* and hence a power *beyond all doubt.* 2) In your enjoyment or use of my product I would have the *direct* enjoyment both of being conscious of having satisfied a *human* need by my work, that is, of having objectified *man's* essential nature, and of having thus created an object corresponding to the need of another *man's* essential nature. 3) I would have been for you the *mediator* between you and the species, and therefore would become recognised and felt by you yourself as a completion of your own essential nature and as a necessary part of yourself, and consequently would know myself to be confirmed both in your thought and your love. 4) In the individual expression of my life I would have directly created your expression of your life, and therefore in my individual activity I would have directly *confirmed* and *realised* my true nature, my *human* nature, my *communal nature.* (Marx 1844b, III)

Marx's utopia, however, comes forward as a grand narrative. In fact, though, this way of metaphorizing creativity has given rise to a great many schools of thought and by implication quite a few—but also different—theories on education and workplace learning: The sociopolitical philosophies of the neo-Marxist Frankfurt school, for example, have promoted educational programs aiming at social critique and reform through a fostering of the child's self-determination, co-determination, and solidarity (i.e., Klafki 1998). A Hegelian, neo-Marxist philosophy of praxis is apparent in Paulo Freire's critical pedagogy, which aims at democratic literacy (Freire 1996). Furthermore, the neo-Marxist notion of creative transformations of culture and self is clearly evident in the more cognitive oriented learning theories of, for example, Vygotsky (1926/1995), Bruner (1990), or

Scandinavian activity theory (i.e., Engestrøm and Miettinen 1999). Overall, all these theories on education and workplace learning seem to save creativity from the narrow domain of aesthetics. The notion of creativity here thus goes beyond the vital human compulsion to produce, as creativity is seen as manifest in the concrete object-related activity that brings into being something new in the world.

The creative labor of a group of Norwegian auditors—all affiliated with different workplaces—is an example. They all use a particular software program designed for public auditing. When using the program, they do not only interact with it as a ready-made tool, they also build, develop, expand, and assess the program continuously while using it. The program is now so well developed and has reached such a high standard that it will be exported and used in transnational auditing. Overall, there are several creative products of the labor of these auditors: First, they have participated in developing a sophisticated tool for auditing to be used within and beyond national borders. Next, their daily interaction on the same software, which they have developed together, has produced a productive community of auditors, reaching across and beyond the local workplaces. These auditors have now a collective experience and expertise that will help to further develop the methods and efficiency of national and transnational auditing procedures. Third, the auditors seem to have acquired a new way of understanding and performing their daily work:

> Now I want to contribute to growth instead of just assessment and control. I have a different outlook. For example I now look for the difference between effective and non-effecting ways of organizing the auditing procedures.
>
> *So you use different knowledge resources than before?*
>
> No. Not really. But I'm more competent in using the resources. I use the same: The government, the ministries, the national assembly, Norwegian laws and regulations. However, I no longer search for information on how to do things
>
> *No?*
>
> Because now the methodology is part of me. My task now is to develop and renew that methodology.

In brief, following the tradition of Marx, creativity is metaphorized as production. Creativity is associated with the concrete object-related activities of human labor, which brings forward something new into the world. Since creativity is seen as the essential spirit of human life, human labor is not only the process of creating values and commodities, but also the manifestation of human beings' vital powers: We create ourselves through labor. Moreover, the driving force is an innate desire to work, to labor, to learn, or to create. Marx shows how I objectify my individuality through my human labor since I in my production enjoy my life and its visible products. I will also enjoy the fact that you use and enjoy my products. Thus, I will receive recognition when you are recognizing the products of my work. Consequently, in my individual labor, I will confirm and realize my human nature, which is a communal, social nature. By implication, a product of human labor is also a productive and pleasant society, a creative community, or fruitful learning culture.

This way of metaphorizing creativity, education, and workplace learning can therefore help to portray and explore the shifting epistementalities now appearing in professional work and learning. It may also help to assess the fruitfulness of the new epistementalities now attained; are they creative or repetitive? However, since action here equates only one particular form of action, namely the object-related productive labor, this way of metaphorizing creativity seems to invite a theoretic distinction between the local and the global and thus to interpret the intersection of local/global epistemologies as a disturbing event. The preliminary findings of the Norwegian study, however, imply that the ways in which global/local epistemologies interact, converge and convert carry an "immanent creation"; a "radical otherness"; a "non-trivial novelty" (i.e., Castoriadis 1987), coming forward, not only as something unforeseen, but also as the *creative* epistementalities now characterizing professional work and learning.

Creativity as Reconstruction

In contrast to the continental philosophical discourses, American pragmatism metaphorizes creativity as reconstruction in terms of a radical remaking of our *common sense*, which not only promotes a reorientation but also contributes to deep-seated transformations that restructure and remake our worldviews, experiences, and habits of thought and action. The creative act is here seen as a reconstruction that affects our ways of seeing the world, our ways of making the world, and by implication the ways of the world themselves. The focus here is not so much the human consciousness but rather the *act* of creation.

In his remarkable theory of signs, Charles Sanders Peirce (1839–1914) depicts this action as "semeiosis": "By semeiosis I mean, an action, or influence, which is, or involves, a cooperation of *three* subjects, such as a sign, its object, and its interpretant, this tri-relative influence not being in any way resolvable into actions between pairs" (Peirce 1907/1998, p. 411). In short, the act of creation is in signs, or rather in the flows of signs, which come forward as a "series of surprises," i.e., one novelty after another that "presses upon every one of us daily and hourly."

Semeiosis thus educates:

> In all the works of pedagogy that ever I read—and that have been many, big, and heavy—I don't remember that any one has advocated a system of teaching by practical jokes, mostly cruel. That, however, describes the method of our great teacher, Experience. She says,
>
> Open your mouth and shut your eyes
>
> And I'll give you something to make wise;
>
> And thereupon she keeps her promise, and seems to take her pay in the fun of tormenting us. (Peirce 1903/1998, p. 154)

Peirce here quotes a folklore children's rhyme, cited when giving the child a gift of sweets. He thus seems to imply that a practical joke is a sweet thing because it conveys learning. This happens because of—as he says—"the action of experience." *Experience* here denotes *semeiosis,* meaning the flows of signs mediating actual lived experiences and our ways of seeing and perceiving the world, thus producing "a series of surprises": "The phenomenon of surprise in itself is highly instructive . . . because of the emphasis it puts upon a mode of consciousness which can be detected in all perception, namely, a double consciousness at once of an *ego* and a *non-ego,* directly acting upon each other" (Peirce 1903/1998, p. 154). In short, experience teaches through its abrupt entrance that bewilders familiar ways of thinking. These moments of surprise, which indeed jumble our categories of thought, happen because of a double consciousness, which, on the one hand, is aware of the familiar and vivid representations *and,* on the other hand, is aware of the new and unexpected. Consequently, the surprise is not so much in the abrupt and unexpected experience. The surprise is rather in the *relationship* between the known and the unknown, between familiar ways of thinking, and something totally new and unexpected, or between the "expected idea" and the "strange intruder."

In the Norwegian study, the group of Norwegian nurses portrays the short-lived epistementalities of a professional field of "fast knowledge." One of the nurses says:

> I'm constantly searching for new research results. So accessing the net, searching databases, updating our knowledge, and validating the flow of information, that's part of our daily routine. And we change our procedures according to the new information. Only since last year, we've changed a lot . . .
>
> *Yes?*
>
> . . . because we constantly learn about new research results, new ways of doing things, or new procedures implemented at another department;—"should we try that"?

> *Hmm. Will you describe yourself as a participant or a bystander to these fast-moving changes?*
>
> Both (laughter)
>
> *Yes?*
>
> You're . . . you're mostly a bystander, but also a participant. For example, a colleague of mine now studies post-operative procedures for patients on heart-lung machines. To his research, which is part of his doctorate, I am a passive bystander. At the same time I have a lot of expertise within that particular area. I'm actually the one doing the labour. And I'm very interested. So I report to him, so therefore . . . I mean . . . that's only one example on how you're always concurrently a participant and a bystander.

This nurse depicts the acts of creation through semeiosis, the flows of signs, which are mediated through the "double consciousness—at once of an ego and a non-ego—directly acting upon each other" (Peirce 1903/1998, p. 154). When she depicts herself as concurrently an observer to and a participant in the fast-moving changes now happening, she describes her "double consciousness": On the one hand, she recognizes the amount and quality of the research work carried out within her field; on the other hand, she acknowledges her own labor and expertise. In other words, she simultaneously keeps an eye on the "non-ego" (the flows of information) and the "ego" (her expertise). These "acts of creation"—constituted through her double consciousness—happen continuously, for example, when she searches the net, discusses with her professional team, attends lectures, gets to learn about different routines carried out at other hospitals, or when she interacts with her patients. In sum, when reading this interview in light of Peirce's theory of signs, it helps to illustrate how the shifting epistementalities of contemporary work life are closely related to the ways in which global/local epistemologies unavoidably interact, converge, convert, and offer new instruments of knowing, acting, and constructing the world of objects. However, it is not yet clear how this way of metaphorizing creativity may help to portray the *creative* epistementalities now characterizing professional work and learning.

Peirce's theory of signs shows a metaphysical realism of a unique kind. To him, the creative "action of experience" is a mediated experience, a flow of dynamic sign systems that acts upon us and thus contributes to a *continuous* reconstruction of both our ways of being in the world and the world itself. However, Peirce also perceives creativity as a *discontinuous* reconstruction. The discontinuous creativity comes forth as spontaneity, a series of guesses, or a play of chance that promotes radical epistemic ruptures. This dynamic occurs through, as he says, abduction or abductive guesses. Peirce describes abduction as an intersubjective creation of new ideas, which he saw as the only kind of reasoning that could generate genuinely new ideas. Abduction appears when we are confronted with a surprise, observe something out of the ordinary, or experience an odd event that bewilders and jumbles our earlier categories of thought, for example, when taught by "practical jokes, mostly cruel"; when being confronted with an impossible task at the workplace; or realizing that one's competency is outdated. Again, the bewilderment is not so much in the surprise itself but rather in the relation between familiar ways of thinking and the totally new and unexpected. Since it is generated by this impossible contrast between our ways of seeing the world and how it actually appears, the purpose of our abductive reasoning is to put forward a somewhat reasonable assumption or working hypothesis, which we later use, test, and act upon in order to attain a plausible assumption: "Abduction is where we find some very curious circumstance, which would be explained by the supposition that it was a case of a general rule and thereupon adopt the supposition" (Peirce 1904/1998, p. 227). The abductive process can thus be described as a series of guesses in order to eventually reach a plausible hypothesis, which we next assess by its likeliness. Our intention is simply to make a working assertion, which we can use to act or test or investigate further and thus to change the *meaning* of the world. However, the product is in fact a creation of something genuinely new (Anderson 1987; Garrison 2005; Kevelson 1998; Liszka 1996; Paavola and Hakkarainen 2005).

The interviews with the Norwegian nurses indicate that abductive reasoning is integrated in their epistementality. When describing their epistementality, the nurses in this group use the term "evidence-based practice." They portray an epistemic culture that can be paralleled with the cultures found within professional research groups, based on a shared interest of inquiry within a specialized field of knowledge. Furthermore, they describe an epistemic practice characterized by highly developed and consistent routines for a systematic inquiry, for sharing their expertise, developing and updating their procedures, and for an uninterrupted and systematic validation of their labor, their disciplined inquiry, and their ways of evaluation. When describing the epistemic culture and the epistemic forms of practice, they seem to have adopted the rhetoric of research, not only speaking of an "evidence-based practice," but also of "systematic investigation," "the research frontier," and "research tools." Overall, the group of nurses portrays an epistementality based on a common research interest and carried out through a highly specialized and orderly set of systematic routines. When one of the interviewees was asked "what if you make a mistake?" she promptly responded: "No, no. No, I can never make a mistake. If I do, the patient will die."

Nevertheless, these nurses seem to have integrated an abductive way of reasoning in their epistementality. In some of the interviews, the interviewees describe how this way of reasoning happens when they are confronted with a surprise, observe something abnormal, or experience an odd event that bewilders and jumbles their habits of thought and action. Here are two examples: One nurse described how she was caught by surprise when learning about another hospital using sterile water when tube-feeding the patients. Another nurse describes how, when scrolling through some newly published articles on how the body reacts to a lower temperature, almost chocked herself when she caught the idea that they could cool down the premature babies in order to save their lives. Both nurses emotionally described the event of catching a new idea on how they could contribute to saving more lives. However, even more interesting is how both ideas gave birth to systematic research in their workplace in order to first find out if their ideas were reasonable, and next if they were possible to put into practice. Both nurses told they spent a lot of time—by themselves, in front of the computer and with their team, in workshops and meetings—in order to develop and test the idea. The first nurse confirmed that they now have changed their routines on tube feeding. The second nurse, however, regrets that they were still not able to practice such a procedure, because "we need bags of a particular type of plastic that does not hurt the babies, and such bags are not yet available." Despite intense research, her team still doubted how the cooling down of such premature babies would affect their blood work: "we need more knowledge, but might be able to implement such a procedure before long. . . ."

These examples illustrate remarkable outcomes of the interweaving of the continuous and discontinuous creativity, and thus how baffling moments of something inexplicable can be integrated in and give birth to new epistementalities. The nurses' ways of integrating the *discontinuous* ruptures that gave birth to their ideas in a *continuous* reconstruction of their epistementalities demonstrate how discontinuous and continuous forms of creativity play together in the reconstruction of our ways of being in the world and of the world itself. The impulse, the radical rupture, the spontaneity should thus be highly affirmed since it is the only way of creating new ideas. However, radical new ideas need to be integrated in and validated by our *common sense* since we uninterruptedly come up with new ideas, but just a few of them seem feasible. In other words, discontinuous creativity needs to go hand in hand with continuous creativity in the generation of new intelligible components of reality.

The Work of Metaphors

Overall, these three ways of metaphorizing creativity may help to illuminate our ways of conceptualizing workplace learning within the new era in that all three metaphors picture creativity as the *act* of creating something new. Nevertheless, they offer somewhat contrasting perspectives on educative experiences, creative acts, and epistemic ruptures and will by implication help to model contemporary workplace learning quite differently. So, what do they invite us to see and not to see?

The philosophical discourse metaphorizing creativity as *expression* pictures creativity as a universal disposition, innate to all human activities, and expressed through the cultural and political forms of society. Creativity is the dynamic vitality of all forms of human practice—it drives our working life and is immanent in any social, political, and cultural form. Creativity is therefore also at the very dynamic heart of educational processes. Education takes the form of self-education through our activities while simultaneously, we are continuously being educated through our participation in and interaction with social, political, and cultural forms. Since creativity and play seem to complement each other, there is a fine line between play and work, and a deep seriousness to unserious playfulness within and beyond the workplace. Nevertheless, this metaphor carries some limitations since this way of metaphorizing creativity only says that creation actually happens without further portraying the specific dynamics of the playful creation.

The philosophical discourse metaphorizing creativity as *production* sees creativity as manifest in the concrete object-related activities of human labor, which brings forward something new into the world. Since creativity is here seen as the essential spirit of human life, human labor is not only the process of creating values and commodities but also the manifestation of human beings' vital powers: We create ourselves through labor with an innate desire to work, to labor, to learn, or to create as a driving force. I objectify my individuality through my labor since I in my production enjoy both my life and its visible products. I will also enjoy the fact that you use and enjoy my products. Thus, I will receive recognition when you are recognizing the products of my work. Consequently, in my individual labor, I confirm and realize my human nature, which is a communal, social nature. By implication, a product of human labor is also a productive society, a creative community, or fruitful learning culture. This way of metaphorizing creativity can therefore help to explore and assess the shifting epistementalities now appearing. Are they creative or repetitive? However, since action here equates only one particular form of action, namely, the concrete object-related productive labour, this way of metaphorizing creativity may invite a theoretic distinction between the local and the global.

The philosophical discourse of metaphorizing creativity as *reconstruction*, however, represents a shift of focus from the human consciousness to the *act* of creation, which is here portrayed as a deep-seated reconstruction that affects our ways of seeing the world, our ways of making the world, and by implication the ways of the world themselves. Here, continuous and discontinuous forms of creativity play together in the reconstruction of reality. More specifically creative action is portrayed as a mediated experience, a flow of dynamic sign systems that acts upon us, contributing to a *continuous* reconstruction of both our ways of being in the world and the world itself. Conversely, creativity is seen as a *discontinuous* reconstruction that comes forward as a series of guesses or play of chance promoting radical epistemic ruptures. This kind of creativity appears when we are confronted with a surprise, observe something out of the ordinary, or experience an odd event that jumbles our earlier categories of thought. The purpose is simply to put forward a somewhat reasonable assumption or working hypothesis, which we later use, test, and act upon in order to arrive at a plausible assumption. This metaphor may thus help to illustrate how the shifting epistementalities of contemporary work life are closely related to the ways in which global/local epistemologies unavoidably interact, converge, convert, and offer new instruments of knowing, acting, and constructing the world of objects. In addition, it helps to portray the *creative* epistementalities now characterizing professional work and learning.

Taken together, the three metaphors offer a thick description on the epistemic ruptures now happening within and beyond working life. Nevertheless, the third metaphor seems to be the only metaphor carrying a rich potential for embracing the ways in which global/local epistemologies now unavoidably interact, converge, and convert. It thus opens possibilities for conceptualizing the epistementalities now characterizing professional work and learning as *creative* shifts generated by and parallel to the extraordinary newness of the phase of the global knowledge economy we are now experiencing.

References

Anderson, D. R. (1987). *Creativity and the Philosophy of C. S. Peirce.* Dordrecht: Martinus Nijhoff.

Banaji, S., Burn, A. and Buckingham, D. (2006). The Rhetoric of Creativity: A Review of the Literature. *A Report for Creative Partnerships.* Centre for the Study of Children, Youth and Partnerships. Institute of Education: University of London.

Beck, U. and Szneider, N. (2006). Unpacking Cosmopolitanism for the Social Sciences: A Research Agenda. *The British Journal of Sociology,* Vol. 57, No. 1, pp 1–23.

Bernard, F. M. (1983). National Culture and Political Legitimacy. *Journal of the History of Ideas,. 44*(2), 231–253.

Bruner, J. (1990). *Acts of Meaning*. Cambridge Mass.: Harvard University Press.

Burawoy, M. (2000). Grounding Globalization. In M. Burawoy et al. (Eds.), *Global Ethnography: Forces, Connections, and Imaginations in a Postmodern World* (pp. 337–350). Berkeley: University of California Press.

Castells, M. (2001). *The Internet Galaxy*. Oxford: Oxford University Press.

Castells, M. (2004). Informationalism, Networks, and the Network Society: A Theoretical Blueprint. In M. Castells (Ed.), *The Network Society: A Cross-Cultural Perspective* (pp. 3–45). Cheltenham: Edward Elgar.

Castoriadis, C. (1987). *The Imaginary Institution of Society*. Cambridge Mass.: The MIT Press.

Engestrøm, Y. and Miettinen, R. (1999). *Perspectives on Action Theory*. Cambridge: Cambridge University Press.

Florida, R. (2002). *The Rise of the Creative Class: And How It's Transforming Work, Leisure, Community and Everyday Life*. New York: Basic Books.

Freire, P. (1996). *Pedagogy of the Oppressed*. London: Penguin.

Froebel, F. (1826/1887). *The Education of Man*. New York and London: D. Appleton and Company.

Garrison, J. (2005). Curriculum, Critical Common-Sensism, Scholasticism, and the Growth of Democratic Character. *Studies in Philosophy and Education, 24,* 174–211.

Herder, J. G. (1772). *Abhandlung über den Ursprung der Sprache (Treatise on the Origin of Language)*. Retrieved January 2010 from http://gutenberg.spiegel.de/?id=5&xid=1162&kapitel=1#gb_found

Howkins, J. (2007). *The Creative Economy: How People Make Money from Ideas*. London: Penguin

Howkins, J. (2009). *Creative Ecologies: Where Thinking Is a Proper Job*. Queensland: University of Queensland Press.

Huizinga, J. (1938/1950). *Homo Ludens: A Study of the Play Element in Culture*. Boston: Beacon Press.

Jensen, K. (2007). The Desire to Learn: An Analysis of Knowledge-Seeking Practices Among Professionals. *Oxford Review of Education, 33*(4), 489–502.

Jensen, K. et al. (2008). Profesjonslæring i endring. [Professional learning in transition]. *Research report*. Oslo: The Norwegian Research Council.

Joas, H. (1996). *The Creativity of Action*. Chicago: Chicago University Press.

Kevelson, R. (1998). *Peirce's Pragmatism: The Medium as Method*. New York: Peter Lang.

Klafki, W. (1998). Characteristics of a Critical-Constructive Didaktik. In B. Gundem and S. Hopmann (Eds.), *Didaktik and/or Curriculum: An International Dialogue*. New York: Peter Lang.

Knorr-Cetina, K. (1997). Sociality with Objects: Social Relations in Postsocial Knowledge societies. *Theory, Culture & Society, 14*(4), 1–30.

Knorr-Cetina, K. (2007). Culture in Global Knowledge Societies: Knowledge Cultures and Epistemic Cultures. *Interdisciplinary Science Reviews, 32*(4), 361–375.

Knorr-Cetina, K. (2009). The Synthetic Situation: Interactionism for a Global World. *Symbolic Interaction, 32*(1), 61–87.

Knorr-Cetina, K. and Bruegger, U. (2005). Complex Global Microstructures. The New Terrorist Society. *Theory, Culture and Society, 22*(5), 213–234.

Lahn, L. (2009). Professional Learning as Epistemic Trajectories. In S. Ludvigsen, A. Lund, I. Rasmussen and R. Säljö (Eds.), *Learning Across Sites: New Tools, Infrastructures and Practices.* Oxford: Pergamon Press.

Landry, C. (2000). *The Creative City: A Toolkit for Urban Innovators.* London: Earthscan Ltd.

Lewis, T. (2004). *The Creative Age.* Cirencester: Management Books 2000 Ltd.

LiKE (2008–2010). Learning Trajectories in Knowledge Economies. Retrieved January 2010 from http://www.pfi.uio.no/forskning/prosjekt/like/index.html

Liszka, J. (1996). *A General Introduction to the Semeiotic of Charles Sanders Peirce.* Bloomington: Indiana University Press.

Lovlie, L. and Standish, P. (2003). Bildung and the Idea of Liberal Education. In L. Løvlie, K. P. Mortensen and S. E. Nordenbo. (Eds.), *Educating Humanity: Bildung in Postmodernity* (pp. 1–24). London: Blackwell.

Marx, K. (1844a). *Critique of Hegel's Philosophy in General.* Retrieved October 2009 from http://www.marxists.org/archive/marx/works/1844/manuscripts/hegel.htm

Marx, K (1844b). *Comments on James Mill, Éléments D'économie Politique.* Retrieved October 2009 from http://www.marxists.org/archive/marx/works/1844/james-mill/index.htm

Nerland, M. (2008). Knowledge Cultures and the Shaping of Work-Based Learning: The Case of Computer Engineering. *Vocations and Learning: Studies in Vocational and Professional Education, 1,* 49–69.

Nerland, M. (2010). Transnational Discourses of Knowledge and Learning in Professional Work: Examples from Computer Engineering. *Studies in Philosophy and Education, 28*(1).

Nerland, M. and Jensen, K. (2007). The Construction of a New Professional Self: A Critical Reading of the Curricula for Nurses and Computer Engineers in Norway. In A. Brown, S. Kirpal and F. Rauner (Eds.), *Identities at Work* (pp. 339–360). Dordrecht: Kluwer Academic Publishers.

Paavola, S. and Hakkarainen, K. (2005). Three Abductive Solutions to the Learning Paradox—With Instincts, Inference and Distributed Cognition. *Studies in Philosophy and Education, 24,* 235–253.

Peirce, C. S. (1903/1998). On Phenomenology. In N. Houser et al. (Eds.), *The Essential Peirce. Selected Philosophical Writings. Vol. 2 (1893–1913)* (pp. 145–159). Bloomington and Indianapolis: Indiana University Press.

Peirce, C. S. (1904/1998). Pragmatism as the Logic of Abduction. In N. Houser et al. (Eds.), *The Essential Peirce. Selected Philosophical Writings. Vol. 2 (1893–1913)* (pp. 226–241). Bloomington and Indianapolis: Indiana University Press.

Peirce, C. S. (1907/1998). Pragmatism. In N. Houser et al. *The Essential Peirce: Selected Philosophical Writings. Vol. 2 (1893–1913)* (pp. 398–433). Bloomington and Indianapolis: Indiana University Press.

Peters, M. A., Marginson, S. and Murphy, P. (2009). *Creativity and the Global Knowledge Economy*. New York: Peter Lang.

ProLearn. (2003–2008). Professional Learning in a Changing Society. Retrieved January 2010 from http://www.pfi.uio.no/prolearn/index.html

Schiller, F. (1794/2004). *On the Aesthetic Education of Man*. New York: Dover.

Seltzer, K. and Bentley, T. (2002). *The Creative Age: Knowledge and Skills for the New Economy*. London: Demos.

Strand, T. (2009). *The Epistemology of Early Childhood Education: The Case of Norway*. Saarbrucken: VDM Verlag.

Strand, T. et al. (2010). Emerging Global Epistemologies in Professional Work and Learning. *Paper*. AERA 2010

Urry, J. (2002). *Global Complexity*. Cambridge: Cambridge University Press.

Vygotsky, L. (1926/1995). *Educational Psychology*. Oxford: Blackwell.

Afterword
Play, the Net, and the Perils of Educating for a Creative Economy

Pat Kane

As complex mammals forged through play, from the beginning to the end of our lifespan, we humans are fated to be lifelong learners. Even so, the crucial question for the effectiveness of education in a creative economy is whether the generativity at the heart of our species-being is a route leading toward autonomy or heteronomy? As the talented writers in this volume have demonstrated, we are living in an age in which our education systems are out of synch with the realities of the twenty-first century. The issue of play is at the heart of many of the dynamic changes we are witnessing today. As many of the authors in this volume have pointed out, present-day educational systems are too rigid for the kind of "play ethic" that creativity and a creative economy require.

The play ethic is what comes after the obsolescence of the work ethic. The work ethic is an ideology or belief-system that asserts that any job has dignity and worth, regardless of how alienated it makes a worker feel or how different it is from a person's desires and aspirations because society recognizes this submission to the job as the basis of social order.

The play ethic is an alternative belief-system that asserts that in an age of mass higher education, continuing advances in personal and social autonomy, and ubiquitous digital networks (and their associated devices), there exists a surplus of human potential and energy that will not be satisfied by the old workplace routines of duty and submission.

One might think that rising affluence, improving health indicators, and the cognitive surplus represented by the Internet would provide the optimum conditions for a post-work/players' identity in the developed world—a spreading "ground-of-play," where our potentiating faculties find ever greater zones in which imagination can enchant and infuse our lives.[1] However, the story may not be so blithely heading in the direction of playful liberation.

The distinct neoteny of our species—that is, the extension of youthful characteristics far into our maturity, by comparison even with other simians—keeps us always, as the Italian autonomist thinker Paolo Virno says, in a state of "permanent formation."[2] We have kept this endemic and anxiety-inducing openness to the world under control, says Virno, by means of what he calls "cultural and social devices"—religions, castes, class identities, civic values, regional and national traditions, foremost among the last of these (as Ernest Gellner might say[3]) being the nation's education system.

However, Virno's warning is that the regime of flexible production and informationalized management that typifies contemporary Western capitalism is now uniquely exploiting our neoteny. Post-Fordism (should we bite the bullet and call it Googlism?) deliberately accelerates this indeterminacy—the faculties that open us up to endemic flexibility and openness—to make it the very fuel of the social and economic order: "The death of specialized instincts and the lack of a definite environment, which have been the same from the Cro-Magnons onwards, today appear as noteworthy economic resources." Virno moves through our natural faculties of potentiality and lashes them methodically to the flexible personality required by informational capitalism.

Our biological non-specialization? The grounding for the "universal flexibility" of labor services: "The only professional talent that really counts in post-Fordist production is the habit not to acquire lasting habits, that is, the capacity to react promptly to the unusual." Does our neotenic forever-youngness keep us always ready to learn and adapt? We are now subject to "permanent formation . . . what matters is not what is progressively learning (roles, techniques, etc) but the display of the pure power to learn." Is it a fact then that we are not determined by our environment but make and construct our worlds? This is mirrored by the "permanent precarity of jobs," where we wander nomadically from one cloud in the nebulous world of labor markets to another.

With a sardonic gloominess worthy of Theodor Adorno, Virno denies that this intrinsically unstable system necessarily leads to unruliness—"far from it." In traditional societies with less pervasive markets (which one presumes includes Fordism), our deep ontological anxiety could be contained by "protective cultural niches." The "omnilateral potentiality" of flexible capitalism shakes those niches to fragments. Yet However, even though this disembeddedness allows for an "unlimited variability of rules," when those rules are applied, they are much more "tremendously rigid" than the Fordist workplace. Each productive instance is like the tight rules of a competitive game, easily entered into but severely binding when the play begins.

When commanded by our managements to respond to today's ad hoc list of tasks and projects, in a world of frazzling openness and potentiality, we display "a compulsive reliance on stereotyped formulae." It is via these formulae that we "contain and dilute" the pervasive indeterminacy of the human condition. Virno characterizes them as reaction-halting behaviors, obsessive tics, the drastic impoverishments of the *ars combinatoria*, the inflation of transient but harsh norms. . . . Though on the one hand, permanent formation and the precarity of employments guarantee the full exposure to the world, on the other they instigate the latter's reduction to a spectral or mawkish dollhouse.

So an education for the creative economy that aims to maintain our potentiating flexibility, which morphs its curricula and pedagogy in the plural energies of the playful self, may not be—according to Virno—as progressive as it thinks it might be. The outcome of our creative industries might all-too-easily be rendered as a "spectral or mawkish dollhouse." Moreover, what might it profit a generation of media studies, liberal arts, or cultural studies graduates to gain a full facility in the *ars combinatoria* yet deploy them in the "drastic impoverishments" of the reality TV show or the taste-marketing analytics consultancy?

If the playfulness that has always been a subterranean touchstone for educators since the Romantic period—from Rousseau to Froebel, Steiner to Montessori, Reggio-Emilia to Summerhill—has now become the Achilles Heel of productive subjectivity, the point of susceptible engagement with processes of miasmic exploitation (or at least expropriation) of human creativity,[4] then the ethical telos of contemporary education would enter a real moment of crisis.

However, if we can question the baroque mechanisms of psychological capture that Virno so mordantly describes, all their fine-grained capitalization of our playful natures, we might find a new foundation for a progressive education. Certainly, Virno's is not the only available social-scientific reading of our wide-open, neotenic natures. Brian Sutton-Smith identifies one evolutionary function of play as the continuation of "neonatal optimism" throughout the life-span. The "unrealistic optimism, egocentricity and reactivity" of the growing child, all of them "guarantors of persistence in the face of adversity," characterize many of our adult play behaviors. Play brings a sense of joyful indefatigability and energetic resilience, which—like the pleasure of sex for procreation—is evolution's "salute" to the human animal for maintaining a "general liveliness," in the face of the challenges of existence.

Sutton-Smith sees play forms as an expression of reflective "secondary" emotion, as ways to deal constructively with the "primary" emotions located in the more reactive parts of the human brain. The amygdala that generates shock, anger, fear, disgust, sadness is mediated by a frontal lobe that trades in pride, empathy, envy, embarrassment, guilt, and shame, with happiness as the emotion that operates across both brain areas. The "secondary" emotions are much more "rule-based" or "situation-based"—our play, games and simulations operate as the medium whereby our basic emotions can be translated into manageable interpersonal and social phenomena.[5]

Compared to Virno's "potential human," fated to indecision in its very constitution, Sutton-Smith's "adaptive potentiator" has a healthy dynamic in its use of play. For the latter, play is "a fortification against the disabilities of life. It transcends life's distresses and boredoms and, in general, allows the individual or the group to substitute their own enjoyable, fun-filled, theatrics for other representations of reality in a tacit attempt to feel that life is worth living."

In the Sutton-Smith vision, play is not the soft spot whereby we are made passive "dividuals" by hyper-capitalism, but the resilient optimism out of which the very possibilities of societal difference are generated. An education for players, founded in this socio-biological vision, becomes a constructive exercise in building forms of simulation, combination and gaming that rehearse that "neonatal optimism."

Even so, isn't such a constitutive "optimism" just what the desiring-machines of Virno's info-capitalism most wish to exploit? The answer returns power to the educator and pupil—but not in the institutions we have inherited from the industrial age. Education has to build those rich "grounds of play" in which the optimism of our species can flourish in a way that outflanks and surpasses any dominion that a powerfully calibrating control-society might assert. It could do worse than to attend to the peculiarly persistent linking of commons and dynamism that characterizes the internet.

For neoteny's generation of play and play forms throughout the human lifespan is one of the deeply constitutive processes shaping the design, functionality, and culture of the Internet. One epochal answer to our potentiating faculties that the Internet could represent is that of an extension of the "ground of play" that we see across the higher complex mammals—that open but distantly monitored developmental zone of time, space, and resource, where potentiating risks are taken by explorative, energetic organisms, in conditions where scarcity is held at bay.

Lion cubs or chimps compelled to play diversively, risk injury and predation, but in a delimited zone with ultimate defenses; children in their local playground, enjoy their rough-and-tumble with solid equipment and open space, under some kind of municipal governance; all of us on the Internet, improvising our sociality and extending our conviviality with powerful communication tools, resting on a complex but (so far) resilient infrastructure. All of these can be cast as complex-mammalian "grounds of play," sharing three conditions—they are, first, loosely but robustly governed; second, a surplus of time, space, and materials is ensured; third, failure, risk, and mess are treated as necessary for development.

So the "constitutive" power of play in humanity—that neoteny-driven potentiation that excites both autonomists and socio-biologists—seems to also require a "constitutional" dimension: a protocol of governance securing certain material and emotional conditions, to enable a rich plurality of playforms. When Lessig speaks of the Net as an "innovation commons," the resonance with a socio-biological vision of the ground of play is clear. His idea that the Internet represents an "architecture of value" is also homologous with these conditions for play: both are discernable zones of rough-and-tumble activity in which our social-ethical identities are forged.

That our schools and colleges could be "innovation commons" and "architectures of value"—could be "constitutional" as much as "institutional"—is a future that many of the writers in this volume are striving to build. Nevertheless, they should realize that play is their deep and elemental ally in such activism. Indeed, educational moments that cleave as closely as possible to the generative structures of the Net will also tap the constitutive power of play.

Notes

1. http://www.theplayethic.com/2009/11/play-potentiality-and-the-constitution-of-the-net.html
2. Paulo Virno, 'Natural-Historical Diagrams: the 'New Global' Movement and the Biological Invariant," in *The Italian Difference,* edited by Alberto Toscano, Re-Press, 2009. http://3.ly/virno
3. http://www.timeshighereducation.co.uk/story.asp?storyCode=96039§ioncode=26
4. http://journal.fibreculture.org/issue5/kucklich.html
5. http://americanjournalofplay.press.illinois.edu/1/1/sutton-smith.html

Contributors

Philippe Aigrain

Philippe Aigrain is the Founder and CEO of Sopinspace (Society for Public Information Spaces) a company that provides free software-based solutions for public debate by citizens of policy issues and for collaborative work over the Internet. He acts at the international level as an advocate for the information and knowledge commons and tries to address challenges in making commons-based cooperation sustainable. He was head of the 'Software technology' sector within the European Commission research programmes, where he initiated the policy in support of free/open source software innovation. He has authored many papers in computer and information science, sociology and history of technology, most of which are accessible under Creative Commons licences from his blog at http://paigrain.publicdebate.net.

Daniel Araya

Daniel Araya is a doctoral candidate in Educational Policy Studies at the University of Illinois (Urbana-Champaign). The focus of his research is the confluence of digital technologies and cultural globalization on systems of education. He has published widely on subjects related to the knowledge economy and peer-to-peer collaboration and is currently editing two books exploring the socioeconomic impact of digital technologies. Daniel has worked with the National Center for Supercomputing Applications (NCSA) and the UIUC Global Studies in Education program and was selected for the 2009 Oxford Internet Institute's Summer Doctoral Programme (SDP). He has a Bachelors degree from McMaster University and a Masters degree from the University of Toronto.

Anne Balsamo

Anne Balsamo is Professor of Interactive Media in the School of Cinematic Arts at the University of Southern California's Annenberg School of Communications. From 2004–2007, she served as the Director of the Institute for Multimedia Literacy. Her work focuses on the relationship between the culture and technology. This focus informs her practice as a scholar, researcher, new media designer and entrepreneur. In 2002, she co-founded Onomy Labs, Inc. a Silicon Valley technology design and fabrication company that builds cultural technologies. Previously she was a member of RED (Research on Experimental Documents), a collaborative research group at Xerox PARC who created experimental reading devices and new media genres. She served as project manager and new media designer for the development of RED's interactive museum exhibit, XFR: Experiments in the Future of Reading. Her first book, *Technologies of the Gendered Body: Reading Cyborg Women* (Duke UP, 1996) investigated the social and cultural implications of emergent bio-technologies. Her new transmedia book project, *Designing Culture: The Technological Imagination at Work* examines the relationship between cultural reproduction and technological innovation (Duke UP, forthcoming).

Michel Bauwens

Michel Bauwens is an active writer, researcher and conference speaker on the subject of technology, culture and business innovation. He is the founder of the Foundation for Peer-to-Peer Alternatives and works in collaboration with a global group of researchers in the exploration of peer production, governance, and property. He has worked as an analyst for the United States Information Agency, knowledge manager for British Petroleum, eBusiness Strategy Manager for Belgacom, as well as an internet entrepreneur in his home country of Belgium. Most recently, he co-produced the 3-hour TV documentary *Technocalyps* with Frank Theys, and co-edited a two-volume book on anthropology of digital society with Salvino Salvaggio. Michel is a Primavera Research Fellow at the University of Amsterdam and external expert at the Pontifical Academy of Social Sciences (2008). He currently lives in Bangkok, Thailand, where he is a lecturer at the

Dhurakij Pundit University's International College, assisting with the development of the Asian Foresight Institute.

Yochai Benkler

Yochai Benkler is the Berkman Professor of Entrepreneurial Legal Studies at Harvard, and faculty co-director of the Berkman Center for Internet and Society. Before joining the faculty at Harvard Law School, he was Joseph M. Field '55 Professor of Law at Yale. He writes about the Internet and the emergence of networked economy and society. His articles include "Overcoming Agoraphobia" (1997/98, initiating the debate over spectrum commons); "Commons as Neglected Factor of Information Production" (1998) and "Free as the Air to Common Use" (1998, characterizing the role of the commons in information production and its relation to freedom); "From Consumers to Users" (2000, characterizing the need to preserve commons as a core policy goal, across all layers of the information environment); "Coase's Penguin, or Linux and the Nature of the Firm" (characterizing peer production as a basic phenomenon of the networked economy) and "Sharing Nicely" (2002, characterizing shareable goods and explaining sharing of material resources online).

Tina (A. C.) Besley

Tina Besley is a Research Professor in Educational Policy Studies at the University of Illinois, Urbana Champaign, currently working in Global Studies in Education. She has been Professor of Counseling in Educational Psychology and Counseling at California State University, San Bernardino and Research Fellow and Lecturer in the Department of Educational Studies at the University of Glasgow, Scotland, UK. Tina's research interests include: youth issues, in particular, notions of self and identity in a globalised world; educational policy and philosophy, especially poststructuralism and the work of Michel Foucault. Tina's book, *Counseling Youth: Foucault, Power and the Ethics of Subjectivity* (Praeger, 2002) is now in a paperback edition (Sense Publishers). With Michael A. Peters, she has written *Building Knowledge Cultures: Education and Development in the Age of Knowledge Capitalism* (Rowman & Littlefield, 2006), *Subjectivity and Truth: Foucault, Education and the Culture of the Self* (Peter Lang, 2007) and is co-editor of *Why Foucault? New Directions in Educational Research* (Peter Lang, 2007). She edited *Assessing the Quality of Educational Research in Higher Education: International Perspectives* (Sense, 2009).

Ruth Bridgstock

Ruth Bridgstock is Vice Chancellor's Postdoctoral Research Fellow at Queensland University of Technology. She is affiliated with the Creative Workforce Programme within the ARC Centre of Excellence in Creative Industries and Innovation. Ruth's research interests relate to lifelong individual career and organisational development in the post-compulsory education and business contexts of the 21st century. Her PhD research took a longitudinal approach to the investigation of individual and contextual predictors of graduate employability and early career success. Her current fellowship project is concerned with building and testing a theory of effective university education for the development of innovators, based on the trajectories of high flyers in the fields of science, technology, engineering, mathematics, and the creative industries.

John Seely Brown

John Seely Brown is a visiting scholar and advisor to the Provost at the University of Southern California (USC) and the Independent Co-Chairman of the Deloitte Center for the Edge. Prior to that he was the Chief Scientist of Xerox Corporation and the director of its Palo Alto Research Center (PARC)—a position he held for nearly two decades. While head of PARC, Brown expanded the role of corporate research to include such topics as organizational learning, knowledge management, complex adaptive systems, and nano/mems technologies. He was a cofounder of the Institute for Research on Learning (IRL). He is a member of the American Academy of Arts and Sciences, the National Academy of Education, a Fellow of the American Association for Artificial Intelligence and of AAAS and a Trustee of the MacArthur Foundation. He serves on numerous public boards (Amazon, Corning, and Varian Medical Systems) and private boards of directors. He has published over 100 papers in scientific journals and was awarded the *Harvard Business Review's* 1991 McKinsey Award for his article, "Research that Reinvents the Corporation" and again in 2002 for his article "Your Next IT Strategy."

Phillip Brown

Phillip Brown is a Distinguished Research Professor in the School of Social Sciences at Cardiff University. He has written, co-authored and co-edited thirteen books, the most recent are *Education, Globalization and Social Change* with Hugh Lauder (2006, Oxford University Press); and *The Mismanagement of Talent: Employability and Jobs in the Knowledge Economy* with A. Hesketh (2004, Oxford

University Press). *The Global Auction: The Broken Promises of Opportunity, Jobs and Rewards*, written with Hugh Lauder and David Ashton was published by Oxford University Press, New York in March 2010.

Leslie Chan

Leslie Chan is a Senior Lecturer in the Department of Social Sciences at the University ofToronto Scarborough, where he is the Programme Supervisor for the New Media Studies.Programme and the International Studies Programme. Since 2000, Leslie has served as the Associate Director of Bioline International, a non-profit, multi-partners electronic service that provides an open access platform for some 60 journals from various developing countries, many of them in the areas of health and food sciences. Leslie was one of the original signatories of the Budapest Open Access Initiative, and he has been active in research and practices on new models of scholarly communication and knowledge dissemination. Leslie serves on a number of boards, including the Advisory Board of the Canadian Research Knowledge Network and the Canadian Association of Learned Journals.

Bill Cope

Bill Cope is a Research Professor in the Department of Educational Policy Studies, University of Illinois, Urbana-Champaign and an Adjunct Professor in the Globalism Institute at RMIT University, Melbourne. He is also a director of Common Ground Publishing, developing and applying new publishing technologies. He is a former First Assistant Secretary in the Department of the Prime Minister and Cabinet and Director of the Office of Multicultural Affairs in Australia. His current research interests include theories and practices of pedagogy, cultural and linguistic diversity, and new technologies of representation and communication. He is currently working on a project investigating the next generation of 'semantic web' technologies.

Dave Cormier

Dave Cormier is a web projects lead at the University of Prince Edward Island, cofounder of Edtechtalk, and president of Edactive Technologies, a social software consulting firm. He teaches academic courses in writing, emerging technology and culture. He is a member of several research communities and has participated in several web based research projects including the Open Habitat project and the Living Archives project. Cormier has produced, designed, or participated in over 300 online webcasts in the past four years and speaks regularly at conferences on topics including rhizomatics, effective use of new technologies, and educational project management and design. His major research interests include the placing of educational technology in a 'postdigital' context, the examination of planned and unplanned communities, rhizomes as a model for knowledge creation, and open-source multiuser virtual environments (MUVEs).

Stuart Cunningham

Stuart Cunningham is Distinguished Professor at Queensland University of Technology, and Director of the Australian Research Council Centre of Excellence for Creative Industries and Innovation. He is author or editor of several books and major reports, most recently: *The Media and Communications in Australia,* 3rd ed (with Graeme Turner, Allen & Unwin, 2010), *Beyond the Creative Industries: Mapping the Creative Economy in the United Kingdom* (with Peter Higgs and Hasan Bakhshi, NESTA, 2008) and *In the Vernacular: A Generation of Australian Culture and Controversy* (University of Queensland Press, 2008).

Shane Dawson

Shane Dawson is Academic Leader: Educational Development, Quality Assurance and Evaluation with the Graduate School of Medicine, University of Wollongong. His research focuses on the application of quantitative data derived from student online activity to inform teaching and learning practice. His research has demonstrated the use of student online communication data to provide lead indicators of student sense of community and course satisfaction. Shane is also interested in the development of social network visualization tools as a resource for teaching staff to better understand, identify and evaluate student engagement, academic performance and creative capacity.

Eduardo de la Fuente

Eduardo de la Fuente is a lecturer in Communications and Media Studies at Monash University and Faculty Fellow with the Centre for Cultural Sociology at Yale University. He has an interdisciplinary background in the fields of communication studies, sociology and social theory. He has held positions at the University of Tasmania (1998–2001) and Macquarie University (2002–7). With Brad West of Flinders University, de la Fuente co-convenes the TASA Cultural Sociology Thematic Group. He has published in various journals including *Sociological Theory, Cultural Sociology, Journal of Classical Sociology, Journal of Sociology, European Journal of Social Theory, Thesis Eleven* and *Distinktion: The Scandinavian Journal of Social Theory.* He is currently completing a scholarly monograph for Routledge on twentieth century music and the question of cultural modernity.

Brian Fitzgerald

Brian Fitzgerald is an internationally recognised scholar specialising in Intellectual Property and Cyberlaw. He holds postgraduate degrees in law from Oxford University and Harvard University, and his recent publications include *Cyberlaw: Cases and Materials on the Internet, Digital Intellectual Property and E Commerce* (2002); *Jurisdiction and the Internet* (2004); *Intellectual Property in Principle* (2004) and *Internet and Ecommerce Law* (2007). Brian is a Chief Investigator and Program Leader for Law in the ARC Centre of Excellence on Creative Industries and Innovation and Project Leader for the Australian Government funded Open Access to Knowledge Law Project (OAK Law) and Legal Framework for e-Research Project. He is also a Program Leader for the CRC Spatial Information. His current projects include work on intellectual property issues across the areas of Copyright, Digital Content and the Internet, Copyright and the Creative Industries in China, Open Content Licensing and the Creative Commons. From January 2002–January 2007 he was Head of the School of Law at QUT in Brisbane. He is currently a specialist Research Professor in Intellectual Property and Innovation at QUT. He is also a Barrister of the High Court of Australia.

Terry Flew

Terry Flew is Professor of Media and Communication in the Creative Industries Faculty, Queensland University of Technology, Brisbane, Australia. He is the author of *New Media: An Introduction* (Oxford, 2008, third edition), *Understanding Global Media* (Palgrave, 2007) and *Creative Industries, Culture and Policy* (Sage, 2010). He has been a chief investigator on Australian Research Council projects on citizen journalism, creative suburbia and creative industries development in China.

Richard Florida

Richard Florida is Professor of Business and Creativity at the Rotman School and Director of the Martin Prosperity Institute, University of Toronto. Florida has held professorships at George Mason University and Carnegie Mellon University and taught as a visiting professor at Harvard and MIT. He earned his Bachelor's degree from Rutgers College and his Ph.D. from Columbia University. His research provides unique, data-driven insight into the social, economic and demographic factors that drive the 21st century world economy. He is author of the best-sellers *The Rise of the Creative Class, The Flight of the Creative Class* and *Who's Your City.*

Jean-Claude Guédon

Jean-Claude Guédon was trained in history of science and now teaches in the Comparative Literature Department of the University of Montreal. With long-standing interests in communication technologies, he has written a small book on the Internet published in 1996 by Gallimard in Paris with a second edition in 2000: *Internet: le monde en réseau.* From 1994 until 2002, he was involved with the Internet Society (ISOC) and chaired the Inet Program Committee in 1996, 1998 and 2000. For the last 10 years or so, he has been involved in the Open Access Movement and has also published on that topic, particularly a paper called "In Oldenburg's Long Shadow," which has been translated into several languages and published as a small book in Italy under the title *Per la Pubblicità del Sapere* (University of Pisa Press, 2004). He is also involved in the Free/Libre Software Movement and is completing a second (and last) term as a member of The French Association for Free Software Users (AFUL). He is a member of the Board of 'Electronic Information for Libraries' (eIFL), an NGO dealing with libraries in poor and transition countries.

Greg Hearn

Greg Hearn is Research Professor in the Creative Industries Faculty at QUT. His work focuses on applications development and evaluation of new technologies and services in the creative industries. He has been involved in high level consulting and applied research examining new media and industry/ organisational forms for more than two decades, with organisations including British Airways, Hewlett Packard, and Australian and international agencies. He was a consultant to the Broadband Services Expert Group, the national policy group that formulated Australia's foundational framework for the internet in 1994. In 2005 he was an invited member of a working party examining the role of creativity in the innovation economy for the Australian Prime Minister's Science Engineering and Innovation Council. He has authored or co-authored over 20 major research reports and a number of books including *The Knowledge Economy Handbook* (2005: Edward Elgar) and *Knowledge Policy: Challenges for the 21st Century* (2008: Edward Elgar).

John Holden

John Holden is an Associate at the independent think-tank Demos, where he was Head of Culture for 8 years, and a visiting Professor at City University. A former investment banker with Masters Degrees in law and in art history, John's interests are in the development of people and organisations in the cultural sector. He has been involved in numerous major projects with the cultural sector ranging across heritage, libraries, music, museums, the performing arts, and the moving image. He has addressed issues of learning, leadership, education, the creative industries, cultural policy and evaluation working with organisations such as DCMS, ACE, HLF, MLA, Creative Partnerships, the Reading Agency and Screen England, as well as individual organisations including Glasgow School of Art, the Sage Gateshead, the British Museum, and the Royal Shakespeare Company. John was a principal organiser of the influential Valuing Culture conference in June 2003, and is the author of numerous works.

John Howkins

John Howkins is Chairman of BOP Consultants and is a member of the United Nations UNDP Advisory Committee on the Creative Economy. He is Deputy Chairman of the British Screen Advisory Council (BSAC) and has worked in over 30 countries including Australia, Canada, China, France, Greece, India, Italy, Japan, Poland, Singapore, UK and USA. He is a Director of HandMade plc, a films and rights owner listed on London's AIM market, and Hotbed Media Ltd. He has advised global corporations, international organizations, governments, and individuals and was associated with HBO and Time Warner from 1982 to 1996 with responsibilities for TV and broadcast businesses in Europe. John is a former Chairman of the London Film School and is a former Executive Director of the International Institute of Communications (IIC). He is currently a Visiting Professor at Lincoln University, England, and Vice Dean and Visiting Professor at the Shanghai School of Creativity, Shanghai Theatre Academy, China. In addition to this, John is the Director of the Adelphi Charter on Creativity, Innovation and Intellectual Property. His books include *Four Global Scenarios on Information* (1997), *The Creative Economy* (2001), and *Creative Ecologies* (2009).

Luke Jaaniste

Luke Jaaniste is a postdoctoral researcher at the Creative Industries Faculty, Queensland University of Technology. His research interests focus around the dynamic of arts and innovation, and the way this connects to the broader creative sector, to public policy, to higher research (in terms of 'practice-led' methodologies), and to tertiary-level pedagogies. In 2008 he co-authored the report on 'The Arts and Australia's National Innovation System 1994–2008' for the Council of Humanities, Arts and Social Science, and is currently a research consultant for the 'Growing Future Innovators' scheme based in Perth, Western Australia. Luke is also a visual and sonic artist, with work exhibited, performed and broadcast nationally, and in 2006 he was selected to represent Australia at the International Rostrum of Composers in Paris.

Mary Kalantzis

Mary Kalantzis is Dean of the College of Education at the University of Illinois, Urbana-Champaign. Before this, she was Dean of the Faculty of Education, Language and Community Services at RMIT University, Melbourne, Australia, and President of the Australian Council of Deans of Education. She has been a Board Member of Teaching Australia: The National Institute for Quality Teaching and School Leadership, a Commissioner of the Australian Human Rights and Equal Opportunity Commission, Chair of the Queensland Ethnic Affairs Ministerial Advisory Committee, Vice President of the National Languages and Literacy Institute of Australia and a member of the Australia Council's Community Cultural Development Board.

Pat Kane

Pat Kane is a musician, writer, consultant and activist. His book *The Play Ethic* (Macmillan 2004) has been highly praised for its originality. Pat writes for the *Guardian* and *Independent,* and was a founding editor of *The Sunday Herald.* He has consulted for organizations as diverse as Lego, Nokia, the BBC, the Cabinet Office, the Scottish Government and Bartle Bogle Hegarty about the power and potential of play. His band Hue And Cry (www.hueandcry.co.uk) has supported artists such as Madonna, U2, James Brown, Van Morrison and Al Green, and their thirteenth album *Open Soul* was released in 2008. He is a regular global keynote speaker on the topic of play and his writing can be found at www.theplayethic.com.

Brian Knudsen

Brian Knudsen is a PhD candidate in Public Policy and Management at the H. John Heinz III College, Carnegie Mellon University. His dissertation explores urban contexts for new social movements. Specifically, it considers the role of walkability, density, and connectivity in undergirding environmental, human rights, and other social advocacy organizations. Previously, he studied the relationship between rates of technological innovation and urban density.

Charles Landry

Charles Landry has written several books including *The Art of City Making* (2006); *The Intercultural City: Planning for Diversity Advantage* (2007) with Phil

Wood; *The Creative City: A Toolkit for Urban Innovators* (2000); *Riding the Rapids: Urban Life in an Age of Complexity* (2004) and, with Marc Pachter, *Culture @ the Crossroads* (2001). He has lectured widely around the world and has presented over 250 keynote addresses on topics including "Risk and creativity," "Creative cities and beyond," "Art and its role in city life," "Complexity and city making," and "Diverse cultures, diverse creativities." With his company Comedia, Charles has worked on several hundred projects and given talks in 45 countries from the wealthy to those less fortunate. His work has ranged from projects to revitalise public, social and economic life through cultural activity, to visionary city and regional strategies.

Hugh Lauder

Hugh Lauder is Professor of Education and Political Economy at the University of Bath (1996–to present). His books include: Brown, P. Lauder, H. Ashton, D. (2010) *The Global Auction: The Broken Promises of Opportunities, Jobs and Rewards*, Oxford University Press, New York; Lauder, H. Brown, P. Dillabough. J-A. and Halsey, A.H. (eds.) (2006) *Education, Globalization and Social Change*, Oxford, Oxford University Press; Brown, P. Green, A and Lauder, H., (2001) *High Skills: Globalisation, Competitiveness and Skill Formation*, Oxford, Oxford University Press; Brown, P. and Lauder, H. (2001) *Capitalism and Social Progress: The Future of Society in a Global Economy*, Basingstoke, Palgrave Press. Reprinted in Chinese, 2007. He has published many academic papers including on international education and globalization, and is editor of the *Journal of Education and Work*. He is a Visiting Professor at the Institute of Education and a member of the ESRC Virtual College.

Edward Lorenz

Edward Lorenz is Professor of Economics at the University of Nice-Sophia Antipolis. His research interests include the development of empirical indicators of organisational innovation and the comparative analysis of national innovation systems. He is Assigned Professor (2003–2008) at Aalborg University.

Bengt-Åke Lundvall

Bengt-Åke Lundvall is Professor in Economics at Aalborg University. His research is organised around innovation systems and learning economies. He coordinates the worldwide network on innovation research, Globelics. Since 2007 he has served as special term professor at Science Po, Paris. He is a former member of the Danish Social Science Research Council and functions as consultant to OECD/DSTI, Unctad, the World Bank, the European Commission as well as to governmental bodies in Denmark and several other countries. He is associate editor of the journals *Innovations, Revue d'Economie Industriel, Research Policy* and *Journal of Industrial and Corporate Change.* He serves as a referee for several other journals, including *Cambridge Journal of Economics* and *Futures.* From 1995–2001 Bengt-Åke Lundvall coordinated Danish Research Unit for Industrial Dynamics (DRUID) and he is a member of the IKE-group. Together with Luc Soete he is the initiator of the worldwide research network GLOBELICS.

Ellen McCulloch-Lovell

Ellen McCulloch-Lovell is Marlboro College's first woman president. She has strong ties to Vermont: she was chief of staff to U.S. Senator Patrick Leahy (D-VT) from 1983 to 1994 and executive director of the Vermont Arts Council (1975 to 1983). McCulloch-Lovell spent seven years in the Clinton administration from 1994 to 2001, serving as executive director of the President's Committee on the Arts and Humanities and deputy assistant to the President and advisor to the First Lady. A believer in civic engagement, she serves on a number of organizations, including as a member of the National Science Foundation's BIO Advisory Council, a regent of the American Architectural Foundation, an advisory council member for the Center for Folklife and Cultural Heritage at the Smithsonian Institution and several local boards. She just completed her sixth year as Marlboro's president.

Erica McWilliam

Erica McWilliam is Professor of Education in the Centre for Research into Pedagogy and Practice, NIE, Singapore. She is also an Adjunct Professor in the ARC Centre of Excellence in Creative Industries and Innovation in Australia. Erica has had a long and distinguished career as a teacher and an educator of teachers in Queensland. Her scholarship covers a wide spectrum, as is evidenced in her numerous publications on creative capacity building, innovative teaching and learning, research methodology and training and educational leadership and management. Her latest book, *The Creative Workforce* (2008), is published with UNSW Press in Sydney.

Peter Murphy

Peter Murphy is Associate Professor in Communications and Director of the Social Aesthetics Research Unit (SARU) at Monash University. His books (some co-authored) include *Global Creation* (Peter Lang, 2009), *Creativity and the Global Knowledge Economy* (Peter Lang, 2009) *Dialectic of Romanticism: A Critique of Modernism* (Continuum, 2004) and *Civic Justice: From Greek Antiquity to the*

Modern World (Prometheus/Humanity Books, 2001). He is co-editor of *Philosophical and Cultural Theories of Music* (forthcoming Brill), *Agon, Logos, Polis* (Franz Steiner, 2000), *The Left in Search of a Center* (University of Illinois Press, 1996) as well as the editor of the special issue of *South Atlantic Quarterly* on *Friendship* (Duke University Press, 1998) and coordinating editor of the international critical theory and historical sociology journal *Thesis Eleven: Critical Theory and Historical Sociology* (Sage). From 1998 to 2001 Murphy worked in senior editorial roles for Australia's most successful Internet start-up company, Looksmart.

Jan Nederveen Pieterse

Jan Nederveen Pieterse is Mellichamp Professor of global studies and sociology in the Global and International Studies Program at the University of California, Santa Barbara. He specializes in globalization, development studies and cultural studies. His current research focuses on new trends in twenty-first century globalization and the implications of economic crisis. Nederveen Pieterse has been visiting professor in Brazil, China, Germany, India, Indonesia, Japan, Pakistan, South Africa, Sri Lanka, Sweden, and Thailand. He is an editor of Clarity Press and associate editor of *Futures, Globalizations, Encounters, European Journal of Social Theory, Ethnicities, Third Text* and *Journal of Social Affairs.*

Michael Parsons

Michael Parsons has been Research Professor of Art Education at the University of Illinois since 2006. Previously he taught at and was chair of the department of Art Education at the Ohio State University. He was born in England and moved to the USA in 1973 for doctoral studies in the history and philosophy of education. He is a Distinguished Fellow of the NAEA, has edited *Studies in Art Education,* has published widely in the art education literature and has taught, lectured and consulted in a number of countries. His recent interests are in globalization and art, metaphors as vehicles for meaning in art, integrated curriculum and the assessment of student learning in art and Lacanian approaches to interpretation.

Matteo Pasquinelli

Matteo Pasquinelli is a writer, curator and researcher at Queen Mary University of London. He has written *Animal Spirits: A Bestiary of the Commons* (2008) and edited the collections *Media Activism* (2002) and *C'Lick Me: A Netporn Studies Reader* (2007). He writes frequently at the intersection of French philosophy, media culture and Italian post-operaismo. His current project is a book about the history of the notion of surplus from biology to knowledge economy and the environmental discourse. He is a member of the Edufactory collective and since 2000 has been editor of the mailing list Rekombinant. Together with Katrien Jacobs and the Institute of Network Cultures, he organised the Art and Politics of Netporn conference (2005) and the C'Lick Me festival (2007). At Queen Mary University of London he co-organised the series of seminars The Art of Rent and the research network The Factory of the Common (2008–2009). Together with Wietske Maas, he developed the art project Urbanibalism. He lives in Amsterdam.

Michael A. Peters

Michael A. Peters is Professor of Education at the University of Illinois (Urbana-Champaign). He is the executive editor of *Educational Philosophy and Theory* (Blackwell) and editor of two international ejournals, *Policy Futures in Education* and *E-Learning* (both with Symposium) and sits on the editorial board of over fifteen international journals. He has written over thirty-five books and three hundred articles and chapters, including most recently: *Global Citizenship Education* (Sense, 2008); *Global Knowledge Cultures* (Sense, 2007); *Subjectivity and Truth: Foucault, Education and the Culture of Self* (Peter Lang, 2007); *Why Foucault? New Directions in Educational Research* (Peter Lang, 2007), *Building Knowledge Cultures: Educational and Development in the Age of Knowledge Capitalism* (Rowman & Littlefield, 2006), and *Knowledge Economy, Development and the Future of the University* (Sense, 2007). He has a strong research interests in distributed knowledge systems, digital scholarship and elearning systems and has acted as an advisor to government on these and related matters in Scotland, NZ, South Africa and the EU.

Sam Pitroda

Sam Pitroda is an internationally respected development thinker, telecom inventor and entrepreneur who has spent 44 years in Information and Communications Technology and related human and national developments. Credited with having laid the foundation for and ushered India's technology and telecommunications revolution in the 1980s, Pitroda has been a leading campaigner to help bridge the global digital divide. During his tenure as Advisor to Prime Minister Rajiv Gandhi in the 1980s Pitroda headed six technology missions related to telecommunications, water, literacy, immunization, dairy and oil seeds. He was also the founder and first chairman of India's Telecom Commission. He was Chairman of India's National Knowledge Commission (2005–2009), an advisory body to the Prime Minister of India, set up to provide a blueprint of reform of the knowledge related institutions and infrastructure in the country. The Commission has offered a series of recommendations on various aspects of the knowledge paradigm to help India meet the challenges of the 21st century. He has recently been appointed as Advisor to the Prime Minister of India on Public Information Infrastructure and Innovations. He holds close to 100 worldwide patents and has published and lectured widely in the US, Europe, Latin America and Asia.

Palle Rasmussen

Palle Rasmussen is Professor of Education and Learning Research at Aalborg University. His research interests include learning in educational institutions and work contexts and educational policy. He is a member of the EU expert network in the social sciences of education, NESSE.

David Rooney

David Rooney is Senior Lecturer in Knowledge Management, University of Queensland Business School. He has researched, taught and published widely in the areas of the knowledge-based economy, knowledge management, wisdom, change management and economic structure of the creative industries. His books include *Public Policy in the Knowledge-Based Economy, The Handbook on the Knowledge Economy, Knowledge Policy: Challenges for the Twenty First Century,* and the forthcoming *Wisdom and Management in the Knowledge Economy.* Rooney has published in many leading academic journals including *The Leadership Quarterly, Public Administration Review, Human Relations* and *Management Communication Quarterly.* Before entering academia he worked in the music and insurance industries.

Sampsung Xiaoxiang Shi

Sampsung Xiaoxiang Shi is a Research Assistant and PhD candidate at the QUT Faculty of Law. His PhD research focuses on copyright law and innovation in the networked information economy in Australia and China. His research interests cover the law of copyright, media,entertainment, and especially the social and legal implications of the internet and ICT. Sampsung was accepted into the annual Summer Doctoral Programme (SDP) run by the Oxford Internet Institute in partnership with The Berkman Center for Internet and Society at Harvard Law School in 2007. He received degrees of Bachelor in Law (2001) and Master in Civil and Commercial Law (2006) from the East China University of Political Science and Law (ECUPL).

Patrick Whitney

Patrick Whitney is Steelcase/Robert C. Pew Professor of Design and the dean of the Institute of Design, at the Illinois Institute of Technology. He has published and lectured throughout the world on the subject of making technological innovations more humane, the link between design and business strategy, and methods for designing interactive communications and products. His work has received support from the John D. and Catherine T. MacArthur Foundation, the Joyce Foundation, the National Endowment for the Arts, and numerous corporations. He has been on the jury of numerous award programs, including the U.S. Presidential Design Awards and was a member of the White House Council on Design. Professor Whitney was the president of the American Center for Design (ACD) and the editor of *Design Journal,* its annual publication. He is on several advisory boards in the U.S. and abroad and is a trustee of the Global Heritage Fund.

John Willinsky

John Willinsky is Professor Education at Stanford University and is a Fellow of the Royal Society of Canada. Until 2007 he was the Pacific Press Professor of Literacy and Technology and Distinguished University Scholar in the Department of Language and Literacy Education at the University of British Columbia (UBC). In addition to this, Willinsky directs the Public Knowledge Project, which is researching systems that hold promise for improving the scholarly and public quality of academic research. He was recently awarded an honorary Doctor of Laws degree from Simon Fraser University for his contribution to scholarly communication.

Index

A

E

F

G

J

K

L

M

Q

R

S

U

V